# Tropical Stream Ecology

# AQUATIC ECOLOGY Series

Series Editor

*James H. Thorp*
Kansas Biological Survey
University of Kansas
Lawrence, Kansas

Editorial Advisory Board
*Alan P. Covich, Jack A. Stanford, Roy Stein* and *Robert G. Wetzel*

Other titles in the series:

*Groundwater Ecology*
Janine Gilbert, Dan L. Danielopol, Jack A. Stanford

*Algal Ecology*
R. Jan Stevenson, Max L. Bothwell, Rex L. Lowe

*Streams and Ground Waters*
Jeremy B. Jones, Patrick J. Mulholland

*Freshwater Ecology*
Walter K. Dodds

*Aquatic Ecosystems*
Stuart E. G. Findlay, Robert L. Sinsabaugh

# *Tropical Stream Ecology*

Edited by

*David Dudgeon*
Department of Ecology and Biodiversity
The University of Hong Kong
Hong Kong SAR
China

Amsterdam • Boston • Heidelberg • London • New York • Oxford
Paris • San Diego • San Francisco • Singapore • Sydney • Tokyo

Academic Press is an imprint of Elsevier

Academic Press is an imprint of Elsevier
84 Theobald's Road, London WC1X 8RR, UK
Radarweg 29, PO Box 211, 1000 AE Amsterdam, The Netherlands
Linacre House, Jordan Hill, Oxford OX2 8DP, UK
30 Corporate Drive, Suite 400 Burlington, MA 01803, USA
525 B Street, Suite 1900, San Diego, CA 92101-4495, USA

First edition 2008

ISBN: 978-0-12-088449-0

For information on all Academic Press publications
visit our website at books.elsevier.com

Printed and bound by CPI Group (UK) Ltd, Croydon, CR0 4YY
Transferred to Digital Printing, 2013

Front cover credit: Turner Creek, far northern Queensland, Australia.
(Photo courtesy of P. Davies)

Back cover credit: Prince Regent River in north-western Australia.
(Photo courtesy of R. Stone)

This book is dedicated to Claudia Cressa and Bill Williams

# Contents

## 3  *Organic Matter Processing in Tropical Streams*

Karl M. Wantzen, Catherine M. Yule, Jude M. Mathooko, and Catherine M. Pringle

## 4  *Macroinvertebrates: Composition, Life Histories and Production*

Dean Jacobsen, Claudia Cressa, Jude M. Mathooko, and David Dudgeon

## 5   *Fish Ecology in Tropical Streams*

Kirk O. Winemiller, Angelo A. Agostinho, and Érica Pellegrini Caramaschi

## 6   *Aquatic, Semi-Aquatic and Riparian Vertebrates*

Nic Pacini and David M. Harper

## 7   *Riparian Wetlands of Tropical Streams*

Karl M. Wantzen, Catherine M. Yule, Klement Tockner, and Wolfgang J. Junk

## 8   *Tropical High-Altitude Streams*

Dean Jacobsen

## 9  Are Tropical Streams Ecologically Different from Temperate Streams?

Andrew J. Boulton, Luz Boyero, Alan P. Covich, Michael Dobson, Sam Lake, and Richard Pearson

## 10  Tropical Stream Conservation

Alonso Ramírez, Catherine M. Pringle, and Karl M. Wantzen

# Contributors

*Numbers in parenthesis indicate the chapter(s) the contributor has written.*

**Angelo A. Agostinho,** (5), Nucleo de Pesquisas em Limnologia, Ictiologia e Aquicultura (NUPELIA), Universidade Estadual de Maringa, Av. Colombo 5790, Bloco H-90, CEP 87.020-900, Maringa, Parana, Brazil

**Andrew J. Boulton,** (9), Department of Ecosystem Management, University of New England, Armidale, NSW 2350, Australia

**Luz Boyero,** (9), School of Tropical Biology, James Cook University, Townsville, Queensland 4811, Australia

**Stuart E. Bunn,** (2), Australian Rivers Institute, Griffith University, Nathan, Queensland 4111, Australia

**Érica Pellegrini Caramaschi,** (5), Universidade Federal do Rio de Janeiro, Departamento de Ecologia, IB, CCS, Ilha do Funao, Caixa Postal 68020, 21941-970, Rio de Janeiro, RJ, Brazil

**Alan P. Covich,** (9), Director, Institute of Ecology, University of Georgia, Ecology Bldg., Athens, GA 30602-2202, USA

**Claudia Cressa,** (4), Instituto de Zoologia Tropical, Facultad de Ciencias, Universidad Central de Venezuela, Apartado Postal 47058, Caracas 1041-A, Venezuela

**Peter M. Davies,** (2), Centre of Excellence in Natural Resource Management, The University of Western Australia, Albany 6330, Australia

**Michael Dobson,** (9), Freshwater Biological Association, The Ferry Landing, Far Sawrey, Ambleside, Cumbria LA22 0LP, UK

**David Dudgeon,** (4), The University of Hong Kong, Department of Ecology & Biodiversity, Pokfulam Road, Hong Kong SAR, China

**Stephen K. Hamilton,** (2), W.K. Kellogg Biological Station and Department of Zoology, Michigan State University, Hickory Corners, MI 49060, USA

**David M. Harper,** (6), Department of Biology, University of Leicester, Leicester LE1 7RH, UK

**Dean Jacobsen,** (4,8), Freshwater Biological Laboratory, University of Copenhagen, Helsingørsgade 51, DK-3400 Hillerød, Denmark

**Wolfgang J. Junk,** (7), Max-Planck Institute of Limnology, Tropical Ecology Working Group, Ploen 24302, Germany

**Sam Lake,** (9), School of Biological Sciences, Monash University, Clayton, VIC 3800, Australia

**William M. Lewis Jr,** (1), Director, Center for Limnology, Cooperative Institute for Research in Environmental Sciences, University of Colorado, Boulder, CO 80309-0216, USA

**Jude M. Mathooko,** (3,4), Department of Zoology, Egerton University, PO Box 536, Njoro, Kenya

**Nic Pacini,** (6), Via Burali d'Arezzo 96, 0420 Itri (LT), Italy

**Richard Pearson,** (9), School of Tropical Biology, James Cook University, Townsville, Queensland 4811, Australia

**Catherine M. Pringle,** (3,10), Institute of Ecology, University of Georgia, Athens, GA 30602, USA

**Alonso Ramírez,** (10), Institute for Tropical Ecosystems Studies, University of Puerto Rico, P.O. Box 21910, San Juan PR 00931, USA

**Klement Tockner,** (7), Department of Limnology, EAWAG/ETH, Überlandstrasse 133, CH-8600 Dübendorf, Switzerland

**Karl M. Wantzen,** (3,7,10), Institute of Limnology, University of Konstanz, Postfach M 659, 78457 Konstanz, Germany

**Kirk O. Winemiller,** (5), Department of Wildlife & Fisheries Sciences, Texas A & M University, College Station, TX 77843, USA

**Catherine M. Yule,** (3,7), Monash University Malaysia, School of Arts and Sciences, 2 Jalan Kolej, Bandar Sunway, Petaling Jaya 46150, Malaysia

# Preface

"Nothing should satisfy short of the best, and the best should always seem a little ahead of the actual."

C. P. Scott, 1921

"The perfect is the enemy of the good."

Voltaire, 1764

The origin of this book lie in a conversation between Claudia Cressa and I at the 2001 Congress of the *Societas Internationalis Limnologiae* in Melbourne. She lamented that the scientific literature lacked a synthetic treatment to tropical stream ecology and felt something should be done about it. This book is an outcome of an attempt to provide that synthesis. In some respects, it was an attempt doomed to failure. The literature available on tropical streams runs into volumes and has grown substantially in recent years. Much of it represents highly descriptive, short-term studies that do not address hypotheses or processes, and thus generates findings that may only be of local interest (Dudgeon, 2003). A good deal of work is published in obscure periodicals, where it often languishes unavailable and unread. Collecting all such publications would be a Sysiphean task resulting in an unwieldy tome. Instead, the approach taken was to invite authors to provide an overview of a topic highlighting what is known about it, and what is unknown or needs to be known about it, and how that knowledge gap might be filled. The original intention was to include authors from all parts of the tropics, but this proved more difficult than expected, and representation from Africa is particularly limited.

The term 'stream' has been interpreted liberally herein, and is taken to include rivers also, although very large rivers have not been considered. Decisions on which topics to be included were made based, in part, on the availability of authors and also by what had been published recently. For instance, recent publications on the ecology, exploitation, and conservation of large river fishes (Allan *et al.*, 2005), the potential effects of climate change on rivers (e.g. Xenopoulos *et al.*, 2005; Brown *et al.*, 2007), and conservation of fresh waters in general (Dudgeon *et al.*, 2006; Postel, 2005) and rivers in particular (e.g. Nilsson *et al.*, 2005; Moulton and Wantzen, 2006; Linke *et al.*, 2007; Nel *et al.*, 2007), meant that these topics have not been covered in depth here. Similarly, the important topic of environmental flows likewise received limited attention, because a major series of articles on this topic is forthcoming in *Freshwater Biology* (see also Arthington *et al.*, 2006). This is not to say that the topic of conservation has been omitted; examples of conservation are included in almost every chapter of this book. Consequently, the final chapter of this book, which is devoted to conservation issues, takes a rather general approach and summarizes a number of matters raised in various guises in other chapters. One potentially important topic, the ecology of birds along tropical rivers and streams, has been omitted here; this decision was made because the literature on these animals is already capacious. I suspect that every reader of this book will find some omission that frustrates or puzzles him or her.

My friend and mentor, the late Bill Williams, was fond of the term 'temperate intellectual hegemony' (Williams, 1988, 1994), which he used to describe the mindset of scientists from northwest Europe and North America when they were faced with observations about inland waters in the tropics and Australia. He felt that research undertaken in the tropics was incorrectly considered as little more than an *ad hoc*, fragmented, and derivative area of limnology that is based upon concepts derived in the temperate realm (Williams, 1994). Concepts and paradigms developed in north-temperate latitudes do not necessarily conform to the reality of ecosystems further south, and this is addressed in various chapters of the book. In particular, Chapter 9 discusses whether tropical and temperate streams differ fundamentally. The answer to this question is confounded by the fact that there is no such thing as a 'typical' tropical stream that can be compared with its temperate counterpart; regional differences in the tropics (and also further north) are quite substantial as is made evident in many chapters (see also Moulton and Wantzen, 2006).

Making generalizations about tropical streams is problematic because much of information about them is derived from a small number of relatively well-studied locations. For example, fairly detailed information is available from parts of the Neotropics (e.g. Costa Rica, Puerto Rico) and northeastern Australia, but data from Africa are generally sparse. The availability of research findings is a related issue. In instances where investigations of tropical streams involves collaboration with scientists based in the temperate zone (mainly North America or Europe), the results are usually published in widely read and well-cited international journals that can be accessed easily (Wishart and Davies, 2002). Of course, this is a good thing. The converse is, however, less good. When the research involves only workers based in a tropical country (which is often a developing nation), for a variety of important reasons (see Williams, 1994), the work has less chance of being published in an international journal (Wishart and Davies, 1998; Dudgeon, 2003), and hence is likely to be overlooked. The combination of papers dealing with Neotropical streams compiled in a recent (2007) special issue of the *Journal of the North American Benthological Society*, which came out when this book was going to press, provides a fitting example of the first of these tendencies, and the mixture of contributors to this volume offers another such instance.

One reason why many indigenous scientists in tropical countries are not publishing their work internationally may be constraints on accessing relevant literature, and thus difficulties with placing the results of their investigations into a wider scientific context. A major aim of this book is to provide a useful aid in that regard. 'Temperate intellectual hegemony' might also bias some referees toward viewing manuscripts that report tropical studies as interesting regional studies first and contributions to the main body of international literature second, whereas studies of temperate waters are viewed in the opposite way. The contents of this book may go some way toward redressing this imbalance.

A book of this type depends on the assistance and goodwill of many people – especially the authors. I am also grateful to the referees who reviewed the chapters and provided the authors with constructive comments and suggestions. My thanks are also due to Claudia Cressa who suggested this task in the first place. While external circumstances conspired to limit her eventual contribution to this book, it may have never been written without her initial involvement. I have benefited greatly from the patience of my graduate students during my periods of inattention to their needs. In addition, I appreciate the support and collegiality provided by my co-workers in the Department of Ecology and Biodiversity at The University of Hong Kong, particularly Richard Corlett, Billy Hau, Nancy Karraker, Kenneth Leung, Lily Ng, Yvonne Sadovy, and Gray Williams.

*David Dudgeon*

# REFERENCES

Allan, J.D., Abell, R., Hogan, Z., Revegna, C., Taylor, B., Welcome, R.L., and Winemiller, K. (2005). Overfishing of inland waters. *BioScience* 55, 1041–1052.

Arthington, A.H., Bunn, S.E., Poff, N.L., and Naiman, R.J. (2006). The challenge of providing environmental flow rules to sustain river ecosystems. *Ecol. Appl.* 16, 1131–1138.

Brown, L.E., Hannah, D.M., and Milner, A.M. (2007). Vulnerability of alpine stream biodiversity to shrinking glaciers. *Global Change Biol.* 13, 958–966.

Dudgeon, D. (2003). The contribution of scientific information to the conservation and management of freshwater biodiversity in tropical Asia. *Hydrobiologia* 500, 295–314.

Dudgeon, D., Arthington, A.H., Gessner, M.O., Kawabata, Z., Knowler, D., Lévêque, C., Naiman, R.J., Prieur-Richard, A.-H., Soto, D., Stiassny, M.L.J., and Sullivan, C.A. (2006). Freshwater biodiversity: importance, threats, status and conservation challenges. *Biol. Rev.* 81, 163–182.

Linke, S., Pressey, R.L., Bailey, R.C., and Norris, R.H. (2007). Management options for river conservation planning: condition and conservation re-visited. *Freshwat. Biol.* 52, 918–938.

Moulton, T.P., and Wantzen, K.M. (2006). Conservation of tropical streams – special questions or conventional paradigms. *Aquat. Conserv.* 16, 659–663.

Nel, J.L., Roux, D.J., Maree, G., Kleynhans, C.J., Moolman, J., Reyers, B., Rouget, M., and Cowling, R.M. (2007). Rivers in peril inside and outside protected areas: a systematic approach to conservation assessment of river ecosystems. *Diversity Distrib.* 13, 341–352.

Nilsson, C., Reidy, C.A., Dynesius, M., and Revenga, C. (2005). Regulation of the world's large river systems. *Science* 308, 405–408.

Postel, S. (2005). "Liquid Assets: The Critical Need to Safeguard Freshwater Ecosystems". Worldwatch Institute, Amherst, USA.

Williams, W.D. (1988). Limnological imbalances: an Antipodean viewpoint. *Freshwat. Biol.* 20, 407–420.

Williams, W.D. (1994). Constraints to the conservation and management of tropical inland waters. *Mitt. Internat. Verein. Limnol.* 24, 357–363.

Wishart, M.J., and Davies, B.R. (1998). The increasing divide between First and Third Worlds: science, collaboration and conservation of Third World aquatic ecosystems. *Freshwat. Biol.* 39, 557–567.

Wishart, M.J., and Davies, B.R. (2002). Collaboration, conservation and the changing face of limnology. *Aquat. Conserv.* 12, 567–575.

Xenopoulos, M. A., Lodge, D. M., Alcano, J., Märker, M., Shulze,. K., and van Vuuren, D. P. (2005). Scenarios of freshwater fish extinction from climate change and water withdrawal. *Global Change Biol.* 11, 57–64.

# *1*

## *Physical and Chemical Features of Tropical Flowing Waters*

William M. Lewis, Jr

Tropical latitudes offer an abundance of well watered landscapes because the Hadley circulation favors heavy seasonal rainfall, particularly within the migration area of the Intertropical Convergence Zone. Hydrographic seasonality typifies tropical streams and rivers, and to a large extent is predictable on the basis of latitude within the tropics. Because thunderstorms play an important role in the delivery of precipitation, hydrographs for small streams often show numerous intra-annual spikes in discharge, but these patterns are obscured through hydrographic averaging at the regional scale. Tropical flowing waters typically show well-defined seasonality in depth and velocity of flow, water chemistry, and metabolic rates, but seasonality is based primarily on hydrology alone rather than hydrology in conjunction with temperature, as would be more typical of temperate latitudes.

Total annual irradiance at tropical latitudes does not differ greatly from that of latitudes as high as 40° because of the high moisture content of tropical air. Because of high mean water temperature, the oxygen reserve at saturation is substantially lower for tropical waters at low elevations than for temperate waters averaged over the growing season. Metabolic rates of rivers within the tropics are affected by thermal variation related to elevation and percent moisture in the atmosphere.

Topography is the dominant control on suspended and dissolved solids of tropical streams and rivers. Concentrations of dissolved solids range from a few mg $L^{-1}$ in wet areas of low gradient and resistant lithology to 1000 mg $L^{-1}$ or more on high gradients with readily eroded lithologies, especially in the presence of disturbance. Seasonality in discharge is accompanied by seasonality in concentrations of dissolved and suspended solids, organic matter, and nutrients; seasonal ranges in concentration often reach an order of magnitude.

Dissolved forms of phosphorus and nitrogen in tropical streams and rivers are present in quantities sufficient to support moderate to high biomass of autotrophs even under pristine conditions. Concentrations of phosphorus (P), nitrogen (N), and dissolved organic carbon (DOC) show no obvious categorical contrast with those of higher latitude at similar elevations, but mechanisms for the delivery of P, N, and DOC from tropical watersheds to streams and rivers, and their subsequent processing within the aquatic environment, may differ in important ways that are not yet well understood.

## I. INTRODUCTION

Generalizations about tropical streams and rivers are most easily approached through the influences of climate. A climatic perspective on tropical flowing waters also establishes a useful framework for comparisons between tropical waters and more familiar temperate waters and highlights the differences that are related to latitude by way of climate. Therefore, this chapter begins with an overview of tropical climatology and follows the climatic connections through hydrology and water temperature. Next is a presentation of water chemistry with a focus on suspended and dissolved solids as well as P, N, and C, which sets the stage for later chapters that deal with ecosystem functions and biotic communities.

Tropical climatology is a subject for books rather than chapters. Only a sketch of tropical climates can be given here, but a substantial narrowing of scope is justified by special relevance of the connections between climate and flowing waters, which are a subset of connections between climate and all ecosystem types combined.

Table I shows a list of climatic variables that are most directly connected to the ecosystem functions of tropical flowing waters. The list differs from one that would be applicable to lakes or to terrestrial ecosystems. For example, lakes are much affected by wind, whereas flowing waters are much less so. The same can be said for heat budgets, which have striking effects on lakes but are less important for flowing waters except insofar as they control the seasonal and diel ranges of temperature.

## II. CLIMATIC ORGANIZING PRINCIPLES

Tropical climates are heterogeneous spatially, but they show a certain amount of order that can be understood in terms of the global distribution and movement of heat and moisture through the atmosphere. For present purposes, tropical latitudes are construed as spanning the tropics of Cancer and Capricorn, although in a climatic sense tropical phenomena may spill beyond or withdraw from these margins at certain times and places.

### A. Hadley Circulation

Within the tropics, one organizing feature is Hadley circulation, which can be depicted as an immensely wide rotation of air roughly spanning each of the hemispheric ranges of tropical latitudes (Fig. 1). The rotation is generated by rising air at the lowest latitudes in response to a combination of heating and convergence of air flows from the two hemispheres. Rising air near the equator carries moisture that was transferred to it as it passed over the ocean surface en route higher to lower latitudes. The rising air releases moisture as precipitation, which falls in copious amounts beneath the convergence of north and south Hadley cells, but not necessarily

*TABLE I*  Climatic Features of the Tropics Most Relevant to Ecosystem Functions of Tropical Streams and Rivers, along with their Direct Ecosystem Connections

| *Climatic variable* | *Ecosystem connection for flowing waters* |
| --- | --- |
| Insolation | Control of photosynthesis |
| Temperature | Regulation of all metabolic rates |
| Precipitation | Control of mass transport and physical habitat through depth and velocity of water |

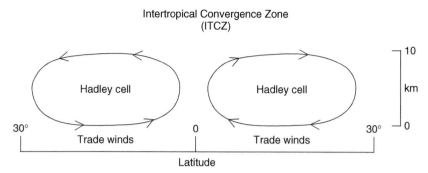

FIGURE 1   Cross-sectional diagrammatic view of the adjoining Hadley cells from the northern and southern hemispheres. Note the vertical exaggeration of the diagram (the cells in reality are approximately 1200 times as wide as they are high). The position of the convergence between the two adjoining (north and south) Hadley cells, although shown here as being on the equator, typically is displaced somewhat from the equator and moves seasonally.

directly over the margins of the cells. The constant push of air from below moves the formerly ascending air mass, now drier, laterally, toward higher latitudes, where it sinks to the earth's surface, causing a constant flow of dry air, or trade winds, back toward the equator, thus completing the cycle.

Hadley circulation is disrupted by other atmospheric forcing factors and through the interruption of its mainly oceanic drivers by land masses. Even so, Hadley circulation causes a band of arid landscapes near the margins of the tropics and abundant precipitation over much of the equatorial zone.

## B. The Intertropical Convergence Zone

Another organizing feature of tropical climate related to Hadley circulation is the intertropical convergence zone (ITCZ). Hadley cells on either side of the equator come together as a convergence of upward-moving, moisture-laden flows (Fig. 1). The convergence between Hadley cells from the two hemispheres is marked by constantly low pressure corresponding to the constant rise of air within the zone of convergence. Either this low-pressure zone (the 'equatorial trough') or the underlying convergence and uplift of air masses marks the ITCZ, which also has been called the 'meteorological equator' (McGregor and Nieuwolt, 1998).

The ITCZ is not aligned with the geographical equator (Fig. 2). It shows seasonal variation that is loosely correlated with the path of the sun, which facilitates the constant rise of air where and when irradiance is most direct. Often there is a notable lag (a few weeks: McGregor and Nieuwolt, 1998) between seasonal movement of the sun and change in position of the ITCZ, presumably reflecting time over which solar heat must accumulate.

The position of the ITCZ, while generalized in Fig. 2 for the entire earth, varies in width and spatial distribution from one continent to the next. In addition, the ITCZ varies from one year to the next in its location and its rate of cross-latitudinal progress. Cloud cover and rainfall associated with the ITCZ therefore are to some extent irregular on a river-basin scale.

## C. Solar Irradiance

Another organizing feature of tropical climates is the seasonal pattern of solar irradiance, which shows latitudinal gradients in total annual amount and seasonality over the tropics (Fig. 3). The gradient in total annual solar irradiance at ground level is the reverse of the expected gradient at the top of the atmosphere because of the high moisture content of air near

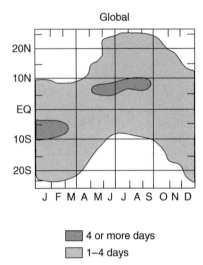

FIGURE 2  Composite global monthly movement of the ITCZ within the tropics. Heavy shading indicates four days or more of strong cloudiness, and light shading indicates one to four days of strong cloudiness. Although virtually all specific locations around the earth reflect the composite pattern, the details differ from one location to another. Redrawn from McGregor and Nieuwolt (1998) after Waliser and Gautier (1993). Copyright John Wiley & Sons Limited. Reproduced with permission.

the equator. In fact, average annual irradiance for land masses within 10° of the equator is little different than it would be at 45° latitude because of the strong suppression of total irradiance at the earth's surface due to high moisture content of the air near the equator (Fig. 3).

Latitude has a strong effect on annual range in solar irradiance per day, as shown in Fig. 3. Annual maximum daily irradiance at the top of the atmosphere varies little with latitude, even beyond the tropics. In contrast, annual minimum irradiance, uncorrected for cloudiness, decreases steeply with latitude; at the equator, it is almost 50% higher than at the margin of the tropics. Latitudinal variation in cloudiness obscures these trends to some degree, however, by suppressing the annual maximum wherever moisture is abundant in the hemispheric summer months.

## D. Monsoons

Movement of the ITCZ and its associated atmospheric moisture is one cause of annual monsoon cycles. Monsoons, which have in common a persistent seasonal reversal of wind direction, are found throughout tropical and southeast Asia as well as northern Australia. They also affect most of tropical Africa north of the equator, but are not pronounced in South America. In general, the northern movement of the ITCZ and its associated low-pressure zone leads to a movement of air from the southern to the northern hemisphere across the equator. If this air is moisture laden, as is often the case, it brings heavy rainfall during the northern hemispheric summer. At a given location, monsoons tend to be oscillatory in intensity, and also show considerable variation from one year to the next (McGregor and Nieuwolt, 1998). East Asia also has a winter monsoon caused by a persistent high-pressure center over the Tibetan plateau.

## E. El Niño Southern Oscillation

Organizing features that are temporally irregular include variations in the pressure gradient of the equatorial Pacific. Fluctuation in the status of this gradient is called 'the southern

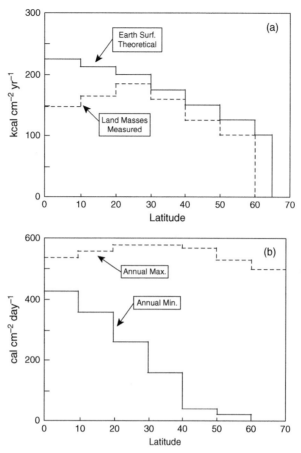

FIGURE 3   (a) Comparison of the amount of irradiance that would reach the surface of the earth at a fixed uniform attenuation coefficient of 0.8, as compared with the actual amount of irradiance observed over the tropical land masses. (b) Gradients in the annual maximum daily irradiance and the annual minimum daily irradiance as a function of latitude, assuming a uniform attenuation coefficient of 0.8. Redrawn from Lewis (1987), based on data from List (1951). Reproduced from Lewis Jr (1987), with permission from the *Annual Review of Ecology and Systematics.* Copyright 1987, Annual Reviews (www.annualreviews.org).

oscillation'; extremes are referred to as 'El Niño Southern Oscillation' (ENSO) conditions. Most often pressure is notably higher in the eastern than in the western Pacific; this condition is termed a 'positive oscillation gradient.' Unusual strengthening of this gradient, which is called 'El Niño,' is associated with an unusual degree of warmth in or around December in marine waters along the western coast of South America. The opposite extreme is designated 'La Niña.' Upwelling caused by extreme thermocline (density gradient) tilt brings up cold, deep ocean water in La Niña years, as contrasted with a Kelvin wave (a large pool of warm water moving across the upper ocean: Kessler *et al.*, 1995) return flow of warm surface waters under El Niño conditions, when the thermocline tilt is relaxed. Most years (about 70%) belong to neither category, but interannual clustering of extreme conditions (El Niño at a frequency of about 20% or La Niña at a frequency of about 10%: website of the National Oceanic and Atmospheric Administration, 2004) is typical (Barry and Chorley, 2003).

The extreme conditions of El Niño and La Niña in the western Pacific are associated with specific climatic variations in various parts of the tropics (see maps given by McGregor and Nieuwolt, 1998). El Niño events correspond to unusually dry conditions (even severe drought) from December through February in Southeast Asia, northern Australia, the eastern Amazon basin, and southeastern Africa; equatorial western Africa is unusually wet. Conditions in these months under La Niña for the most part are opposite (including flooding in some locations) and, in addition, tropical western Africa tends to be cooler than usual. From June to August, El Niño years bring drought to much of India, and southern Southeast Asia as well as northern Australia, warmth to western South America, and warmth plus drought to lower central America and upper South America. La Niña in these months is nearly the opposite, with the addition of a cool tendency in western tropical Africa.

## F. Spatially Irregular Phenomena

A number of other climatic phenomena of the tropics, including tropical thunderstorms, cyclones, jets, and orographic effects, deserve consideration in the analysis of climate at any particular location. For example, the frequency and severity of thunderstorms determines the frequency and magnitude of significant non-seasonal irregularities in the hydrograph, which in turn have ecological effects through mobilization of sediment, movement of biofilms, and facilitation of transport from land surfaces to streams and rivers. Such phenomena do not lend themselves well to generalization because they are spatially even more heterogeneous than the above-mentioned organizing features of tropical climate.

## III. TEMPERATURE

Mean daily air temperatures in the tropics are affected primarily by moisture content of the atmosphere and elevation. The timing of the annual thermal maximum, however, is determined primarily by latitude. Within 10° of the equator, most of the tropical land mass at sea level has a daily mean temperature of 25–30°C in July (northern hemispheric summer). Mean air temperatures near the equator in December–January also are 25–30°C, although there is a seasonal shift in the isotherms that reflects cooling of a few degrees north of the equator and warming of a few degrees south of the equator in the northern hemispheric winter (Fig. 4). Toward the margin of the tropics, temperatures are much higher (30–35° at sea level) in July for the northern hemisphere and in January for the southern hemisphere. The higher maxima at the margins of the tropics reflect the much lower moisture content of air there, as explained mostly by the Hadley circulation (McGregor and Nieuwolt, 1998).

## A. Air Temperature

Near the equator, the annual range of mean daily air temperature is 1–4°C (mean near 2°C). At the margin of the tropics, the annual amplitude reaches 12–16°C. In coastal zones, the annual amplitude is reduced by as much as 50% through oceanic influences. Diel air-temperature variations are approximately 10°C near the equator, where they are moderated by suppression of back radiation due to humidity. At the tropical margins, where humidity is low, they are 15–20°C. Close proximity to the coast reduces diel variations by as much as 50%.

## B. Effects of Elevation on Air and Water Temperature

The mean normal lapse rate (rate of decrease in temperature with elevation) within the tropics is approximately 0.65°C per 100 m of elevation (McGregor and Nieuwolt, 1998). Therefore, the air temperature at high elevations can be quite low (Fig. 5). Elevation-related suppression of air temperature ranging from a few degrees to 20°C or more affects about 20% of the tropical land mass (Fig. 5).

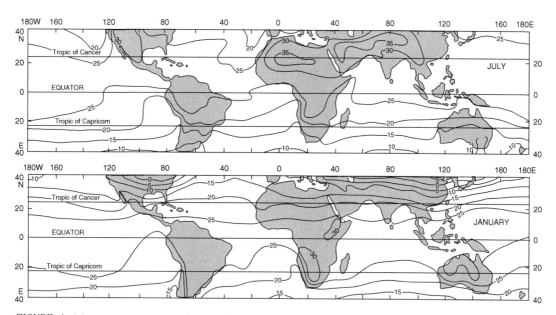

FIGURE 4 Mean air temperatures during July and January at sea level. Redrawn from McGregor and Nieuwolt (1998), with permission from John Wiley & Sons Limited.

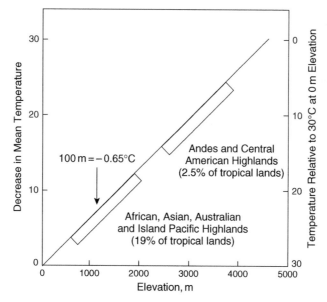

FIGURE 5 Expected decrease in mean air temperature with elevation at mean normal lapse rate of 0.65°C per 100 m of elevation. The brackets show the elevation range encompassing most tropical lands with elevations above 500 m.

Figure 6 gives information on water temperatures for selected stations covering a range of latitudes and elevations. Maximum water temperatures for stations of low to moderate elevation range between 25 and 35°C, reflecting the range of mean daily air temperatures within the tropics. As expected from air temperatures, the highest maximum water temperatures are associated with low humidity (e.g. Black Alice River, Australia).

At elevations above 750 m, the effect of elevation is distinguishable from other factors that influence maximum temperature. The maximum temperatures at the highest elevations

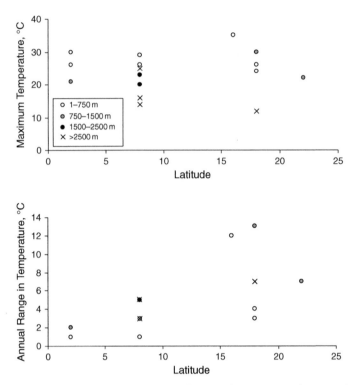

*FIGURE 6* Examples of maximum temperature and annual range of temperatures for tropical running waters over a range of elevations and latitudes. All data except for the Mameyes are derived from a series of individual daytime measurements at a specific site, rather than true daily averages. Data for sites 1 through 8 are from a review by Ward (1985), who gives the original sources. Locations 9 through 14 are from Petr (1983), and locations 15 and 16 are from Ortiz-Zayas (1998). 1 = Black Alice River, Queensland, Australia; 2 = Amazon River main stem; 3 = Orinoco River main stem; 4 = forest streams of the central Amazon; 5 = the Luhoho River, equatorial Africa; 6 = Zambezi River, southern Africa; 7 = Mogi Guassu River, headwaters, Brazil; 8 = Matadoro headwaters, Ecuadorian Amazon; 9–13 = Curare, New Guinea; 15–16 = Mameyes, Puerto Rico (coastal).

(above 3000 m) are not so low, however, as would be expected from the application of the mean normal lapse rate to a base temperature of 30°C or more for sea level. Streams at high elevation are small, and thus are subject to a large diel fluctuation in temperature (see also Chapter 8 of this volume). Because Fig. 6 gives random daytime temperatures rather than 24-hour average temperatures, except for the Rio Mameyes, the observations exceed the daily mean temperature. There may be other reasons as well for the discrepancy between expected and observed temperatures, but they could only be elucidated if the bias associated with daytime sampling were first eliminated.

## C. Overview of Range in Water Temperature

Annual range of water temperature follows patterns related to latitude, humidity, size of water body, and proximity to coastal waters. Annual range of temperature is lowest for the largest water bodies within 10° of the equator (approximate annual range of daily means: 1°C, as in the Orinoco and Amazon main stems). The highest ranges (above 7°C) are for waters located above 15° latitude, but the range is substantially suppressed for waters that are coastal (e.g. Rio Mameyes, Puerto Rico).

Because reliable, continuous temperature-recording instruments now are inexpensive, more detailed data on temperature should become available in the future. The influence of 24-hour

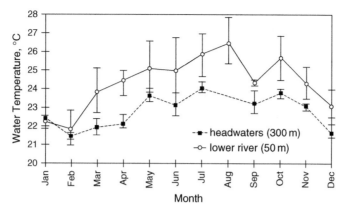

FIGURE 7 Seasonal variation in water temperature in the headwaters and in the lower part of Rio Mameyes. Bars indicate the range of diel temperatures.

records on the interpretation of temperature can be inferred from data on the Rio Mameyes, Puerto Rico, as reported by Ortiz-Zayas (1998: Fig. 7): a daytime maximum could easily be 2°C higher than a daily average.

## D. Effects of Temperature on Metabolism and Oxygen Saturation

The main significance of temperature for flowing waters is through its effect on metabolism and the capacity of water to hold dissolved oxygen. Drastic suppression of metabolism by temperatures approaching 0°C is very rare in the tropics (it can occur in very small flowing waters at the highest elevations). At sea level, for equatorial waters in humid portions of the tropics, the annual and daily fluctuations are so minor that they hardly need to be considered. Over the full range of latitudes, elevations, and humidities within the tropics, however, there is sufficient thermal variation to affect metabolism substantially (Table II). The range in mean temperatures with latitude, elevation, and humidity correspond to an approximate twofold range in metabolic rate that must be taken into account.

Much of the tropics is characterized by water temperatures that correspond to low oxygen content at saturation. This is an important practical consideration, especially when coupled with high metabolic rates. Tropical flowing waters as a whole have a lower oxygen reserve and a higher potential oxygen demand for a given amount of organic loading than temperate waters,

TABLE II  Approximate Effects of Temperature on Aquatic Metabolism and Oxygen Availability (Shown as Concentration Equal to 100% Saturation) for a Representative Range of Conditions within the Tropics

| Location | Temperature, °C | | Metabolic rate* | | $O_2$ concentration, mg $L^{-1}$ | |
|---|---|---|---|---|---|---|
| | Mean | Range | Mean | Range | Mean | Range |
| Low elevation | | | | | | |
| 0–10° latitude, humid, 0 m | 27 | 25–29 | 81 | 70–93 | 7.9 | 7.6–8.1 |
| 15–25° latitude, dry, 0 m | 29 | 23–35 | 93 | 61–142 | 7.6 | 7.0–8.4 |
| Montane | | | | | | |
| 0–20° latitude, 1000 m | 21 | 18–24 | 52 | 43–65 | 7.7 | 7.3–8.1 |
| 0–20° latitude, 2000 m | 14 | 10–18 | 33 | 25–43 | 7.8 | 7.2–8.5 |

* Given as a percent of the rate at 30°C, assuming $Q_{10} = 2.0$

except when temperate waters briefly reach the height of summer warmth. Thus, tropical waters are more vulnerable to organic loading (Lewis, 1998).

The geographic and seasonal mean oxygen saturation concentrations corresponding to expected water temperature, discounting unusual extremes, spans a relatively narrow range $(7.5–8.5 \, \text{mg L}^{-1})$. The narrowness of this range, which represents about 90% of tropical flowing waters, is explained by the inverse effects of elevation and temperature on oxygen saturation. While much higher saturation concentrations would be expected at the lower temperatures of tropical waters at high elevation, the temperature effect is largely offset by reduced barometric pressure at higher elevations (see also Chapter 8 of this volume).

In contrast with conditions in the tropics, saturation concentrations for oxygen over a range of geographic and elevation conditions representing the bulk of flowing waters in the temperate zone would span the range $7.5–14 \, \text{mg L}^{-1}$. Thus, the range for temperate latitudes is not only much broader, but also centers around a considerably higher oxygen concentration, which provides a much larger oxygen reserve to offset other respiratory losses at night or under other conditions when respiration exceeds photosynthesis.

## IV. PRECIPITATION

Precipitation is more abundant within the tropics than at higher latitudes as a whole. Factors contributing to high annual precipitation in the tropics include the high moisture-holding capacity of warm air and the constant uplift of moist air in conjunction with the ITCZ. The amount and seasonality of precipitation at a given latitude can be predicted to some degree simply from the position of the ITCZ. In the winter of the northern hemisphere, the precipitation maximum lies below the equator, but in July it is north of the Equator (Fig. 8). Mean annual precipitation is highest near the equator, but especially so in the northern hemisphere, reflecting a bias in the position of the ITCZ toward a northern-hemisphere position. Aggregate precipitation (Fig. 8) fails to indicate orographic effects of precipitation or variations in precipitation with distance from the coast. These two factors and other regional phenomena cause significant basin-scale variability in precipitation at any particular latitude within the tropics.

The degree and pattern of seasonality are dictated to some extent by movement of the ITCZ. Within the equatorial zone, each year there are two wet seasons and two dry seasons corresponding to the twice-annual passage of the ITCZ over this area. At the margins of the tropics, the two wet seasons and the two dry seasons are merged because the ITCZ pivots around these latitudes during its reversal of direction rather than passing over them twice. In general, the degree of seasonality increases with latitude, reflecting the greater annual range in distance of the ITCZ for higher latitudes within the tropics. There are numerous deviations from these patterns. Jackson (1989) and Talling and Lemoalle (1998) provide maps that give more details on the geographic distribution of rainfall in the tropics.

## V. RUNOFF

Runoff is about 60% of precipitation in the moist portions of the tropics (Lewis *et al.*, 1995); under semi-arid conditions, this percentage may be as low as 20%. Runoff generally reflects the seasonality of precipitation, but the spatial heterogeneity of precipitation in large basins smooths the extremes that might be observed in small basins. Much of the precipitation within the tropics is accounted for by convective mechanisms producing thunderstorms (McGregor and Nieuwolt, 1998), which are local rather than regional. Therefore, the

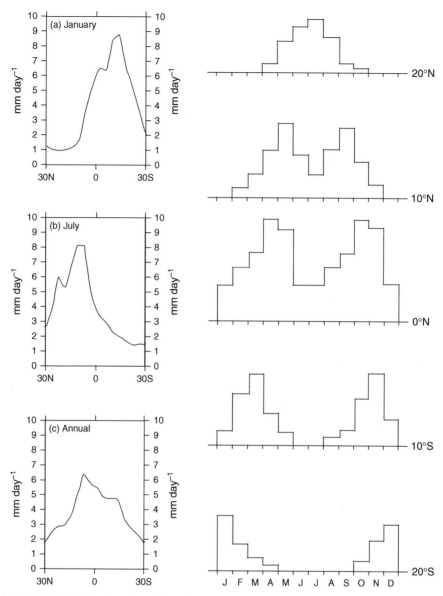

**FIGURE 8** Latitudinal distribution of rainfall in the tropics (left panels) and expected seasonal patterns (right panels). There are numerous deviations from these patterns at specific locations, as explained in the text. Redrawn and slightly modified from McGregor and Nieuwolt (1998).

hydrograph of a small stream in the moist tropics may show numerous spikes in discharge that are temporally uncorrelated with similar spikes in discharge at other locations in the drainage network (Fig. 9). Through combination of discharges showing different temporal patterns, there is a progressive suppression of short-term extremes in discharge extending from streams of lower order to streams of high order. In streams of the highest order, the smoothing effect on discharge is extended further by latitudinal breadth of coverage for the drainage basin. Even though hydrographs of large rivers tend to be smooth, the seasonal contrasts may be quite extreme, including seasonal alternation between drought and massive flooding (Fig. 10). Because the ITCZ at a given longitude is centered over specific latitude bands at specific times, movement

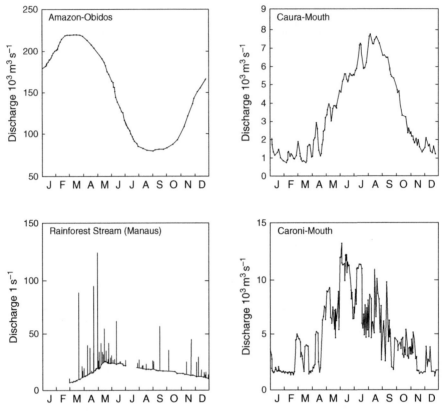

FIGURE 9   Tropical hydrographs illustrating seasonal and short-term variations in discharge. The Amazon shows the smoothing effects of area-based and latitudinal averaging because of its great size. The Caura, a tributary of the Orinoco, shows an intermediate degree of smoothing, and the Caroni, adjacent to the Caura, shows the effects of hydroelectric regulation superimposed on a regime essentially identical to that of the Caura. Extreme short-term variation is illustrated by an Amazonian rainforest stream. Modified from Lewis *et al.* (1995), with permission from Elsevier; rainforest stream data from Lesack (1988).

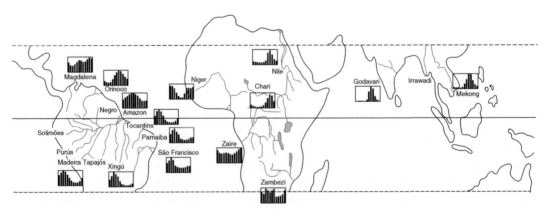

FIGURE 10   Multi-year average hydrographs for large rivers at tropical latitudes showing latitudinal tendencies in seasonal discharge. Redrawn from Vorosmarty *et al.* (1998).

*TABLE III*   Characteristics of 20 of the Largest Tropical Rivers (Amazon Tributaries are Indented)

| Location | Latitudinal span | Orientation | Area, $10^3 km^2$ | Discharge (Q), $m^3 s^{-1}$ | Q/Area, $mm y^{-1}$ | TSS, $mg L^{-1}$ | TDS, $mg L^{-1}$ |
|---|---|---|---|---|---|---|---|
| Magdalena | 2 N–11 N | N–S | 270 | 7000 | 830 | 780 | 120 |
| Orinoco | 2 N–10 N | E–W | 950 | 38000 | 1300 | 80 | 25 |
| Amazon | 15 S–2 N | E–W | 6100 | 220000 | 1200 | 220 | 41 |
| Solimões | 12 S–0 | E–W | 1200 | 43000 | 1200 | 380 | 82 |
| Purús | 12 S–6 N | N–S | 370 | 11000 | 1000 | 74 | – |
| Negro | 4 S–3 N | E–W | 620 | 30000 | 1600 | 7 | 6.5 |
| Madeira | 20 S–4 S | N–S | 1300 | 30000 | 740 | 540 | 60 |
| Tapajós | 13 S–3 S | N–S | 500 | 14000 | 910 | – | 18 |
| Xingú | 15 S–2 S | N–S | 510 | 17000 | 1200 | – | 28 |
| Tocantins | 16 S–2 S | N–S | 820 | 18000 | 680 | – | – |
| São Francisco | 20 S–8 S | N–S | 640 | 2800 | 140 | 60 | – |
| Parnaiba | 10 S–3 S | N–S | 320 | 2300 | 220 | – | – |
| Nile[2,3] | 0–31 N | N–S | 2960 | 950 | 10 | 0 | 208 |
| Niger[2,3] | 5–15 N | E–W | 1210 | 6090 | 159 | 208 | 53 |
| Zaire[2,3] | 10 S–10 N | E–W | 3820 | 39600 | 327 | 34 | 38 |
| Zambezi[2,3] | 20 S–10 S | E–W | 1200 | 7070 | 186 | 148 | 70 |
| Chari[3] | 5 N–15 N | E–W | 600 | 1320 | 69 | – | 64 |
| Godavari[2] | 17 N–20 N | E–W | 310 | 2660 | 271 | 1143 | – |
| Mekong[1,2,3] | 10 N–33 N | N–S | 790 | 14900 | 595 | 340 | 105 |
| Irrawaddy[1,2] | 15 N–29 N | N–S | 430 | 13600 | 995 | 619 | – |

Source for South America is Lewis *et al.* (1995); source for discharge and area outside South America is Vorosmarty *et al.* (1998) except as noted. Hyphens show missing data. See Fig. 10 for hydrographs. TSS = total suspended solids, TDS = total dissolved solids.
[1] Watershed narrow above tropic of Cancer.
[2] Sediment from Milliman and Meade (1983).
[3] Meybeck (1976).

of the ITCZ across a basin of great latitudinal breadth smooths the seasonal precipitation signal for the basin as a whole.

Table III shows total runoff and specific runoff for 20 of the largest tropical rivers. In aggregate, these drainage basins account for 45% of the total land area of the tropics. All are large enough to reflect the regional averaging of thunderstorms as well as the latitudinal averaging of seasonal effects mentioned earlier. Hydrographs for these rivers generally reflect, however, the expected pattern of precipitation in relation to latitude in the tropics (Fig. 10), with a few exceptions that reflect well-known irregularities in the latitudinal distribution of precipitation.

## VI. SUSPENDED AND DISSOLVED SOLIDS

Sediment load has been studied extensively in connection with erosion and continental denudation (Milliman and Meade, 1983). Although the emphasis of such studies is on transport, which is only indirectly related to ecological processes in streams and rivers, sediment concentration, which can be calculated from load and discharge if bedload is ignored (which it usually is), affects transparency, and thus photosynthesis, in flowing waters.

Transport and concentration of total dissolved solids is loosely related to that of suspended solids, but the overall ratio of suspended to dissolved solids is about 5 : 1 (Meybeck, 1976). Inorganic dissolved solids (mostly salts) are much less significant ecologically than suspended solids except at high concentrations ($>2000$ mg L$^{-1}$) (Wetzel, 2001), but can be generally equated with the availability of specific nutrients, such as phosphorus and silicon.

Factors affecting both suspended and dissolved load and concentration include slope (topography), annual precipitation, lithology, disturbance, and natural or anthropogenic impoundment. A useful frame of reference proposed by Carson and Kirkby (1972) and used by Stallard and Edmond (1983) in explaining variations in denudation rates for tropical Latin America is based on the contrast between weathering and transport in controlling denudation rates. Watersheds with limited weathering produce rock-weathering byproducts (dissolved and suspended solids) at rates that are below the potential for transport. As a result, such watersheds have thin soils, reflecting weak or temporary accumulation of weathering byproducts, are wet, and have steep slopes, thus showing efficient removal of weathering byproducts. In contrast, transport-limited landscapes receive or produce weathering byproducts at a rate that exceeds the removal rate. Such landscapes have thick soils containing a large inventory of weathering byproducts. Watersheds may be mainly transport-limited or weathering-limited, but may also present a mosaic of types, as in the case of a steep terrain draining to a flat valley floor.

Topography is of overriding importance at the continental scale in determining the rate of transport for suspended and dissolved solids. In the Amazon drainage, for example, over 80% of both dissolved and suspended solids for the entire drainage derive from the Andes, which account for only 12% of the total drainage area (Drever, 1997, as obtained from Gibbs, 1967); Callède *et al.* (1997) put the percentage Andean contribution of suspended solids as high as 97% for the Amazon. Stallard (1985) makes a similar point in a different way, through a combination of topography and rock type in relation to load of dissolved solids (Fig. 11). The main factor differentiating physiographic regions with respect to dissolved solids content of water is topographic, but erodibility of rock plays an important secondary role within a given physiographic region, as is evident particularly for the Andes. Worldwide precipitation would account for substantial variation as well (Meybeck, 1976), but in the tropics the diversity of moisture regimes is lower than it is worldwide.

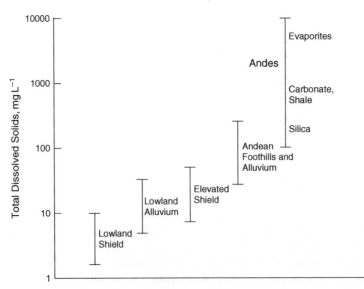

FIGURE 11   Amounts of total dissolved solids characteristic of physiographic regions of tropical South America showing the dominance of topography and secondary role of lithology in controlling export of total dissolved solids. Modified from Stallard (1985), with permission from Kluwer Academic Publishers.

The 20 major rivers listed in Table III illustrate the degree of variation in concentrations of suspended and dissolved solids, some of which can be explained by the main variables that control sediment transport. For perspective, Meybeck and Helmer (1989) give the most typical natural concentrations as $150\,mg\,L^{-1}$ (suspended) and $65\,mg\,L^{-1}$ (dissolved); weighted averages are much higher for suspended solids and slightly higher for dissolved solids because of some very potent sources (e.g. large Asian rivers, such as the Yangtze and Ganges). Among the 20 rivers of Table III, the Rio Negro is extreme in its low concentration of dissolved and suspended solids, as explained by absence of high relief in the watershed, with the exception of some elevated shield that does not weather as efficiently as montane topography because of its rectangular cross-section (Stallard and Edmond, 1983). In addition, shield lithology is extremely resistant to weathering, thus producing only small amounts of chemical weathering byproducts. For similar reasons, tributaries coming to the Orinoco main stem from the south are poor in suspended solids. The Orinoco main stem, which is very strongly influenced by the large contribution of runoff from the Guyana Shield, carries a low concentration of suspended solids among the rivers listed in Table III, even though it receives enough suspended solids from the Andes and the Andean alluvial plain to bring its suspended solids above the baseline reflected by the Rio Negro and other adjacent shield rivers lacking major montane influences. Similarly, the Amazon main stem shows the mixed influences of extensive shield drainage in the center of the watershed and mountainous headwaters to the west.

Of the rivers listed for the American tropics, the Magdalena carries the highest concentrations of dissolved and suspended solids because of the high proportion of steep terrain within its watershed and probably some effects of land cover disturbance. The São Francisco shows relatively low suspended solids because of impoundment.

The African rivers show surprisingly low concentrations of suspended solids, as explained mainly by low population densities, but also by impoundments (the Zambezi) or natural lakes (the Zaire). The Nile once carried $110 \times 10^6\,t\,yr^{-1}$ of sediment, corresponding to a concentration of over $3000\,mg\,L^{-1}$ (Milliman and Meade, 1983), but virtually all of its sediment now is captured behind the Aswan Dam. In Asia, the Godavari, Mekong, and Irrawaddy all derive sediment from Himalayan sources, but also are influenced by substantial augmentation from intensive land use, particularly for the Godavari and Irrawaddy.

Although suspended and dissolved load are presented most often as discharge-weighted annual averages, seasonal variability is ecologically important. For example, concentration of total suspended solids in the Orinoco main stem varies by a factor of almost 10 (Fig. 12). The minimum coincides with lowest discharge and the maximum appears on the rising limb of the hydrograph, but prior to the hydrographic peak (the same is true of the Amazon: Seyler and Boaventura, 2001).

Because much of the ecological significance of suspended solids is related to the interception of light, it is useful to relate concentrations of suspended solids to the penetration depth for 1% photosynthetically available radiation (PAR), which is the approximate threshold for positive net photosynthesis. This relationship is shown in Fig. 13 on the basis of an equation provided by Owen Lind (Baylor University: personal communication) and derived from measurements in Lake Chapala, Mexico, which has a wide seasonal and interannual range of suspended solids ($\eta = 2.018 + 0.0605 \times TSS$ where $\eta$ is given as units/m$^2$ and TSS is given as $mg\,L^{-1}$: $n = 76$; $r^2 = 0.80$). While the relationship may differ slightly from one source of suspended solids to the next, the general indication of Fig. 13 is that photosynthesis over most of a stream or river channel is virtually impossible at concentrations of suspended solids exceeding $100\,mg\,L^{-1}$. Penetration of PAR can be matched to rivers of varied size by use the of detached axis to the right, which shows river size (as discharge) matching the penetration depths on the Y axis as derived from the equation of Church (1996). Where substantial dissolved color (mainly humic and fulvic acids derived from degradation of plant biomass in soils: Wetzel, 2001) is present,

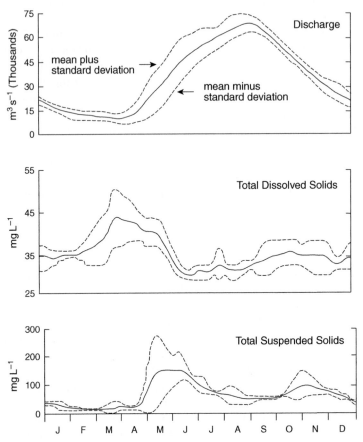

**FIGURE 12** Illustration of seasonal changes in concentrations of dissolved and suspended solids in a large tropical river (the Orinoco). Standard deviations indicate interannual variation for a given time of the year. Redrawn from Lewis and Saunders (1989), with permission from Springer-Verlag.

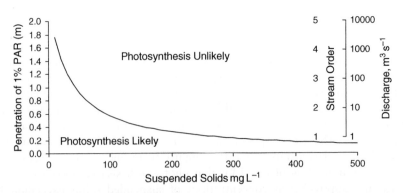

**FIGURE 13** Relationship between suspended solids and depth of penetration for 1% of photosynthetically available radiation (PAR assumed here to be the threshold for positive net photosynthesis). Approximate mean depth for streams of various order indicate the approximate range of suspended solids consistent with net photosynthesis for streams of a given order. The detached axis to the right shows river size (expressed as discharge) corresponding to light penetration on the Y-axis on the left.

even the smaller amounts of suspended solids may not be consistent with photosynthesis in deep water columns because of the additional removal of PAR by dissolved color. Comparison of Fig. 13 with Table III shows that benthic photosynthesis would not be expected in large rivers of the tropics.

## VII. PHOSPHORUS, NITROGEN, AND CARBON

The amounts of phosphorus, nitrogen, and carbon in tropical streams and rivers are of special interest because of the regulatory role that these elements play in aquatic ecosystems. Extremes in concentrations of phosphorus, nitrogen, and dissolved organic carbon would follow from unregulated water pollution, which has already occurred in the most populous regions of the tropics. Of greater fundamental interest is the range of concentrations of these key elements to be expected under natural conditions, and a general view of the factors that control natural concentrations. For present purposes, the natural range of concentrations will be presented for various tributary drainages in the Amazon and Orinoco basins, as reported from the literature (Lewis *et al.*, 1995). Not only do these basins reflect natural conditions, they also offer a wide range of geologic and physiographic conditions for comparison, and are represented by a reasonable amount of field data spanning periods of more than one year.

### A. Phosphorus

Particulate and total dissolved phosphorus, which together make up total phosphorus, are subject to very different kinds of control mechanisms in the watersheds from which they originate. In world rivers, particulate phosphorus accounts for over 90% of total phosphorus (Meybeck, 1982). The dissolved fraction, although present in much smaller amounts, cannot be discounted because it is more readily available to autotrophs.

Within the minimally disturbed watersheds of South America represented in Fig. 14, concentrations of particulate phosphorus range over two orders of magnitude and are in all cases lower than the world average for rivers as reported by Meybeck (1982). The relatively lower concentrations of particulate phosphorus are explained by the absence of water pollution and by the lower extremes of suspended solids in the Amazon and Orinoco basins as compared with the basins of Asiatic rivers, particularly where land disturbance plays a major factor in liberating suspended solids to surface waters (Downing *et al.*, 1999).

The wide variation in particulate phosphorus reflects variation in total suspended solids (Meybeck, 1982); the relationship between these two variables is quite close for undisturbed watersheds. Therefore, particulate phosphorus can be categorized by a scheme similar to the one that is used for suspended solids, i.e. according to weathering rate as determined by slope and rock type. In general, Andean streams and rivers carry a few hundred $\mu$g L$^{-1}$ of particulate phosphorus; Andean alluvium carries less than $100\,\mu$g L$^{-1}$; and streams running on flat terrain over shield or continental alluvium carry $10\,\mu$g L$^{-1}$ or less.

Total dissolved phosphorus falls within a much narrower range of concentrations than particulate phosphorus (Fig. 14). The shield drainages and continental alluvium, which share slow weathering rates, tend to have low concentrations, although the Rio Negro shows as much as $19\,\mu$g L$^{-1}$. Montane and Andean alluvial watersheds have concentrations that range mostly between 10 and $50\,\mu$g L$^{-1}$. There is little understanding at present of the reasons for differences in TDP concentrations within these groups of rivers. Given the importance of phosphorous to autotrophs in flowing water (Wetzel, 2001), studies of the mechanisms controlling TDP in tropical rivers and streams are needed.

One counterintuitive conclusion from phosphorus concentrations in undisturbed tropical rivers and streams is that they contain sufficient TDP to support a substantial amount of

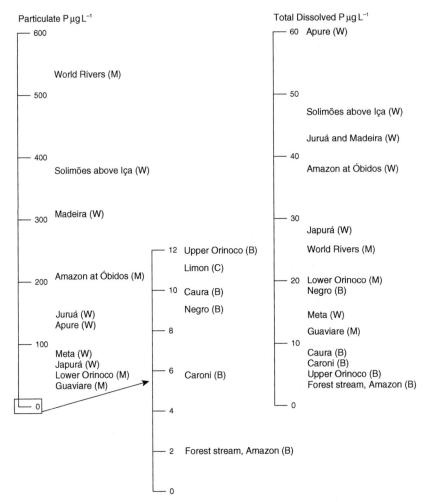

*FIGURE 14* Concentrations of particulate phosphorus and total dissolved phosphorus in undisturbed or minimally disturbed rivers of South America (central and northern South America). Apure, Meta, Guaviare, Caura, and Caroni are the main tributaries of the Orinoco main stem. Juruá, Negro, and Japurá are tributaries of the Amazon/Solimóes. Limon is a second order montane drainage near the Caribbean coast of Venezuela. Letters indicate general classification (W = white water, B = black water, C = clear water, M = mixed). All data are from Lewis *et al.* (1995). Data on world rivers are from Meybeck (1979).

autotrophic growth, as judged by nutrient responses in temperate streams and rivers (Dodds *et al.*, 2002). Even on continental alluvium and shield, there is sufficient TDP to support moderate growth of periphyton although canopy conditions in these areas may minimize photosynthesis. The presence of TDP in the amounts as shown in Fig. 14 is contradictory to the notion that nutrient cycling in tropical forests is especially tight (literature summarized by Schlesinger, 1997). It appears that there is sufficient leakage of TDP from tropical forest nutrient cycles to sustain substantial growth of autotrophs (Lewis, 1986; Lewis *et al.*, 1990; Ortiz-Zayas *et al.*, 2005).

## B. Nitrogen

Transport of total nitrogen and all nitrogen fractions increases steeply as a function of runoff from tropical watersheds (Lewis *et al.*, 1999). The rate of increase in transport is somewhat

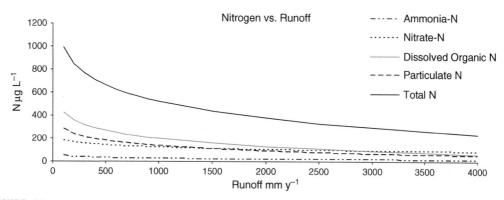

FIGURE 15   Concentrations of total nitrogen and nitrogen fractions (discharge-weighted) over a range of annual runoff as determined by Lewis *et al.* (1999).

lower, however, than the rate of increase in discharge. Thus, there is certain amount of dilution of total nitrogen and nitrogen fractions between the lowest and the highest rates of runoff (Fig. 15). At runoff between 400 and 2000 mm yr$^{-1}$, which would encompass most tropical watersheds, the amounts of all nitrogen fractions, except for ammonia, are substantial and could support autotrophic growth under appropriate light conditions, as judged by response of temperate streams and rivers to nitrogen (Dodds *et al.*, 2002). Here, as in the case of phosphorus, the ability of tropical forests to retain and recycle nitrogen has probably been overstated, as shown by the consistent presence of substantial amounts of nitrogen (all fractions), where anthropogenic disturbance has been minimal or non-existent (Lewis *et al.*, 1999).

There appears to be little difference across latitudes in the concentrations of total nitrogen and nitrogen fractions for a given amount of runoff (Lewis, 2002). The underlying mechanisms are not well understood. In the tropics, moisture stimulates nitrogen fixation, which allows constant export of nitrogen. At temperate latitudes, nitrogen fixation is partially suppressed by cool weather, but the demand for nitrogen also is suppressed by the same mechanism, thus resulting in nitrogen export.

Nitrogen delivery to streams and rivers is controlled hydrographically. The time course of nitrogen concentrations has not been extensively studied, but the existing information suggests peaks of nitrate and dissolved organic nitrogen concentrations occur on the rising limb of the hydrograph (Lewis and Saunders, 1989), as would be typical also of undisturbed temperate watersheds.

## C. Organic Carbon

Dissolved organic carbon (DOC) is one basis for distinguishing different basic water types of the tropics (Sioli, 1984). In the Amazon and elsewhere, waters that are darkly colored with humic and fulvic acids (limnohumic acids) are designated as 'black waters.' Waters of high turbidity are 'white waters,' and waters with neither characteristic are 'clear waters' (Sioli, 1984). In general, black waters emanate from sandy soils or shield, and white waters emanate from mountains or montane alluvium. Clear waters are much scarcer than the other two types, and their deficiency of dissolved color remains poorly explained.

The visual contrast between black waters and white waters has led to the widespread notion that black waters contain large amounts of DOC in the form of humic acids, whereas white waters do not. This notion is misleading, as the amount of DOC in most tropical waters falls between 3 and 10 mg L$^{-1}$, regardless of the visual classification (Lewis *et al.*, 1995). While it is possible to find specific examples of black waters with unusually high DOC, this is not typical.

Most black waters owe their striking appearance to the absence of inorganic turbidity, which would offset some of the absorbance of light through backscatter, combined with impressive absorption of short wavelengths of light by DOC within an otherwise clear water column.

In undisturbed tropical watersheds, rivers and streams carry modest amounts of particulate carbon. In the lower range, which is typical for shield waters, the concentrations of particulate organic carbon (POC) may be less than $1 \, mg \, L^{-1}$. Concentrations often approach $10 \, mg \, L^{-1}$ where slopes are higher. The ratio of POC to DOC is highly variable (Meybeck, 1982; Lewis *et al.*, 1995).

As is typical of temperate watersheds, DOC often shows peak concentrations on the rising limb of the hydrograph, rather than at peak discharge or at minimum discharge (Lewis and Saunders, 1989; Saunders and Lewis, 1989), but some rivers show nearly static DOC concentrations (Lewis *et al.*, 1986). The same is true of POC, but probably for different reasons (flushing of soils for DOC, resuspension of particles for POC). Most POC is accounted for as byproducts of vascular plants rather than suspended organisms (Devol and Hedges, 2001).

## VIII. CONCLUSIONS

Tropical flowing waters in some ways differ as a group from their counterparts at higher latitudes, but in other respects latitudinal differences are weak or nil (for further discussion of such differences see Chapter 9 of this volume). Tropical streams and rivers are stable thermally but should show seasonality driven by hydrology. Their metabolic potential is higher and more stable than that of their temperate counterparts, but supplies of suspended and dissolved solids, as well as nutrients, span the same ranges and show the same responses to hydrology as would be expected at higher latitudes. The rivers and streams of the tropics still offer opportunities to observe and analyze natural phenomena over a wide range of geologic and physiographic conditions, whereas those opportunities are all but lost in the highly modified flowing waters of temperate latitudes.

## REFERENCES

Barry, R.G., and Chorley, R.J. (2003). "Atmosphere, Weather, and Climate." 8th Edition. Routledge, London.

Callède, J., Guyot, J.L., Ronchail, J., L'Hôte, Y., Niel, H., and De Oliviera, E. (1997). La variabilité des débits de l'Amazone à Obidos (Amazonas, Brèsil). *In:* "Sustainability of Water Resources under Increasing Uncertainty" (D. Rosbjerg, N.E. Boutayeb, Z.W. Kundzewicz, A. Gustard, and P.F. Rasmussen, Eds.), Vol. 240, pp. 163–172. IAHS, Wallingford.

Carson, M.A., and Kirkby, M.J. (1972). "Hillslope, Form and Process." Cambridge, New York.

Church, M. (1996). Channel morphology and typology. *In:* "River Flows and Channel Forms: Selected Extracts from the Rivers Handbook" (G. Petts and P. Calow, Eds.), pp. 185–203. Blackwell Science, Oxford.

Devol, A.H., and Hedges, J.I. (2001). Organic matter and nutrients in the mainstem Amazon River. *In:* "The Biogeochemistry of the Amazon Basin" (M.E. McClain, R.L. Victoria, and J.E. Richey, Eds.), pp. 275–306. Oxford University Press, Oxford.

Dodds, W.K., Smith, V.H., and Lohman, K. (2002). Nitrogen and phosphorus relationships to benthic algal biomass in temperate streams. *Can. J. Fish. Aquat. Sci.* 59 (5), 865–874.

Downing, J.A., McClain, M., Twilley, R., Melack, J.M., Elser, J., Rabalais, N.N., Lewis, W.M. Jr, Turner, R.E., Corredor, J., Soto, D., Yanez-Arancibia, A., Kopaska, J.A., and Howarth, R.W. (1999). The impact of accelerating land-use change on the N-cycle of tropical aquatic ecosystems: current conditions and projected changes. *Biogeochemistry* 47, 109–148.

Drever, J.I. (1997). "The Geochemistry of Natural Waters: Surface and Groundwater Environments." 3rd Edition. Prentice-Hall, Upper Saddle River.

Gibbs, R.J. (1967). The geochemistry of the Amazon River System: I. The factors that control the salinity and the composition and concentration of the suspended solids. *Bull. Geol. Soc. Am.* 78, 1203–1232.

Jackson, I.J. (1989). "Climate, Water and Agriculture in the Tropics." Longman Group, England.

Kessler, W.S., McPhaden, M.J., and Weickmann, K.M. (1995). Forcing of intraseasonal Kelvin waves in the equatorial Pacific. *J. Geophys. Res.* **100**, 10613–10631.

Lesack, L.F.W. (1988). "Mass Balance of Nutrients, Major Solutes and Water in an Amazon Floodplain Lake and Biogeochemical Implications for the Amazon Basin". Ph.D. Thesis. University of California, Santa Barbara, CA, 484 pp.

Lewis, W.M. Jr (1986). Nitrogen and phosphorus runoff losses from a nutrient-poor tropical moist forest. *Ecology* **67**, 1275–1282.

Lewis, W.M. Jr (1987). Tropical limnology. *Annu. Rev. Ecol. Syst.* **18**, 159–184.

Lewis, W.M. Jr (1998). Aquatic environments of the Americas: basis for rational use and management. *Proc. 4th Int. Congr. Environ. Issues* **4**, 250–257.

Lewis, W.M. Jr (2002). Yield of nitrogen from minimally disturbed watersheds of the United States. *Biogeochemistry* **57/58**, 375–385.

Lewis, W.M. Jr, and Saunders, J.F. III. (1989). Concentration and transport of dissolved and suspended substances in the Orinoco River. *Biogeochemistry* **7**, 203–240.

Lewis, W.M. Jr, Saunders, J.F. III, Levine, S.N., and Weibezhan, F.H. (1986). Organic carbon in the Caura River, Venezuela. *Limnol. Oceanogr.* **31**, 653–656.

Lewis, W.M. Jr, Weibezahn, F.H., Saunders, J.F. III, and Hamilton, S.K. (1990). The Orinoco River as an ecological system. *Interciencia* **15**, 346–357.

Lewis, W.M. Jr, Hamilton, S.K., and Saunders, J.F. III. (1995). Rivers of Northern South America. *In:* "Ecosystems of the World: River and Stream Ecosystems" (C.E. Cushing, K.W. Cummins, and G.W. Minshall, Eds.), Vol. 22, pp. 219–256. Elsevier Science B.V., Amsterdam.

Lewis, W.M. Jr, Melack, J.M., McDowell, W.H., McClain, M., and Richey, J.E. (1999). Nitrogen yields from undisturbed watersheds in the Americas. *Biogeochemistry* **46**, 149–162.

List, R.J. (1951). Smithsonian meteorological tables. *Smith. Misc. Coll.* **114**, 411–447.

McGregor, G.R., and Nieuwolt, S. (1998). "Tropical Climatology: An Introduction to the Climates of the Low Latitudes." 2nd Edition. Wiley, Chichester, UK.

Meybeck, M. (1976). Total dissolved transport by world major rivers. *Hydrol. Sci. Bull.* **21**, 265–284.

Meybeck, M. (1979). Concentrations des eaux fluviales en elements majeurs et apports en solution aux oceans. *Rev. Geol. Dyn. Geogr. Phys.* **21**, 215–246.

Meybeck, M. (1982). Carbon nitrogen and phosphorus transport by world rivers. *Am. J. Sci.* **287**, 401–450.

Meybeck, M., and Helmer, R. (1989). The quality of rivers: from pristine state to global pollution. *Palaeogeogr. Palaeocl. Palaeoclim. (Global Planet. Change Sect.)* **75**, 283–309.

Milliman, J.D., and Meade, R.H. (1983). World-wide delivery of river sediment to the oceans. *J. Geol.* **91**, 1–21.

National Oceanic and Atmospheric Administration website: http://www.noaa.gov/ (2004).

Ortiz-Zayas, J.R. (1998). "The Metabolism of the Rio Mameyes, Puerto Rico: Carbon Fluxes in a Tropical Rain Forest River". Ph.D. Thesis. University of Colorado, Boulder, CO. 344 pp.

Ortiz-Zayas, J.R., Lewis, W.M. Jr, Saunders, J.F. III, McCutchan, J.H. Jr, and Scatena, F. (2005). Metabolism of a tropical rainforest stream. *J. N. Am. Benthol. Soc.* **24**(4), 769–783.

Petr, T. (1983). Limnology of the Purari basin. *In:* "The Purari – Tropical Environment of a High Rainfall River Basin." (T. Petr, Ed.), pp. 141–178. Dr. W. Junk Publishers, The Hague.

Saunders, J.F. III, and Lewis, W.M. Jr (1989). Transport of major solutes and the relationship between solute concentrations and discharge in the Apure River, Venezuela. *Biogeochemistry* **8**, 101–113.

Schlesinger, W.H. (1997). "Biogeochemistry: An Analysis of Global Change." 2nd Edition. Academic Press, San Diego.

Seyler, P.T., and Boaventura, G. (2001). Trace elements in the mainstem Amazon River. *In:* "The Biogeochemistry of the Amazon Basin." (M.E. McClain, R.L. Victoria, and J.E. Richey, Eds.), pp. 307–327. Oxford University Press, New York.

Sioli, H. (1984). The Amazon and its main affluents: Hydrography, morphology of the river courses, and river types. *In:* "The Amazon Limnology and Landscape Ecology of a Mighty Tropical River and its Basin" (S. Sioli, Ed.), pp. 127–165. Dr. W. Junk Publishers, The Hague.

Stallard, R.F. (1985). River chemistry, geology, geomorphology, and soils in the Amazon and Orinoco Basins. *In:* "The Chemistry of Weathering." (J.I. Drever, Ed.), pp. 293–316. Reidel, Dordrecht.

Stallard, R.F., and Edmond, J.M. (1983). Geochemistry of the Amazon, 2. The influence of geology and weathering environment on the dissolved load. *J. Geophys. Res.* **88**, 9671–9688.

Talling, J.F., and Lemoalle, J. (1998). "Ecological Dynamics of Tropical Inland Waters." Cambridge University Press, Cambridge.

Vorosmarty, C.J., Fekete, B.M., and Tucker, B.A. (1998). Global River Discharge, 1807–1991, Version 1.1 (RivDis). Oak Ridge National Laboratory Distributed Archive Data Center, Oak Ridge, TN, USA (http://www.daac.ornl.gov).

Waliser, D.E., and Gautier, C. (1993). A satellite derived climatology of the ITCZ. *J. Clim.* **6**, 2162–2174.

Ward, J.V. (1985). Thermal characteristics of running waters. *Hydrobiologia* **125**, 31–46.

Wezel, R.G. (2001). "Limnology." 3rd Edition. Academic Press, New York.

# 2

# Primary Production in Tropical Streams and Rivers

Peter M. Davies, Stuart E. Bunn, and Stephen K. Hamilton

Net primary production is a fundamental ecological process that reflects the amount of carbon synthesized within an ecosystem, which is ultimately available to consumers. Although current ecosystem models of streams and rivers have placed variable emphasis on the importance of in-stream primary production to aquatic food webs, recent research indicates that aquatic algae are a significant contributor to food webs in tropical rivers and streams. This is in contrast to many well-studied north temperate latitude streams in deciduous forest ecosystems, which are thought to depend mainly upon terrestrial leaf litter and detritus-based food webs.

A review of available literature suggests that rates of in-stream primary production in tropical regions are typically at least an order of magnitude greater than comparable temperate systems. Although nutrient status can significantly modify rates, the ultimate driver of aquatic primary production is light availability. Rates of benthic gross primary productivity in tropical streams range from 100 to 200 $mg\,C\,m^{-2}\,d^{-1}$ under shaded conditions to much higher values associated with open canopies. Light inputs to the channel can be controlled by stream orientation, with east-west channels receiving much more light compared to those orientated north-south. Rates of production for large tropical rivers are similar to those for streams, although factors that regulate production are different and hence they respond differently to human impact. Values for rivers range from 10 to 200 $mg\,C\,m^{-2}\,d^{-1}$ to more than 1000 $mg\,C\,m^{-2}\,d^{-1}$. Production is often limited by turbidity, which tends to be at a maximum after high flow events. In polluted tropical rivers, productivity responds to nutrient enrichment and can attain rates of 6000 $mg\,C\,m^{-2}\,d^{-1}$.

The highest rates of production in tropical river systems typically occur in floodplains subject to seasonal inundation, where aquatic vascular plants dominate total productivity. These macrophytes (herbaceous vascular plants that can be primarily terrestrial or aquatic) can proliferate *in situ* or be transported from upstream. Rooted aquatic plants with emergent or floating leaves respond to the rising water level, sometimes elongating their stems at a rate of 20 $cm\,d^{-1}$, and many terrestrial plants tolerate prolonged submergence. These ecosystems can attain very high rates of primary production that rival those of intensively managed agro-ecosystems. Floodplain forest can also be a productive component of these ecosystems. Both attached algae (periphyton) and phytoplankton contribute substantially to algal production in floodplain waters. Floodplains are important for fodder and for nursery habitat for fish, which re-invade main channels when floods recede.

Tropical rivers may flow into coastal mangrove ecosystems, where rates of productivity are variable and often dependent of methodologies of measurement. Rates of mangrove production range from $1300\,mg\,C\,m^{-2}\,d^{-1}$ in the Términos Lagoon, Mexico to $1900-2700\,mg\,C\,m^{-2}\,d^{-1}$ in the Fly River estuary (Papua New Guinea). However, rates of phytoplankton growth within mangrove forests are low, probably controlled by shading and turbidity, and are comparable to those of tropical streams.

As pressures for water resource development intensify, tropical fluvial ecosystems are coming under increasing pressure. It is important to understand how these ecosystems function and to ensure problems of developing water resources in temperate regions are not repeated in the tropics.

## I. INTRODUCTION

Carbon is the essential building block of living tissue and the primary production of organic carbon fuels ecosystems and sustains populations of consumers. In stream and river ecosystems, photosynthesis by either terrestrial or aquatic primary producers can provide the initial carbon source, through the conversion of light energy into plant biomass. The total amount of carbon fixed is termed gross primary production (GPP) and is usually expressed per unit area and time. Net primary production (NPP) is the amount of carbon allocated to plant biomass after the losses from respiration (R) associated with maintenance of cellular processes (Bott, 1983). This primary production can occur above or below the water surface within the channel, or as terrestrial production on land adjacent to the water that is subsequently transported to the water, or as aquatic or terrestrial production in seasonally inundated floodplains fringing large rivers.

Streams and rivers have long been regarded as somewhat unusual in the way they function as ecosystems. They are considered to be open systems, in the sense that energy and essential nutrients constantly 'leak' downstream and are effectively lost to primary producers or consumers at any particular stream or river reach, with little opportunity for return or recycling. They are also considered strongly heterotrophic; that is, often far more carbon is present and consumed than is produced within the system (i.e. $GPP \ll R$) (Webster and Meyer, 1997; Bunn *et al.*, 1998a; Thorp and Delong, 2002). Consequently, much of the carbon 'available' in particulate or dissolved form in a stream or river may be derived from the surrounding catchment. This is particularly true for small temperate forest streams, which have been the subject of much study and upon which many of our prevailing ecological models have been derived (Fisher and Likens, 1973; Cummins, 1974; Gregory *et al.*, 1991). The strong riparian control of the light regime in such streams effectively limits aquatic production and food webs are considered to be strongly dependent on terrestrial sources of carbon, supplied largely from vascular plant growth in the riparian zone.

However, not all stream and river systems show such a strong dependence on catchment-derived carbon (see also Chapter 1 of this volume). Where sunlight reaches the channel, in-stream sources can be significant and, in the extreme case of some desert streams, autochthonous production can significantly exceed that derived from the catchment (Bunn *et al.*, 2005). There are also marked shifts in the importance of autotrophic production moving down the stream hierarchy (Rosemond *et al.*, 2002), as the degree of direct riparian canopy shading over the river diminishes (e.g. Lamberti and Steinman, 1997).

There is still considerable debate about the relative importance of aquatic versus catchment-derived sources of organic production to food webs in larger river systems (see Bunn *et al.*, 1998b; Bunn *et al.*, 2003). Although terrestrial sources of carbon derived either from upstream processes (Vannote *et al.*, 1980) or from lateral exchange during or following floods (Junk *et al.*, 1989) are considered to be a major contributor to the food webs of large rivers, there is a growing view that these earlier models of ecosystem function have understated the role of autochthonous sources (Lewis *et al.*, 2001; Thorp and Delong, 2002; Bunn *et al.*, 2003; Winemiller, 2004).

Tropical river systems undoubtedly sit at the high end of the global productivity spectrum and the wet and dry forest catchments they drain have the highest rates of primary production of any terrestrial biome (e.g. 1000–3500 g dry wt $m^{-2} yr^{-1}$) (Whittaker and Likens, 1975). They are characterized by climatic conditions that often promote year-round growth, with less seasonal variation in solar irradiance than temperate latitudes. However, greater short-term variation can occur in tropical systems due to changes in atmospheric moisture (Lewis *et al.*, 1995). Seasonal changes in flow regimes and hydrological connectivity clearly regulate how tropical systems function (Douglas *et al.*, 2005). Australian rivers represent the extreme in hydrologic variability, with flows that are always strongly seasonal (McMahon *et al.*, 1991); high inter-annual variability also can occur as a consequence of the El Niño Southern Oscillation (Hamilton and Gehrke, 2005).

In undisturbed tropical streams and rivers, dissolved forms of phosphorus and nitrogen are low but present in sufficient concentrations to support moderate to high biomass of autotrophs (see Chapter 1 of this volume). Much of the ecosystem production that supports aquatic food webs in floodplain rivers may occur outside the main channel, in backwaters and floodplains. Given these features, it is perhaps not surprising that recent studies suggest a far greater dependence of tropical river food webs on aquatic primary production than their temperate equivalents (Lewis *et al.*, 2001; Douglas *et al.*, 2005). This highlights the importance of understanding the factors that influence composition and production of aquatic plants in tropical systems.

Previous reviews of data on the primary productivity of streams and rivers included only a few examples of tropical systems (e.g. Lamberti and Steinman, 1997), in contrast to the relatively large number from other biomes (Cushing *et al.*, 1995). An additional constraint is that the use of different methodologies and units of measurement have made it difficult to derive robust generalizations from existing data sets. In this chapter, we consider the major groups of aquatic primary producers in tropical stream and river systems and the factors that influence their composition and production. We have structured this around different catchment settings – to compare and contrast the differences between small streams, large rivers, floodplains, and mangrove estuaries that occur at the mouths of many tropical rivers. Additional information on tropical floodplains and their wetlands can be found in Chapter 7 of this volume. In each catchment setting, we also comment, where information is available, on the relative importance of aquatic primary production to tropical stream and river food webs.

## II. SMALL STREAMS

### A. Primary Producers

Much of the primary production within the channels of streams and rivers, particularly those with fast flowing water, is typically restricted to benthic habitats involving algae and cyanobacteria (Canfield Jr and Hoyer, 1988; Lamberti and Steinman, 1997). The distribution, biomass and production of aquatic plants in small streams are controlled by a number of factors, among which light availability is the most important (Boston and Hill, 1991; Hill, 1996). Optimum light requirements differ for various aquatic plant groups and there is evidence that light intensity is a major factor determining the composition of algal communities (Ghosh and Gaur, 1994; Hill, 1996; Larned and Santos, 2000; Mosisch *et al.*, 2001; Bixby *et al.*, 2003). For example, green algae (chlorophytes), as a group, require much higher light intensities than diatoms or cyanobacteria (Langdon, 1988) and some, e.g. the filamentous chlorophyte *Spirogyra*, are unable to tolerate low light conditions (Graham *et al.*, 1995).

## B. Factors Controlling Productivity in Small Streams

Although solar radiation in the tropics is higher than in temperate regions, moist forest canopies over streams can reduce irradiance at the valley floor to 1% of that in the upper canopy (Brinkman, 1985). The high angle of the sun often leads to greater inputs of light to the stream channel than in temperate regions, especially in channels with an east-west orientation (Fig. 1). In these situations, the sun can track along an open channel with little direct interception by the riparian vegetation and consequently below-canopy light intensities remain high for many months of the year. In higher latitudes, the path of the sun is effectively screened by riparian vegetation for much of the year and below-canopy light intensities rarely attain the high levels of the tropics. These latitudinal differences in light regime can have a marked effect on the spatial and temporal dynamics of aquatic plants. For example, below-canopy light intensity in the tropical stream in Fig. 1 would exceed the likely threshold for abundant growth of filamentous chlorophytes (approximately $13 \, mol \, m^{-2} \, d^{-1}$) (Langdon, 1988) for almost 180 days of the year, whereas the same stream at higher latitudes would experience this for only about half that period (Fig. 1).

As well as channel orientation, the structural characteristics of the riparian vegetation can also have an influence on the below-canopy light regime in forest streams (Davies *et al.*, 2004). The three rainforest streams in Fig. 2 are part of the same subcatchment in south-east Queensland, Australia and have similar catchment sizes, channel dimensions and total percent canopy cover (Bunn *et al.*, 1999a,b). The east-west flowing stream (Peters Creek) experiences much higher below-canopy light levels than the other two, which rarely exceed levels known to stimulate growth of filamentous green algae.

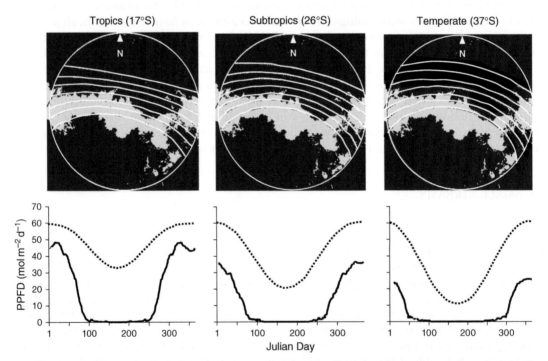

FIGURE 1   Hemispherical images of the riparian canopy above a small rainforest stream in eastern Australia (13 m wide) showing the effect of latitude (tropics, subtropics, and temperate zone). The white curves indicate the trajectory of the sun during each month of the year (Bunn *et al.*, 1999b). The associated graphs show estimated photosynthetically active photon flux density above canopy (dotted line) and below canopy (solid line) throughout the year.

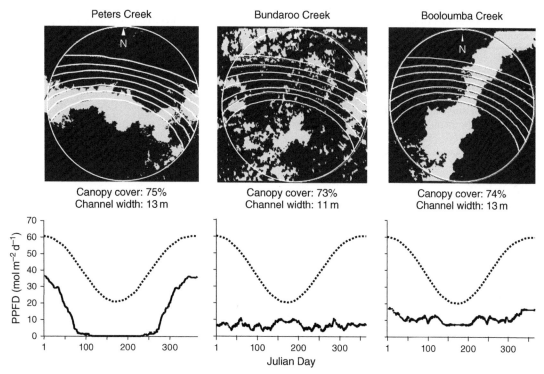

**Peters Creek**     **Bundaroo Creek**     **Booloumba Creek**

Canopy cover: 75%
Channel width: 13 m

Canopy cover: 73%
Channel width: 11 m

Canopy cover: 74%
Channel width: 13 m

*FIGURE 2*   Hemispherical images of the riparian canopy above three small subtropical rainforest streams in southeast Queensland, Australia showing the influence of channel orientation and canopy characteristics on potential light inputs. The white curves indicate the trajectory of the sun during each month of the year (Bunn *et al.*, 1999b). Canopies here are evergreen. The associated graphs show estimated photosynthetically active photon flux density above canopy (dotted line) and below canopy (solid line) throughout the year.

Given the higher solar intensity and angle of the sun, it is not surprising that rates of benthic primary production in small tropical streams are at the high end of the range recorded for forest stream ecosystems (Table I; see also Lamberti and Steinman, 1997). Tropical forest streams typically have rates of benthic GPP in the range of 100–200 mg C m$^{-2}$ d$^{-1}$, with much higher values in the dry tropical streams characterized by more open canopies (e.g. tributaries of the Ord River in north-western Australia; Table I). In some pristine seasonal dry forests and montane forests the riparian vegetation may not be sufficiently dense to fully shade streams (Lewis *et al.*, 1995).

The strong control of riparian shading on primary production in tropical and subtropical streams is evident in disturbed systems where the canopy has been wholly or partially removed. Significant differences in the standing stock of periphyton in Hong Kong streams were regulated by differences in canopy cover, with the lowest standing stocks occurring in shaded streams (Dudgeon, 1988). Similarly, Mosisch *et al.* (2001) showed light availability was the overriding factor controlling the accrual of algal biomass in subtropical streams in Australia, even in the presence of elevated nutrient concentrations. Bunn *et al.* (1999a) also found that much of the observed variation in benthic primary production on cobbles in small streams in the same region was explained by riparian canopy cover (68%), with land use (% crops and pasture) accounting for only an additional 14%. The measured increase in benthic primary production with decreasing riparian canopy cover was accompanied by a shift in aquatic plant composition, from light-sensitive diatoms and cyanobacteria in low-light conditions to filamentous green algae and ultimately macrophytes in systems flowing through cleared riparian zones.

*TABLE I* Rates of Primary Production (mg C m$^{-2}$ d$^{-1}$ Unless Noted Otherwise) in Selected Tropical Streams and Rivers

| Site | Comment | Rate | Source |
|---|---|---|---|
| Small streams | | | |
| Tributaries of the Ord River, northwestern Australia | Tropical dry savannah, seasonal means from three streams | 280 ± 30 | Douglas *et al.* (2005) |
| Tributaries, Mary River, southeast Queensland, Australia | Subtropical forest, ~70% canopy cover, ~10 km$^2$ catchment area (winter measurements only) | 152 ± 58 | Bunn *et al.* (1999a) |
| Tributaries, Johnstone River, far north Queensland, Australia | Tropical rainforest, ~80% canopy cover, ~2 km$^2$ catchment area (S = summer, W = winter) | S = 198 ± 31; W = 164 ± 24 | Bunn *et al.* (1999a) |
| Tributaries of Rio Mameyes Basin, Puerto Rico | Tropical rainforest, undisturbed | 150 (variance = 22.5) | Ortiz-Zayas *et al.* (2003) |
| Frijoles River, Panama | Tropical rainforest, small streams: open pools | 1130 | Power (1984) |
| | Tropical rainforest, small streams: shaded pools | 970 | |
| Limón River, Venezuela | Second order, undisturbed (diatoms only) | 1200 | Lewis and Weibezahn (1976) |
| Sungai Gombak, Malaysia | Periphyton productivity (range of five sites), small river system | 5–113 | Bishop (1973) |
| Itaqueri and Lobo Rivers, São Paulo State, Brazil | | 34–90 | de Oliveira and Calijuri (1996) |
| Tai Po Kau Stream, Hong Kong | Upland streams | | Dudgeon (1983) |
| | Unshaded | 81.2 | |
| | Shaded | 33.2 | |
| River Shatt Al-Arab, Iraq | Arid zone planktonic production | 145–888 | Huq *et al.* (1981) |
| Quebrada Bisley, Puerto Rico | First order, intact tropical forest; GPP based on diel oxygen fluxes | 26 | Mulholland *et al.* (2001) |
| *Hugh White Creek (Cowetta Hydrologic Laboratory)* | *Second order, mature deciduous forest* | *15.9* | *Webster et al. (1997)* |

(*continued*)

TABLE I    (*continued*)

| Site | Comment | Rate | Source |
|------|---------|------|--------|
| *Unnamed stream, H.J. Andrews Experimental Forest, Oregon, USA* | *First order, montane coniferous forest, logged in 1972–1974.* | 211 | *Triska et al. (1982)* |
| Bear Brook, New Hampshire, USA | *Second order, deciduous forest* | 9.6 | *Findlay et al. (1997)* |
| Rivers | | | |
| Ord River, north western Australia | Tropical dry savannah. Regulated sites, seasonal means, benthic production | 340 ± 40 | Douglas *et al.* (2005) |
| Lower river reaches (Puente Roto, La Vega), Rio Mameyas basin, Puerto Rico | Third order, intact tropical forest | 638 (variance = 103) | Ortiz-Zayas (2003) |
| Orinoco River, Venezuela | Phytoplankton in main stem (unregulated; moderate turbidity) | 19–43 | Lewis (1988) |
| | Phytoplankton in tributaries (variable turbidity) | 4–26 | |
| Paraná River, Argentina/Paraguay | Upper Paraná (200 km upstream of the Paraguay River confluence); phytoplankton production (high turbidity) | 10 (floods) to 1 000 | Bonetto (1986) |
| Negro River (Amazon Basin) | Acidic blackwater river | 141 | Lewis *et al.* (1995) |
| Tapajós River (Pará, Brazil) | Clearwater river with lacustrine conditions near mouth | 1370 | Schmidt (1982) |
| River Champanala, India | Side spill channel of the Ganges at Bhagalpur. Macrophyte production | 3900 | Hasan (1988) |
| River Mahanadi at Sambalpur in Orissa, India | Degraded river system | 1030–1780 (mean = 1 419) | Patra *et al.* (1984), cited in Dudgeon (1995) |
| Kahn in India | Sewage and industrial discharges | 6070 | Rama Rao *et al.* (1979) |
| Testa River system in Sikkim Himalayas, India | Phytoplankton production | 440–2180 (per m³) | Venu *et al.* (1985) |

(*continued*)

TABLE I    (*continued*)

| Site | Comment | Rate | Source |
|------|---------|------|--------|
| Ganges, India | GPP limited by turbidity | 910–1020 | Natarjan (1989), cited in Dudgeon (1995) |
| Nile River | Phytoplankton production | 1500–4000 | Talling (1976) |
| *Matamek River, Quebec, Canada* | *Sixth order, boreal coniferous forest* | 726 | *Naiman (1983)* |
| **Floodplains** | | | |
| Orinoco River fringing floodplain | Annual NPP divided by 365 days $yr^{-1}$ | | |
| | $C_4$ grasses (based on maximum biomass) | 1000 | Hamilton and Lewis (1987) |
| | Phytoplankton | 904 | Lewis *et al.* (2001) |
| | Periphyton on floating macrophytes | 685 | Lewis *et al.* (2001) |
| Amazon River fringing floodplain | Emergent vascular plants (some with both terrestrial and aquatic growth phases); annual NPP multiplied by 50% C and divided by 365 days $yr^{-1}$ | 50–100 t $ha^{-1}$ $yr^{-1}$ = 7000–14000 mg C $m^{-2}$ $d^{-1}$ | Junk and Piedade (1993); Morison *et al.* (2000) |
| | Floodplain forest; annual NPP multiplied by 50% C and divided by 365 days $yr^{-1}$ | 27–37 t $ha^{-1}$ $yr^{-1}$ = 3700–5000 mg C $m^{-2}$ $d^{-1}$ | Worbes (1997) |
| | Phytoplankton (floodplain lakes) | 800 | Schmidt (1973) |
| | | 680 | Melack and Fisher (1990) |
| | Periphyton: On floating macrophytes | 1200 | |
| | On submersed forest leaves | 760 | Melack and Forsberg (2001) |
| **Mangroves** | | | |
| Daintree River, *Rhizophora* forest | Biomass accumulation (net above ground dry biomass accrual) | 25.9 t $ha^{-1}$ $yr^{-1}$ = 3550 mg C $m^{-2}$ $d^{-1}$ | Clough (1992) |
| Thai Mangrove forest | Benthic community metabolism | 110–180 | Kristensen *et al.* (1988) cited in Alongi (1994) |
| | Mangrove production | 1900–2750 | |

(*continued*)

TABLE I   *(continued)*

| Site | Comment | Rate | Source |
|------|---------|------|--------|
| Fly River Estuary, Papua New Guinea | Phytoplankton Mangrove | 180 1900–2700 | Robertson *et al.* (1991, 1993) |
| Terminos Lagoon, Mexico | Phytoplankton Mangrove | 1200 1300 | Day *et al.* (1982) |

Some examples from temperate forest biomes are also included for comparison (shown in italics). Unless noted otherwise, these estimated represent either gross primary production (GPP) for benthic algae and phytoplankton, or net primary production (NPP) for vascular plants (often based on maximum standing plant biomass). To facilitate comparisons, annual values reported by some authors have been scaled to daily rates (or vice versa), but include the assumption that production rates were stable throughout the year.

In extreme cases, the lack of riparian vegetation can lead to explosive growths of aquatic and semi-aquatic plants. For example, in a highly disturbed stream flowing through sugar cane fields in the wet tropics of northern Australia, Bunn *et al.* (1998a) reported that the invasive grass *Urochloa mutica* achieved a growth rate (above ground) of $2.8 \, g$ dry wt $m^{-2} \, d^{-1}$, reaching a standing biomass of over $1.3 \, kg$ dry wt $m^{-2}$. This not only had a major influence on aquatic habitat but also trapped sufficient sediment to reduce channel capacity and subsequently increase flooding frequency (Bunn *et al.*, 1998a).

Such human-induced shifts in the production and composition of aquatic plants have important consequences for higher order consumers in tropical streams. There is growing evidence that food webs in tropical forest streams are strongly dependent on algal sources of carbon (see Douglas *et al.*, 2005). Consequently, disturbance of the riparian zone, accompanied by increased nutrient inputs (e.g. Ramos-Escobedo and Vázquez, 2001; Rosemond *et al.*, 2002), may stimulate higher primary production. This extra energy is not necessarily incorporated into secondary production because increases in light intensity tend to shift the aquatic plant composition from palatable microalgae to filamentous algae and macrophytes, which are not readily grazed by many primary consumers. In the case of invasive macrophytes, such as the African Para grass *Urochloa mutica*, there was no evidence of incorporation of its distinctive $C_4$ carbon isotope signature into the local aquatic food web in Australia (Bunn *et al.*, 1997a; Clapcott and Bunn, 2003), which accords with the results of work in the Orinoco and Amazon floodplains on the minor importance of native aquatic $C_4$ grasses in aquatic food webs (Hamilton *et al.*, 1992; Forsberg *et al.*, 1993).

Although shading by riparian vegetation and the resulting below-canopy light regime have the greatest influence on the composition and production of aquatic plants in tropical streams, other factors can also play an important role. In the absence of pollution, nutrient levels in tropical streams and rivers are low (Pringle *et al.*, 1986; Douglas *et al.*, 2005; but see Chapter 1 of this volume) and nutrient limitation can have a significant influence on aquatic primary production in situations where light is not limiting. In disturbed subtropical streams with low riparian shading, Mosisch *et al.* (2001) found that nitrogen was a primary limiting nutrient for algal growth (on artificial substrates). Similar evidence of primary nitrogen limitation has been found in about 70% of the streams in south-eastern Queensland (Udy and Dennison, 2005) and in other tropical (e.g. Downing *et al.*, 1999; Flecker *et al.*, 2002) and temperate streams (Hill and Knight, 1988).

Physical disturbance of the stream bed from episodic or seasonal high flow events can have a major effect on algal biomass in subtropical and tropical streams (Pringle *et al.*, 1986; Mosisch and Bunn, 1997; Pringle and Hamazaki, 1997; Townsend and Padovan, 2005). Similarly, Dudgeon (1982) recorded a substantial reduction in standing stock of periphyton from 323.4

to $9.6\,mg\,m^{-2}$ in a Hong Kong forest stream during 1 month in the wet season. High flow events are often associated with increased turbidity, also limiting in-stream primary production (Lewis *et al.*, 1995).

Top-down control by fish and shrimps in tropical streams and rivers has been shown to exert a strong influence on standing stocks of benthic organic matter, nutrients and algal communities (e.g. Wootton and Oemke, 1992; Pringle and Blake, 1994; Kent, 2001; Flecker *et al.*, 2002; Wirf, 2003; de Souza and Moulton, 2005). Pringle and Hamazaki (1997) found that fish played a key role in maintaining the stability of benthic algal assemblages and influenced resistance to storm events. Although these conspicuous macroconsumers appear to have a disproportionately important role in some tropical stream ecosystems (Douglas *et al.*, 2005), aquatic insects (e.g. Moulton *et al.*, 2004; Barbee, 2005) and tadpoles (Ranvestel *et al.*, 2004) can also have a strong influence on algal stocks. Power (1987) showed the distribution of algae in Panamanian streams was regulated by armoured catfish (Loricariidae), and a distinctive 'bath-tub ring' of algae developed in shallower regions beyond the grazing range of these fish. The impacts of grazing on rates of primary production, as opposed to standing stocks of biomass, have been less frequently investigated due to methodological challenges. It is possible that some grazers actually stimulate primary production by removing senescent material and sediment (Pringle *et al.*, 1993a) and by regenerating nutrients. However, the high turnover in algal biomass often obscures this effect (Allan 1995).

## III. RIVERS

### A. Primary Producers: Macrophytes and Phytoplankton

Tropical rivers are poorly understood compared to their temperate counterparts. Generally, rivers draining undisturbed tropical catchments are low in nutrients and consequently aquatic primary production is low (Douglas *et al.*, 2005), although moderate levels of algal biomass can accumulate (see also Chapter 1 of this volume). For example, phytoplankton biomass measured as chlorophyll *a* concentrations is low in the main stem of the major rivers and tributaries of rivers such as the Orinoco and Amazon (Hamilton *et al.*, 1992; Lewis *et al.*, 1995) and the Daly in Australia (Webster *et al.*, 2005). Diatoms (e.g. *Melosira*) can dominate the phytoplankton of large South American rivers (Lewis *et al.*, 1995), particularly in systems with relatively high reactive silica. A low diversity of chlorophytes seems typical of tropical rivers but is unusual given their importance in oligotrophic waters of higher latitudes, since chlorophytes typically require higher light intensities than diatoms (Richardson *et al.*, 1983; Langdon, 1988). The open channels of large tropical rivers are generally unsuitable habitats for aquatic macrophytes due to a range of factors including the depth and often highly variable water level, high turbidities and associated suspended sediment loads, high current velocities and mobile bed materials (Lewis *et al.*, 1995; Vieira and Necchi, 2003). However, their shorelines and floodplains can support luxuriant growths of macrophytes, as will be discussed below.

Generally, highest phytoplankton densities in tropical rivers occur during low-water periods when residence time is longest and turbidity may be diminished (di Persia and Neiff, 1986). For example, phytoplankton densities recorded in rivers in Côte d'Ivoire during floods were very low contrasting the maximum density observed during low flow periods (Lévêque, 1995). Similarly, seasonal studies of riverine phytoplankton in rivers and streams of tropical Asia showed highest densities of phytoplankton occurred during the dry season which declined during the monsoon due to the combination of dilution and increased turbidity (Dudgeon, 1995).

Given the low background levels of nutrients, eutrophication of tropical rivers and estuaries is usually a consequence of run-off from agricultural activities (Shehata and Bader, 1985; Finlayson *et al.*, 2005). Rates of production for large tropical rivers range from lows of

$10$–$200\,\text{mg}\,\text{C}\,\text{m}^{-2}\,\text{d}^{-1}$ to $>1000\,\text{mg}\,\text{C}\,\text{m}^{-2}\,\text{d}^{-1}$ in relatively unpolluted systems (Table I), and may exceed $6000\,\text{mg}\,\text{C}\,\text{m}^{-2}\,\text{d}^{-1}$ in polluted systems (e.g. the Kahn in India receiving both raw sewage and industrial discharge).

## B. Factors Controlling Productivity in Tropical Rivers

High turbidity is an important regulatory factor for phytoplankton growth in large tropical rivers (Lewis, 1988; Bonetto and Wais, 1995; Chapter 1 of this volume). For example, Bonetto (1986) recorded peak phytoplankton production in the upper Paraná River (about $200\,\text{km}$ upstream of Paraguay River confluence) occurring at lower discharges ($\sim1000\,\text{mg}\,\text{C}\,\text{m}^{-2}\,\text{d}^{-1}$), with much lower rates during high discharge ($\sim10\,\text{mg}\,\text{C}\,\text{m}^{-2}\,\text{d}^{-1}$) due to high turbidity. Similarly, low phytoplankton productivity in the Orinoco River main stem is a function of low light penetration in a relatively deep water column (Lewis, 1988).

Flows have also been shown to regulate algal standing biomass in tropical rivers (Pringle *et al.*, 1993b). For example, Townsend and Padovan (2005) showed that high flows in the Daly River, in the Northern Territory of Australia, substantially reduced the biomass of *Spirogyra*. Flow is also important because hydrological connectivity between the main channel and back-waters can determine the standing biomass of algae in rivers (Pringle *et al.*, 1993b). Most suspended algae in the lower reaches of the Orinoco River has been transported from upstream pools and anabranches rather than from the floodplain (Lewis, 1988), consequently even minor changes to hydrology and connectivity can substantially regulate the amount of algal biomass in the main stem of the river.

## IV. FLOODPLAINS AND WETLANDS

### A. Primary Producers Associated with Floodplains and Wetlands

Globally, tropical floodplains are among the most productive landscapes, owing to their continual enrichment by import and retention of nutrient-rich sediments from the headwaters and from lateral sources (Tockner *et al.*, in press; Chapter 7 of this volume). For example, in the Inner Delta of the Niger River, over 550,000 people use the floodplain for grazing of about two million sheep and goats during the post-flood dry season (Dugan, 1990).

The seasonal pattern of inundation is clearly the main driver of productivity in floodplain rivers (e.g. Junk *et al.*, 1989; Chapter 7 of this volume) and rates of primary production typically show a high degree of spatial and temporal variation (see Table I). For example, in the Niger River, phytoplankton biomass is significantly elevated in floodplains compared to the main channel (Imevbore and Bakare, 1974 cited in Welcomme, 1986). The effects of flooding and water level fluctuations play an important role in maintaining the diversity of floodplain vegetation in the Amazon (Junk, 1986). The Flood-Pulse Concept identifies seasonal inundation and drainage as the primary driver of ecological processes in large tropical rivers (Junk *et al.*, 1989; Tockner *et al.*, 2000; Chapter 7 of this volume), particularly when the flood pulse is both predictable and of long duration (Junk and Welcomme, 1990) and co-occurs with warm temperatures (Winemiller, 2004). In Australia's wet-dry tropics, as in other tropical floodplain rivers, these seasonal changes in water levels are associated with dramatic shifts in communities of aquatic primary producers (Finlayson *et al.*, 1990; Finlayson, 1993).

Macrophytes appear to be the major primary producers within floodplains (Bayley, 1989; Junk and Piedade, 1993; Lewis *et al.*, 2001; Melack and Forsberg, 2001). On the Central Amazon floodplain, for example, four main groups of primary producers contribute to primary production. Herbaceous macrophytes are most important accounting for an estimated 65% of NPP, followed by flooded forest trees (28%), periphyton (5%), and phytoplankton (2%)

(Melack and Forsberg, 2001; estimates scaled by the relative areas of each habitat across the region). In spite of their modest contribution to overall production, algae (periphyton and phytoplankton) appear to be the main drivers of aquatic food webs in tropical river systems (Lewis *et al.*, 2001; Winemiller, 2004; Douglas *et al.*, 2005). However, this does not mean that macrophytes are unimportant: in addition to providing essential habitat for aquatic invertebrates and fish, they also offer a large surface area of substrate in the photic zone for epiphytic algae (Douglas, Bunn and Davies, unpublished data). It is also clear that native macrophytes play an important role in the diets of some semi-aquatic and terrestrial fauna on the floodplains of rivers in tropical Australia (and elsewhere), particularly for water birds (Whitehead and Tschirner, 1992; Whitehead and Saalfeld, 2000; Kingsford *et al.*, 2004) and mammals (e.g. Redhead, 1979; Madsen and Shine, 1996; Wurm, 1998; for more details, see Chapter 6 of this volume).

On tropical floodplains, rooted aquatic plants with floating stems and emergent leaves (e.g. grasses such as *Oryza*, *Echinochloa*, and *Paspalum*) respond rapidly to rising water levels, sometimes elongating their stems at a rate of $20 \, cm \, d^{-1}$ (Junk, 1986). Many of the rooted species may adopt a temporary free-floating habit – either directly or by colonizing other floating plants – whilst other species are entirely free-floating (e.g. successful and highly-invasive species such as Water lettuce, *Pistia stratiotes*, and Water hyacinth, *Eichhornia crassipes*). Free-floating plants can take advantage of expanding habitat during flood phase and drift with wind and water current onto the flooded areas, which facilitates rapid colonization (Junk, 1986). Seasonal drainage of these areas can result in nearly complete turnover of the herbaceous plant communities, although some species exhibit alternating aquatic and terrestrial growth forms (Junk and Piedade, 1993).

Emergent vascular plants growing along river banks and in floodplain environments of turbid, seasonally fluctuating rivers such as the Amazon, Orinoco, and Paraguay in South America can attain very high rates of primary production (Junk and Piedade, 1993; Morison *et al.*, 2000). Accounting for turnover of biomass, these kinds of plants can attain rates of gross primary production as high as $50–100 \, t \, ha^{-1} \, yr^{-1}$, although their maximum standing biomass is generally much lower, in the range of $4–23 \, t \, ha^{-1}$ (Junk and Piedade, 1993). Among the most productive is *Echinochloa polystachya*, one of several species of $C_4$ grasses estimated to cover ~$20\,000 \, km^2$ of the Central Amazon floodplain (Table I). The primary production of $C_4$ grasses in these floodplain environments rivals that of intensively-fertilized maize crops in warm temperate regions.

Other terrestrial plants are able to tolerate prolonged submergence: for example, the dry-season annual *Ludwigia densiflora* can achieve standing crops of $7–8 \, t$ dry wt $ha^{-1}$ (Junk, 1986). Dense forests are found on many tropical floodplains including the Amazon, where they are somewhat less productive on an areal basis than the floating macrophyte stands but still more productive than nearby upland forests (Worbes, 1997). Floodplains of the River Niger are mostly of the savanna type, where the active growth phase of aquatic grasses (e.g. *Oryza barthii*, *Echinochloa* spp.) is associated with the beginning of the flood period (Welcomme, 1986). On the Middle Niger, *Echinochloa stagnina* can cover up to 40% of the floodplain, and is important for livestock fodder as well as a nursery for juvenile fish.

## B. Factors Controlling Productivity of Floodplains and Wetlands

Water chemistry plays an important role in regulating aquatic primary production in tropical floodplains. The most dominant macrophytes of the Amazon River main stem and its floodplain are virtually absent from its major tributary, the Rio Negro, and its floodplain as a consequence of water chemistry: the Rio Negro is poor in dissolved minerals and nutrients and has a low pH (Lewis *et al.*, 1995; see also Chapter 1 of this volume). Low dissolved inorganic nitrogen concentrations are typical of the floodplain lakes of the Paraná River (Carignan and

Neiff, 1994), where Water hyacinth is the dominant primary producer (Carignan and Neiff, 1992). Many tropical rivers are characterized by nitrogen limitation (Downing *et al.*, 1999), and may be attributable to a relatively high phosphorus supply (e.g. from Andean rivers) and/or high losses of macrophytic nitrogen during floods (Saunders and Lewis, 1988). Algae in Central Amazon floodplain waters appear to shift seasonally between limitation by phosphorus and nitrogen depending on the concentrations of suspended inorganic material that carries available phosphorus (Melack and Forsberg, 2001; see also Chapter 1 of this volume).

There is some evidence that nitrogen may be a limiting nutrient for algal production on the inundated floodplain of the East Alligator River, northern Australia and in small seasonal streams at nearby Kapalga (Kakadu National Park) (Douglas *et al.*, 2005). In both cases, low (negative) stable nitrogen isotope values of epiphytes were indicative of nitrogen-fixation and the periphyton was dominated by cyanobacteria (e.g. *Nostoc*).

## V. MANGROVES

### A. Primary Producers

Mangrove ecosystems form as a consequence of geomorphologically-defined habitats that are clearly a function of catchment processes (Woodroffe, 1992); for that reason, they deserve some consideration in this volume. They often dominate low-energy, muddy, tropical shorelines (typically river deltas), and inhabit a physically evolving habitat that can change at timescales shorter than the lifespan of the vegetation. The most extensive mangroves are found in the Bay of Bengal, the deltas of the Ganges and Bramhaputra rivers and the deltas of the Fly and Putari rivers in Papua New Guinea (Woodroffe, 1992).

Mangroves were once considered productive sources of organic matter, which subsidized near-shore food webs (Robertson and Blaber, 1992). However, increasing evidence has shown this may not always be the case (Loneragan *et al.*, 1997). Nevertheless, like the macrophytes of river floodplains, they are important determinants of habitat even if they do not contribute significantly to aquatic food webs directly through their primary production. Rates of productivity in mangroves are highly variable between systems (Table I) and undoubtedly the function of different methodologies and measurement of different productivity components. Rates of mangrove production range from $1300\,mg\,C\,m^{-2}\,d^{-1}$ in the Terminos Lagoon, Mexico to $1900–2700\,mg\,C\,m^{-2}\,d^{-1}$ in the Fly River Estuary (Table I). However, rates of phytoplankton growth within mangrove forests are low and comparable to tropical streams (Table I).

### B. Factors Controlling Mangrove Production

Mangrove production is controlled, to a large extent, by the distribution of suitable habitat. Factors including turbidity and light attenuation, which regulate primary production in tropical streams and rivers, appear to be unimportant for controlling mangrove productivity. Mangrove habitat is controlled by landform patterns and physical processes, particularly the sedimentary setting (Woodroffe, 1992). Consequently, flow regulation of rivers and subsequent changes to sediment delivery can influence the distribution and productivity of mangroves. In Australia, the erratic behaviour of tropical cyclones and the inherent variability of rainfall as a consequence of the El Niño Southern Oscillation results in highly episodic delivery of water and sediments to mangrove systems (Robertson and Blaber, 1992).

Alterations to the downstream fluxes of nitrogen and cycling within drainage basins can lead to substantial changes in the community structure of mangroves (Alongi, 1994; Downing *et al.*, 1999). Increased levels of catchment-derived enhances primary production, particularly around river deltas and often, in calcareous marine systems, causes a switch from phosphorus to nitrogen limitation (Downing *et al.*, 1999).

## VI. ECOSYSTEM COMPARISONS

Tropical rivers and their associated floodplains are globally the most productive of ecosystems. Rates of primary production in tropical systems are substantially greater by at least an order of magnitude than their temperate counterparts (Fig. 3). Values for Australian tropical stream systems, at about $200\,\mathrm{mg\,C\,m^{-2}\,d^{-1}}$, are less than their South American counterparts ($\sim 1000\,\mathrm{mg\,C\,m^{-2}\,d^{-1}}$; Fig. 3). These differences may be a consequence of larger channels of South American streams, and consequential increases in light, or possibly methodological differences.

Primary production in tropical rivers is variable and typically only a component of overall productivity (e.g. phytoplankton) is measured. With more open channels, the productivity of tropical rivers is elevated over nearby streams (Fig. 3), although turbidity, typically following high-flow events, can limit production (e.g. the Ganges River; Table I). Many of these large systems are densely settled and vulnerable to nutrient enrichment; under such conditions, primary production can attain levels of $4000$–$6000\,\mathrm{mg\,C\,m^{-2}\,d^{-1}}$ (Table I).

Floodplains are the most productive components of tropical river ecosystems (see also Chapter 7 of this volume). Macrophytes dominate overall productivity and, in South American floodplains, may contribute in excess of $10\,000\,\mathrm{mg\,C\,m^{-2}\,d^{-1}}$ (Table I). Some of these plants are transported downstream where they proliferate in newly inundated areas of the floodplain. Rates of algal production in floodplain waters are lower than those of aquatic macrophytes and floodplain forest, and depend on the river water having a residence time of sufficient duration to permit development of a phytoplankton community. The productivity of mangroves is around $1000$–$3000\,\mathrm{mg\,C\,m^{-2}\,d^{-1}}$ (Table I; Fig. 3), and results in a shading effect that may inhibit the growth of phytoplankton.

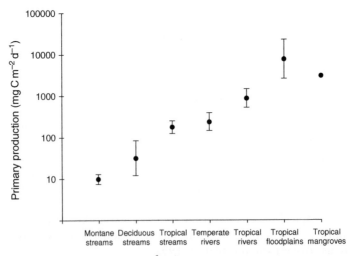

FIGURE 3   Rates of primary production (in mg C m$^{-2}$ d$^{-1}$) for a range of temperate streams, tropical streams, tropical rivers, tropical floodplains and mangrove systems. Sites only included where methodologies for measuring GPP were comparable. Data are from Table I and, for temperate systems largely from Webster and Meyer (1997) and Bunn and Davies (unpublished Australian data). Note, the temperate river data includes some nutrient-enriched sites. For the Amazon and Orinoco floodplains, data represent overall means accounting for the relative areas of each habitat (Lewis et al., 2001; Melack and Forsberg, 2001).

## VII. THREATS

Threats to tropical water resources will continue until there is broad recognition of the ecological values and inherent ecosystem services provided by intact freshwater ecosystems (see also Chapter 10 of this volume). Globalization will undoubtedly increase pressure on water resources, and may be exacerbated by political instability in some regions. In addition, as long as tropical water resources are considered 'free' to human populations (e.g. Junk, 2002), there is little to stimulate development of local capacity to manage these resources wisely.

Currently, the largest remaining topical floodplains are in South America where about 20% of lowlands are flooded annually (for more information, see Chapter 7 of this volume). In Africa also, many large floodplains are still relatively intact (e.g. the Sudd and Congo Basin); however, they are being transformed at an accelerating rate as a result of water management activities; in particular, large-scale irrigation schemes and dam construction (Gordon, 2003). In tropical Asia, many large river floodplains are already densely settled and dominated by humans, with the Ganges being an obvious example, whereas others (such as the Mekong) are less degraded. Downstream in mangroves, threats arise from a variety of human activities including mariculture, forestry and timber cutting, and reclamation (e.g. Pannier, 1979; Ong, 1982).

In Australia, about 70% of the country's freshwater resources are in tropical rivers and wetland systems (Hamilton and Gehrke, 2005). At present, most of them have largely unmodified flow regimes and are generally free from the impacts associated with intensive land uses characteristic of the temperate regions or parts of tropical Asia (Dudgeon, 2000; Douglas *et al.*, 2005). However, invasive aquatic and floodplain weeds, grazing by feral animals, altered fire regimes, and irrigation (Hart, 2004) are major threats that have substantially degraded some tropical Australian systems (Storrs and Finlayson, 1997). For example, 14 of the top 18 weeds in Australia occur in floodplains and wetlands (Humphries *et al.*, 1991; Bunn *et al.*, 1997b), and species such as *Mimosa pigra* and *Urochloa mutica* have had major a impact in undeveloped floodplain river systems of the Australian wet-dry tropics (Douglas *et al.*, 1998).

Tropical river systems are often characterized by strongly seasonal flow regimes, and alterations to these (see Pettit *et al.*, 2001; Chapter 10 of this volume) affect material transfer from the catchment including carbon, sediment and nutrients (see Pearson and Connolly, 2000). At present, we do not know enough to predict the impacts of these changes on primary productivity, and this represents a critical gap in our knowledge and a research priority. Adequate understanding of how tropical rivers systems function and how they may respond to disturbance is lacking. Such fundamental information is not only required urgently, but must be applied in such a manner as to ensure that problems associated with water-resource developments in temperate latitudes are not repeated in the tropics (Australian Tropical Rivers Group, 2004).

## REFERENCES

Allan, J.D. 1995. Stream ecology: structure and function of running waters. Chapman and Hall, London.

Alongi, D.M. 1994. The role of bacteria in nutrient recycling in tropical mangrove and other coastal benthic ecosystems. *Hydrobiologia* 285: 19–32.

Australian Tropical Rivers Group. 2004. *Securing the North: Australia's Tropical Rivers*. A statement by the Australian Tropical Rivers Group. World Wildlife Fund Australia, Sydney, Australia.

Barbee, N.C. 2005. Grazing insects reduce algal biomass in a neotropical stream. *Hydrobiologia* 532: 153–165.

Bayley, P.B. 1989. Aquatic environments in the Amazon Basin, with an analysis of carbon sources, fish production, and yield. *In* D.P. Dodge (ed.) *Proceedings of the International Large Rivers Symposium. Can. J. Fish. Aquat. Sci. Spec. Publ.* 106: 399–408.

Bishop, J.E. 1973. *Limnology of a Small Malayan River Sungai Gombak*. W. Junk, The Hague. 485 pp.

Bixby, R.J., Ramírez, A., and Pringle, C.M. 2003. Effects of phosphorus and light on algal communities in lowland neotropical streams, Costa Rica. *J. North Am. Benthol. Soc.* 20: 205–226.

Bonetto, A.A. 1986. The Parana River system. *In* B.R. Davies and K.F. Walker (eds.) *The Ecology of River Systems.* pp. 541–555. W. Junk, Dordrecht, Netherlands.

Bonetto, A.A., and Wais, I.R. 1995. Southern South American streams and rivers. *In* C.E. Cushing, K.W. Cummins and G.W. Minshall (eds.) *River and Stream Ecosystems.* pp. 257–293. Elsevier Science B.V., Amsterdam.

Boston, H.L., and Hill, W.R. 1991. Photosynthesis-light relations of stream periphyton communities. *Limnol. Oceanogr.* 36: 644–656.

Bott, T.L. 1983. Primary productivity in streams. *In* J.R. Barnes and G.W Minshall (eds.) *Stream Ecology: Application and Testing of General Ecological Theory.* Plenum Press, New York. pp. 29–53.

Brinkman, W.L.F. 1985. Studies on hydrobiogeochemistry of a tropical lowland river system. *GeoJournal* 11: 89–101.

Bunn, S.E., Davies, P.M., and Kellaway, D.M. 1997a. Contributions of sugar cane and invasive pasture grass to the aquatic food web of a tropical lowland stream. *Mar. Freshw. Res.* 48: 173–179.

Bunn, S.E., Boon, P.I., Brock, M.A., and Schofield, N.J. (eds.) 1997b. *National Wetlands R&D Program Scoping Review.* Land and Water Resources R&D Corporation. Special Publication 1/97. Land and Water, Canberra, ACT, Australia.

Bunn, S.E., Davies, P.M., Kellaway, D.M., and Prosser, I. 1998a. Influence of invasive macrophytes on channel morphology and hydrology in an open tropical lowland stream, and potential control by riparian shading. *Freshw. Biol.* 39: 171–178.

Bunn, S.E., Davies, P.M., and Mosisch, T.D. 1998b. Contribution of algal carbon to stream food webs. *J. Phycol.* 34: 10–11.

Bunn, S.E., Davies, P.M., and Mosisch, T.D. 1999a. Ecosystem measures of river health and their response to riparian and catchment degradation. *Freshw. Biol.* 41: 333–345.

Bunn, S., Mosisch, T., and Davies, P.M. 1999b. Temperature and light. *In*: S. Lovett and P. Price (eds.) *Riparian Land Management Technical Guidelines Volume One: Principles of Sound Management*, pp. 17–24. Land and Water Resources R&D Corporation, Canberra.

Bunn, S.E., Davies, P.M., and Winning, M. 2003. Sources of organic carbon supporting the food web of an arid zone floodplain river. *Freshw. Biol.* 48: 619–635.

Bunn, S.E., Balcombe, S.R., Davies, P.M., Fellows, C.S., and McKenzie-Smith, F.J. 2005. Aquatic productivity and food webs of desert river ecosystems. *In* R. Kingsford (ed.) *Changeable, Changed, Changing: the Ecology of Desert Rivers.* pp. 76–99. Cambridge University Press, United Kingdom.

Canfield, D.E. Jr, and Hoyer, M.V. 1988. Influence of nutrient enrichment and light availability on the abundance of aquatic macrophytes in Florida streams. *Can. J. Fish. Aquat. Sci.* 45: 1467–1472.

Carignan, R., and Neiff, J.J. 1992. Nutrient dynamics in the floodplain ponds of the Parana River (Argentina) dominated by the water hyacinth *Eichhornia crassipes.* *Biogeochemistry* 17: 85–121.

Carignan, R., and Neiff, J.J. 1994. Limitation of water hyacinth by nitrogen in subtropical lakes of the Parana Floodplain (Argentina). *Limnol. Oceanogr.* 39: 439–443.

Clapcott, J.E., and Bunn, S.E. 2003. Can $C_4$ plants contribute to the aquatic food webs of subtropical streams? *Freshw. Biol.* 48: 1105–1116.

Clough, B.F. 1992. Primary productivity and growth of mangrove forests. *In* A.I. Robertson and D.M. Alongi (eds.) *Coastal and Estuarine Studies: Tropical Mangrove Systems.* pp. 225–249. American Geophysical Union, Washington, DC, USA.

Cummins, K.W. 1974. Structure and function of stream ecosystems. *Bioscience* 24: 631–641.

Cushing, C.E., Cummins, K.W., and Minshall, G.W. 1995. *River and Stream Ecosystems.* Elsevier Science B.V., Amsterdam, Netherlands. 817 pp.

Davies, P.M., Cook, B., Rutherford, J.C., and Walshe, T. 2004. Managing high in-stream temperatures using riparian vegetation. *Rip Rap* 22: 12–20. ISSN 1445-3924.

Day, J.W., Day, R.H., Barreiro, M.T., Ley-Lon, F., and Madden, C.J. 1982. Primary production of the Laguna de Terminos, a tropical estuary in the southern Gulf of Mexico. *Oceanol. Acta* 5: 269–276.

de Souza, M.L., and Moulton, T.P. 2005. The effects of shrimps on benthic material in a Brazilian island stream. *Freshw. Biol.* 50: 592–602.

de Oliveira, M.D., and Calijuri, M.D.C. 1996. Estimate of the rate of primary production in two lotic systems, based on hourly change of dissolved oxygen – Itaqueri and Lobo Rivers (Sao Paulo State). *An. Acad. Bras. Ciênc.* 68: 103–111.

di Persia, D.H., and Neiff, J.J. 1986. The Uruguay River system. *In* B.R. Davies and K.F. Walker (eds.) *The Ecology of River Systems.* pp. 599–621. W. Junk, Dordrecht, Netherlands.

Douglas, M., Finlayson, C.M., and Storrs, M.J. 1998. Weed management in tropical wetlands of the Northern Territory, Australia. *In* W.D. Williams (ed.) *Wetlands in a Dry Land: Understanding for Management.* pp. 76–99. Environment Australia, Canberra, Australia.

Douglas, M.M., Bunn, S.E., and Davies, P.M. 2005. River and wetland food webs in Australia's wet-dry tropics: general principles and implications for management. *Mar. Freshw. Res.* 56: 329–342.

Downing, J.A., McClain, M., Twilley, R., Melack, J.M., Elser, J., Rabalais, N.N., Lewis, W.M., Turner, R.E., Corredor, J., Soto, D., Yanez-Arancibia, A., Kopaska, J.A., and Howarth, R.W. 1999. The impact of accelerating land-use change on the N-cycle of tropical aquatic ecosystems: current conditions and projected changes. *Biogeochemistry* 46: 109–148.

Dudgeon, D. 1982. Spatial and seasonal variations in the standing crop of periphyton and allochthonous detritus in a forest stream in Hong Kong, with notes on the magnitude and fate of riparian leaf fall. *Arch. Hydrobiol. Suppl.* 64: 189–220.

Dudgeon, D. 1983. Preliminary measurements of primary production and community respiration in a forest stream in Hong Kong. *Arch. Hydrobiol.* 98: 287–298.

Dudgeon, D. 1988. The influence of riparian vegetation on macroinvertebrate community structure in four Hong Kong streams. *J. Zool. Lond.* 216: 609–627.

Dudgeon, D. 1995. The ecology of rivers and streams in tropical Asia. *In* C.E. Cushing, K.W. Cummins and G.W. Minshall (eds.) *River and stream ecosystems.* pp. 615–657. Elsevier Science B.V., Amsterdam.

Dudgeon, D. 2000. The ecology of tropical Asian rivers and streams in relation to biodiversity conservation. *Annu. Rev. Ecol. Syst.* 31: 239–263.

Dugan, P.J. (ed.) 1990. *Wetland Conservation: a Review of Current Issues and Required Action.* IUCN, Gland, Switzerland.

Findlay, S., Likens, G.E., Hedin, L., Fisher, S.G., and McDowell, W.H. 1997. Organic matter dynamics in Bear Brook, Hubbard Brook Experimental Forest, New Hampshire, USA. *J. North Am. Benthol. Soc.* 16: 43–46.

Finlayson, C.M. 1993. Vegetation changes and biomass on an Australian monsoonal floodplain. *In* B. Gopal, A. Hillbricht-Ilkowska and R.G. Wetzel (eds.) *Wetlands and Ecotones: Studies on Land-Water Interactions.* pp. 157–171. International Scientific Publications: New Delhi.

Finlayson, C.M., Bailey, B.J., and Cowie, I.D. 1990. Characteristics of a seasonally flooded freshwater system in monsoonal Australia. *In* D.F. Whigham, D.F. Good and J. Kvet (eds.) *Wetland Ecology and Management: Case Studies.* pp. 141–162. Kluwer Academic, Dordrecht.

Finlayson, C.M., Bellio, M.G., and Lowry, J. 2005. A conceptual basis for the wise use of wetlands in northern Australia – linking information needs, integrated analysis, drivers of change and human well-being. *Mar. Freshw. Res.* 56: 269–277.

Fisher, S.G., and Likens, G.E. 1973. Energy flow in Bear Brook, New Hampshire: an integrative approach to stream ecosystem metabolism. *Ecol. Monogr.* 43: 421–439.

Flecker, A.S., Taylor, B.W., Bernhardt, E.S., Hood, J.M., Cornwell, W.K., Cassatt, S.R., Vanni, M.J., and Altman, N.S. 2002. Interactions between herbivorous fishes and limiting nutrients in a tropical stream ecosystem. *Ecology* 83: 1831–1844.

Forsberg, B.R., Araujolima, C., Martinelli, L.A., Victoria, R.L., and Bonassi, J.A. 1993. Autotrophic carbon-sources for fish of the Central Amazon. *Ecology* 74: 643–652.

Ghosh, M., and Gaur, J.P. 1994. Algal periphyton of an unshaded stream in relation to *in situ* nutrient enrichment and current velocity. *Aquat. Bot.* 47: 185–189.

Graham, J.M., Lembi, C.A., Adrian, H.L., and Spencer, D.F. 1995. Physiological responses to temperature and irradiance in *Spirogyra* (Zygnematales, Charophyceae). *J. Phycol.* 31: 531–540.

Gregory, S.V., Swanson, F.J., McKee, W.A., and Cummins, K.W. 1991. An ecosystem perspective of riparian zones. *Bioscience* 41: 540–551.

Hamilton, S.K., and Gehrke, P.C. 2005. Australia's tropical river systems: current scientific understanding and critical knowledge gaps for sustainable management. *Mar. Freshw. Res.* 56: 243–252.

Hamilton, S.K., and Lewis, W.M. Jr 1987. Causes of seasonality in the chemistry of a lake on the Orinoco River floodplain, Venezuela. *Limnol. Oceanogr.* 32: 1277–1290.

Hamilton, S.K., Lewis, W.M. Jr, and Sippel, S.J. 1992. Energy sources for aquatic animals in the Orinoco River floodplain: evidence from stable isotopes. *Oecologia* 89: 324–330.

Hart, B.T. 2004. Environmental risks associated with new irrigation schemes in northern Australia. *Ecol. Manage. Rest.* 5: 106–115.

Hasan, R. 1988. Annual production and productivity of the macrophytes of the River Champanala, a side spill channel of the Ganges at Bhagalpur. *Acta Hyrdochim. Hydrobiol.* 16: 573–578.

Hill, W.R. 1996. Factors affecting benthic algae – effects of light. *In* R.J. Stevenson, M.L. Bothwell and R.L. Lowe (eds.) *Algal ecology – freshwater benthic ecosystems.* pp. 121–148. Academic Press, San Diego, California, USA.

Hill, W.R., and Knight, A.W. 1988. Nutrient and light limitation of algae in two northern California streams. *J. Phycol.* 24: 125–132.

Humphries, S.E., Groves, R.H., and Mitchell, D.S. 1991. *Plant Invasions of Australian Ecosystems – a Status Review and Management Directions.* Kowari 2, Australian National Parks and Wildlife Service, Canberra, Australia.

Huq, M.F., Al-Saadi, H.A., and Hameed, H.A. 1981. Studies on the primary production of the River Shatt-al-Arab at Basrah, Iraq. *Hydrobiologia* 77: 25–29.

Junk, W.J. 1986. Aquatic plants of the Amazon system. *In* B.R. Davies and K.F. Walker (eds.) *The Ecology of River Systems.* pp. 319–337. W. Junk, Dordrecht, Netherlands.

Junk, W.J. 2002. Long-term environmental trends and the future of tropical wetlands. *Environ. Conserv.* 29: 414–435.

Junk, W.J., and Piedade, M.T.F. 1993. Biomass and primary-production of herbaceous plant communities in the Amazon floodplain. *Hydrobiologia* 263: 155–162.

Junk, W.J., and Welcomme, R.L. 1990. Floodplains. *In* B.C. Patten (ed.) *Wetlands and Shallow Continental Water Bodies.* pp. 491–524. SPB Academic Publishing, The Hague, Netherlands.

Junk, W.J., Bayley, P.B., and Sparks, R.E. 1989. The flood pulse concept in river-floodplain systems. *In* D.P. Dodge (ed.) *Proceedings of the International Large River Symposium. Can. Spec. Publ. Fish. Aquat. Sci.* 106: 110–127.

Kent, S. 2001. *Top-down Control in a Stream Community Under Contrasting Flow Regimes.* Honours Thesis. Northern Territory University, Darwin, Australia.

Kingsford, R.T., Jenkins, K.M., and Porter, J.L. 2004. Imposed hydrological stability on lakes in arid Australia and effects on waterbirds. *Ecology* 85: 2478–2492.

Lamberti, G.A., and Steinman, A.D. 1997. A comparison of primary production in stream ecosystems. *J. North Am. Benthol. Soc.* 16: 95–104.

Langdon, C. 1988. On the causes of interspecific differences in the growth-irradiance relationship for phytoplankton. II: a general review. *J. Plankton Res.* 10: 1291–1312.

Larned, S.T., and Santos, S.R. 2000. Light- and nutrient-limited periphyton in low order streams of Oahu, Hawaii, *Hydrobiologia* 432: 101–111.

Lévêque, C. 1995. River and stream ecosystems of northwestern Africa. *In* C.E. Cushing, K.W. Cummins and G.W. Minshall (eds.) *River and Stream Ecosystems.* pp. 519–536. Elsevier Science B.V., Amsterdam.

Lewis, W.M. Jr 1988. Primary production in the Orinoco River. *Ecology* 69: 679–692.

Lewis, W.M. Jr, and Weibezahn, F.H. 1976. Chemistry, energy flow, and community structure in some Venezuelan freshwaters. *Arch. Hydrobiol. Suppl.* 50: 145–207.

Lewis, W.M. Jr, Hamilton, S.K., and Saunders, J.F. III. 1995. Rivers of northern South America. *In* C.E. Cushing, K.W. Cummins and G.W. Minshall (eds.) *River and Stream Ecosystems.* pp. 219–256. Elsevier Science B.V., Amsterdam.

Lewis, W.M. Jr, Hamilton, S.K., Rodríguez, M.A., Saunders, J.F. III, and Lasi, M.A. 2001. Foodweb analysis of the Orinoco floodplain based on production estimates and stable isotope data. *J. North Am. Benthol. Soc.* 20: 241–254.

Loneragan, N.R., Bunn, S.E., and Kellaway, D.M. 1997. Are mangrove and seagrasses sources of organic carbon for penaeid prawns in a tropical Australian estuary? A multiple stable isotope study. *Mar. Biol.* 130: 289–300.

Madsen, T., and Shine, R. 1996. Seasonal migration of predators and prey: pythons and rats in tropical Australia. *Ecology* 77: 149–56.

McMahon, T.A., Finlayson, B.L., Haines, A.T., and Srikanthan, R. 1991. *Global Runoff: Continental Comparisons of Annual Flow and Peak Discharges.* Catena Verlag. Cremlingen, Germany.

Melack, J.M., and Fisher, T.R. 1990. Comparative limnology of tropical floodplain lakes with an emphasis on the Central Amazon. *Acta Limnologica Brasiliensia* 3: 1–48.

Melack, J.M., and Forsberg, B.R. 2001. Biogeochemistry of Amazon Floodplain Lakes and Associated Wetlands. *In* M.E. McClain, R.L. Victoria, and J.E. Richey (eds.) *The Biogeochemistry of the Amazon Basin.* pp. 235–274. Oxford University Press, Oxford U.K.

Morison, J.I.L., Piedade, M.T.F., Müller, E., Long, S.P., Junk, W.J., and Jones, M.B. 2000. Very high productivity of the $C_4$ aquatic grass *Echinochloa polystachya* in the Amazon floodplain confirmed by net ecosystem $CO_2$ flux measurements. *Oecologia* 125: 400–411.

Morton, S.R., and Brennan, K.G. 1991. Birds. *In* C.D. Haynes, M.G. Ridpath, and M.A.J. Williams (eds.) *Monsoonal Australia Landscape, Ecology and Man in the Northern Lowlands.* pp. 133–151. A. A. Balkema, Rotterdam.

Mosisch, T., and Bunn, S.E. 1997. Temporal patterns in stream epilithic algae in response to discharge regime. *Aquat. Bot.* 5: 181–193.

Mosisch, T., Bunn, S.E., and Davies, P.M. 2001. The relative importance of shading and nutrients on algal production in subtropical streams. *Freshw. Biol.* 46: 1269–1278.

Moulton, T.P., Souza, M.L., Silveira, R.M.L., and Krsulović, F.A.M. 2004. Effects of ephemeropterans and shrimps on periphyton and sediments in a coastal stream (Atlantic forest, Rio de Janeiro, Brazil). *J. North Am. Benthol. Soc.* 23: 868–881.

Mulholland, P.J., Fellows, C.S., Tank, J.L., Grimm, N.B., Webster, J.R., Hamilton, S.K., Marti, E., Ashkenas, L., Bowden, W.B., Dodds, W.K., McDowell, W.H., Paul, M.J., and Peterson, B.J. 2001. Inter-biome comparison of factors controlling stream metabolism. *Freshw. Biol.* 6: 1503–1517.

Ong, J.E. 1982. Mangroves and mariculture. *Ambio* 11: 252–257.

Ortiz-Zayas, J.R., Lewis, W.M. Jr, Saunders, J.F. III, and Scatena, F.N. 2003. Metabolism of a tropical rainforest stream. *J. North Am. Benthol. Soc.* 24: 769–783.

Pannier, F. 1979. Mangroves impacted by human-induced disturbances: a case study of the Orinoco Delta mangrove ecosystem. *Environ. Manage.* 3: 205–216.

Pearson, R.G., and Connolly, N.M. 2000. Nutrient enhancement, food quality and community dynamics in a tropical rainforest stream. *Freshw. Biol.* 43: 31–42.

Pettit, N.E., Froend, R.H., and Davies, P.M. 2001. Identifying the natural flow regime and the relationship with riparian vegetation for two contrasting Western Australian rivers. *Regulated Rivers: Research and Management* 17: 201–215.

Power, M. 1984. Habitat quality and the distribution of algae-grazing catfish in a Panamanian stream. *J. Anim. Ecol.* 53: 357–374.

Power, M. 1987. Predator avoidance by grazing fishes in temperate and tropical streams: importance of stream depth and size. *In* W.C. Kerfoot and A. Sih (eds.) *Predation: Direct and Indirect Impacts in Aquatic Communities.* University Press of New England, Dartmouth, USA.

Pringle, C.M., and Hamazaki, T. 1997. Effects of fishes on algal response to storms in a tropical stream. *Ecology* 78: 2432–2442.

Pringle, C.M., Paaby-Hansen, P., Vaux, P.D., and Goldman, C.R. 1986. *In situ* assays of periphyton growth in a lowland Costa Rican stream. *Hydrobiologia* 134: 207–213.

Pringle, C.M., and Blake, G.A. 1994. Quantitative effects of atyid shrimp (Decapoda: Atyidae) on the depositional environment in a tropical stream: use of electricity for experimental exclusion. *Can. J. Fish. Aquat. Sci.* 51: 1443–1450.

Pringle, C.M., Blake, G.A., Covich, A.P., Buzby, K.M., and Finley, A. 1993a. Effects of omnivorous shrimp in a montane tropical stream: sediment removal, disturbance of sessile invertebrates and enhancement of understory algal biomass. *Oecologia* 93: 1–11.

Pringle, C.M., Rowe, G.L., Triska, F.J., Fernandez, J., and West, J. 1993b. Landscape linkages between geothermal activity and solute composition and ecological response in surface waters draining the Atlantic slope of Costa Rica. *Limnol. Oceanogr.* 38: 753–774.

Rama Rao, S.V., Singh, V.P., and Mall, L.P. 1979. The effect of sewage and industrial waste discharges on the primary production of a shallow turbulent river. *Water Res.* 13: 1017–1021.

Ramos-Escobedo, M.G., and Vázquez, G. 2001. Major ions, nutrients and primary productivity in volcanic neotropical streams draining rainforest and pasture catchments at Los Tuxtlas, Vera Cruz, Mexico. *Hydrobiologia* 445: 67–76.

Ranvestel, A.W., Lips, K.R., Pringle, C.M., Whiles, M.R., and Bixby, R.J. 2004. Neotropical tadpoles influence stream benthos: evidence for the ecological consequences of decline in amphibian populations. *Freshw. Biol.* 49: 274–285.

Redhead, T.D. 1979. On the demography of *Rattus sordidus colletti* in monsoonal Australia. *Aust. J. Ecol.* 4: 115–136.

Richardson, K., Beradall, J., and Raven, J.A. 1983. Adaptation of unicellular algae to irradiance: an analysis of strategies. *New Phytol.* 93: 157–191.

Robertson, A.I., and Blaber, S.J.M. 1992. Plankton, epibenthos and fish communities. *In* A.I. Robertson and D.M. Alongi (eds.) *Coastal and Estuarine Studies: Tropical Mangrove Systems.* pp. 173–234. American Geophysical Union, Washington, DC, USA.

Robertson, A.I., Daniel, P.A., and Dixon, P. 1991. Mangrove forest structure and productivity in the Fly River Estuary, Papua New Guinea. *Mar. Biol.* 111: 147–155.

Robertson, A.I., Daniel, P.A., Dixon, P., and Alongi, D.M. 1993. Pelagic biological processes along a salinity gradient in the Fly River estuary and adjacent river plume (Papua New Guinea). *Cont. Shelf Res.* 13: 205–224.

Rosemond, A.D., Pringle, C.M., Ramírez, A., Paul, M.J., and Meyer, J.L. 2002. Landscape variation in phosphorus concentration and effects on detritus-based tropical streams. *Limnol. Oceanogr.* 47: 278–289.

Saunders, J.F. III, and Lewis, W.M. Jr 1988. Transport of phosphorus, nitrogen, and carbon by the Apure River, Venezuela. *Biogeochemistry* 5: 323–342.

Schmidt, G.W. 1973. Primary production of phytoplankton in the three types of Amazonian water, 3: primary productivity of phytoplankton in a tropical floodplain lake of Central Amazonia, Lago do Castanho, Amazonas, Brazil. *Amazoniana* 4: 379–404.

Schmidt, G.W. 1982. Primary production of phytoplankton in three types of Amazonian waters. V. Some investigations on the phytoplankton and its primary productivity in the clear water of the lower Rio Tapajós (Pará, Brazil). *Amazoniana* 7: 335–348.

Shehata, S.A., and Bader, S.A. 1985. Effect of Nile River water quality on algal distribution at Cairo, Egypt. *Environ. Int.* 11: 465–474.

Storrs, M.J., and Finlayson, C.M. 1997. *A Review of Wetland Conservation Issues in the Northern Territory.* Supervising Scientist, Darwin, Northern Territory.

Talling, J.F. 1976. Phytoplankton: composition, development and productivity. *In* J. Rzóska (ed.) *The Nile: Biology of an Ancient River.* pp. 385–402. W. Junk, The Hague.

Thorp, J.H., and Delong, AM.D. 2002. Dominance of autochthonous autotrophic carbon in food webs of heterotrophic rivers. *Oikos* 96: 543–550.

Tockner, K., Malard, F., and Ward, J.V. 2000. An extension to the flood pulse concept. *Hydrobiol. Proc.* 14: 2861–2883.

Tockner, K., Bunn, S.E., Gordon, C., Naiman, R.J., Quinn, G.P., and Stanford, J.A. in press. Floodplains: critically threatened ecosystems. *In:* N. Polunin (ed.) *Future of Aquatic Ecosystems.* Cambridge University Press, United Kingdom.

Townsend, S.A., and Padovan, A.V. 2005. The seasonal accrual and loss of benthic algae (*Spirogyra*) in the Daly River, an oligotrophic river in tropical Australia. *Mar. Freshw. Res.* 56: 317–327.

Triska, F.J., Sedell, J.R., Cromack, S.V., and Gregory, S.V. 1982. Coniferous forest streams. *In* R.L. Edmonds (ed.) *Analysis of Coniferous Forest Ecosystems in the Western United States*. pp. 292–322. Hutchinson Ross, Pennsylvania, USA.

Udy, J., and Dennison, W.C. 2005. Nutrients. *In*: E.G. Abal, S.E. Bunn and W.C. Dennison, *Healthy Waterways, Healthy Catchments: Making the Connection in South East Queensland*. pp. 93–118. Moreton Bay and Catchments Partnership, Brisbane, Queensland, Australia.

Vannote, R.L., Minshall, G.W., Cummins, K.W., Sedell, J.R., and Cushing, C.E. 1980. The river continuum concept. *Can. J. Fish. Aquat. Sci.* 37: 130–137.

Venu, P., Kumar, B., and Bhasin, M.K. 1985. Water chemistry and production studies on Testa River and its two tributaries in Sikkim Himalayas, India. *Acta Bot. Ind.* 13: 158–164.

Vieira, J. Jr, and Necchi, O. Jr 2003. Photosynthetic characteristics of charophytes from tropical lotic ecosystems. *Phycol. Res.* 1: 51–60.

Webster, J.R., and Meyer, J.L. 1997. Stream organic matter budgets. *J. North Am. Benthol. Soc*: 16: 3–26.

Webster, J.R., Meyer, J.L., Wallace, J.B., and Benfield, E.F. 1997. Organic matter dynamics in Hugh White Creek, Coweeta Hydrologic Laboratory, North Carolina, USA. *J. North Am. Benthol. Soc*: 16: 74–78.

Webster, I.T., Rea, N., Padovan, A.V., Dostine, P., Townsend, S.A., and Cook, S. 2005. An analysis of primary production in the Daly River, a relatively unimpacted tropical river in northern Australia. *Mar. Freshw. Res.* 56: 303–316.

Welcomme, R.L. 1986. The Niger River system. *In* B.R. Davies and K.F. Walker (eds.) *The Ecology of River Systems*. pp. 9–23. W. Junk, Dordrecht.

Whitehead, P.J., and Saalfeld, K. 2000. Nesting phenology of magpie geese (*Anseranas semipalmata*) in monsoonal northern Australia: responses to antecedent rainfall. *J. Zool.* 251: 495–508.

Whitehead, P.J., and Tschirner, K. 1992. Sex and age-related variation in foraging strategies of magpie geese *Anseranas semipalmata*. *Emu* 92: 28–32.

Whittaker, R.H., and Likens, G.E. 1975. The biosphere and man. *In*: H. Leith and R.H. Whittaker (eds.) *Primary Productivity of the Biosphere*. Springer-Verlag, New York, USA.

Winemiller, K.O. 2004. Floodplain river food webs: generalizations and implications for fisheries management. *In* R.L. Welcomme and T. Petr (eds.) *Proceedings of the Second International Symposium on the Management of Large Rivers for Fisheries: Sustaining Livelihoods and Biodiversity in the New Millennium, Volume 1*. pp. 285–309. Food and Agriculture Organization of the United Nations and Mekong River Commission, Texas, USA.

Wirf, L. 2003. Spatial variation in top-down control in an Australian tropical stream. Honours thesis. Charles Darwin University, Darwin.

Woodroffe, C. 1992. Mangrove sediments and geomorphology. *In* A.I. Robertson and D.M. Alongi (eds.) *Coastal and Estuarine Studies: Tropical Mangrove Ecosystems*. pp. 251–292. American Geophysical Union, Washington, DC, USA.

Wootton, J.T., and Oemke, M.P. 1992. Latitudinal differences in fish community trophic structure, and the role of fish herbivory in a Costa Rican stream. *Environ. Biol. Fishes* 35: 311–319.

Worbes, M. 1997. The forest ecosystem of the floodplains. *In* W.J. Junk (ed.) *The Central Amazon Floodplain: Ecology of a Pulsing System*. pp. 223–266. Springer, Berlin.

Wurm, P.A.S. 1998. A surplus of seeds: high rates of post-dispersal seed predation in a flooded grassland in monsoonal Australia. *Aust. J. Ecol.* 23: 385–392.

# 3

# Organic Matter Processing in Tropical Streams

Karl M. Wantzen, Catherine M. Yule, Jude M. Mathooko, and Catherine M. Pringle

Organic matter derived from many sources provides a basis for stream food webs. In terms of weight, leaves from the surrounding land constitute the largest allochthonous source of energy for stream consumers, but other items, including fruits, flowers, wood and twigs, and terrestrial insects, are also important. Timing of allochthonous inputs can vary markedly due to the phenology of the riparian vegetation, retention mechanisms in the aquatic-terrestrial transition zone, and local climate (especially the incidence of high-rainfall events), but seasonality of litter inputs is different, and often much less marked, than is typical of streams in temperate latitudes. As in such streams, litter decomposition rates depend on the interaction of physical factors (flow, temperature), water chemistry (dissolved nutrients), and biological agents (micro-organisms and detritivores – especially shredding invertebrates). Because vascular plant biodiversity in the tropics is high, varied leaf characteristics (hardness, phenolic content, and other aspects of leaf chemistry) contribute to great variability in breakdown rate: fast-decomposing leaves persist for a few days only, whereas highly recalcitrant species take well over a year to decompose. In all the above cases, the decomposition process includes an initial rapid leaching phase when water-soluble compounds are lost, followed by colonization by micro-organisms (fungi and bacteria), and subsequent mechanical breakdown of the leaf structure by invertebrate shredder and hydraulic forces. Undecomposed leaves are sometimes exported downstream during flood events, and thence deposited in water-logged riparian zones or, in some cases, forming dense accumulations of peat that are important as carbon sinks and as habitat for specialized biota. Recent research indicates that the role of invertebrate shredders in processing organic matter in tropical streams is less than in temperate latitudes, and there may be a higher proportion of material that is recalcitrant and/or exported from streams (or stored as peat) before it is decomposed completely. Autochthonous energy sources may be particularly important to consumers in tropical streams, and there is some evidence of a lesser reliance on allochthonous organic matter than in temperate streams.

## I. INTRODUCTION

Decomposition is a central process in the organic matter budgets of stream ecosystems (Kaushik and Hynes, 1971; Webster and Benfield, 1986; Boulton and Boon, 1991; Abelho, 2001). The terms 'decomposition' and 'organic matter processing' are rather general terms that refer to the metabolism of a wide range of organic substances (dead plant and animal material) through an assortment of processes that include a variety of physical, chemical, and biological players. In addition, metabolic processes resulting in decomposition occur at different sites within stream ecosystems: for example, in pools of the riparian zone, on floodplains, within debris dams, and on or among the sediments in riffles and streams. Any comparison between latitudes, habitats, or types of organic substances can be confounded by a multiplicity of environmental variables and site characteristics, and generalizations must be made with caution. Here, we synthesize information on decomposition in tropical streams and make some comparisons with temperate systems. We focus mainly on particulate organic matter (POM).

## II. ORGANIC MATTER DYNAMICS

### A. Diversity of Particulate Organic Material

The most intensively studied source of organic matter in streams is plant litter – especially senescent leaves and dead wood – that has been shed from terrestrial and riparian plants. Other allochthonous energy sources include flowers, fruits and seeds, pollen, carcasses, and feces from terrestrial organisms but are much less apparent, and generally less studied, due to their relatively low biomass, local occurrence or high temporal variability. Nonetheless, their contribution to in-stream food webs can be significant due to a relatively high energy and/or nutrient contents compared to leaves and wood. Significantly, the types of organic matter that are most conspicuous in streams are precisely those that have not yet been consumed, degraded or processed by biota – either because these entered the water only recently, or because they are recalcitrant (i.e. resistant to decomposition). This may give rise to a false impression of the relative importance of different energy sources to stream consumers. For instance, fruits, pollen, and fleshy petals (e.g. of bat-pollinated flowers) are nutritious and eaten readily by terrestrial consumers; and because of these characteristics, they can be expected to decompose quickly in streams. Conversely, leaves, bark, and wood have been selected to resist terrestrial herbivores and forces of degradation while they are on the living plant, and nutrients may be translocated from leaves during senescence. For these reasons, such plant litter decomposes slowly in streams.

Allochthonous organic inputs derived from animals include the rain of honeydew from aphids or feces and frass produced by herbivores in the riparian zone. Large numbers of terrestrial insects fall into streams and, depending on the structure of the surrounding vegetation and thus the magnitude of this 'rain', stream drift may contain considerable amounts of terrestrial insects that can enhance food availability for fishes and other predators (Mathooko and Mavuti, 1992; Chan and Dudgeon, 2006). A manipulative study in a temperate Japanese stream showed a cascading effect whereby fish predation pressure on aquatic insects increased considerably when terrestrial insects falling from the canopy were excluded (Nakano *et al.*, 1999).

In some cases, tropical streams flow underneath bird or bat colonies and feces and animal carcasses can considerably raise the organic matter inputs. Although direct studies are lacking, analogous human impacts provide an idea of their probable effects. Thus, in an extremely nutrient-poor erosion gully in Mato Grosso, Brazil, dead chickens dumped from a nearby farm caused a 10-fold increase in benthic invertebrate densities on artificial substrates (Wantzen and Junk, 2006; Fig. 1). The importance of fruits in the diet of fishes has been well documented in

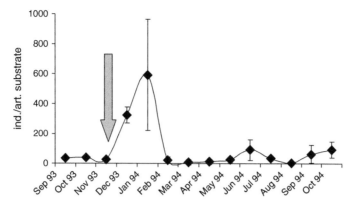

FIGURE 1   Effects of artificial increase of food sources for benthic insect larvae in an erosion-gully stream 'Formoso' in Mato Grosso, Brazil. Dumping of dead chickens from a chicken farm (about 200-week-old chicken bodies, black arrow) into the small rivulet (discharge: 20 L s$^{-1}$) increased the abundance of aquatic insects colonizing artificial substrates by a factor of 10–100 within 10 weeks. When the corpses were fully consumed, insect densities declined to the levels similar to those prevailing before dumping. Modified after Wantzen and Junk (2006).

Amazonian rainforest streams (Knöppel, 1970; Goulding *et al.*, 1988) and elsewhere (Dudgeon, 1999), and Larned (2000) and Larned *et al.* (2001) have drawn attention to their potential importance as a food source for stream invertebrates. Estimates of the magnitude of such inputs are scarce, but fruits contributed about 45 g ash-free dry weight (AFDW) m$^{-2}$yr$^{-1}$ to the total litter fall into a closed-canopy site on the Njoro River, Kenya (Magana, 2000), and riparian fruit input to tropical streams is sometimes high and continuous (Larned *et al.*, 2001). The importance of these and other nutrient- and energy-rich allochthonous inputs for fish and invertebrates in tropical streams has probably been underestimated (Wantzen and Junk, 2000; Larned *et al.*, 2001).

The largest proportion of allochthonous organic matter entering most streams comprises leaves, bark, and wood – especially twigs. The variety of these inputs depends on the biodiversity of the riparian ecosystem, which is the source of most of this organic matter. Since vascular plant biodiversity increases with decreasing latitude (Barthlott *et al.*, 1996), there are substantial regional differences in the composition of litter entering streams. Riparian vegetation in temperate latitudes is often species poor, and in Central Europe, for example, generally no more than 10 tree species contribute to the bulk of the litter input in a particular stream. Few species shed bark or twigs in a regular manner, or produce energy-rich flowers or fruit parts. In tropical latitudes, diversity is higher, e.g. gallery forests in Brazilian Cerrado streams have about 50 species of trees per hectare (Wantzen, 2003). Such diversity is typical; rainforests in Costa Rica's Caribbean lowlands support more than 320 tree species (Hartshorn, 1983) and about 1000 species occur on the whitewater floodplain of the Amazon (Wittmann *et al.*, 2004). In the Old-World topics, dipterocarp forests in lowland Borneo have an estimated 3000 tree species (IUCN, 1991), at densities more than 100 tree species per hectare (Kiew, 1998), whereas 186 species were recorded along a rift valley stream in the highlands of Kenya (Mathooko and Kariuki, 2000). The implications of this diversity for carbon processing in streams are yet unclear, and their link between riparian plant community composition and energy flow or in-stream food webs needs has to be investigated.

Given the diversity of potential allochthonous energy sources in tropical streams, analyses of carbon budgets should not only include studies of the decomposition pattern of leaves from important or dominant species, but also assessment of other POM sources and their dynamics. This will involve a variety of samplers (e.g. drift nets for POM > 200 μm; water-filled pan traps for falling insects; litter traps of various designs as in, for example, Fig. 2). Since tropical

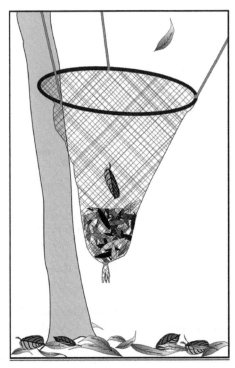

FIGURE 2  Standard leaf trap for measurement of litter inputs (diameter: 60 cm). Construction is simple, using flexible plastic canes and mosquito mesh.

latitudes differ in seasonality from the temperate zone, which will have implications for the timing of allochthonous inputs into streams, due consideration must be given to sampling effort in order to account for and adequately document temporal and spatial variations in inputs. Estimate of the larger wood fractions require long-term studies and a larger sampling area that may involve marking whole trees on standardized plots. There is certainly no 'one-size-fits-all' sampling strategy for tropical regions, as quantities and types (or quality) of litter inputs vary strongly between sites and years. In the lowland dipterocarp forests of Southeast Asia, for instance, the simultaneous massive flowering and mast fruiting of dozens of tree species occurs at irregular, multiyear intervals, across thousands of kilometres (Janzen, 1974; Sakai, 2002). The affects of this phenomenon are to create a 'hot moment' for stream energy budgets (Wantzen and Junk, 2006), and may have important consequences for consumers who are able to respond quickly to the increased availability of allochthonous sources. Given the occurrence of such marked interannual variation, some of the published literature on litter inputs to tropical streams, and particularly data accumulated over a year or less (such as most of those summarized in Tables I and II), may need to be treated as rather conservative or indicative of the lower end of the scale of temporal variability.

## B. Timing of Litter Inputs

In north temperate vegetation is mainly made up of either evergreen coniferous forests or deciduous broad-leaf forests that shed leaves when day length and temperature decrease in autumn each year (e.g. Dudgeon and Bretschko, 1996). The timing of leaf loss in tropical riparian forests is more variable (Fig. 3), in part because they contain a broad range of deciduous

*TABLE I*  Annual Input of Different Litter Types and their Carbon to Nitrogen Ratios at Open- and Closed-Canopy Sites along the Nyoro River, Kenya [Data from Magana (2000)]

| Litter type | Site | kg m$^{-2}$ yr$^{-1}$ | C : N |
|---|---|---|---|
| Wood and bark | Open canopy | 0.036 | |
| | Closed canopy | 0.271 | 51.9 |
| Fruits | Open canopy | 0.061 | |
| | Closed canopy | 0.045 | 39.9 |
| FPOM | Open canopy | 0.014 | |
| | Closed canopy | 0.143 | 19.4 |
| Unidentified fragments | Open canopy | 0.196 | |
| | Closed canopy | 0.415 | 58.1 |

*TABLE II*  Summary of Data on Litter Fall (kg m$^{-2}$ yr$^{-1}$) at a Range of Tropical Sites

| Forest type | Location | Total litter | Leaf litter | %N | %p | Source |
|---|---|---|---|---|---|---|
| Montane (2550 m asl) | Amazonia | 0.70 | 0.46 | 1.2 | 0.09 | Veneklaas (1991)[†] |
| Montane (3370 m asl) | Amazonia | 0.43 | 0.28 | 0.7 | 0.04 | Veneklaas (1991)[†] |
| Upland | Amazonia | 0.80 | | 1.3 | 0.04 | Dantas and Phillipson (1989)[†] |
| Terra firme | Amazonia | 0.74 | 0.56 | 1.4 | 0.03 | Klinge and Rodrigues (1968)[†] |
| Terra firme | Amazonia | 0.79 | 0.64 | – | – | Franken (1979)[†] |
| Terra firme | Amazonia | 0.83 | 0.54 | 1.8* | 0.02* | Luizao (1989)[†] |
| Terra firme | Amazonia | 1.02 | 0.76 | 1.6* | 0.03* | Cuevas and Medina (1986)[†] |
| Riparian | Amazonia | 0.64 | 0.43 | 1.2 | 0.02 | Franken (1979)[†] |
| Terra firme | Amazonia | 0.74 | 0.47 | 1.4* | 0.03* | Luizao (1989)[†] |
| Campina | Amazonia | – | – | 1.0 | 0.05 | Klinge (1985)[†] |
| Tall caatinga | Amazonia | 0.56 | 0.40 | 0.7 | 0.05 | Cuevas and Medina (1986)[†] |
| Bana | Amazonia | 0.24 | 0.21 | 0.6 | 0.02 | Cuevas and Medina (1986)[†] |
| Igapó | Amazonia | 0.68 | 0.53 | – | – | Adis *et al.* (1979)[†] |
| Igapó | Amazonia | 0.67 | – | 1.4 | – | Irmler (1982)[†] |
| Average | Amazonia | 0.68 | 0.48 | 1.2 | 0.04 | McClain and Richey (1996)[†] |
| Gallery forest | Cerrado, Central Brazil | 0.82 | | | | Wantzen and Wagner (2006) |
| Evergreen savanna | Pantanal, Central Brazil | 0.75 – 1.02 | | | | Haase (1999) |
| Semi-deciduous savanna | Pantanal, Central Brazil | 0.48 – 0.75 | | | | Haase (1999) |
| Tropical rainforest | Pasoh, Malaysia | 1.06 | 0.63 – 0.75 | | | Ogawa (1978) |
| *Shorea* plantation | India | | 0.59 | | | Puri (1953)[‡] |
| *Tectona* plantation | India | | 0.53 | | | Seth *et al.* (1963)[‡] |
| Deciduous forest | Sagar, India | | 0.26 – 0.93 | | | Upadhyaya (1955)[‡] |
| Deciduous forest | Varansi, India | | 0.10 – 0.62 | | | Singh (1968)[‡] |
| Deciduous forest | Udaipur, India | | 0.4 | | | Garg and Vyas (1975)[‡] |
| Average | 10°S–10°N | 0.68 | | | | Bray and Gorham (1964) |

*Calculated as percent of leaf litter.
[†]Cited from McClain and Richey (1996).
[‡]Cited from Garg and Vyas (1975).

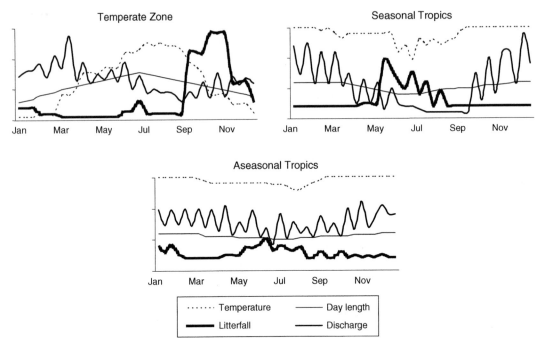

*FIGURE 3* Idealization of seasonal changes in temperature, day length, discharge, and timing of leaf litter input in temperate and tropical latitudes.

to semi-deciduous tree species that occur in highly mixed associations. Even when seasonal leaf fall does occur, it may not occur as synchronously as in the temperate latitudes, and may reflect a response to water scarcity (or other factors) rather than temperature and day length. In some parts of the tropics, the constant supply of water in the vicinity of streams and rivers provides conditions for tree growth that may change the timing of litter fall relative to that in more upland or drier sites (Wantzen and Junk, in press). For example, gallery forests along streams running through seasonal savannas (e.g. the Brazilian Cerrado, or the African *Andropogon* savannas) appear as permanently green belts compared to the much more seasonal vegetation further away. In wetter, less-seasonal tropical regions, the riparian tree phenology follows more or less closely the patterns of the surrounding vegetation and the appearance of the whole community (e.g. in tropical rainforests) is evergreen even if leaf fall is actually occurring all year round.

Contrary to most temperate species, the lifespan of some tropical tree leaves can extend over several years (Coley, 1988). Others lose their leaves on a seasonal basis or more erratically during the year. Because of the variety of species-specific patterns, distinct seasonal trends in allochthonous inputs to streams are lacking, especially in equatorial latitudes (e.g. Yule and Pearson, 1996). In more seasonal parts of the tropics, especially toward the northern and southern limits, triggers for phenological events like flowering, leaf flushing, and fruiting, or shedding leaves and bark, include monsoonal rainfall and drought (Dudgeon and Bretschko, 1996), floodplain inundation (Worbes *et al.*, 1992), and fire (Oliveira-Filho *et al.*, 1989). Moreover, torrential rainstorms, typhoons, and cyclones, as well as human activities such as fruit harvesting, contribute large amounts of fresh green leaves and wood to streams. All of these factors combine to make the predictability of timing and magnitude of leaf fall in the tropics considerably less pronounced than in most parts of the temperate zone.

## C. Accession Pathways

Litter input is composed of variable proportions of shed leaves falling directly into the stream channel, and lateral transport of fallen leaves that are blown or flushed into the channel. Lateral transport depends on the bank steepness, soil surface heterogeneity, understorey vegetation, wind and surface runoff, as well as human modification of the riparian vegetation, and may be at least half of that falling into the stream directly (Dudgeon, 1992) depending on the type and stature of riparian vegetation. In temperate zones, synchronous shedding of tree leaves in autumn combined with die-back of riparian herbs and strengthening winds enhance the transport of dry litter into the streams during snow-free periods in winter (Wantzen and Wagner, 2006). Litter accession into tropical streams is reduced by plants, which act like a mechanical filter to retain leaves and retard lateral transport. Retention of fallen litter by rapidly growing basidiomycetous fungi and moss has been reported on forested upland slopes in Puerto Rico (Covich, 1988) and in Kenya (Mathooko *et al.*, 2001), and basidiomycetes reduce downhill leaf transport rates to Puerto Rican streams by ∼40% (Lodge and Asbury, 1988). These fungi contribute to the maintenance of litter mats that retard soil erosion, and tree roots as well as associated mycorrhizae tend to retain the intact leaves with the result that they break down *in situ* and 'tighten' nutrient cycles. This can be seen from a study of streams in Cerrado of Mato Grosso, Brazil, where deployment of leaf traps two distances (1 m and 30 m) from the channel indicated negligible horizontal transport perpendicular to the stream channel at sites that did not flood seasonally (K. M. Wantzen, unpublished observations).

Topical rainstorms transport large quantities of litter into and along streams and rivers. They flush the floodplain and surrounding hillslopes bringing in leaves and wood that may already have been subject to some decomposition. Short intensive flow pulses may also flush out litter accumulations from deep pools and carry them downstream or deposit them in the riparian zone. In streams with very 'flashy' hydrographs, litter may be alternately picked up and deposited in terrestrial, aquatic, or marshy environments where it may be subject to differing decomposition processes.

## D. Decomposition and Storage of Organic Matter in Riparian Zones

Wetting and drying has an important influence on litter decomposition and the distribution of in-stream and riparian POM, especially in the large areas of the tropics where streams experience alternation of distinct dry and wet seasons or monsoon cycles. As rains and runoff fluctuate seasonally, the wetted area of the stream channel expands or contracts. Consequently, two zones can be identified: first, a wet zone defined as 'the sediment area wetted by water flow at the time of observation' and, second, a dry zone defined as 'the dry area on both sides of a stream bordered by the edge of the flowing water at the time of observation and the highest extent reached by the stream flow in its history' (Mathooko, 1995). Decomposition of *Dombeya goetzenii* (Sterculiaceae) leaves was four times faster in the wet zone than in the dry zone in the Njoro River, Kenya (Mathooko *et al.*, 2000a; see also Table III). Similar results have been reported from Australia (Boulton, 1991) and Amazonia (Furch and Junk, 1997), and an experimental study in which litter was moved from the dry to the wet zone confirmed this effect (Table III). Transport of leaves from dry riparian sites into the wetted stream channel appears to promote nutrient release and fragmentation. Boulton (1991) demonstrated that microbial enrichment (measured as ATP) of submerged leaves was higher than that of leaves that were exposed to air, and enhancement of microbial activity certainly contributes significantly to increased decomposition of submerged litter. Small streamside pools or fringing wetlands in riparian regions of swamp forests can serve as important sites or 'hotspots' of decomposition (Wantzen and Junk, 2000; see also Chapter 7), but only as long as there is connectivity between

TABLE III   Comparison of the Processing Rates ($-k$ per day) of Leaves of *Dombeya goetzenii* in a Wet–Dry Interchange Experiment (Data from Mathooko *et al.*, 2000a)

| Treatment of the litterbags | $-k$ |
| --- | --- |
| Litterbags in dry zone | 0.171 |
| Litter bags interchanged from dry to wet zone | 0.711 |
| Litter bags in wet zone | 0.789 |
| Litter bags interchanged from wet to dry zone | 0.004 |

the lateral water body and the stream channel. Isolated floodplain and wetland pools may serve as traps for organic matter (Wantzen *et al.*, 2005a).

   In some instances, the riparian zone is so retentive for organic matter that the export of coarse material to streams seems to be virtually non-existent. Much of the organic material in the Malaysian, Indonesian, and Papua New Guinea streams is broken down to fine particulate matter prior to entering the streams (C. M. Yule, personal observation; Yule and Pearson, 1996). Generally, the only intact leaves are those that fall directly from the canopy overhead. In Konaiano Creek on Bougainville Island, Papua New Guinea, the mean standing stock of detritus was 0.04 to 0.23 kg m$^{-2}$ (Yule, 1996), whereas in North American streams at a similar altitude stocks ranged from 0.2 to 0.9 kg m$^{-2}$ (Cummins *et al.*, 1989); see Dudgeon and Bretschko (1996) for the European–Asian comparison. While these values are not exceptionally low, most of the detritus in Konaiano Creek comprised sticks and twigs as well as fine organic material rather than intact leaves or leaf parts (Yule and Pearson, 1996).

## E. In-stream Decomposition Processes

   Decomposition of organic matter in streams is caused by a number of interacting processes (Fig. 4), and their joint effects are usually studied by measuring loss of detrital mass over time. This is not wholly satisfactory, since weight loss [or changes in ash-free dry weight (AFDW)]

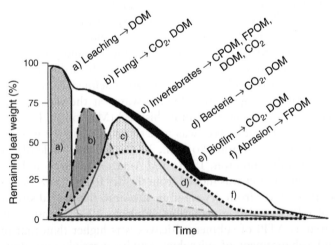

FIGURE 4   Synthesis of processes acting during decomposition of plant litter in fresh water and their effects on weight loss. Note that a temporary increase of litter dry weight may be caused by growth of biofilms on the leaf surface (e). The relative contribution of individual processes (a)–(f) may differ between streams. Highly modified after Suberkropp (1998).

does not provide direct information on the fate of this material or its uptake and assimilation by consumers. Gessner *et al.* (1999) propose a more organism-centered perspective on leaf–litter breakdown, which acknowledges that 'degradation' may begin even before leaves are shed. In tropical rainforests, many living leaves are colonized by epiphyllic algae and mosses. The surface characteristics and the chemistry of these leaves are also influenced by colonization of fungi and activity of herbivores. While parts of individual leaves may be killed when infected by pathogenic fungi, some fungi appear to maintain leaf activity in order to profit by the products of photosynthesis even after the onset of senescence of infected leaves (Butin, 1995). All these processes interfere with the quality of the leaves before they are shed and eventually reach the water. The term 'leaching' describes the extraction of soluble compounds by water. Leaching rates are affected by the integrity of the leaf surface, and leaching may occur during rain when senescing leaves are still attached to the tree. Once fallen leaves enter the stream, osmotic breakage of dead cell walls, penetration by fungal hyphae, and softening of the structural elements by microbial enzymes combined with feeding by invertebrate shredders enhance leaching. Leaching rates generally peak 24–48 h after immersion (Fig. 4) but some leaching continues for weeks (France *et al.*, 1997). The leachates (sugars, amino acids, etc.) are generally energy-rich and easily absorbed by bacteria (Strauss and Lamberti, 2002). Studies in temperate waters have shown that benthic decomposition of leaves enhanced microalgal biomass in the water column, demonstrating the role of allochthonous detritus as a nutrient source for primary production (Fazi and Rossi, 2000). The importance of epilithic algae growing on submerged litter in tropical streams has not been studied, but they could provide a significant enrichment of the food value of leaves consumed by invertebrates. As Fig. 4 shows, a temporary increase of dry weight during decomposition may be caused by the growth of biofilms on the leaf surface (as well as increased endophytic microbial biomass), as has been reported for *Syzygium cordatum* (Myrtaceae) litter by Mathooko *et al.* (2000b). Laboratory studies show that light favors biofilm quality on litter and thus the growth of temperate-zone invertebrate shredders (Franken *et al.*, 2005), and there is no reason to assume that this effect does not occur in tropical streams.

Fungi and bacteria growing on the leaf surface and inside the mesophyll produce enzymes that degrade structural polysaccharides, such as cellulose, resulting in a softening of leaf structure and an increase in food value for shredders (Kaushik and Hynes, 1971). Fungal biomass and reproduction generally peaks 1 or 2 weeks after immersion in temperate streams (Gessner and Chauvet, 1994). A few existing studies on tropical streams confirm this pattern (e.g. 10–20 days in Columbia; Mathuriau and Chauvet, 2002), or indicate that it may occur even more quickly (e.g. within 7 days in Costa Rica; Stallcup *et al.*, 2006). Invertebrate shredders and large benthic omnivores (decapods, crabs, and fish) contribute to the comminution and consumption of the litter and associated microbes (see Section V) and, together with physical degradation by the water current, reduce the leaf particles to tiny fragments and fibres. Although the retention of coarse litter in some tropical headwaters appears generally high (Mathooko, 1995; Morara *et al.*, 2003), large amounts of leaf material and fine fragments of organic material are transported to the lower course and floodplains especially, as described earlier, during spates and high-flow events. This organic material forms large accumulations in the deposition zone of rivers, often alternatively layered as sandy-loamy layers within 'sand/debris dunes' (Fittkau, 1982; Wantzen *et al.*, 2005).

## F. Abiotic Factors Affecting In-stream Decomposition

The influence of submergence and wetting on decomposition has been described earlier (see Section D) but processing of allochthonous detritus in streams can be affected by other aspects of stream hydrology. Foremost among these is retentiveness, which determines where decomposition of allochthonous organic material actually takes place once it has entered the stream. Roots, fallen branches, and stones act as obstacles that retain drifting litter (e.g. Mathooko *et al.*, 2001; Morara *et al.*, 2003), and coarse gravel riffles are especially efficient retainers of organic matter (Hynes, 1970). While the form and extent of stony retention structures depend on the geological setting, the amount of wood in the stream reflects the species composition, stature, and condition of the riparian forest. Forest streams in undisturbed catchments contain large amounts of logs and have heterogeneous channels that contribute to efficient litter retention and may allow formation of debris dams. On the other hand, stream and river management and flow regulation (or channelization) is often accompanied by wood removal (Diez *et al.*, 2000), and aggressive forestry practices or vegetation clearance may alter wood inputs and can cause bank erosion leading ultimately to reduced retention capacity (Wantzen, 2003; 2006). Where streams suffer from siltation, increased shear stress favors physical disintegration of the leaf structure, rather than processing of litter by microbes and invertebrates.

In seasonal tropical streams, strong rainstorm events at the onset of the wet season flush out much of the litter that might have accumulated during lower-flow periods (Franken, 1979; Pearson *et al.*, 1989). Frequent recurrence of spates of flood events may 'reset' the system (*sensu* Fisher, 1983) and carry away leaves before degradation can occur. In such cases, large quantities of leaves may be transported out of smaller tropical streams (Mathooko *et al.*, 2001; Morara *et al.*, 2003) or decomposed in the wetlands and along floodplains of larger rivers (Wantzen and Junk, 2000). For example, some southern temperate streams New Zealand have natural 'flashy' flow regimes and hence tend to have low retention capacity for leaves. Rounick and Winterbourn (1983) have suggested that low retentiveness was the cause of scarcity of specialised invertebrate shredders in such systems, and that much litter breakdown (fragmentation) was probably accomplished by physical processes. A similar pattern of high leaf export has been observed by Schwarz and Schwoerbel (1997) in Mediterranean streams on Corsica, and may be anticipated in tropical streams (especially those with steep courses) that experience high rainfall.

Sediment transport during spates may bury leaf litter. Studies in temperate streams indicate that buried leaves decomposed more slowly than those on the surface of the stream bed (Metzler and Smock, 1990), and were subject to less feeding by invertebrates although they did not differ in protein content (Herbst, 1980). By contrast, Mayack *et al.* (1989) reported increased breakdown of buried leaves due to feeding by tipulid (Diptera) larvae. Observations from Neotropical streams and rivers suggest that burial of leaves by sediment layers reduces decomposition and may lead to accumulation of layers of organic matter in floodplain sand dunes (Wantzen *et al.*, 2005b; Rueda-Delgado *et al.*, 2006).

While spates transport litter from streams, water scarcity or lack of flow also negatively affects decomposition. In seasonal eucalypt forest streams in Australia, periods of flow cessation during the dry season result in the accumulation of leaves, because plant leachates, high temperatures, and reduced oxygen are inimical to the activities of detritivorous invertebrates (Bunn, 1988). In floodplain areas, where elevated flows in the river mainstem cause backflooding of tributary streams, reduced flows in tributaries result in litter accumulations with characteristic faunal assemblages (e.g. Henderson and Walker, 1986). Substantial reductions in rates of leaf decomposition occurred during the backflooding phase in an Amazonian floodplain tributary, although dissolved oxygen was still present (Rueda-Delgado *et al.*, 2006; see also Table IV). In deep or isolated water bodies on floodplains, there may be sufficient microbial activity associated with litter accumulations to deplete dissolved oxygen levels entirely; this and associated hydrogen sulphide production commonly causes mass mortality of aquatic invertebrates during the inundation phase (Junk and Robertson, 1997) with consequential declines in metazoan-mediated litter processing.

TABLE IV  Summary of the Results of Recent Studies of the Processing Rates ($-k$ per day) of Tree Leaves in Tropical Streams

| Tree species | Location | Method | Duration (days) | $-k$ | Source |
|---|---|---|---|---|---|
| Vangueria madagascariensis | Njoro River, Kenya | Wire cages | 70 | 0.047 | Dobson et al. (2003) |
| Dombeya goetzenii | Njoro River, Kenya | Wire cages | 70 | 0.010 | Dobson et al. (2003) |
| Dombeya goetzenii | Njoro River, Kenya | Litter bag | 70 | 0.711–0.789 | Mathooko et al. (2000a) |
| Syzygium cordatum | Njoro River, Kenya | Wire cages | 70 | 0.022 | Dobson et al. (2003) |
| Syzygium cordatum | Njoro River, Kenya | Litter bag | 56 | 0.001 | Mathooko et al. (2000b) |
| Rhus natalensis | Njoro River, Kenya | Wire cages | 70 | 0.026 | Dobson et al. (2003) |
| Cecropia latiloba | Arenosa FDP, Colombia | Litter bag | 56 | 0.031 | Rueda-Delgado et al. (2006) |
| Cecropia latiloba | Arenosa BFP, Colombia | Litterbag | 56 | 0.009 | Rueda-Delgado et al. (2006) |
| Tessaria integrifolia | Arenosa FDP, Colombia | Litter bag | 56 | 0.029 | Rueda-Delgado et al. (2006) |
| Tessaria integrifolia | Arenosa BFP, Colombia | Litter bag | 56 | 0.009 | Rueda-Delgado et al. (2006) |
| Symmeria paniculata | Arenosa FDP, Colombia | Litter bag | 56 | 0.010 | Rueda-Delgado et al. (2006) |
| Symmeria paniculata | Arenosa BFP, Colombia | Litter bag | 56 | 0.001 | Rueda-Delgado et al. (2006) |
| Salix humboldtiana | Paraná, Argentina | Litter bag | 56 | 0.0119 | Capello et al. (2004) |
| Croton gossypifolius | Cabuyal, Colombia | Litter bag | 43 | 0.0651 | Mathuriau and Chauvet (2002) |
| Clidemia sp. | Cabuyal, Colombia | Litter bag | 43 | 0.0235 | Mathuriau and Chauvet (2002) |
| Cecropia schreberiana | Puerto Rico (micro-organisms excluded) | Laboratory microcosms | 84 | 0.00083 | Wright and Covich (2005) |
| Cecropia schreberiana | Puerto Rico (micro-organisms present) | Laboratory microcosms | 84 | 0.0035 | Wright and Covich (2005) |
| Dacryodes excelsa | Puerto Rico (micro-organisms excluded) | Laboratory microcosms | 84 | 0.0014 | Wright and Covich (2005) |
| Dacryodes excelsa | Puerto Rico (micro-organisms present) | Laboratory microcosms | 84 | 0.0073 | Wright and Covich (2005) |
| Trema integerrima | La Selva Biological Station, Costa Rica | Litter bag | 21 | 0.0451 | Ardón et al. (2006) |
| Castilla elastica | La Selva Biological Station, Costa Rica | Litter bag | 80 | 0.0064 | Ardón et al. (2006) |

(continued)

TABLE IV    (*continued*)

| Tree species | Location | Method | Duration (days) | −k | Source |
|---|---|---|---|---|---|
| *Zygia longifolia* | La Selva Biological Station, Costa Rica | Litter bag | 80 | 0.0020 | Ardón *et al.* (2006) |
| *Radermachera glandulosa* | Second-order streams, Thailand | Leaf packs | 35 | 0.0413 | Parnrong *et al.* (2002) |
| *Pometia pinnata* | Second-order streams, Thailand | Leaf packs | 35 | 0.0236 | Parnrong *et al.* (2002) |
| *Hevea brasiliensis* | Second-order streams, Thailand | Leaf packs | 35 | 0.0636 | Parnrong *et al.* (2002) |
| *Nephelium lappaceum* | Second-order streams, Thailand | Leaf packs | 35 | 0.0380 | Parnrong *et al.* (2002) |
| *Eucalyptus camaldulensis* | Second-order streams, Thailand | Leaf packs | 35 | 0.0747 | Parnrong *et al.* (2002) |
| *Acacia mangium* | Second-order streams, Thailand | Leaf packs | 35 | 0.0682 | Parnrong *et al.* (2002) |

## III. THE SIGNIFICANCE OF PHYSICAL AND CHEMICAL COMPOSITION OF LEAVES

High vascular plant biodiversity in the tropics has prompted suggestions that leaf quality (especially phytochemistry) has a more important influence on decomposition dynamics than in the case of temperate streams (e.g. Stout, 1989; Irons *et al.*, 1994; Aerts, 1997). Leaf chemical properties known to affect litter decomposition in aquatic ecosystems include tannins and other phenolics (Stout, 1989; Ostrofsky, 1997; Wantzen *et al.*, 2002), lignin (Gessner and Chauvet, 1994), nitrogen content (e.g. Melillo *et al.*, 1983), and carbon-to-nitrogen or carbon-to-phosphorus ratios (Enríquez *et al.*, 1993). Leaves have evolved not only to optimize photosynthesis but also to be defended against terrestrial herbivores, especially insects, and there is evidence that the amounts of defensive compounds that they contain increase with leaf age (Coley, 1988). Chemicals remaining in fallen leaves, especially proanthocyanidins, may retard decomposition by inhibiting activities of microbes and invertebrates (Wantzen *et al.*, 2002). Acid waters (typically pH 3.5–3.7), low dissolved oxygen, and tough leaves high in tannins and lignin retard the decomposition of litter in Malaysian peat swamps where even fruits and flowers do not break down readily (Fig. 5; C. M. Yule, unpublished observations); in such situations, partly decomposed litter can build up layers of peat up to 20 m thick. Leaf recalcitrance is a key contributor to this build up, as species of leaves that lack defensive compounds break down quite rapidly in the peat swamp. There are no invertebrate shredders, and the peat-swamp food web seems to be based on bacteria that utilize dissolved organic matter leached from leaves and other detritus (C. M. Yule *et al.*, unpublished observations). Elsewhere, leachates from litter may have rather different effects; for instance, soluble polyphenols from eucalypt leaves inhibit microbial activity during low-flow conditions in Australian streams (Bunn, 1988), and several periods of inundation and leaching appear necessary before microbes can fully utilize eucalypt leaves in intermittent streams (Boulton and Boon, 1991). Likewise, Walker (1986, 1995) has shown that leachates from leaves in Amazonian blackwaters inhibit bacterial growth and, in turn, influence the occurrence of mosquitoes (whose larvae feed on planktonic bacteria) in Amazonian blackwater and whitewater areas.

In addition to potential 'antifeedant' compounds such as polyphenols and tannins (for a discussion of terminology, see Duffey and Stout, 1996), structural compounds (i.e. lignin and

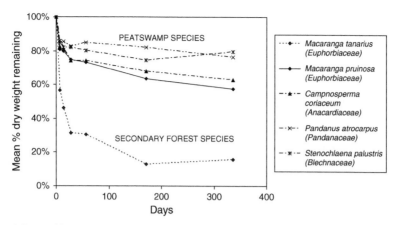

FIGURE 5   Breakdown of leaves within litter bags in a Malaysian peat swamp over a year. Leaves of peat-swamp trees hardly decomposed, but leaves of a secondary forest species broke down relatively rapidly (C. M. Yule and L. Gomez, unpublished observations).

cellulose) tend to be higher in tropical leaves than in their temperate counterparts (Coley and Barone, 1996). They remain in the leaf matrix until the end of the decomposition process that and, therefore, have longer lasting effects than other defensive compounds are leached relatively quickly. Indeed, Gessner and Chauvet (1994) considered that lignin played a key role in controlling hyphomycete fungal growth on litter. In lowland Costa Rican streams, it appears that the lignin and cellulose content of litter limit the extent of phosphorus stimulation of leaf breakdown, fungal biomass, and microbial respiration (Ardón *et al.*, 2006).

## IV. MICROBIAL CONTRIBUTIONS TO ORGANIC MATTER PROCESSING

The interplay and relative importance of microbes that contribute to the decomposition of organic matter streams is poorly understood, especially so in the tropics (Wright and Covich, 2005). Studies in temperate streams indicate that there is a functional overlap among different groups (e.g. Webster and Benfield, 1986; Graça, 1993; Gessner and Chauvet, 1994; Gessner *et al.*, 1997; Raviraja *et al.*, 1998). For instance, aquatic hyphomycetes perform well at low water temperatures (Bärlocher and Kendrick, 1974) and dominate fungal assemblages in temperate streams (Gessner *et al.*, 1997), but they may lose their 'advantage' over other fungal taxa in warmer waters (Graça, 1993). In a seminal paper, Irons *et al.* (1994) hypothesized that the rate of microbially mediated processing of leaf litter will rise with decreasing latitude, i.e. that the role of microbes will be relatively more important in the tropics than in the temperate zone. Given that decomposition rates tend to increase with temperature, as does microbial metabolism, there is certainly a biological basis for this hypothesis, but the activity of other consumers might rise with temperature also, and it is not clear how microbial assemblage composition or functional organization might change with latitude. There is a large body of taxonomic information on aquatic hyphomycetes in tropical streams (e.g. Marvanova, 1997), but insufficient data to draw any robust conclusions about the relative contribution of hyphomycetes, other fungi and bacteria to litter decomposition in the tropics or to generalize about changes in their representation as decomposition proceeds. The results of the existing literature are inconclusive; for instance, litter breakdown rate did not change between streams with high and low hyphomycete diversity and productivity (Raviraja *et al.*, 1998). In Western Australian streams, actinomycete bacteria seem to contribute more than that of hyphomycetes to decomposition of *Eucalyptus obliqua* (Bunn, 1988). While dense growths of fungal hyphae have been reported on submerged litter in Central Amazonia (Henderson and Walker, 1986), the presence of fungal biomass may not necessarily be correlated with high levels of growth or metabolic activity (Gessner and Chauvet, 1994). Chemical exclosure experiments in Puerto Rico indicated that bacteria and fungi perform different functions in litter processing since selective poisoning of either one or both groups indicated the presence of non-additive effects (Wright and Covich, 2005). A particularly interesting study by Rueda-Delgado *et al.* (2006) in a Columbian floodplain stream showed that although fungal hyphae were scarce on decomposing litter, leaf ergosterol content (a biomarker of fungal biomass) was comparable to that reported from litter in European streams (e.g. Mathuriau and Chauvet, 2002). In lowland Costa Rican streams, high hyphomycete fungal biomass (measured as ergosterol) on leaves in lowland Costa Rican streams was associated with fast breakdown rate of high-quality leaves (Rosemond *et al.*, 2002, Ardón *et al.*, 2006), but the leaf characteristics and the high ambient levels of phosphorus in these streams limit our ability to generalize from these results. Clearly more data are needed, particularly on the role of bacteria in the latter stages of leaf breakdown when the substrate is highly fragmented. Improvement of techniques to estimate bulk microbial biomass is also needed, and must take account of the fact that ergosterol does not occur in all fungi.

## V. THE ROLE OF SHREDDERS IN ORGANIC MATTER PROCESSING

There are many reports of a paucity or a conspicuous lack of shredders in tropical streams in Africa, Asia, and the Neotropics systems (e.g. Walker, 1987; Dudgeon, 1992, 1994, 1999, 2000; Irons *et al.*, 1994; Dudgeon and Bretschko, 1996; Yule, 1996; Rosemond *et al.*, 1998; Dudgeon and Wu, 1999; Dobson *et al.*, 2002; Mathuriau and Chauvet, 2002; Rueda-Delgado *et al.*, 2006; Wantzen and Wagner, 2006). Far fewer researchers record either similar shredder representation as in north temperate streams (in tropical Queensland, Australia: Cheshire *et al.*, 2005) or even high litter breakdown rates due to the presence of crabs acting as macro-shredders (Moss, 2007). Elsewhere in the tropics, there is anecdotal information of leaf-processing proceeded by microbial activity in the virtual absence of invertebrate shredding (e.g. in Malaysian peat swamps; C. M. Yule, unpublished observations). The body of evidence from the tropics, thus far, is that the importance of invertebrate shredders for the comminution of allochthonous inputs and facilitation of fine-particle-feeding detritivores (filters and gatherer–collectors), and their dominance or co-dominance of headstream benthic communities, does not coincide with the models of community structure and function embodied in the River Continuum Concept (Vannote *et al.*, 1980). Finally, this concept to tropical streams and rivers must be reconsidered (e.g. Wantzen and Wagner, 2006), and its applicability to temperate running waters has also been questioned (e.g. Heard and Richardson, 1995; Schwarz and Schwoerbel, 1997) but especially so in the tropics. Certainly, the concept needs to be adjusted to take account of tropical peat swamps, and organic matter dynamics on floodplains (see Chapter 7 of this volume).

There is evidence that omnivorous fishes and decapod crustaceans play key roles in detrital processing in some tropical streams (Wootton and Oemke, 1992; Irons *et al.*, 1994; Crowl *et al.*, 2001; March *et al.*, 2001). Fishes that feed on allochthonous material are much more common in tropical than in temperate streams (e.g. Bowen, 1983; Wootton and Oemke, 1992; Dudgeon, 1999). Island streams can be dominated by omnivorous decapod shrimps which play a key role in organic matter processing and mobilization (Covich and McDowell, 1996; Pringle *et al.*, 1999; Crowl *et al.*, 2001; Larned *et al.* 2001; March *et al.*, 2001; Souza and Moulton, 2005). These omnivores are generally larger than insect shredders and rapidly process considerable volumes of coarse and fine organic matter. Rates of detrital processing varied as a function of the nature of the shrimp assemblage along an elevational gradient in a Puerto Rican stream (March *et al.*, 2001); fastest rates were found at high-elevation sites dominated by *Xiphocaris* and *Atya* (Atyidae). Total insect densities were low and insect shredders were rare, but shrimp densities approached 30 individuals per square meter. At lower elevations, the role of atyids in leaf processing was suppressed by the activity of predatory *Macrobrachium* shrimps (Palaemonidae). In lowland Costa Rica, Rosemond *et al.* (1998) found that omnivorous fish and shrimps caused loss of weight from leaf packs although the cause (feeding and/or physical disruption) was not identified. In a Kenyan stream, freshwater crabs comminuted large amounts of leaf litter within the space of a few days (Moss, 2007). While our understanding of the role and importance of shredders in tropical streams is evolving, the functional replacement of insects and small crustaceans by larger decapod shredders (shrimps and crabs) appears to be a general feature in the tropics (Dudgeon, 1999; Dobson *et al.*, 2002; Wantzen and Wagner, 2006), with decapods (especially atyid shrimps) occupying the niches filled by amphipod and isopod detritivores in north temperate. More research on the feeding activities of these creatures is needed, since recent work indicates shrimp contributions to litter processing and fragmentation do not reflect the significance of this energy source to their diet or secondary production (Yam and Dudgeon, 2005).

## VI. METHODOLOGICAL CONSTRAINTS ON DECOMPOSITION STUDIES

There are substantial interspecific differences in the physical and chemical composition of leaves (see Section III). Even within the same species, the toughness, form, and chemical composition varies in response to edaphic and climatic conditions, and can depend also on the degree of leaf senescence and extent of herbivory and pathogen attack. Because of the species richness of tropical trees and the fact that synchronized abscission periods are often lacking, many researchers in the tropics have used green leaves, or dried samples of fresh leaves that were sampled prior to abscission, as a means of investigating litter decomposition. Green leaves have a different physical and chemical structure than senescent, naturally abscised leaves, and this will affect their decomposition and palatability to detritivores. Studies of a feeding by a chironomid (*Stenochironomus*: Diptera) leaf miner in a Neotropical stream has shown that green leaves both from temperate and tropical tree species were consumed rapidly (Wantzen and Wagner, 2006), whereas naturally fallen leaf litter showed extremely slow decomposition (K. M. Wantzen *et al.*, unpublished observations).

The tendency to use fresh leaves in decomposition studies, and the confounding effects of seasonal flow patterns and poorly known life cycles of detritivores in tropical streams, hamper our ability to make latitudinal or regional comparisons of litter processing between latitudes or among regions. Indeed, even within a region or a single stream, there are marked species-specific differences between decomposition rates (see review by Abelho, 2001; Table IV). For instance, Dobson *et al.* (2003) found that rates of mass-specific loss varied as much as five times among leaves in a Kenyan stream, but these researchers and Mathooko *et al.* (2000a) found that leaves in most species had broken down almost entirely within 3–4 months. However, one species (*Syzygium cordatum*) with tough, well-defended leaves was estimated to persist for well over 2 years (Mathooko *et al.*, 2000b).

Despite differences in methods used, some general trends emerge from the array of decomposition studies reviewed by Abelho (2001) and the more recent tropical literature (Table IV). Leaves of rapidly growing tree species from nutrient-rich environments (e.g. whitewater floodplains; Furch and Junk, 1997; Rueda-Delgado *et al.*, 2006) or fertile volcanic soils (Benstead, 1996) have faster decomposition rates than leaves of trees from nutrient-poor rainforest areas (Rueda-Delgado *et al.*, 2006); leaves from peat swamp trees have the slowest decomposition rates (K. M. Wantzen and C. M. Yule, unpublished observations).

The methods used in decomposition studies in the tropics and elsewhere vary greatly, and have been critically reviewed by Boulton and Boon (1991) and, more recently, by Graça *et al.* (2005). Decisions about study methodology are important because they affect estimates of decomposition rate regardless of site and species. Season also plays an influential role; Rueda-Delgado *et al.* (2006) found highly significant differences in leaf decomposition rates between the backflooded phase and the flashy-discharge phase of Arenosa Stream in Colombia (Table IV; see also Section F.). Comparability of future studies of leaf decomposition will depend on the use of a standard methodology across sites/countries as seen, for example, in the WW-DECOEX project (Wantzen and Wagner, 2006). Key components of this protocol include the use of 5–10 g recently shed air-dried litter inside mesh bags (mesh size 5 mm) with sufficient being placed at the start of the study to allow collection of at least five replicates on each sampling interval. Bag retrieval should take place after 1, 7, 14, 28, and 56 days, and after 6 and 12 months (if leaves persist for that long), with decomposition (i.e. weight loss of leaves) measured as ash-free dry matter (AFDM). When large benthic omnivores are present, additional leaf packs (i.e. leaves tethered with thread) should be deployed so that potential shredders can gain access to the litter. The use of electric fences to exclude larger consumers (e.g. Rosemond *et al.*, 1998; March *et al.*, 2001) may give further insights into their role in litter processing.

## VII. AUTOCHTHONOUS PLANT LITTER

A conventional view of stream ecology is that shading by terrestrial vegetation, especially at low-order sites, limits autochthonous primary production so that stream organic matter budgets depend mainly on allochthonous inputs (Vannote *et al.*, 1980). Inevitably, this simple model does not apply in all circumstances, even in temperate latitudes (e.g. Franken *et al.*, 2005). In Neotropical savannas, stretches of closed forest canopy alternate with sections of scattered gallery forest and grassland (Wantzen, 2003). Shallow soils, recurrent fires and water-logged soils in hyperseasonal savannas reduce the growth of tall trees (Oliveira-Filho *et al.*, 1989) and, in such circumstance, in-stream primary production may be high. Human clearance of forest for the development of pastures and agriculture also facilitates autochthonous production (Martinelli *et al.*, 1999; Mathooko and Kariuki, 2000). Even in closed-canopy tropical sites, aquatic macrophyte growth can be considerable if the current is not excessive (e.g. Furch and Junk, 1980). The fate of this autochthonous plant material after death is not clear and requires study. However, isotopic studies of riparian and emergent grasses in tropical Australia and Brazil indicate that, despite high biomass, their contributions to in-stream organic carbon budgets are limited (Bunn *et al.*, 1997; Martinelli *et al.*, 1999; for details, see Chapter 2, Sections II-B and III-A).

## VIII. CONCLUSIONS

Although the general features of organic matter processing in northern-temperate and tropical streams are similar, some differences between them are nonetheless evident (for further consideration of latitudinal differences, see Chapter 9). Leaf litter input into tropical streams generally comprises a greater range of species, and the timing of the input ranges from completely asynchronous in the wet tropics to more-or-less synchronous in savanna climates. Higher ambient temperatures and the lack of winter freezing remove temperature constraints on decomposition rates, and may have the consequence that litter quality has a more influential role on organic matter dynamics in the tropics than in temperate streams. Warmer temperatures directly affect metazoans (e.g. Sweeney, 1984), and may alter the proportionate contributions of metazoans and microbes to organic matter processing (Irons *et al.*, 1994) as well as influencing the relative importance of fungi and bacteria (Graça, 1993).

While seasonality in temperate streams is determined by temperature and day length, tropical streams are either aseasonal (e.g. Yule and Pearson, 1996) or seasonality tends to be linked with flow fluctuations (e.g. Dudgeon, 2000; Wantzen and Junk, 2000; Junk and Wantzen, 2004). Life cycles of detritivores in temperate streams appear synchronized with leaf availability (e.g. Cummins *et al.*, 1989) and, in that sense, consumers seem to be adapted to make efficient use of litter inputs (see Vannote *et al.*, 1980). In seasonal tropical streams, there is little evidence of adaptation of detritivores to make use of dry-season accumulations of leaves (Wantzen and Wagner, 2006), and wet-season spates may flush out litter before it is processed by consumers. Aseasonal tropical streams are particularly 'flashy', and this means that leaves may not be retained in streams long enough for decomposition to be completed. The range of plant secondary and structural compounds in tropical leaves probably limits the importance of shredders in leaf breakdown, and there is evidence that a significant proportion of the litter input may enter the stream as fine particles rather than intact leaves (e.g. Yule and Pearson, 1996). Therefore, it seems possible that collectors and filter feeders that consume physically and microbially degraded leaves play a crucial role in organic matter processing in tropical streams than shredders.

Our ability to generalize about differences between organic matter processing between tropical and temperate streams, and the organisms involved in such processing, is constrained by the fact that we still know little of the diversity of aquatic invertebrates in tropical streams (Vinson and Hawkins, 1998; Wantzen and Junk, 2000). As in temperate latitudes, this biota is likely to be influenced strongly by diversity of the surrounding vegetation, and thus it seems an appropriate precautionary principle to maintain riparian zones as a primary goal for conservation of biodiversity and ecosystem functioning in tropical streams (Dudgeon, 2000; Pringle, 2001; Wantzen *et al.*, 2006; see also Chapter 10). A further constraint on generalization is the extent of biogeographic and regional variation within tropical and temperate latitudes; not all 'topical' streams are alike or represent a single type, and neither are all 'temperate streams' similar to each other; both represent a range of ecosystem types. The data that we have presented here indicate that while the number of tropical streams studied may be too small to permit robust generalizations, they are sufficient to indicate the extent of regional differences within the tropics. Further research is needed to uncover the extent of these differences, their importance, and the factors that underlie them.

Research priorities that need attention include measures of the amounts and condition of litter entering streams, in particular the relative importance of intact versus fragmented, partially decomposed leaves, and possible facilitation of aquatic breakdown by initial processing on land. This needs to be accompanied by estimates of litter retention in streams, especially large accumulations of detritus, and the responses of this material (and associated organisms) to seasonal and aseasonal flow pulses and flood events. Knowledge of the importance of inputs of other allochthonous energy sources aside from leaves (wood, fruit, flowers, frass, carcasses of insects and other animals) will also be required for elaboration of stream energy budgets. Better understanding of the role of plant defensive and structural compounds in litter breakdown are also needed, as this will influence how much litter is processed *in situ* or exported downstream in a relatively undecomposed or refractory condition. In addition, the relative importance of fungi and bacteria, and the way in which they interact to influence organic matter processing is poorly understood, and the relationship between microbial and faunal activities – especially the role of shredders – is still unclear. While tropical shredders appear to be less abundant and consequently less important in organic matter processing than their temperate counterparts, detailed purpose-designed studies will be needed to confirm this premise.

# REFERENCES

Abelho, M. (2001). From litterfall to breakdown in streams: a review. *The Scientific World* 1, 656–680.

Aerts, R. (1997). Climate, leaf litter chemistry and leaf litter decomposition in terrestrial ecosystems: a triangular relationship. *Oikos* 79, 439–449.

Ardón, M., Stallcup, L. A., and Pringle, C. M. (2006). Does leaf quality stimulate the stimulation of leaf breakdown by phosphorus in Neotropical streams? *Freshwater Biology* 51, 618–633.

Bärlocher, F., and Kendrick, B. (1974). Dynamics of the fungal population on leaves in a stream. *Journal of Ecology* 63, 761–791.

Barthlott, W., Lauer, W., and Placke, A. (1996). Global distribution of species diversity in vascular plants: towards world map of phytodiversity. *Erdkunde* 50, 317–327.

Benstead, J. P. (1996). Macroinvertebrates and the processing of leaf litter in a tropical stream. *Biotropica* 28, 367–375.

Boulton, A. J. (1991). Eucalypt leaf decomposition in an intermittent-stream in South-Eastern Australia. *Hydrobiologia* 211, 123–136.

Boulton, A. J., and Boon, P. I. (1991). A review of methodology used to measure leaf litter decomposition in lotic environments: time to turn over an old leaf? *Australian Journal of Marine and Freshwater Research* 42, 1–43.

Bowen, S. H. (1983). Detritivory in Neotropical fish communities. *Environmental Biology of Fishes* 9, 137–144.

Bray, J. R., and Gorham, E. (1964). Litter production in the forests of the world. *Advances in Ecological Research* 2, 101–157.

Bunn, S. E. (1988). Processing of leaf litter in two northern jarrah forest streams, Western Australia: II. The role of macroinvertebrates and the influence of soluble polyphenols and inorganic sediment. *Hydrobiologia* 162, 211–223.

Bunn, S. E., Davies, P. M., and Kellaway, D. M. (1997). Contributions of sugar cane and invasive pasture grass to the aquatic food web of a tropical lowland stream. *Marine and Freshwater Research* 48, 173–179.

Butin, H. (1995). "Tree Diseases and Disorders: Causes, Biology, and Control in Forest and Amenity Trees." Oxford University Press, Oxford.

Capello, S., Marchese, M. R., and Ezcurra de Drago, I. (2004). Descomposición de hojas de Salix humboldtiana y colonización por invertebrados en la llanura de inundación del río Paraná Medio. *Amazoniana* 18, 125–144.

Chan, E. K. W., and Dudgeon, D. (2006). Riparian vegetation affects the food supply of stream fish in Hong Kong. In Jim, C. Y. and Corlett, R.T. (eds.) "Sustainable Management of Protected Areas for Future Generations" pp. 219–231. World Conservation Union (IUCN) and World Commission on Protected Area (WPCA), Gland, Switzerland.

Cheshire, K., Boyero, L., and Pearson, R. G. (2005). Food webs in tropical Australian streams: shredders are not scarce. *Freshwater Biology* 50, 748–769.

Coley, P. D. (1988). Effects of plant growth rate and leaf lifetime on the amount and type of anti-herbivore defense. *Oecologia* 74, 531–536.

Coley, P. D., and Barone, J. A. (1996). Herbivory and plant defenses in tropical forests. *Annual Review of Ecology and Systematics* 27, 305–335.

Covich, A. P. (1988). Geographical and historical comparisons of Neotropical streams: biotic diversity and detrital processing in highly variable habitats. *Journal of the North American Benthological Society* 7, 61–386.

Covich, A. P., and McDowell, W. H. (1996). The stream community. In Reagan, D. P. and Waide, R. B. (eds.) "The Food Web of a Tropical Rain Forest." pp. 433–459. University of Chicago Press, Chicago, IL.

Crowl, T. A., McDowell, W. H., Covich, A. P., and Johnson, S. L. (2001). Freshwater shrimp effects on detrital processing and nutrients in a tropical headwater stream. *Ecology* 82, 775–783.

Cummins, K. W., Wilzbach, M. A., Gates, D. M., Perry, J. B., and Taliaferro, W. B. (1989). Shredders and riparian vegetation. *Bioscience* 39, 24–30.

Diez, J. R., Larrañaga, S., Elosgi, A., and Pozo, J. (2000). Effect of removal of wood on streambed quality and retention of organic matter. *Journal of the North American Benthological Society* 19, 621–632.

Dobson, M., Magana, A., Mathooko, J. M., and Ndegwa, F. K. (2002). Detritivores in Kenyan highland streams: more evidence for the paucity of shredders in the tropics? *Freshwater Biology* 47, 909–919.

Dobson, M., Mathooko, J. M., Ndegwa, F. K., and M'Erimba, C. M. (2003). Leaf litter processing in a Kenyan highland stream, the Njoro River. *Hydrobiologia* 519, 207–210.

Dudgeon, D. (1992). "Patterns and Processes in Stream Ecology. A Synoptic Review of Hong Kong Running Waters." (Die Binnengewässer, Vol. 29). 147 pp. Schweizerbartsche Buchhandlung, Stuttgart.

Dudgeon, D. (1994). The influence of riparian vegetation on macroinvertebrate community structure and functional organization in six New Guinea streams. *Hydrobiologia* 294, 65–85.

Dudgeon, D. (1999). "Tropical Asian Streams: Zoobenthos, Ecology and Conservation." 830 pp. Hong Kong University Press, Hong Kong.

Dudgeon, D. (2000). The ecology of tropical Asian rivers and streams in relation to biodiversity conservation. *Annual Review of Ecology and Systematics* 31, 239–263.

Dudgeon, D., and Bretschko, G. (1996). Allochthonous inputs in land-water interactions in seasonal streams: tropical Asia and temperate Europa. In Schiemer, F. and Boland, K. T. (eds.) "Perspectives in Tropical Limnology." pp. 161–179. SPB Academic Publishing, Amsterdam.

Dudgeon, D., and Wu, K. K. Y. (1999). Leaf litter in a tropical stream: food or substrate for macroinvertebrates? *Archiv für Hydrobiologie* 146, 65–82.

Duffey, D., and Stout, M. J. (1996). Antinutritive and toxic components of plant defensive against insects. *Archives of Insect Biochemistry and Physiology* 32, 3–37.

Enríquez, S., Duarte, C. M., and Sand-Jensen, K. (1993). Patterns in decomposition rates among photosynthetic organisms: the importance of detritus C : N : P content. *Oecologia* 94, 457–471.

Fazi, S., and Rossi, L. (2000). Effects of macro-detritivores density on deaf detritus processing rate: a macrocosm experiment. *Hydrobiologia* 435, 127–134.

Fisher, S. G. (1983). Succession in streams. In Barnes, J. R. and Minshall, G. W. (eds.) "Stream Ecology: Application and Testing of General Ecological Theory." pp. 7–27. Plenum Press, New York.

Fittkau, E. J. (1982). Struktur, Funktion und Diversität zentralamazonischer Ökosysteme. *Archiv für Hydrobiologie* 95, 29–45.

France, R. L., Culbert, H., Freeborough, C., and Peters, R. H. (1997). Leaching and early mass loss of boreal leaves and wood in oligotrophic water. *Hydrobiologia* 345, 209–214.

Franken, W. (1979). Untersuchungen im Einzugsgebiet des zentralamazonischen Urwaldbaches "Barro Branco" auf der "terra firme": I. Abflußverhalten des Baches. *Amazoniana* 6, 459–466.

Franken, R. J. M., Peeters, E. T. H. M., Gardeniers, J. J. P., Beijer, J. A. J., and Scheffer, M. (2005). Growth of shredders on leaf litter biofilms: the effect of light intensity. *Freshwater Biology* 50, 459–466.

Furch, K., and Junk, W. J. (1980). Water chemistry and macrophytes of crecks and rivers in Southern Amazonia and the Central Brazilian Shield. In: Furtado, J. I. (ed.): "Tropical Ecology and Development", pp. 771–796. *The International Society of Tropical Ecology*, Kuala Lumpur.

Furch, K., and Junk, W. J. (1997). The chemical composition, food value and decomposition of herbaceous plants and leaf litter of the floodplain forest. In Junk, W. J. (ed.) "The Central Amazonian Floodplain: Ecology of a Pulsing System." pp. 187–205. Springer, Berlin.

Garg, R. K., and Vyas, L. N. (1975). Litter production in deciduous forest Near Udaipur (South Rajasthan), India. In Golley, F. B. and Medina, E. (eds.) "Tropical Ecological Systems. Trends in Terrestrial and Aquatic Research." pp. 131–135. Springer, New York.

Gessner, M. O. (1991). Differences in processing dynamics of fresh and dried leaf litter in a stream ecosystem. *Freshwater Biology* 26, 387–398.

Gessner, M. O., and Chauvet, E. (1994). Importance of stream microfungi in controlling breakdown rates of leaf litter. *Ecology* 75, 1807–1817.

Gessner, M. O., Suberkropp, K., and Chauvet, E. (1997). Decomposition of plant litter by fungi in marine and freshwater ecosystems. In Wicklow, W. and Söderström, J. (eds.) "The Mycota IV – Environmental and Microbial Relationships." pp. 303–321. Springer, Berlin.

Gessner, M. O., Chauvet, E. and Dobson, M. (1999). A perspective on leaf litter breakdown in streams. *Oikos* 85, 377–384.

Goulding, M., Carvalho, M. L., and Ferreira, E. G. (1988). "Rio Negro, South America – Rich Life in Poor Water: Amazonian Diversity and Foodchain Ecology as Seen Through Fish Communities." 200 pp. SBP Publishing, The Hague.

Graça, M. A. S. (1993). Patterns and processes in detritus-based stream systems. *Limnologica* 23, 107–114.

Graça, M. A. S., Bärlocher, F., and Gessner, M. O. (2005). "Methods to Study Litter Decomposition – a Practical Guide." Springer, Berlin.

Haase, R. (1999). Litterfall and nutrient return in seasonally flooded and non-flooded forest of the Pantanal, Mato Grosso, Brazil. *Forest Ecology and Management* 117, 129–147.

Hartshorn, G. S. (1983). Plants. In Janzen, D. H. (ed.) "Costa Rican Natural History." pp. 118–183. University of Chicago Press, Chicago, IL.

Heard, S. B., and Richardson, J. S. (1995). Shredder-collector facilitation in stream detrital food webs: is there enough evidence? *Oikos* 72, 359–366.

Henderson, P. A., and Walker, I. (1986). On the leaf litter community of the Amazonian blackwater stream Tarumzinho. *Journal of Tropical Ecology* 2, 1–17.

Herbst, G. N. (1980). Effects of burial on food value and consumption of leaf detritus by aquatic invertebrates in a lowland forest stream. *Oikos* 35, 411–424.

Hynes, H. B. N. (1970). "The Ecology of Running Waters." University of Toronto Press, Toronto.

Irons, J. G., Oswood, M. W., Stout, R. J., and Pringle, C. M. (1994). Latitudinal patterns in leaf litter breakdown: is temperature really important? *Freshwater Biology* 32, 401–411.

Janzen, D. H. (1974). Tropical blackwater rivers, animals, and mast fruiting by the Dipterocarpaceae. *Biotropica* 6, 69–103.

Junk, W. J., and Robertson, B. A. (1997). Aquatic invertebrates. In Junk, W. J. (ed.) "The Central Amazonian Floodplain: Ecology of a Pulsing System." pp. 279–298. Springer, Berlin.

Junk, W. J., and Wantzen, K. M. (2004). The flood pulse concept: new aspects, approaches, and applications – an update. In Welcomme, R. and Petr, T. (eds.) "Proceedings of the 2nd Large River Symposium (LARS), Pnom Penh, Cambodia." pp. 117–149. RAP Publication 2004/16. Food and Agriculture Organization and Mekong River Commission. FAO Regional Office for Asia and the Pacific, Bangkok.

Kaushik, N. K., and Hynes, H. B. N. (1971). The fate of dead leaves that fall into streams. *Archiv für Hydrobiologie* 68, 465–515.

Kiew, R. (1998). Species richness and endemism. In Soepadmo, E. (ed.) "The Encyclopedia of Malaysia: Plants." pp. 14–15. Archipelago Press, Singapore.

Knöppel, H.-A. (1970). Food of Central Amazonian fishes. Contribution to the nutrient-ecology of amazonian rain-forest-streams. *Amazoniana* 2, 257–352.

Larned, S. T. (2000). Dynamics of coarse riparian detritus in a Hawaiian stream ecosystem: a comparison of drought and post-drought conditions. *Journal of the North American Benthological Society* 19, 215–234.

Larned, S. T., Chong, C. T., and Punewai, N. (2001). Detrital fruit processing in a Hawaiian stream ecosystem. *Biotropica* 33, 241–248.

Lodge, D. J., and Asbury, C. E. (1988). Basidiomycetes reduce export of organic matter from forest slopes. *Mycologia* 80, 888–890.

March, J. G., Benstead, J. P., Pringle, C. M., and Ruebel, M. W. (2001). Linking shrimp assemblages with rates of detrital processing along an elevational gradient in a tropical stream. *Canadian Journal of Fisheries and Aquatic Sciences* 58, 470–478.

Martinelli, L. A., Ballester, M. V., Krusche, A. V., Victória, R. L., de Camargo, P. B., Bernardes, M., and Ometto, J. P. H. B. (1999). Landcover changes and $\delta^{13}$ C composition of riverine particulate organic matter in the Piracicaba River basin (southeast region of Brazil). *Limnology and Oceanography* 44, 1826–1833.

Marvanova, L. (1997). Freshwater hyphomycetes: a survey with remarks on tropical taxa. In Janardhanan, K.K., Rajendran, K. C., Natarajan, M. and Hawksworth, D. L. (eds.) "Tropical Mycology." pp. 169–226. Science Publishers, Enfield, NH.

Mathooko, J. M. (1995). The retention of plant coarse particulate organic matter (CPOM) at the surface of the wet-store and dry-store zones of the Njoro River, Kenya. *African Journal of Ecology* 33, 159.

Mathooko, J. M., and Kariuki, S. T. (2000). Disturbances and species distribution of the riparian vegetation of a Rift Valley stream. *African Journal of Ecology* 38, 123–129.

Mathooko, J. M., and Mavuti, K. M. (1992). Composition and seasonality of benthic invertebrates, and drift in the Naro Moru River, Kenya. *Hydrobiologia* 232, 47–56.

Mathooko, J. M., M'Erimba, C. M. and Leichtfried, M. (2000a). Decomposition of leaf litter of *Dombeya goetzenii* in the Njoro River, Kenya. *Hydrobiologia* 418, 147–152.

Mathooko, J. M., Magana, A. M., and Nyang'au, I. M. (2000b). Decomposition of *Syzygium cordatum* in a Rift Valley stream ecosystem. *African Journal of Ecology* 38, 365–368.

Mathooko, J. M., Morara, G. O., and Leichtfried, M. (2001). Leaf litter transport and retention in a tropical Rift Valley stream: an experimental approach. *Hydrobiologia* 443, 9–18.

Mathuriau, C., and Chauvet, E. (2002). Breakdown of leaf litter in a Neotropical stream. *Journal of the North American Benthological Society* 2, 384–396.

Mayack, D. T., Thorp, J. H., and Cothran, M. (1989). Effects of burial and floodplain retention on stream processing of allochthonous litter. *Oikos* 54, 378–388.

McClain, M. E., and Richey, J. E. (1996). Regional-scale linkages of terrestrial and lotic ecosystems in the Amazon basin: a conceptual model for organic matter. *Algological Studies* 113, 111–125.

Melillo, J. M., Naiman, R. J., Aber, J. D., and Eshleman, K. N. (1983). The influence of substrate quality and steam size on wood decomposition dynamics. *Oecologia* 58, 281–285.

Metzler, G. M., and Smock, L. A. (1990). Storage and dynamics of subsurface detritus in a sand-bottomed stream. *Canadian Journal of Fisheries and Aquatic Sciences* 47, 588–594.

Morara, G. O., Mathooko, J. M., and Leichtfried, M. (2003). Natural leaf litter transport and retention in a second-order tropical stream: the Njoro River, Kenya. *African Journal of Ecology* 41, 277–279.

Moss, B. (2007). Rapid shredding of leaves by crabs in a tropical African stream. *Verhandlungen Internationale Vereinigung für Limnologie* 29, in press.

Nakano, S., Miyasaka, H., and Naotoshi, K. (1999). Terrestrial-aquatic linkages: riparian arthropod inputs alter trophic cascades in a stream food web. *Ecology* 80, 2435–2441.

Ogawa, H. (1978). Litter production and carbon cycling in Pasoh Forest. *Malaysian Nature Journal* 30, 367–373.

Oliveira-Filho, A. T., Shepherd, J. G., Martins, F. M., and Stubblebine, W. H. (1989). Environmental factors affecting physiognomic and floristic variation in an area of Cerrado in central Brazil. *Journal of Tropical Ecology* 5, 413–431.

Ostrofsky, M. L. (1997). Relationship between chemical characteristics of autumn-shed leaves and aquatic processing rates. *Journal of the North American Benthological Society* 16, 750–759.

Parnrong, S., Buapetch, K., and Buathong, M. (2002). Leaf decomposition rates in three tropical streams of southern Thailand: the influence of land use. *Verhandlungen Internationale Vereinigung für Limnologie* 28, 475–479.

Pearson, R. G., Tobin, R. K., Smith, R. E. W., and Benson, L. J. (1989). Standing crop and processing of rainforest litter in a tropical Australian stream. *Archiv für Hydrobiologie* 115, 481–498.

Pringle, C. M. (2001). River conservation in tropical versus temperate latitudes. In Boon, P. J., Davies, B. R. and Petts, G. E. (eds.) "Global Perspectives on River Conservation: Science, Policy and Practice." pp. 373–383. John Wiley & Sons Ltd., Chichester.

Pringle, C. M., Hemphill, N., McDowell, W. H., Bednarek, A., and March, J. G. (1999). Linking species and ecosystems: different biotic assemblages cause interstream differences in organic matter. *Ecology* 80, 1860–1872.

Raviraja, N. S., Sridhar, K. R., and Bärlocher, F. (1998). Breakdown of *Ficus* and *Eucalyptus* leaves in an organically polluted river in India: fungal diversity and ecological functions. *Freshwater Biology* 39, 537–545.

Rosemond, A. D., Pringle, C. M., and Ramírez, A. (1998). Macroconsumer effects on insect detritivores and detritus processing in a tropical stream. *Freshwater Biology* 39, 515–523.

Rosemond, A. D., Pringle, C. M., Ramírez, A., Paul, M. J., and Meyer, J. (2002). Landscape variation in phosphorus concentration and effect on detritus-based tropical streams. *Limnology and Oceanography* 47, 278–289.

Rounick, J. S., and Winterbourn, M. J. (1983). Leaf processing in two contrasting beech forest streams: effects of physical and biotic factors on litter breakdown. *Archiv für Hydrobiologie* 96, 448–474.

Rueda-Delgado, G., Wantzen, K. M., and Beltrán, M. (2006). Leaf litter decomposition in an Amazonian floodplain stream: impacts of seasonal hydrological changes. *Journal of the North American Benthological Society* 25, 231–247.

Sakai, S. (2002). General flowering in lowland mixed dipterocarp forests of South-east Asia. *Biological Journal of the Linnean Society* 75, 233–247.

Schwarz, A. E., and Schwoerbel, J. (1997). The aquatic processing of sclerophyllous and malacophyllous leaves on a Mediterranean island (Corsica): spatial and temporal pattern. *Annales de Limnologie* 33, 107–119.

Souza, M. L., and Moulton, T. P. (2005). The effects of shrimps on benthic material in a Brazilian island stream. *Freshwater Biology* 50, 592–602.

Stallcup, L. A., Ardón, M., and Pringle, C. M. (2006). Effects of P- and N-enrichment on leaf breakdown in detritus-based tropical streams. *Freshwater Biology* 51, 1515–1526.

Stout, R. J. (1989). Effects of condensed tannins on leaf processing in mid-latitude and tropical streams: a theoretical approach. *Canadian Journal of Fisheries and Aquatic Sciences* 46, 1097–1106.

Strauss, E. A., and Lamberti, G. A. (2002). Effect of dissolved organic carbon quality on microbial decomposition and nitrification rates in stream sediments. *Freshwater Biology* 47, 65–74.

Suberkropp, K. (1998). Microorganisms and organic matter decomposition. In Naiman, R. J. and Bilby, R. E. (eds.) "River Ecology and Management. Lessons from the Pacific Coastal Region." pp. 120–143. Springer, New York.

Sweeney, B. W. (1984). Factors influencing life history patterns of aquatic insects. In Resh, V. H. and Rosenberg, D. M. (eds.) "Ecology of Aquatic Insects." pp. 56–100. Praeger Scientific Publishers, New York.

Vannote, R. L., Minshall, G. W., Cummins, K. W., Sedell, K. W., and Cushing, C. E. (1980). The River Continuum Concept. *Canadian Journal of Fisheries and Aquatic Sciences* 37, 130–137.

Vinson, M. R. and Hawkins, C. P. (1998). Biodiversity of stream insects: variation at local, basin, and regional scales. *Annual Review of Entomology* 43, 271–293.

Walker, I. (1986). Experiments on colonization of small water bodies by Culicidae and Chironomidae as a function of decomposing plant substrates and implications for natural Amazonian Ecosystems. *Amazoniana* 10, 113–125.

Walker, I. (1987). The biology of streams as a part of Amazonian forest ecology. *Experientia* 43, 279–287.

Walker, I. (1995). Amazonian streams and small rivers. In Tundisi, J. G., Bicudo, C. E. M. and Matsamura-Tundisi, T. (eds.) "Limnology in Brazil." pp. 167–194. Brazilian Academy of Sciences, Rio de Janeiro.

Wantzen, K. M. (2003). Cerrado Streams – characteristics of a threatened freshwater ecosystem type on the tertiary shields of South America. *Amazoniana* 17, 485–502.

Wantzen, K. M. (2006). Physical pollution: effects of gully erosion in a tropical clear-water stream. *Aquatic Conservation* 16, 733–749.

Wantzen, K. M., and Junk, W. J. (2000). The importance of stream-wetland-systems for biodiversity: a tropical perspective. In Gopal, B., Junk, W. J. and Davies, J. A. (eds.) "Biodiversity in Wetlands: Assessment, Function and Conservation." pp. 11–34. Backhuys, Leiden, The Netherlands.

Wantzen, K. M., and Junk, W. J. (2006). Aquatic-terrestrial linkages from streams to rivers: biotic hot spots and hot moments. *Archiv für Hydrobiologie/Supplement* 158, 595–611.

Wantzen, K. M., and Junk, W. J. (in press). Riparian wetlands. In Ronan, B. (ed.) "Encyclopedia of Ecology." Elsevier, Amsterdam.

Wantzen, K. M., and Wagner, R. (2006). Detritus processing by shredders: a tropical-temperate comparison. *Journal of the North American Benthological Society* 25, 214–230.

Wantzen, K. M., Wagner, R., Suetfeld, R., and Junk, W. J. (2002). How do plant-herbivore interactions of trees influence coarse detritus processing by shredders in aquatic ecosystems of different latitudes? *Verhandlungen Internationale Vereinigung für Limnologie* 28, 815–821.

Wantzen, K. M., Da Rosa, F. R., Neves, C. O., and Nunes da Cunha, C. (2005a). Leaf litter addition experiments in riparian ponds with different connectivity to a Cerrado Stream in Mato Grosso, Brazil. *Amazoniana* **18**, 387–396.

Wantzen, K. M., Drago, E., and da Silva, C. J. (2005b). Aquatic habitats of the Upper Paraguay River-Floodplain-System and parts of the Pantanal (Brazil). *Ecohydrology and Hydrobiology* **21**, 1–15.

Wantzen, K. M., Sá, M. F. P., Siqueira, A., and Nunes da Cunha, C. (2006). Conservation scheme for forest-stream-ecosystems of the Brazilian Cerrado and similar biomes in the seasonal tropics. *Aquatic Conservation* **16**, 713–732.

Webster, J. R., and Benfield, E. F. (1986). Vascular plant breakdown in freshwater ecosystems. *Annual Review of Ecology and Systematics* **17**, 567–594.

Wittmann, F., Junk, W. J., and Piedade, M. T. F. (2004). The várzea forests in Amazonia: flooding and the highly dynamic geomorphology interact with natural forest succession. *Forest Ecology and Management* **196**, 199–212.

Wootton, J. T., and Oemke, M. P. (1992). Latitudinal differences in fish community trophic structure, and the role of fish herbivory in a Costa Rican stream. *Environmental Biology of Fishes* **35**, 311–319.

Worbes, M., Klinge, H., Revilla, J. D., and Martius, C. (1992). On the dynamics, floristic subdivision and geographical distribution of várzea forests in Central Amazonia. *Journal of Vegetation Sciences* **3**, 553–564.

Wright, M. S., and Covich, A. P. (2005). Relative importance of Bacteria and Fungi in a Tropical Headwater Stream: Leaf decomposition and invertebrate feeding preference. *Microbial Ecology* **49**: 536–546.

Yam, S. W. R., and Dudgeon, D. (2005). Stable isotope investigation of food use by *Caridina* spp. (Decapoda: Atyidae) in Hong Kong streams. *Journal of the North American Benthological Society* **24**, 68–81.

Yule, C. M. (1996). Trophic relationships and food webs of the benthic invertebrate fauna of two aseasonal tropical streams on Bougainville Island, Papua New Guinea. *Journal of Tropical Ecology* **12**, 517–534.

Yule, C. M., and Pearson, R. G. (1996). Aseasonality of benthic invertebrates in a tropical stream on Bougainville Island, Papua New Guinea. *Archiv für Hydrobiologie* **137**, 5–117.

# Macroinvertebrates: Composition, Life Histories and Production

Dean Jacobsen, Claudia Cressa, Jude M. Mathooko, and David Dudgeon

Tropical stream macroinvertebrates are, in general, very incompletely known and in most tropical regions, there are significant taxonomic impediments to species- or genus-level identifications. Despite this, a considerable amount of information about macroinvertebrates at the family or sub-family level is available from tropical streams, and hypotheses, patterns and generalizations about their ecology can be derived from such data. This chapter describes the composition of the macro-invertebrate fauna of tropical streams, especially the insects, with the aim of uncovering the patterns in diversity and richness that they exhibit. No attempt is made to review the primary taxonomic literature. There appears to be a latitudinal trend towards increased richness of families and species of macroinvertebrates in topical streams, but some taxa (e.g. Plecoptera) are largely confined to temperate regions and others (decapod crustaceans) are mainly tropical. There are also some significant differences in faunas among tropical regions. The seasonality of macroinvertebrate populations and assemblages, and the factors driving life-history evolution and their consequences for secondary production are discussed with special emphasis on insects and decapod crustaceans. Many but by no means all tropical stream insects show semi-continuous or year-round reproduction that is almost independent of season, but strong seasonality is seen in some instances and is usually related to stream-flow patterns. The data indicate that insects and other tropical macroinvertebrates are sometimes affected by temperature, but that flow regime and habitat stability have a primary influence on life-history parameters and population dynamics. Published secondary-production estimates from tropical streams are not high, but may be lower than actual values because of the difficulty of determining growth and turnover ratios of populations that exhibit continuous and asynchronous development. Decapods make an important contribution to overall production in some tropical streams, and may offset low production by insects. Brief consideration is given to food webs and the trophic base of macroinvertebrate production in tropical streams: although data are scarce, there is evidence of dependence on autochthonous (algal) energy sources with a minor role for allochthonous foods. The chapter discusses the use of tropical macroinvertebrates in biomonitoring and includes suggestions for future research. The need for manipulative field experiments to elucidate the role of macroinvertebrates in stream ecosystem processes (including predation and competition) is stressed. Studies of production dynamics and life-history parameters should include a broader range of taxa in a wider range of tropical sites since, at present, the literature is dominated by studies from a few locations. More research in Africa is needed.

## I. INTRODUCTION

The biogeography, diversity and ecology of macroinvertebrates in tropical streams are poorly known. 'Macroinvertebrates' are those invertebrates exceeding 0.5 mm body size, or large enough to be seen by the naked eye, and comprise mostly insects as well as decapod crustaceans, molluscs, leeches, oligochaetes and planarians. Most of them live on or among streambed sediments, and hence are often referred to as macrobenthos, although the majority of stream insects have an amphibiotic life cycle and spend their adult stage on land, a fact that is sometimes conveniently ignored in studies of 'aquatic' macrobenthos. 'Tropical', as used here and elsewhere, encompasses the lands lying between the Tropics of Cancer and Carpricorn. Some major tropical rivers flow into (or out) of these latitudes from the temperate zone, and therefore the distribution of stream biota will necessarily coincide with these 'tropical' boundaries. Even within the tropics *sensu stricto*, climatic conditions and elevation determine stream conditions, and can range from relative aseasonality in equatorial headwater streams to truly seasonal conditions with wide lateral flooding in streams within monsoonal regions where there are marked wet and dry seasons. The pattern of climatic seasonality will have an influence on almost all aspects of the ecology of stream invertebrates, particularly their distribution, life histories and timing of life-cycle events. Inter-annual climatic fluctuations and disturbances such as droughts and intense storms may also have important implications for stream macroinvertebrates (e.g. Covich *et al.*, 2003).

The tropical stream biota is complex and biodiverse (see, for example, Chapters 5 and 6 of this volume): much about macroinvertebrate biogeography and ecology is poorly known and data on macroinvertebrate diversity are sparse and disconnected both at the local and regional scale. A few summaries are available (e.g. Dudgeon, 1999a, 2000) but these have the effect of emphasizing how little we know rather than providing robust generalities that apply widely. Although there are certainly exceptions (e.g. streams in Puerto Rico have been well studied), it is nonetheless evident that ecological understanding of the composition, diversity, role and importance of macroinvertebrates in tropical streams lags far behind that in temperate regions, and relatively little is known about the life histories and habits of major taxa.

### A. The Taxonomic Impediment

One reason for the paucity of ecological studies of tropical stream invertebrates in the international literature (for further data and explanations, see Dudgeon, 2003) is the fact that identification of tropical species is difficult for non-specialists, because many lower taxa (especially genera and species) have received limited study and relevant literature is scarce (Boyero, 2002). Manifest errors arise when identification guides from temperate latitudes are used in the tropics, especially if these comprise keys to genera or species. Much of the relevant tropical literature that does exist is made up of isolated descriptions of one or a few new species at one or a handful of locations, and revisions of genera or family at a regional level or other monographic treatments that might have application over a region or biome are very scarce. A notable advance in this regard is the recent set of keys to Southeast Asian aquatic invertebrates collated by Yule and Yong (2004).

The problem of identification is confounded by the *ad hoc* nature of taxonomy itself, where a decision has to be made about the identity and status of a group of organisms based on available data (Oliver, 1979). Additional confusion arises (in the case of freshwater molluscs, for example) where changes in shell form and appearance can be induced by environmental conditions, or where different life stages live in water and on land (most of the aquatic insects) and have different habits and appearance. Taxonomic decisions are subject to continuous revision and reassessment as further information becomes available, thus knowledge of the macroinvertebrate fauna of tropical streams is improved – albeit very gradually. Slow progress reflects a lack of specialist systematists, inadequate training or capacity-building, plus a perception that fundamental taxonomic work is outmoded, lacks status and does not warrant high priority; it is therefore poorly resourced. A great deal is yet to be done: many, perhaps most, tropical stream invertebrates remain undiscovered and/or undescribed (Cressa and Holzenthal, 2003), and even those that are better known have not yet been studied in all parts of the tropics. Perhaps this is not surprising: mayflies (Ephemeroptera) in Europe were recorded as 'day-flies', 'ephemeron' or 'heremobio' by Aristotle (384–322 BC) and his successor Theophrastus, but the first account of their biology, Ougert Cluyt's booklet *De Heremobio* (1634), was not published until 2000 years later (Williams and Feltmate, 1992).

Among the tropical stream macroinvertebrates that are relatively well known, the Simuliidae (Diptera), or blackflies, have received a great deal of attention (Currie and Craig, 1987) because some species are vectors for filarial nematode parasites that attack humans and domestic animals (Walsh, 1985). Onchocerciasis (river blindness) control programs in tropical Africa – especially West Africa – have resulted in the recognition of over 193 species (Shelley, 1988) in habitats ranging from torrential streams to the carapaces of freshwater crabs and prawns (Williams and Hynes, 1971; Burger, 1987). Details of the success of what has been described as the largest blackfly control program ever mounted or, perhaps, the biggest that will ever be mounted, are given by Resh *et al.* (2004).

## B. Challenges and Potential Solutions

Analysis of the biogeography of most tropical stream invertebrates is certainly hampered by taxonomic uncertainty, and by the lack of regional studies of many groups, so that interpretation of any patterns that are uncovered is problematic. In isolated insular streams, for example, some taxa may be 'missing' because of events in the Earth's geological history (Covich, 1988), absence of studies devoted to a particular taxon, or extirpation by anthropogenic impacts. The last two can produce 'false negatives'. Human impacts not only reduce original geographic ranges by eliminating species from previously-favourable habitat. Deliberate or accidental species introductions disrupt once-formidable geographic barriers, provide a basis for further colonization and range extensions of newly-established species, and allow intermixing of previously isolated faunas. Apart from their intrinsic academic interest, understanding of pre-human biogeographic patterns may allow prediction of the consequences of ongoing and future anthropogenic alterations of the biosphere. Climatic changes in particular will have profound implications for precipitation, evaporation and runoff, and hence the seasonality and magnitude of stream flows in the tropics to which the biota are adapted. It is these changes, and the need to predict and manage their consequences, that makes it imperative to improve understanding of the ecology of macroinvertebrates and other tropical stream biota.

How can this goal be achieved? To be realistic, constraints upon species-level identifications will remain for the vast majority of tropical stream macroinvertebrates in the foreseeable future. Most ecological work in these habitats will have to depend on identification to family and/or subfamily, which can be done accurately given some basic training. These identifications can be supplemented by separation into morphospecies, preferably matched across sites, for some or all of the major taxa. Only where regional and local faunas are exceptionally well studied will more accurate identification be possible. For that reason, much of the following account is based upon research conducted at the family or subfamily level, and richness – as used herein – refers richness of these taxa and, sometimes, morphospecies richness. Rarely, if ever, do tropical stream ecologists work at the same level of taxonomic resolution as their temperate counterparts. This fact needs to be kept in mind when comparing the results of studies at different latitudes, but this does not mean that such comparisons are not possible or worthwhile. Much can be learned about ecological processes in streams (e.g. the role of macroinvertebrates in leaf litter breakdown, the impacts of fish predation on invertebrate assemblages, or the effects of grazers on periphyton stocks) without the need for species-level identification of macroinvertebrates. While the taxonomic impediment is significant, it should not be an excuse for second-rate work nor a reason to avoid the many exciting research opportunities and applications presented by tropical stream invertebrates.

## C. Chapter Objectives

An exhaustive review of all aspects of the ecology of tropical stream invertebrates lies beyond the scope of a single chapter. Moreover, aspects of invertebrate ecology relating to organic matter processing (Chapters 3 and 7), assemblage composition at high-altitudes (Chapter 8) and tropical-temperate comparisons (Chapter 9), as well as a variety of other topics, including some relevant to conservation (Chapter 10), are considered elsewhere in this volume. Our objectives in this chapter are to describe the composition of the macroinvertebrate fauna of tropical streams (with the caveats about taxonomic penetration given above) with the aim of uncovering the patterns in diversity and richness that they exhibit. No attempt is made to review the primary taxonomic literature. An account is also given of the seasonality of macroinvertebrate populations and assemblages, and the factors driving life-history evolution and their consequences for secondary production are discussed. Brief consideration is given to the food webs and energy flow in relation to the functional role of macroinvertebrates in tropical stream ecosystems (so elaborating upon some matters raised in Chapters 2, 3 and 9). Finally, we review the current but somewhat limited use of macroinvertebrates in the biological assessment of tropical streams, and evaluate the potential for development of much-needed biomonitoring tools in the tropics. The chapter concludes with an outline of research needs and priorities.

## II. COMPOSITION OF THE FAUNA

### A. General Characteristics

At first glance, the macroinvertebrate fauna of tropical streams is quite similar to that present in temperate streams as most of the same macroinvertebrate orders occur in both (Hynes, 1970). Tropical stream assemblages seem to be just as variable, both within and among regions and continents, as those in temperate streams. Nevertheless, tropical stream faunas have some distinctive features that distinguish them from streams at higher latitudes. These differences occur primarily at lower taxonomic levels (families, genera and species) and in the relative abundance and richness of orders. In very general terms, the fauna of tropical lowland streams is relatively rich in decapod crustaceans (shrimps, prawns and crabs) and prosobranch snails, as well as insects such as Odonata, Heteroptera and perhaps also Trichoptera; by contrast, Plecoptera are depauperate. Details are given below. The richness of the other major insect orders (i.e. Ephemeroptera, Coleoptera and Diptera) seems to be quite similar to that of temperate streams.

### B. Composition of the Non-Insect Fauna

Altitude is a prime factor determining composition and diversity of macroinvertebrates in tropical streams (Jacobsen *et al.*, 1997; Jacobsen 2003, 2004). High-altitude streams (at or above 3000 m above sea level) are discussed in Chapter 8 of this volume, and the following account therefore focuses on tropical streams at lower elevations. Table I provides a general overview of the composition of the zoobenthic fauna sampled by roughly comparable

*TABLE I*   Mean Percent of Total Individuals Attributed to Specific Non-insect Taxa Classes
and Insect Orders in the Stream Macrobenthos of Six Tropical Regions

| | Total individuals (%) | | | | | | |
|---|---|---|---|---|---|---|---|
| | *SU* | *HK* | *PN* | *EP* | *EA* | *BO* | *Mean* |
| Non-insects | | | | | | | |
| Planariidae | <0.1 | 0.5 | – | 3.2 | 1.3 | 0.2 | 0.9 |
| Oligocheata | – | – | – | – | 0.2 | 1.0 | 0.2 |
| Hydracarina | – | – | – | – | <0.1 | 0.3 | <0.1 |
| Decapoda | – | 0.4 | <0.1 | – | – | – | <0.1 |
| Bivalvia | – | <0.1 | – | 0.1 | – | – | <0.1 |
| Gastropoda | – | 3.4 | – | 0.8 | <0.1 | 0.7 | 0.8 |
| Insects | | | | | | | |
| Plecoptera | <0.1 | 3.1 | – | 0.3 | 0.8 | 1.7 | 1.0 |
| Ephemeroptera | 47.3 | 28.2 | 50.9 | 29.9 | 21.2 | 25.4 | 33.8 |
| Odonata | 0.4 | 0.8 | 1.3 | 8.3 | 1.6 | 1.4 | 2.3 |
| Megaloptera | – | 0.1 | – | 0.7 | 0.3 | 0.7 | 0.3 |
| Hemiptera | – | 1.4 | 0.8 | 6.4 | 1.3 | 1.6 | 1.9 |
| Coleoptera | 6.7 | 15.5 | 10.1 | 25.8 | 13.3 | 22.1 | 16.6 |
| Lepidoptera | 0.2 | 0.2 | 1.4 | 0.9 | 2.2 | 0.3 | 0.9 |
| Trichoptera | 23.7 | 13.1 | 14.3 | 9.9 | 8.4 | 28.0 | 16.2 |
| Diptera | 21.6 | 34.3 | 21.1 | 13.6 | 47.9 | 16.6 | 25.2 |

Only taxa comprising >0.1% in at least one region are included.
SU, Sulawesi (number of sites = 5; Dudgeon, 2006 and unpublished observations); HK, Hong Kong (*n* = 36;
D. Dudgeon, unpublished observations); PN, Papua New Guinea (*n* = 6; Dudgeon, 1994); EP, Ecuador Pacific (*n* = 12;
Schultz, 1997); EA, Ecuador Amazon (*n* = 12; Bojsen and Jacobsen, 2003); BO, Bolivia (*n* = 12; Moya, 2006).

methods in unpolluted streams within six different tropical regions, three in Asia and three
in South America (similar quantitative data from tropical African lowland streams were not
available). The percentage of non-insects varies from practically none to 5% of total density
in the six regions, but there are other situations where they may become more important.
Decapod crustaceans and snails seem to be particularly abundant in tropical island streams
or streams close to the coast, as reported from Puerto Rico (Covich and MacDowell, 1996),
St. Vincent (Harrison and Rankin, 1975), Costa Rica (Pringle and Ramírez, 1998), Sri Lanka
(Starmühlner, 1984) and the Philippines (Bright, 1982).

   The non-insect groups are dominated by turbellarian Planariidae, molluscs (especially gastropods) and crustaceans. Many of the snail families present in tropical streams are cosmopolitan, especially pulmonates such as Lymnaeidae, Planorbidae, Ancylidae and Physidae, but they
appear to be more important in temperate streams. In contrast, Thiaridae *sensu lato* (including
Pachychilidae) and Ampullaridae are exclusively tropical and fresh water, whereas the neritid snails are mainly marine but include tropical freshwater representatives. Richness in some
streams is high: for example, the Mekong River supports species-flocks of stenothyrid and
pomatiopsid snails (>110 endemic species) and may have the richest extant freshwater mollusc
fauna (Dudgeon, 1999a; 2000). Other generally less numerous non-insect groups in tropical
streams are Oligochaeta, Hirudinea, Hydracarina and bivalve molluscs, although they may be
abundant in particular circumstances or habitats (e.g. large rivers or floodplain lakes).

   Small crustaceans such as amphipods (Gammaridae and Hyalellidae) and isopods (Asellidae)
are common and abundant in temperate streams, as are the larger decapod crayfishes, but
they generally do not occur in tropical streams. Parastacid crayfish are only found in New
Guinea, Australia and Madagascar while North American Cambaridae crayfish just extend

south to Central America. Instead, these crustaceans are replaced in tropical streams by a variety of other decapods: crabs (a few Grapsidae but mainly Potamidae and allied families) and two circumtropical shrimp families – Palaemonidae (*Macrobrachium* spp.) and Atyidae (mainly *Caridina*). These groups can be highly diverse: over 100 species of shrimps have been recorded in India alone (Kottelat and Whitten, 1996; Dudgeon, 1999a); the Asia freshwater crabs include over 80 genera and at least 185 species occur in China although many more await description by scientists (Kottelat and Whitten, 1996; Ng, 1988). While the data in Table I suggest that decapod densities are low, they may be misleading since decapods are not efficiently collected by quantitative Surber or kick samples, and can be quite diverse (e.g. Victor and Ogbeibu, 1985; Ng, 1988) and abundant although their numbers are often underestimated (Covich, 1988; Ramírez and Pringle, 1998a; Ramírez *et al.*, 1998; Dobson, 2004; Dobson *et al.*, 2007). For instance, Pringle and Ramírez (1998) noted that shrimp larvae constituted the main part of invertebrate drift in a Costa Rican stream, although they were practically absent from quantitative benthos samples; Dudgeon (2006) reported the same phenomenon in Sulawesi streams. By contrast, densities of predatory *Macrobrachium hainanense* (Palaemonidae) in Hong Kong stream pools reached 3–5 individuals per m$^2$ which, given that they can attain 7 cm body length, represents a considerable biomass and secondary production (Mantel and Dudgeon, 2004a). Some *Macrobrachium* may attain lengths of 20–25 cm, and can be of considerable importance for aquaculture and fisheries (e.g. Victor and Ogbeibu, 1985; Dudgeon, 1999a).

As our understanding of tropical streams increases, it is becoming clear that shrimps play important roles as grazers, predators and detritivores (e.g. Pringle *et al.*, 1993; 1999; Pringle, 1996; Pringle and Hamazaki, 1998; Crowl *et al.*, 2001; March *et al.*, 2001, 2002; Mantel and Dudgeon, 2004a,b; De Souza and Moulton, 2005; Yam and Dudgeon, 2005a), perhaps serving as 'ecosystem engineers' (but see Mantel and Dudgeon, 2004c), and they are prey for a variety of riparian and semi-aquatic vertebrates (Arkell, 1979; Butler and du Toit, 1994, Purves *et al.*, 1994; Butler and Marshall, 1996; see also Chapter 6 of this volume). In general, atyid shrimps are primary consumers, whereas *Macrobrachium* spp. are omnivores or predators; consequently, the ecological influence of shrimps is strongly taxon dependent. However, both families are circumtropical and often occur in the same stream.

Since they are confined to the tropics (and parts of the subtropics), the distribution and regional composition of lotic decapod assemblages is not well known, and there have been rather few studies of their population dynamics (but see recent papers by Mantel and Dudgeon, 2004a,b, 2005; Yam and Dudgeon, 2005a,b, 2006; Dobson *et al.*, 2007). This is particularly true of freshwater crabs: nearly 1000 species have been described (Sternberg and Cumberlidge, 2001), but even the most basic ecological information is known for only a handful of them (Ng, 1988; Dobson, 2004; Dobson *et al.*, 2007).

Tropical freshwater crabs have evolved a much higher family richness than the shrimps, with families specific to different continents. Two families of crabs, Trichodactylidae and Grapsidae, dominate Neotropical lowland streams. Tropical Asian streams have higher diversity, hosting the Grapsidae, Gecarcinucidae, Sundathelphusidae, Parathelphusidae, Potamidae, Isolapotamidae and Sinopotamidae, plus a few Hymenosonatidae (Ng, 1988; Dudgeon, 1999a), while tropical Australian streams are inhabited by Grapsidae and Sundathelphusidae. The Potamidae is the only family of freshwater crabs reported from Afrotropical streams (Table II), where *Potamonautes* spp. appear to be the main invertebrate shredders in (at least) Kenyan streams (Dobson *et al.*, 2002).

The biomass of crabs in tropical and southern African streams can exceed the rest of the benthic fauna by an order of magnitude, and a single species characteristically dominates the fauna and, perhaps, ecosystem processes (Turnbull-Kemp, 1960; Hill and O'Keeffe, 1992). In Tanzania, Abdallah *et al.* (2004) estimated crab biomass within a single stream to be 58–94% of total

TABLE II  Population Density and Biomass Estimates for Lotic Freshwater Crabs (*Potamonautes* spp.) in Africa

| Species and locality | Density (m$^{-2}$) | Biomass (dry mass gm$^{-2}$) | Sampling method | Sampling frequency | Reference |
|---|---|---|---|---|---|
| *Potamonautes perlatus* | | | | | |
| Zimbabwe | 0.8–1.3 | 2.2–5.4* | Baited traps | Short period | Turnbull-Kemp (1960) |
| Cape Province | 2.0 | | Hand search | Single collection | Arkell (1979) |
| Cape Province | 0.3–2.7 | | Benthic sample | Monthly for 1 year | King (1983) |
| Cape Province | 1.7–5.2 | 54–136 | Baited traps | Short period | Hill and O'Keeffe (1992) |
| Zimbabwe | 0.1–2.2 | | Baited traps | Short period | Butler and du Toit (1994) |
| Cape Province | 2.9–15.6 | | Hand search | Widely spaced for >1 year | Somers and Nel (1998) |
| *Potamonautes* sp. | | | | | |
| Tanzania – erosive stream | 4.0 | 0.6 | Surber sampler | Single collection | Abdallah *et al.* (2004) |
| Tanzania – flood plain | 24.8 | 11.5 | Pond net | Single collection | Abdallah *et al.* (2004) |
| *Potamonautes obdneri* and *Potamonautes* sp. | | | | | |
| Kenya | 10.7 –16.0 (107.2 if juveniles included) | 6.5–8.2 | Surber sample | Single collection | Dobson et al. (2007, unpublished observations) |
| *Potamonautes subukia* (Chinga crab) | | | | | |
| Kenya | 7.2 | 2.4 | Surber sample | Monthly for 1 year | Dobson et al. (2007, unpublished observations) |
| Kenya | 24.8 | 4.6 | Hand search | Single collection | Dobson et al. (2007, unpublished observations) |

* Estimated dry mass, based on fresh weight data.

macroinvertebrate biomass, the variation depending upon microhabitat, whereas Dobson *et al.* (2007; see Table II) reported that crabs often comprised at least 70% of total macroinvertebrate biomass in streams around Mt Kenya; even in streams where crabs were rare they accounted for nearly 40% of biomass. More than half of East African freshwater crab species may be endangered or vulnerable, but insufficient information is available to make a realistic assessment of their conservation status (Darwall *et al.*, 2005). Threats from habitat loss, pollution and water abstraction are likely to be major risks factors, but threats from competitors, parasites and introduced (non-native) predators may be significant also (Williams *et al.*, 1964; Dobson, 2004; Ogada, 2006).

## C. Composition of the Insect Fauna

In terms of abundance, insects dominate the stream fauna in all tropical regions (Table I), as is the case for most temperate streams. There is, however, variation among streams and regions with respect to the orders that dominate insect assemblages across a sample of non-African streams from different parts of the tropics (Table I). In Africa also, insects dominate the benthos (exceeding 90% of individuals) at forested and cultivated sites along the Sagana River, Kenya (Mwangi, 2000), but their relative proportions vary greatly. Chironomids, simuliids (Diptera) and Ephemeroptera were the most abundant insect taxa in the forested site, constituting over 88% of the macroinvertebrate assemblage, while the abundance of Coleoptera and Trichopterans rarely exceeded 5%, and chironomids made up more than half of the zoobenthos at cultivated sites. More than 60% of the individuals collected from Kalengo Mountain Stream, Democratic Republic of Congo, were Chironomidae followed by Ephemeroptera and Trichoptera (Statzner, 1975). As Table I shows, Trichoptera can vary in their relative abundance across regions by a factor of around three (e.g. EA or EP versus BO or SU), and the same pattern has been reported in other studies. In Sábalo Stream, Costa Rica, Trichoptera comprised a few percent of total zoobenthos (Ramírez and Pringle, 1998b), but reached 27–37% in some Venezuelan streams (Río Orituco and Río Camurí Grande, respectively: Cressa, 1994, 1998).

Despite some inter-regional variability, the four orders that constitute the majority of the tropical stream insect assemblages included in Table I are, in descending order of mean importance (relative abundance), Ephemeroptera, Diptera, Coleoptera and Trichoptera. Streams in Africa follow the same pattern (e.g. Petr, 1970; Hynes, 1975; Mathooko and Mavuti, 1992; Dobson *et al.*, 2002), and this appears to hold true for other tropical streams in continental areas (e.g. Bishop, 1973). The same orders dominate the insect fauna of most streams at higher latitudes (Hynes, 1970). Plecoptera and other insect orders can be locally abundant in tropical streams but, as Table I shows, they generally occur in relatively low numbers (only in EP do they exceed 10% of insects).

Streams in different parts of the tropics are dominated not only by the same orders, but also by the same few insect families. Chironomidae (Diptera), Baetidae and Leptophlebiidae (Ephemeroptera), Hydropsychidae (Trichoptera) and Elmidae (Coleoptera) are among the top-ranked families in all regions included in Table III. They are likewise common and often dominant in streams at high latitudes. However, they may be represented by different species in different tropical regions. For example, Dudgeon (1988a) demonstrated that even though the top-ranked families did not vary much between four Hong Kong streams, each stream was dominated by different species. Stream insect families that are locally common are generally widespread throughout the tropics (with the possible exception of northern Australia). The proportion of families shared between regions on the same continent and among continents is much higher for locally-abundant families than for locally-rare families (Fig. 1). In other words, those families limited to specific regions are also generally the least abundant locally (i.e. they are present at low densities), and the qualitative differences in the composition of stream faunas between different tropical regions occur primarily within the locally least-abundant groups.

*TABLE III*  Dominant Taxa in the Stream Macrobenthos of Six Tropical Regions Listed in Order of Decreasing Density

| Sulawesi | Hong Kong | Papua New Guinea | Ecuador, Pacific | Ecuador, Amazon | Bolivia, Chapare |
|---|---|---|---|---|---|
| Chironomidae (D) | Chironomidae (D) | Baetidae (E) | Elmidae (C) | Chironomidae (D) | Elmidae (C) |
| Baetidae (E) | Leptophlebiidae (E) | Chironomidae (D) | Baetidae (E) | Elmidae (C) | Hydropsychidae (T) |
| Hydropsychidae (T) | Scirtidae (C) | Leptophlebiidae (E) | Chironomidae (D) | Tricorythidae (E) | Tricorythidae (E) |
| Caenidae (E) | Baetidae (E) | Elmidae (C) | Tricorythidae (E) | Leptophlebiidae (E) | Chironomidae (D) |
| Tricorythidae (E) | Heptagenidae (E) | Hydropsychidae (T) | Leptophlebiidae (E) | Baetidae (E) | Baetidae (E) |
| Philopotamidae (T) | Elmidae (C) | Caenidae (E) | Naucoridae (H) | Hydropsychidae (T) | Odontoceridae (T) |
| Leptophlebiidae (E) | Hydropsychidae (T) | Philopotamidae (T) | Hydropsychidae (T) | Simuliidae (D) | Leptophlebiidae (E) |
| Elmidae (C) | Caenidae (E) | Glossosomatidae (T) | Libellulidae (O) | Pyralidae (L) | Xiphocentronidae (T) |
| Psephenidae (C) | Philopotamidae (T) | Pyralidae (L) | Gomphidae (O) | Euthyplociidae (E) | Helicopsychidae (T) |
| Simuliidae (D) | Simuliidae (D) | Libellulidae (O) | Hydroptilidae (T) | Limoniidae (D) | Perlidae (P) |
| Prosopistomatidae (E) | Gastropoda | Psephenidae (C) | Psephenidae (C) | Naucoridae (H) | |
| Pyralidae (L) | Ephemerellidae (E) | Ptilodactylidae (C) | Planariidae | Coenagrionidae (O) | |
| | Perlidae (P) | Tipulidae (D) | Odontoceridae (T) | Ceratopogonidae (D) | |
| | Hydrophilidae (C) | Philopotamidae (T) | Ptilodactylidae (C) | | |
| | Psephenidae (C) | Coenagrionidae (O) | | | |
| | Helotrephidae (H) | | | | |
| | Pseudoneureclipsidae (T) | | | | |
| | Nemouridae (P) | | | | |

Only taxa comprising >1% in at least one region are included.
D, Diptera; E, Ephemeroptera; T, Trichoptera; C, Coleoptera; P, Plecoptera; H, Heteroptera; L, Lepidoptera.
For sources, see Table I.

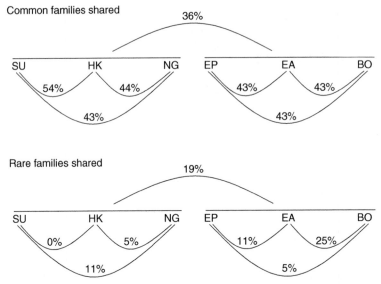

FIGURE 1   The percentage of shared families among the ten most abundant insect families (above) and among the ten least abundant families (below) occurring in streams in each of six tropical regions. SU, Sulawesi; HK, Hong Kong; NG, Papua New Guinea; EP, Ecuador Pacific; EA, Ecuador Amazon; BO, Bolivia (for sources, see Table I).

Some more-or-less cosmopolitan families such as Oligoneuriidae and Prosopistomatidae (Ephemeroptera), Naucoridae (Heteroptera), Philopotamidae, Calamoceratidae and Xiphocentronidae (Trichoptera), Pyralidae (Lepidoptera), Ptilodactylidae, Lampyridae and Psephenidae (Coleoptera), and Perlidae (Plecoptera) seem to be generally more important in tropical than in temperate streams. The Tricorythidae (Leptohyphidae: Ephemeroptera) is especially notable as it is the only predominately tropical family that is sufficiently abundant to be ranked among the top five families in Table III. The Perlidae seems to be the only family of Plecoptera in equatorial streams, represented by *Anacroneuria, Enderleina* and *Macrogynoplax* in the Neotropics (Stark, 2001), *Neoperla* (*sensu lato*) in Asia and Africa (Zwick, 1976, 1986; Dudgeon, 1999a) and *Stenoperla* and *Dinotoperla* in tropical Australia (Pearson *et al.*, 1986), although other families with a mainly temperate distribution (Nemouridae and Leuctridae) may penetrate the northern margins of the tropics (Dudgeon, 1999a). Aquatic moth larvae belonging to the Pyralidae are not often common in temperate streams, but they can be very abundant in tropical lowland streams (e.g. New Guinea and the Ecuadorian Amazon; Table III), particularly in open reaches where they graze periphyton beneath silken retreats. In addition, some cosmopolitan families such as Caenidae and Heptageniidae (Ephemeroptera) are apparently more important in tropical Asian and African streams than in Neotropical streams, although the latter is scare in Sulawesi and does not extend south into Papua New Guinea or Australia. Note that in the majority of cases where the same families occur on different continents, they are represented by different genera in each region.

Surprisingly few families of stream insects are strictly tropical, but examples occur within the Odonata, Heteroptera, Ephemeroptera, Trichoptera and Coleoptera. Curiously, the Diptera contains no exclusively-tropical aquatic families. Many of the exclusively-tropical families are confined to particular regions, such as the Neotropics (Euthyplocidae: Ephemeroptera) or Old World tropics of Africa and Asia (Prosopistomatidae: Ephemeroptera, Pseudoneureclipsidae: Trichoptera, and Heterotrephidae: Heteroptera). Among the Odonata alone, there are some substantial differences between the Old and New World tropics: nine families have been recorded in Hong Kong streams, of which five are unknown from the Neotropics, whereas three out of seven families reported from Ecuador do not occur in Hong Kong (Dudgeon, 1989a; Jacobsen, 2004). The exclusively-tropical families tend to be relatively rare at a local scale. Only the Euthyplociidae is ranked among the top-10 most abundant families in any region in Table III. Most of the 'real' tropical families are among the families that make up less than 1% of individuals, and so are not listed in Table III. No more than 4% of the individuals among the 10 most abundant families in any of the six data sets in Table III are strictly tropical, but 26% of the 10 least-abundant families included in these data sets were tropical.

## D. Diversity of Tropical Stream Faunas

It might seem obvious that the orders which make the greatest contribution to zoobenthos densities in tropical streams would be those that contribute most to taxonomic richness, and in some studies this is certainly the case (e.g. the Río Camurí Grande, Venezuela: Cressa, 1998). In other instances, however, this relationship does not hold true. Among the six tropical data sets used as examples in Tables I and III, the Trichoptera and Diptera are the most diverse orders of zoobenthos in terms of family richness, followed by Coleoptera, Ephemeroptera, Odonata and Heteroptera (Table IV). Plecoptera, Megaloptera and Lepidoptera are usually represented by only one family each (Perlidae, Corydalidae and Pyralidae, respectively), and therefore contribute little to overall richness. Trichoptera and especially Odonata are disproportionately important in terms of family richness relative to their percentage abundance while, despite high population densities, Diptera and Ephemeroptera make rather small contributions to total family richness (Fig. 2).

Contributions by individual orders to total family richness and species richness are not identical because species : family ratios differ among orders (Table V). For example, Odonata in Hong Kong streams constitute 28% of total family richness but only 6% of total species richness, whereas Coleoptera account for 14% of family richness and, because of high diversity of Elmidae, contribute 27% of species richness (Dudgeon, 1988a). Ephemeroptera contributed 16% of family richness and 23% of species richness in Hong Kong streams, and these insects seem to have diversified more at the species level than at the family level in the tropics. For example, slightly more than 85 species of Ephemeroptera are recorded from East Africa (see Appendix 1 in Mathooko, 1998), despite the fact that "... workers on the African fauna in the past have been more conservative in their approach and, especially in the case of nymphs, have either tended to assign taxa with unusual characters to existing genera" (Gillies, 1988).

*TABLE IV*   Percentage of the Total Number of Insect Families Attributed to Individual Orders in Streams within Six Tropical Regions

| | *Percent of total number of insect families* | | | | | | |
|---|---|---|---|---|---|---|---|
| | *SU* | *HK* | *PN* | *EP* | *EA* | *BO* | *Mean* |
| Plecoptera | 2.3 | 5.0 | – | 2.5 | 3.0 | 3.8 | 2.8 |
| Ephemeroptera | 18.6 | 13.3 | 12.5 | 12.5 | 12.1 | 11.5 | 13.4 |
| Odonata | 14.0 | 13.3 | 6.3 | 12.5 | 15.2 | 7.7 | 11.5 |
| Megaloptera | – | 1.7 | – | 2.5 | 3.0 | 3.8 | 1.8 |
| Heteroptera | – | 1.7 | 6.3 | 10.0 | 9.1 | 3.8 | 5.2 |
| Coleoptera | 18.6 | 15.0 | 25.0 | 12.5 | 15.2 | 11.5 | 16.3 |
| Lepidoptera | 2.3 | 1.7 | 3.1 | 2.5 | 3.0 | 3.8 | 2.7 |
| Trichoptera | 27.9 | 26.7 | 18.8 | 25.0 | 21.2 | 30.8 | 25.1 |
| Diptera | 16.3 | 21.7 | 28.1 | 20.0 | 18.2 | 23.1 | 21.2 |

For sources, see Table I.

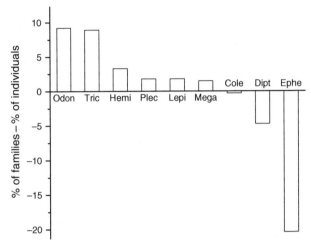

*FIGURE 2*    The contribution to total insect family richness made by each stream insect order relative to its contribution to total insect density. Values are derived from means for the six tropical stream data sets shown in Tables I and IV. Odon, Odonata; Tric, Trichopters; Hemi, Heteroptera; Plec, Plecoptera; Lepi, Lepidoptera; Mega, Megaloptera and Neuroptera; Cole, Coleoptera; Dipt. Diptera; Ephe, Ephemeroptera.

*TABLE V*    Number of families (F), species (S) and the species: family (S : F) ratio for aquatic insect orders on three continents

|  | Tropical South America | | | Europe | | | North America | | |
|---|---|---|---|---|---|---|---|---|---|
|  | F | S | S : F | F | S | S : F | F | S | S : F |
| Ephemeroptera | 10 | 184 | 18.4 | 17 | 217 | 12.8 | 21 | 599 | 28.5 |
| Plecoptera | 2 | 230 | 50.0 | 7 | 387 | 55.3 | 9 | 577 | 64.1 |
| Odonata | 19 | 1491 | 78.5 | 10 | 127 | 12.7 | 10 | 422 | 42.2 |
| Heteroptera | 14 | 715 | 51.1 | 12 | 129 | 10.8 | 18 | 421 | 23.4 |
| Megaloptera | 4 | 42 | 11.5 | 4 | 16 | 4.0 | 4 | 70 | 17.5 |
| Coleoptera | 18 | 1913 | 106.3 | 23 | 967 | 42.0 | 18 | 1214 | 67.4 |
| Trichoptera | 14 | >2500* | 107.1 | 22 | 895 | 40.7 | 23 | 1385 | 60.2 |
| Total | 81 | 6975 | 73.4 | 95 | 2738 | 28.8 | 103 | 4688 | 45.5 |

Data for tropical South America from Hurlbert *et al.* (1981), Flint *et al.* (1999) and Cressa and Stark (2003); for Europe from Illies (1978); and for North America from Merritt and Cummins (1996). Megaloptera as used here includes Neuroptera. Diptera have been excluded because data for the order are very incomplete. * R. Holzenthal, University of Minnesota (personal communication). Modified from Jacobsen *et al.* (1997).

The widespread genus *Baetis* (*sensu* Gillies, 1994) is particularly in need of revision and subdivision into other genera such as *Nigrobaetis* in East Africa (Mathooko, 1997).

Evidence of the species-level diversification of tropical Ephemeroptera is evident from a study of 26 stream sites within a relatively limited area of northwestern Panamá, where Flowers (1991) collected just five families but an astonishing 82 species indicating a high degree of endemism. By contrast, Dudgeon (1988a) collected eight families and 29 species of Ephemeroptera from four Hong Kong streams; recent collections from the 1000-km$^2$ territory have raised the total to 56 species in nine families (D. Dudgeon, unpublished observations). Comparisons between data such as these are, in many instances, confounded by incomplete knowledge of the tropical stream insects, especially at lower taxonomic levels, and obstacles to reliable species identifications. As a result, generalizations about species: family ratios are probably not robust, and comparative studies of taxonomic richness should probably be confined to the family level.

Taxonomic impediments to species identification limit the ability of stream ecologists to decide whether the diversity of macroinvertebrates in tropical streams varies consistently among regions. Although the global tally of aquatic insects is far from complete, there is some evidence that Trichoptera in tropical Asia are exceptionally diverse compared to other tropical areas in terms of both total and relative species richness. It supports 1.22 species kilohectare$^{-1}$, compared to only 0.60 species kilohectare$^{-1}$ in the Neotropics (Morse, 1997). The equivalent figures for Australasia and Africa are 0.72 and 0.25, respectively. While the African caddisfly fauna may be under collected, this is not the case for Australia. Large expanses of arid land may be responsible for the lower per unit area richness of Australia relative to Asia, but it does not explain the difference between Asia and the Neotropics. High diversity in Asia is also a feature of other orders of aquatic insects. For example, Indonesia (1 905 000 km$^2$) supports more dragonfly (Odonata) species (over 660) than any other country, and even tiny Hong Kong (~1000 km$^2$) hosts 111 species (Wilson, 2003).

Some studies indicate high zoobenthos richness in Old World tropical streams: Bishop (1973) identified 204 morphospecies from the Sungai Gombak in Peninsular Malaysia, and Pearson *et al.* (1986) found more than 267 morphospecies in Yuccabine Creek in northern Australia. Similarly, Dudgeon (1988a) recorded 94 morphospecies (excluding Chironomidae) from benthic samples taken on one day in a single riffle of Tai Po Kau Forest Stream in Hong Kong. In the Neotropics, by contrast, Ramírez and Pringle (1998b) identified only 53 morphospecies in Sábalo Stream, Costa Rica, while Cressa (1998) likewise reported 52 morphospecies in the Río Camurí Grande, Venezuela. Similar total richness was apparent in the Baharini Springbrook and Njoro River (51 and 64 taxa, respectively) in the Lake Nakuru drainage, Kenya (Shivoga, 1999), and so do not reflect a straightforward Old World versus New World disparity in richness. Such marked differences among regions may be due, in part, to varying sampling techniques, processing thoroughness and study intensity and duration but they may also indicate real variations in richness of zoobenthos assemblages. One comparison of streams sampled in a broadly similar way, yielded at total of between 70 and 94 species per site (overall total 126) in four Hong Kong streams (Dudgeon, 1994) and only 28–42 species in six northern New Guinea streams (overall total 64). Similar differences are seen between the relatively rich piedmont streams of the Ecuadorian Amazon in Napo region (Bojsen and Jacobsen, 2003) and the more species-poor piedmont-streams in the Chapare region of the Bolivian Amazon (Moya, 2006). These data suggest that there are quite substantial differences in taxon richness among tropical streams, and even within regions of the tropics, such that the latter tend to obscure or override patterns of inter-regional variation.

## E. Latitudinal Patterns in Diversity

Ecologists remain uncertain about whether tropical streams contain more diverse assemblages of macroinvertebrates than temperate streams. A latitudinal gradient in diversity is certainly evident for fishes (see Chapter 5 of this volume) and can be seen among decapod crustaceans (see Section II-B), but is less conspicuous for total richness macroinvertebrate assemblages. The results of investigations in some tropical streams have yielded high species totals, leading to suggestions that such richness is a general feature of tropical streams (e.g. Bishop, 1973; Stout and Vandermeer, 1975; Pearson *et al.*, 1986; Lake *et al.*, 1994). However, this generalization does not take account of the variation in species richness evident among tropical streams. Furthermore, very high numbers of macroinvertebrate species have been recorded in certain temperate streams, such as the total of 642 species known from the Breitenbach in Germany (Allan, 1995). Any latitudinal trends in total species richness may be masked by the

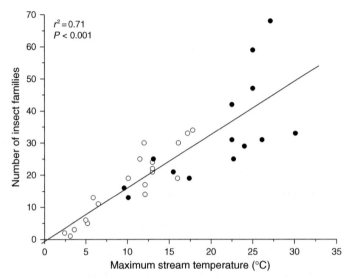

FIGURE 3  Regression of family richness of stream insects versus maximum stream temperature from studies of individual localities reported in the literature ($y = 1.60x - 0.18$, $r^2 = 0.71$, $p < 0.001$). Family richness of aquatic insects ($y$) increased linearly with maximum stream temperature ($x$) for both temperate-arctic streams ($y = 1.93x - 2.96$, $r^2 = 0.84$, $p < 0.001$; open circles) and tropical streams ($y = 1.72x - 2.80$, $r^2 = 0.48$, $p < 0.01$; filled circles) and relationships were not significantly different (for further details, see Jacobsen *et al.*, 1997). Reprinted with kind permission from Blackwell Publishing.

distribution patterns of individual insect orders: most species of Plecoptera, for example, are primarily temperate (Illies, 1969), whereas odonates are mainly tropical (Corbet, 1980). Trichoptera seem to have diversified at temperate as well as at tropical latitudes (Ross, 1967), with the Phryganeidae, Stenopsychidae and Limnephilidae occurring mostly in temperate regions, and other taxa (e.g. macronematine Hydropsychidae) occurring mostly in warmer climates. Despite, or perhaps because of, these taxon-specific trends, Jacobsen *et al.* (1997) noted a common relationship between maximum stream temperature and insect family richness irrespective of latitude and altitude (Fig. 3).

A number of authors have failed to find consistent differences in macroinvertebrate richness between tropical and temperate streams (e.g. Patrick, 1964; Arthington, 1990; Flowers, 1991; Boyero, 2002), and have drawn attention to the potential confounding effects of sampling method, stream type (e.g. tropical highland versus temperate lowland), taxonomic penetration, and the influence of small-scale spatial variability (see also Chapter 9 of this volume). The search for latitudinal patterns in stream macroinvertebrate richness requires application of standardized designs and sampling techniques across latitudes. Unfortunately, this has rarely been done (see Stout and Vandermeer, 1975). In a study of macroinvertebrate richness in a number of physically comparable lowland streams, Jacobsen *et al.* (1997) reported that mean richness of insect families in Ecuadorian coastal streams (29.3) was around 50% higher than in Danish streams (20). The same difference was seen between total family richness in 8 streams in Ecuador (44) and Denmark (30).

While there may be a latitudinal trend in family richness, it is not immediately evident whether this is paralleled by a change in species richness because the number of species per family may not be constant at a global scale. If there are more species per family in the tropics, the latitudinal trend in species richness might even be more marked than appears to be the case for families. Jacobsen *et al.* (1997) compared species: family ratios for aquatic insect orders in South America, North America and Europe. Although knowledge of the Neotropical fauna is relatively incomplete, South American aquatic insect families nevertheless contains the most species (Table V). In addition, there are more species of aquatic insects in the Neotropics than at higher latitudes, with Odonata, Heteroptera, aquatic Coleoptera and perhaps Trichoptera accounting for this difference. Neotropical Trichoptera may be considerably more diverse than indicated by the data in Table V (Flint *et al.*, 1999), and include families (the Anomalopsychidae) which occur only in the Neotropics. This does not mean, however, that Fig. 3 cannot be regarded as representing species richness as well as family richness in relation to stream temperature. The data in Table V concerns whole continents, while Fig. 3 concerns local diversity found at single stream sites, and individual streams in the tropics do not necessarily house more species of each family occurring, even though more species taxonomically may belong to each family.

If individual stream sites do in fact have more species per family closer to the equator the relationship between species richness and maximum stream temperature will be exponential instead of linear, but a global relationship should still persist. The data presented in Fig. 3 suggest that tropical lowland streams have about 1.5–2 times more insect families than their temperate equivalents, and Table V shows that aquatic insect families in South America have, on average, 1.5–2 times more species per family than in North America or Europe. On this basis, Jacobsen *et al.* (1997) have made the extrapolation that, overall, South American lowland streams should contain 2–4 times more species than streams in Europe.

Temperature affects stream invertebrates in a number of ways (Ward and Stanford, 1982; Allan, 1995) and is the abiotic variable most obviously related to latitude (see Chapter 1 for more information). As part of the voluminous debate on the possible causes of latitudinal gradients in biodiversity, Rohde (1992) concluded that temperature may be a key factor as it is associated with higher mutation rates and shorter generation times, thereby speeding up evolution and speciation. Quaternary climatic history may also be responsible, at least in part, since much of the landscape at temperate latitudes was covered by glaciers some 20 000 years ago. Other factors contributing to higher species richness in tropical streams could include greater food availability and habitat diversity (Covich, 1988) and the intensity of predation (Stout and Vandermeer, 1975; Fox, 1977). We can find no evidence to suggest that differences in habitat diversity or food in tropical streams allows them to support more species (see also Flowers, 1991) although it is possible that the greater diversity of tropical plants results in a more variable quality of detrital food (see Chapter 3 of this volume). Fox (1977) suggested that higher proportions of predators in tropical streams could reduce invertebrate densities and increase diversity by maintaining competitively-superior prey species at lower densities, thereby allowing the coexistence of less-competitive species. The two insect orders making a large contribution to species richness of aquatic insects in the Neotropics (Odonata and Heteroptera: Table V) comprise exclusively predatory species, thereby lending support to Fox's hypothesis. In addition, the greater richness of fishes in tropical streams (see Chapter 5) could also increase predation pressures on macroinvertebrates. High species richness but low densities of mayflies in a Costa Rican streams have been attributed to the influence of fish predation (Flowers and Pringle, 1995), and there is experimental evidence of the effects of fish predation on the abundance of zoobenthos at tropical latitudes (e.g. Dudgeon, 1991, 1993, 1996a; Flecker, 1992a,b).

Environmental variability and disturbance (caused by spates, for example) are regarded as key influences on the diversity of stream communities (e.g. Stout and Vandermeer, 1975; Stanford and Ward, 1983; Arthington, 1990; Reice *et al.*, 1990). Ward and Stanford (1983) considered that the predictions of the intermediate-disturbance hypothesis (Connell, 1978) were applicable to stream invertebrate communities, with maximum richness reached at intermediate levels of disturbance. While many tropical streams experience seasonal spates or flow pulses (see Chapter 1 of this volume), defining the intensity of such disturbances, and determining their effects on macroinvertebrates, is by no means straightforward. Dudgeon (1993) found that invertebrate densities and richness increased during the wet season in a Hong Kong stream when spates were frequent; evidently, not all such floods represent disturbance events (see Brewin *et al.*, 2000). Experimental studies of the effects of disturbance on Ephemeroptera in a Kenyan stream showed that disturbance could have positive or negative effects on species diversity, and did not accord with predictions of the intermediate-disturbance hypothesis (Mathooko, 1999). Other research by Flecker and Feifarek (1994) suggests that disturbance plays a seasonally important role in structuring invertebrate assemblages in Venezuelan streams that have severe and unpredictable changes in discharge, but the effects seem to involve depletion of benthic populations rather than changes in species richness. Inter-year variation in rainfall in Hong Kong streams produce similar effects and have taxon-specific consequences on macroinvertebrate standing stocks (densities and biomass) and production (Dudgeon, 1999b; see below) but do not seem to influence species richness.

## III. LIFE HISTORIES AND SECONDARY PRODUCTION

### A. Seasonality

Most studies of the life-history attributes of macroinvertebrates in tropical streams concern the temporal components of voltinism (or frequency of breeding) and phenology (seasonal and synchrony), and most have concerned insects. Conditions in the tropics can range from the relative aseasonality of equatorial streams to marked seasonality imposed by monsoons further away from the climatic equator (see Chapter 1 for details). Much of the literature is based upon studies of insect emergence or flight periodicity of aquatic insects, some of which have been of short-term duration and thus may not represent the extent of seasonal change in tropical streams (Cressa, 2003). However, there have also been numerous investigations of larval growth that have been continued for 12 months or longer and, in some cases, research has included both the larval and the adult insects. Asynchronous emergence and semi-continuous or year-round reproduction and larval growth that is almost independent of season have been reported from all tropical regions (e.g. Froehlich, 1969; Bishop, 1973; Zwick, 1976; Marchant, 1982a,b; Wolf *et al.*, 1988; Cressa, 1990, 2003; Campbell, 1995; Dudgeon, 1995a,b, 1996b,c, 1997, 1999c; Yule, 1995, 1996; Marchant and Yule, 1996; Mathooko, 1996; Ramírez and Pringle, 1998b; Salas and Dudgeon, 2003; Freitag, 2004). These observations are indicative of multivoltine life histories. There are a few instances, as in Konaiano Creek on Bougainville Island, Papua New Guinea, where virtually all species of aquatic insects had aseasonal life histories (Yule and Pearson, 1996), but many examples of synchronous emergence have been reported from streams across the tropics (Table VI; see also Dudgeon, 1999a and references therein) even in areas with apparently aseasonal climates (McElravy *et al.*, 1982; Flowers and Pringle, 1995). Such seasonality can be very marked, with emergence being confined to only a few days each year (e.g. the ephemerid mayfly *Ephemera pictipennis* in Hong Kong) while in others the flight period can extend over weeks (e.g. sympatric *Ephemera spilosa*) or months (e.g. Hong Kong stream odonates: Dudgeon, 1989b,c). The Hong Kong data are of particular interest since they are derived from long-term studies at a single site (Tai Po Kau Forest Stream) where a wide range of insect seasonality is seen, from unsynchronized almost aseasonal growth and adult emergence through somewhat synchronous development to highly synchronized growth and emergence with short flight periods. In the case of Trichoptera and Ephemeroptera, this range of patterns occurs within one order. Clearly, there is no common or uniform response of insect seasonality to a particular combination of climatic conditions.

Seasonal fluctuations in macroinvertebrate densities can occur in populations that undergo continuous growth and reproduction and have multivoltine life histories (Marchant, 1982a; Jacobi and Benke, 1991; Benke, 1998). They result from the fact that rates of growth, hatching and mortality in tropical streams are not constant during the year (e.g. Marchant, 1982a;

*TABLE VI*    Examples of Insect Species from Tropical Streams with Seasonal Adult Emergence or Flight Periods

| Order/species | River | Author |
|---|---|---|
| Diptera | | |
| *Simulium palauensis* | Headwaters of Ngerekiil River, The Philippines | Bright (1982) |
| Plecoptera | | |
| *Anacroneuria* sp. | Headwaters of Río Matadero, Ecuador | Turcotte and Harper (1982) |
| Ephemeroptera | | |
| *Ephemera pictipennis* | Tai Po Kau Forest Stream, Hong Kong | Dudgeon (1996d) |
| *Ephemera spilosa* | Tai Po Kau Forest Stream, Hong Kong | Dudgeon (1996d) |
| *Euthyplocia hecuba* | Río Tempesquito and Quebrada Marilin, Costa Rica | Sweeney *et al.* (1995) |
| Trichoptera | | |
| *Macrostemum fastosum* | Tai Po Kau Forest Stream, Hong Kong | Dudgeon (1988b) |
| *Macrostemum lautum* | Tai Po Kau Forest Stream, Hong Kong | Dudgeon (1988b) |
| *Polymorphanisis astictus* | Tai Po Kau Forest Stream, Hong Kong | Dudgeon (1988b) |
| *Atopsyche majada* | R. Chiriqui, Panama | McElravy *et al.* (1982) |
| *Chimarra* sp. | R. Chiriqui, Panama | McElravy *et al.* (1982) |
| *Leptonema intermedium* | R. Chiriqui, Panama | McElravy *et al.* (1982) |
| *Leptonema simulans* | R. Chiriqui, Panama | McElravy *et al.* (1982) |
| *Leptonema* sp. | R. Chiriqui, Panama | McElravy *et al.* (1982) |
| Coleoptera | | |
| *Eubrianax* sp. | Tai Po Kau Forest Stream, Hong Kong | Dudgeon (1995a) |
| *Sinopsephenus chinensis* | Tai Po Kau Forest Stream, Hong Kong | Dudgeon (1995a) |
| *Psephenoides* sp. | Tai Po Kau Forest Stream, Hong Kong | Dudgeon (1995a) |

Jackson and Sweeney, 1995a; Dudgeon, 1997) as has been widely reported from subtropical and temperate streams (Mackey, 1977a,b; Sweeney *et al.*, 1986; Benke, 1993; Marchant and Hehir, 1999; Huryn and Wallace, 2000). Some of variation in these parameters will be due to spates that cause mortality and will change food availability and other conditions that influence growth. Where climatic seasonality is present in the tropics, it is usually reflected in an uneven distribution of rainfall and is manifested in the occurrence of alternating wet and dry seasons (Walter and Medina, 1971; Walter and Breckle, 1985; see also Chapter 1 of this volume). Intense rainfall events and spates can occur unpredictably during (usually) the wet season, and can have significant effects on macroinvertebrate populations (e.g. Flecker and Feifarek, 1994; Dudgeon, 2000; but see Brewin *et al.*, 2000). One possible outcome is that macroinvertebrate densities will tend to peak during the dry season when flows are stable and are depleted by spates during the wet season (e.g. Dudgeon, 1996b,c,d, 1999c; Cressa, 1998; Maldonado *et al.*, 2001). Density differences between the wet and dry seasons can be large: in the Neotropics, fluctuations from 250 to 1250 individuals per m$^2$ have been reported in Río Sábalo, Costa Rica (Ramírez and Pringle, 1998b), from 785 to 1672 individuals per m$^2$ in Río Orituco, Venezuala (Cressa, 1990), and from 0 to >64 500 individuals per square metre in Río Las Marías, Venezuaela (Flecker and Feifarek, 1994). By contrast, macroinvertebrate densities in Magela Creek, northern Australia, were greatest during the late wet season and early dry season (Marchant, 1982a,b). A bimodal pattern has also been reported from small Venezuelan streams where benthic macroinvertebrates almost vanish as the dry season advances but populations re-establish quickly at the start of the wet season (Rincón and Cressa, 2000; Flecker and Feifarek, 1994).

There is unlikely to be a general seasonal pattern in abundance of all macroinvertebrate taxa in all tropical streams. This is evident from the fact that Flowers and Pringle (1995) reported a lack of any seasonal pattern among the Ephemeroptera fauna of Río Sábalo in Costa Rica. Both composition and densities exhibited large fluctuations over a three-year study period, that were not related with rainfall nor were they consistent among years. By contrast, Ramírez and Pringle (1998b) recorded a bimodal pattern in total macroinvertebrate abundance from the same stream. A trimodal seasonal pattern among Elmidae (Coleoptera), Chironomidae, and various Trichoptera and Ephemeroptera has been reported from a stream in the Ecuadorian Andes (Turcotte and Harper, 1982). Further complications that seem to rule out the presence of a general pattern of seasonality are evident from a study of Ephemeroptera in the Río Bento Gomes, Brazil: peaks in abundance occurred during both the wet and dry seasons but the composition of these two assemblages was different (Nolte *et al.*, 1997).

While the intensity and exact timing of spates will vary within and between years, spate-induced disturbances occur only during the wet season and are thus, to some extent, predictable. They therefore allow the possibility of life-history adaptations or ecological responses by macroinvertebrates that represent adjustments to predictable flow patterns (Brewin *et al.*, 2000; Dudgeon, 2000). In this context, it is striking that where emergence and/or reproduction by stream macroinvertebrates in tropical Asian streams is more-or-less synchronized or shows seasonal periodicity, it coincides with or precede the onset of wet-season monsoons. The recruitment of juveniles and the emergence of mature insect larvae as adults has been interpreted as an evolutionary response to predictable flow increases that can result in spate-induced mortality of large larvae (hence emergence is timed to minimize this risk) but make available new habitat and feeding opportunities for new recruits (Dudgeon, 1999a, 2000).

## B. Life Spans and Voltinism

The association of multivoltine life histories with asynchronous growth and near-continuous emergence and recruitment has generally been assumed for tropical stream insects. However, unequivocal demonstration of multivoltinism requires estimation of insect life spans since it is, at least in theory, possible to have a population of insects with continuous, asynchronous growth and year-round recruitment in which each individual lives for 365 days and thus are univoltine. Reports of life spans of tropical stream insects are usually based on interpretations of size-frequency histograms of larval populations sampled over a year (e.g. Benke 1993; Dudgeon, 1996b,c,d) but other approaches (e.g. Marchant and Yule 1996) or laboratory growth experiments (Jackson and Sweeney, 1995a; Cressa and Barrios, 2002; Salas and Dudgeon, 2002) have been employed in a few instances.

The difficulty of using field samples to trace a discrete cohort from birth to maturity in species that show asynchronous growth limits the availability of life-span estimates for tropical macroinvertebrates. Using laboratory reared larvae under simulated field conditions, Jackson and Sweeney (1995a) were able to determine development times of 35 species of stream insects in Costa Rica. The durations ranged from 17 to 197 days: i.e. all species were multivoltine or at least (approximately) bivoltine. Chironomids generally had the shortest life spans (17–73 days) and the stonefly *Anacroneuria* sp. (Perlidae) had the longest (135–197 days). Considerable variability was present among Ephemeroptera, with life spans ranging from 26–34 days (*Acerpenna* sp.: Baetidae) to 131–165 days (*Thraulodes* sp.; Leptophlebiidae), to Trichoptera, with life spans ranging from 50–54 days (*Oecetis* nr. *prolongata*: Leptoceridae) to 70–171 days (*Phylloicus ornatus*; Calamoceratidae). Life spans (and growth rates) of individual species may vary between habitats or in different parts of their geographic range: life-span estimates for *Nectopsyche gemmoides* (Leptoceridae) in the Río Camurí Grande, Venezuela (Cressa and Barrios, 2002), were less (66 days) than those recorded for *N. gemmoides* in Costa

Rica (87 days; Jackson and Sweeney (1995a). Life spans of congeneric helicopsychid caddisflies were more similar: 69 days for *Helicopsyche camuriensis* in Venezuela versus 77 days for Costa Rican *Helicopsyche dampfi*.

One limitation on the use of laboratory-reared animals to estimate life spans and growth is the possibility that conditions in nature are not simulated exactly. In a comparison of 7 species of mayflies reared in the laboratory and in small field enclosures in three Hong Kong streams, Salas and Dudgeon (2002) found that growth rates in the field were the same as or (more often) less than those in laboratory (range: 53–100% of laboratory rates) where ample, high-quality food was provided. As in other tropical studies, all species were multivoltine with baetids completing over eight generations per year, heptageniids at least four and *Choroterpes* sp. (Leptophlebiidae) up to eight. Growth rates varied between streams for five of the seven species studied, and seemed to be reduced by canopy cover. Elsewhere in the tropics, estimated life spans of stream insects on Bougainville Island, Papua New Guinea Life, ranged between 40 and 250 days (Marchant and Yule, 1996). Ephemeroptera were shortest (40–110 days), Trichoptera were intermediate (95–185 days) and Odonata were longest (250 days). In tropical Australia, life spans of around 1 month have been reported for the baetid *Cloeon fluviatile* (Marchant, 1982a) and 75–120 days for *Jappa* spp. (Leptophlebiidae; Campbell, 1995). *Centroptilum* sp. (Baetidae) in Ghana has a larval development time of only 18 days (Hynes, 1975), but equally fast growth rates for baetids in the subtropical south of the United States have been reported (Benke and Jacobi, 1986; see also Benke, 1998).

Marchant and Yule (1996) noted a positive correlation between life span duration and maximum larval dry weight (Marchant and Yule, 1996), and a similar relationship (with adult dry weight) was reported by Jackson and Sweeney (1995a) in Costa Rica. Such relationships probably depend on the fact that, all other things being equal, larger insects will take longer to reach adulthood than smaller ones, and may offer a short-cut for estimation of approximate life spans. Interestingly, the Costa Rican burrowing mayfly *Euthyplocia hecuba* (Euthyplociidae) has a life-cycle duration of 22 months (Sweeney *et al.*, 1995) and appears to be the only semivoltine tropical stream insect confirmed thus far. It is an unusually large mayfly, and females can reach almost 150 mg dry mass.

While the majority of the tropical stream insects appear to be multivoltine, this is not always the case (see above), and the factors affecting their life histories are not well known. The possibility of adaptive responses to spate-induced disturbance have been discussed above, but temperature (Sweeney and Vannote, 1978; Vannote and Sweeney, 1980; Sweeney, 1984; Jacobi and Benke, 1991), food availability (Anderson and Sedell, 1979; Ward and Cummins, 1979; Sweeney, 1984; Benke and Jacobi, 1986), predation and competition (Kohler and Mcpeek, 1989; Ball and Baker, 1996; Peckarsky *et al.*, 2001) acting singly or in combination have proximate influences on life-history parameters of stream insects. Their effects are likely to be similar in the tropics, where food availability (especially algae and cyanobacteria) plus temperature have been shown to affect growth rates of Hong Kong mayflies (Salas and Dudgeon, 2002), with temperature explaining 23–77% of the total variation in field growth rates throughout the year (see also Benke *et al.*, 1992). More research on the relative importance of such proximate factors is needed, together with investigations to determine the generality of the size versus life span relationship for tropical stream insects (Jackson and Sweeney, 1995a; Marchant and Yule, 1996). Sampling programs of sufficient duration to include inter-year variations in weather (Walter and Breckle, 1985) should be established so that their possible consequences on long-term or inter-annual change on life histories and seasonality can be assessed.

## C. Breeding by Other Macroinvertebrates

Because insects constitute the majority of species of stream macroinvertebrates, their life histories and population dynamics have received more attention than those of other taxa.

Considerable attention has been paid to the dynamics of freshwater snails (mainly pulmonates) in the context of their role in transmitting parasites of medical importance (e.g. Brown, 1980), but the most of them are not stream dwellers and so are not considered further. Pulmonates can occasionally become numerous in tropical streams when flows are low and food is abundant, but such conditions are usually short-lived and these snails do not persist when normal flow conditions return (e.g. Dudgeon, 1983). Reproduction by the prosobranch snail *Brotia hainanensis* (Pachychilidae) in Hong Kong streams occurs twice each year occurring immediately before and immediately following the wet season, and has been interpreted as a life-history adaptation to seasonal flows (Dudgeon, 1982, 1989d). The life cycle of these snails involves brooding young within a pouch and release of juvenile snails so that there is no egg or planktonic larval stage (unlike the life cycle of freshwater Neritidae: see Blanco and Scatena, 2006).

The same suppression of the planktonic stage is evident in certain species of *Macrobrachium* and atyid shrimps, especially those that occur in stony upland streams (Jalihal *et al.*, 1993; Dudgeon, 1999a and references therein), but some shrimps that have not gone through the full process of 'freshwaterization' (*sensu* Jalihal *et al.*, 1993) retain a planktonic larval stage that develops in estuaries or coastal waters and requires inclusion of a migratory stage in the life cycle (March *et al.*, 2003; the conservation implications are discussed in Chapter 10 of this volume and in Greathouse et al., 2006). A downstream migration to estuaries is required for breeding in freshwater grapsid crabs (e.g. *Eriocheir* spp.), but most other stream-dwelling crabs do not migrate nor do they have planktonic larvae. Instead, juveniles that resemble miniature adults hatch from large, yolky eggs brooded beneath the abdomen. In some cases, freshwater crabs show aseasonality in breeding (ovigerous females occur all year round), as in *Potamonautes* sp. in Kenya (Dobson *et al.*, 2007). Other tropical freshwater crabs seem to exhibit a variety of patterns, with release of juveniles during high-flow and low-flow periods depending on species and habitat (Dudgeon, 1999a; Dobson *et al.*, 2007 and references therein).

Shrimps breed throughout the year in some parts of the tropics (Mizuno, 1982; Walker and Ferreira, 1985; Wowor, 1985; Mossolin and Bueno, 2002), whereas, in more seasonal areas, the proportion of ovigerous females in shrimp populations tends to be positively related to flow or temperature (Dudgeon, 1985, 1999a; Collart and Megalhães, 1994; Johnson *et al.*, 1998; Mantel and Dudgeon, 2004a; Yam and Dudgeon, 2006). Studies of individually tagged *M. hainanense* in Hong Kong streams indicated that growth rates vary seasonally in response to stream temperature (Mantel and Dudgeon, 2004a). These palaemonid shrimps are larger and more longer lived than stream insects (see above) and smaller *Caridina* (life span 17–22 months; Yam and Dudgeon, 2005b) with females surviving for 3 years and males, which grow larger, living for 4 years or longer (Mantel and Dudgeon, 2004a, 2005). Competition for females may be responsible for the larger size of males, and sexual dimorphism in chelipeds and/or body size is common in other freshwater decapods.

Periods of high flow and spate-induced mortality may influence shrimp population sizes and limit recruitment and production (Benzie and de Silva, 1988; Hancock and Bunn, 1997; Mantel and Dudgeon, 2004a; Yam and Dudgeon, 2005b; 2006), especially in parts of the tropics with a marked wet season. However, even within a particular region, life-history patterns may differ in response to local factors. Inter-stream differences in growth rates and breeding frequency of *Caridina* shrimps in Hong Kong are affected by environmental factors, such as the extent of stream shading (which influences food resources and tends to increase shrimp production) and the intensity of spates in individual streams (Yam and Dudgeon, 2005b, 2006). The number of cohorts and the frequency of recruitment of *C. cantonensis* (1–3 times each year) were highly site-specific, but consistent between years, as were growth-rate differences with faster-growing animals reaching sexual maturity earlier and at a smaller size (Yam and Dudgeon, 2005b). Life-history variation can be associated with genetic differences among *C. cantonensis* populations in Hong Kong (Yam and Dudgeon, 2005c), but a causal link has yet to be established. Brood size varied among sites also, and broods included greater numbers of eggs at sites where algal food was abundant (Yam and Dudgeon, 2005a,b). Elsewhere, Johnson *et al.* (1998) concluded that the reproductive patterns of Neotropical atyids are strongly influenced by the stability of their habitats: *Atya lanipes* and *Xiphocaris elongata* in Puerto Rico showed reproductive seasonality in a large stable pool but bred throughout the year in a smaller less stable headwater pool in the same drainage. These findings suggest that, as is the case with stream insects, populations of other tropical macroinvertebrates are sometimes affected by temperature, but that flow regime and habitat stability have a primary influence on their life-history parameters and population dynamics.

## D. Secondary Production

Macroinvertebrate secondary production estimates integrate information on population dynamics (growth, survivorship, mortality) and energy flow that are of considerable ecological importance (Benke, 1984, 1996). Despite this, a review of macroinvertebrate secondary production estimates by Benke (1993) reported only two papers from the tropics, and a subsequent synthesis by Huryn and Wallace (2000) revealed a similar scarcity of information. Only one production estimate has been reported from the Neotropics (Ramírez and Pringle, 1998b). More estimates are available from mainland tropical Asia (Hong Kong: Dudgeon, 1999b and references therein; Salas and Dudgeon, 2003; Mantel and Dudgeon, 2004a; Yam and Dudgeon, 2006), the Philippines (Bright, 1982) and Papua New Guinea (Yule and Pearson, 1996), but these data are derived from a very limited number of sites. We know of no secondary production estimates for macroinvertebrates in tropical African streams.

The limited information indicates that secondary production of aquatic insects in tropical streams is not especially high. Habitat-weighted secondary production by the entire stream insect assemblage in a Costa Rican stream (373 mg ash-free dry mass m$^{-2}$ yr$^{-1}$; no more than 500 mg dry mass) was low compared with temperate streams (Ramírez and Pringle, 1998b), most likely because of the dominance of non-insect invertebrates such as shrimps which were not included in the estimate but are abundant in Costa Rica (see Pringle and Hamazaki, 1998; Ramírez *et al.*, 1998). Dominance of the biomass of Puerto Rican streams by shrimps (with few insects) has also been reported (March *et al.*, 2001, 2002).

Other estimates of production by an entire macroinvertebrate assemblage (either as all insects or all taxa combined) are unavailable from any tropical stream. However, if we sum the production estimates made for a range of taxa in Tai Po Kau Forest Stream, Hong Kong, the dry mass total ranges from 5.64–7.64 g m$^{-2}$ yr$^{-1}$ (depending on the particular year; see Dudgeon, 1999b) which is over 10 times higher than the Costa Rican estimate despite including only part of the Hong Kong macroinvertebrate assemblage. Separate estimates of production of the palaemonid shrimp *Macrobrachium hainanense* in pools in Tai Po Kau Forest Stream and at another site in Hong Kong where densities where more than two-fold higher indicated that, surprisingly, production was similar in both streams (1.34–1.51 g m$^{-2}$ yr$^{-1}$) with relatively minor inter-year fluctuations (Mantel and Dudgeon, 2004a). If production by *M. hainanensis* in Tai Po Kau Forest Stream is added to the other macroinvertebrate data, the estimate of annual macroinvertebrate production in this Hong Kong stream totals about 9 g m$^{-2}$.

There are no other productivity estimates for field populations of *Macrobrachium* in the literature, so direct comparison with other studies of congeneric species is not possible. Densities in the Hong Kong sites (2.2–5.7 individuals m$^{-2}$) were within the range for *Macrobrachium* spp. recorded in Puerto Rico (1.3–6.2 m$^{-2}$; March *et al.*, 2001, 2002), Costa Rica (2–6 individuals m$^{-2}$; Pringle and Hamazaki, 1998) and the Philippines (3 m$^{-2}$; Bright, 1982). However, the Costa Rican shrimps were larger than Hong Kong *Macrobrachium*, so standing stocks in terms of biomass were higher (Ramírez and Pringle, 1998a,b). Production estimates for *Caridina* shrimps in Hong Kong are not comparable with those of *Macrobrachium* since these shrimp genera tend not to be abundant in the same streams: however, in four streams where they were studied, *Caridina* were estimated to contribute around 10% of total macroinvertebrate production (Yam and Dudgeon, 2006).

Comparing production estimates for other components of the macroinvertebrate fauna, values for filter-feeding Trichoptera range from 0.19 mg m$^{-2}$ yr$^{-1}$ (*Wormaldia*: Philopotamidae) to 2.27 g m$^{-2}$ (*Stenopsyche angustata*: Stenopsychidae) (see Table VII; all data are in dry mass units for purposes of comparison), which are generally lower than the range of values for filter-feeding insects in subtropical streams (3–300 g m$^{-2}$ yr$^{-1}$; Huryn and Wallace, 2000). However, the value for individual hydropsychid species in Table VII fall within the range of those reported for these filter-feeders in temperate latitudes (for further discussion, see Dudgeon, 1997), but the total production of hydropsychids in Tai Po Kau Forest Stream, Hong Kong (3.67–3.74 g m$^{-2}$ yr$^{-1}$), was at the lower end of estimates for their temperate equivalents. Inter-year variations in production of individual species in this stream was substantial, with the percentage variation ranging from 3–774% (mean = 206%, $n = 19$) equivalent to a difference in production that, depending on species, ranged from 0.21 to 8.44 times between years. Given such variation at a single site, the lack of data from most parts of the tropics, and the incomplete representation of different taxa, it is not possible to draw robust conclusions about latitudinal trends in insect production from the limited tropical data set. Nor is it possible to make an assessment of the importance of decapod production within and among tropical regions, although it seems clear that they make a significant contribution to total production (Ramírez and Pringle, 1998b; Mantel and Dudgeon, 2004a; Dobson *et al.*, 2007). Suffice to state at present that values for insects appear low compared to estimates reported for equivalent temperate taxa.

Benke (1993) reported that secondary production is highly correlated with and dependent on biomass ($r^2 = 0.70$, $p < 0.001$) as well as growth rate and any factors affecting them (Huryn and Wallace, 2000). Insect body size and density are the main determinants of biomass and neither are exceptionally high in tropical streams (e.g. Cressa, 1998; Jacobsen and Encalada, 1998; Ramírez and Pringle, 1998b). Indeed, a glance at samples of the insects from stony streams in the lowland tropics reveals that they appear, or average, to be smaller than their temperate counterparts. Benke (1993) considered that substrate stability and food availability are the main factors affecting macroinvertebrate biomass in streams in general, and some authors (Flecker and Feifarek, 1994; Uieda and Gajardo, 1996; Jacobsen and Encalada, 1998; Maldonado *et al.*, 2001) have suggested that dynamic flow regimes and disturbance may play an important role in structuring tropical stream macroinvertebrate communities. Regular depletion of biomass by spates could deplete standing stocks accounting for low productivity, but evidence for this is scant. The inter-year variations in secondary production in Tai Po Kau Forest Stream, Hong Kong, described above coincide with differences in rainfall between years, with production (and biomass) of more rheophilic taxa increasing substantially during years with higher flow when densities and production of species typical of depositional habitats are reduced (Dudgeon, 1999b; Salas and Dudgeon, 2003). By contrast, production by *Caridina* shrimps in 4 Hong Kong streams did not vary between years over a 2-year study, but spate-induced mortality during the wet season led to substantial seasonal differences in productivity during both years (Yam and Dudgeon, 2006). There were also large site-specific differences in production among sites, with greater productivity in two unshaded sites streams where shrimps depended on algal food (Yam and Dudgeon, 2005a).

Even if biomass of macroinvertebrates in tropical streams is low, growth rates are strongly influenced by temperature and could be expected to be high thereby contributing to greater rates of production. Given that mean annual temperature of tropical streams are generally above 20°C (Pringle, 1991; Jackson and Sweeney, 1995a; Cressa and Barrios, 2002), we should

*TABLE VII* Mean Density (Individuals m$^{-2}$), Mean Biomass (B; mg m$^{-2}$), Cohort Production Interval (CPI), Secondary Production (P; mg dry mass m$^{-2}$ yr$^{-1}$) and P/B of Tropical Stream Insects

| Taxon | D | B | CPI | P | P/B | Stream, Location | Reference |
|---|---|---|---|---|---|---|---|
| Ephemeroptera | | | | | | | |
| Pseudocloeon sp. | 239.1 | 43.5 | 86 | 899.5 | 20.7 | Konaiano Stream, Bougainville Island, Papua New Guinea | Marchant and Yule (1996) |
| Ephemera spilosa | 11.1 | 10.2 | 365 | 37.3 | 7.3 | Tai Po Kau Forest Stream, Hong Kong | Dudgeon (1996d) |
| Afronurus sp. | 99.6 | 24.0 | 365 | 143.8 | 6.0 | Tai Po Kau Forest Stream, Hong Kong | Dudgeon (1996b) |
| Cinygmina sp. | 31.4 | 24.0 | 365 | 116.8 | 4.9 | Tai Po Kau Forest Stream, Hong Kong | Dudgeon (1996b) |
| Epeorus sp. | 18.2 | 15.2 | 365 | 100.0 | 6.6 | Tai Po Kau Forest Stream, Hong Kong | Dudgeon (1996b) |
| Iron sp. | 8.8 | 10.6 | 365 | 95.6 | 9.1 | Tai Po Kau Forest Stream, Hong Kong | Dudgeon (1996b) |
| Paegniodes cupulatus | 7.8 | 7.6 | 365 | 42.2 | 5.5 | Tai Po Kau Forest Stream, Hong Kong | Dudgeon (1996b) |
| Afronurus sp. | 80.3 | 41.7 | 91 | 1481.9 | 39.1 | Tai Po Kau Forest Stream, Hong Kong | Salas and Dudgeon (2003) |
| Cinygmina sp. | 13.8 | 3.4 | 91 | 107.2 | 27.2 | Tai Po Kau Forest Stream, Hong Kong | Salas and Dudgeon (2003) |
| Procloeon sp. | 37.6 | 1.2 | 46 | 69.5 | 94.6 | Tai Po Kau Forest Stream, Hong Kong | Salas and Dudgeon (2003) |
| Baetiella pseudofrequenta | 8.5 | 0.3 | 46 | 15.7 | 77.3 | Tai Po Kau Forest Stream, Hong Kong | Salas and Dudgeon (2003) |
| Choroterpes sp. | 9.1 | 2.7 | 61 | 75.3 | 62.8 | Tai Po Kau Forest Stream, Hong Kong | Salas and Dudgeon (2003) |
| Afronurus sp. | 58.1 | 37.7 | 91 | 1210.8 | 33.2 | Shing Mun Stream, Hong Kong | Salas and Dudgeon (2003) |
| Cinygmina sp. | 67.4 | 17.7 | 91 | 606.3 | 31.8 | Shing Mun Stream, Hong Kong | Salas and Dudgeon (2003) |
| Procloeon sp. | 32.5 | 1.1 | 46 | 61.6 | 99.3 | Shing Mun Stream, Hong Kong | Salas and Dudgeon (2003) |
| Baetiella pseudofrequenta | 31.1 | 1.6 | 46 | 73.1 | 109.8 | Shing Mun Stream, Hong Kong | Salas and Dudgeon (2003) |
| Choroterpes sp. | 17.0 | 4.1 | 61 | 128.6 | 49.4 | Shing Mun Stream, Hong Kong | Salas and Dudgeon (2003) |
| Thraulodes sp.* | 4.1 | 0.7 | 159 | 4.2 | 6.5 | Sábalo River – Riffle, Costa Rica | Ramírez and Pringle (1998b) |
| Thraulodes sp.* | 1.5 | 0.2 | 159 | 2.3 | 10.5 | Sábalo River – Pool, Costa Rica | Ramírez and Pringle (1998b) |
| Leptohyphes sp.* | 6.2 | 0.4 | 82 | 2.2 | 5.1 | Sábalo River – Riffle, Costa Rica | Ramírez and Pringle (1998b) |
| Leptohyphes sp.* | 2.6 | 0.3 | 82 | 4.4 | 16.3 | Sábalo River – Pool, Costa Rica | Ramírez and Pringle (1998b) |
| Tricorythodes sp.* | 124.8 | 0.3 | 86 | 138.8 | 19.1 | Sábalo River – Riffle, Costa Rica | Ramírez and Pringle (1998b) |
| Tricorythodes sp.* | 46.0 | 2.7 | 86 | 64.9 | 24.3 | Sábalo River – Pool, Costa Rica | Ramírez and Pringle (1998b) |
| Thraulodes sp. | 27.5 | 49.7 | 242 | 452.9 | 18.9 | Camurí Grande, Venezuela | Cressa (2003) |
| Leptohyphes spp. | 11.3 | 18.0 | 115 | 237.2 | 9.9 | Camurí Grande, Venezuela | Cressa (2003) |
| Haplohyphes sp. | 6.4 | 25.5 | 110 | 308.8 | 20.3 | Camurí Grande, Venezuela | Cressa (2003) |

(continued)

TABLE VII (continued)

| Taxon | D | B | CPI | P | P/B | Stream, Location | Reference |
|---|---|---|---|---|---|---|---|
| **Plecoptera** | | | | | | | |
| *Anacroneuria bifasciata* | 7.2 | 14.3 | 194.5 | 94.5 | 6.6 | Camuri Grande, Venezuela | Cressa (2003) |
| **Trichoptera** | | | | | | | |
| *Macrostemum fastosum* | 31.4 | 79.8 | 365 | 383.6 | 4.8 | Tai Po Kau Forest Stream, Hong Kong | Dudgeon (1997) |
| *Polymorphanisus astictus* | 3.8 | 7.8 | 365 | 37.8 | 4.8 | Tai Po Kau Forest Stream, Hong Kong | Dudgeon (1997) |
| *Cheumatopsyche spinosa* | 319.8 | 31.6 | 182.5 | 265.2 | 8.4 | Tai Po Kau Forest Stream, Hong Kong | Dudgeon (1997) |
| *Cheumatopsyhe ventricosa* | 75.6 | 15.4 | 182.5 | 150.0 | 9.7 | Tai Po Kau Forest Stream, Hong Kong | Dudgeon (1997) |
| *Herbertorossia quadrata* | 12.2 | 26.2 | 182.5 | 387.0 | 14.8 | Tai Po Kau Forest Stream, Hong Kong | Dudgeon (1997) |
| *Hydatopsyche melli* | 13.2 | 40.2 | 365 | 147.4 | 3.7 | Tai Po Kau Forest Stream, Hong Kong | Dudgeon (1997) |
| *Hydropsyche chekiangana* | 7.6 | 8.6 | 365 | 38.0 | 4.4 | Tai Po Kau Forest Stream, Hong Kong | Dudgeon (1997) |
| *Stenopsyche angustata* | 77.2 | 260.2 | 365 | 2265.4 | 8.7 | Tai Po Kau Forest Stream, Hong Kong | Dudgeon (1996c) |
| *Leptonema* sp. | 7.4 | 1.0 | 103 | 12.4 | 12.7 | Sábalo River – Riffle, Costa Rica | Ramírez and Pringle (1998b) |
| *Leptonema* sp. | 0.3 | 0.03 | 103 | 1.1 | 33.3 | Sábalo River – Pool, Costa Rica | Ramírez and Pringle (1998b) |
| *Polyplectropus* sp. | 0.5 | 0.03 | 96 | 0.8 | 23.3 | Sábalo River – Riffle, Costa Rica | Ramírez and Pringle (1998b) |
| *Polyplectropus* sp. | 1.1 | 0.1 | 96 | 0.8 | 7.0 | Sábalo River – Pool, Costa Rica | Ramírez and Pringle (1998b) |
| *Wormaldia* sp. | 0.8 | 0.9 | 45 | 1.2 | 1.4 | Sábalo River – Riffle, Costa Rica | Ramírez and Pringle (1998b) |
| *Wormaldia* sp. | 0.2 | 0.02 | 45 | 0.2 | 9.5 | Sábalo River – Pool, Costa Rica | Ramírez and Pringle (1998b) |
| *Nectopsyche gemmoides* | 3.1 | 0.3 | 149 | 4.4 | 17.7 | Camuri Grande, Venezuela | Cressa (2003) |
| **Coleoptera** | | | | | | | |
| *Eubrianax* sp. | 52.4 | 23.6 | 182.5 | 304.4 | 12.9 | Tai Po Kau Forest Stream, Hong Kong | Dudgeon (1995a) |
| *Sinopsephenus chinensis* | 59.6 | 102.4 | 182.5 | 1718.4 | 16.8 | Tai Po Kau Forest Stream, Hong Kong | Dudgeon (1995a) |
| *Mataeopsephus* sp. | 1.4 | 2.0 | 182.5 | 33.6 | 16.8 | Tai Po Kau Forest Stream, Hong Kong | Dudgeon (1995a) |
| *Psephenoides* sp. | 11.9 | 0.5 | 182.5 | 4.1 | 7.47 | Tai Po Kau Forest Stream, Hong Kong | Dudgeon (1995a) |
| *Hydrocyphon* sp. | 153.2 | 8.6 | 91.3 | 107.5 | 12.1 | Tai Po Kau Forest Stream, Hong Kong | Dudgeon (1995b) |
| **Diptera** | | | | | | | |
| *Simulium palauensis* | 381.8 | 25.0 | 182.5 | 50.0 | 2.0 | Ngerekiil River, the Philippines | Bright (1982) |
| *Tanypodinae*\* | 15.6 | 0.1 | – | 10.5 | 85.6 | Sábalo River – Riffle, Costa Rica | Ramírez and Pringle (1998b) |
| *Tanypodinae*\* | 14.4 | 0.2 | – | 11.5 | 69.0 | Sábalo River – Pool, Costa Rica | Ramírez and Pringle (1998b) |
| *Non-Tanypodinae*\* | 208.7 | 1.1 | – | 116.3 | 103.6 | Sábalo River – Riffle, Costa Rica | Ramírez and Pringle (1998b) |
| *Non-Tanypodinae*\* | 117.9 | 0.5 | – | 52.4 | 96.2 | Sábalo River – Pool, Costa Rica | Ramírez and Pringle (1998b) |

Secondary production was estimated by the size–frequency method or (indicated by \*) the instantaneous-growth method; original data were standardized to dry mass m$^{-2}$ units where necessary.

expect higher growth rates of organisms living in tropical streams than for temperate ones (Sweeney and Vannote, 1978; Vannote and Sweeney, 1980; Sweeney, 1984). High growth rates are certainly indicated by the short time taken by most tropical insects to grow to maturity (see Section III-B.) but could be a result of their apparent smaller body sizes rather than faster growth. Comparisons of mass-specific growth rates can correct for this size difference. Data for a variety of mayflies in a Hong Kong indicate maximum growth rates of up to $0.23$ mg mg$^{-1}$ d$^{-1}$ at around $22°C$ (Salas and Dudgeon, 2001). These are close to the fastest growth rates in the literature ($0.26$ and $0.35$ mg mg$^{-1}$ d$^{-1}$) and, although they were reported from subtropical southern United States, water temperatures were warmer than in the Hong Kong study ($27.5°C$ and $30°C$, respectively; Benke and Jacobi, 1986; Benke et al., 1992). It thus seems likely that the short life spans of macroinvertebrates in tropical streams is due to fast growth rates that result from high prevailing temperatures at these latitudes.

Among other invertebrates, *Caridina cantonensis* shrimps in Hong Kong grew quickly, reaching sexual maturity in 4–8 months depending on site, while *Caridina serrata* could mature in 3 months (Yam and Dudgeon 2005b). Mass specific growth rate in both species is size dependent and slows after maturity, although these shrimps have a maximum life span of 17–22 months. Growth of larger and more long-lived (over 3 years) *Macrobrachium hainanense* in Hong Kong likewise varied with size (and temperature) but mean rates were less than *M. sintangense* in Indonesia (Wowor, 1985) and *M. amazonicum* and *M. acanthurus* in Brazil (Guest, 1979; Valenti et al., 1987) at lower, warmer latitudes. Growth rates of all *Macrobrachium* spp. exceed those reported for temperate crayfish, which are relatively large and more long-lived (for references, see Mantel and Dudgeon, 2004a).

One implication of the fast growth and short development time of tropical stream invertebrates, especially insects, and their asynchronous growth is the difficulty of applying the size-frequency method for estimating secondary production. This approach depends on the investigators ability to identify and follow discrete cohorts of animals in field samples or, at the very least, to determine the cohort production interval (CPI) with accuracy (see Benke, 1984, 1996). The CPI of insects is approximated by the larval development time but this is problematic in species where growth is asynchronous and larvae of all size groups are present throughout all or much of the year. If CPI is overestimated (i.e. larval growth is faster than assumed), then production tends to be underestimated. For example, Dudgeon (1996b) estimated the voltinism and CPI of the heptageniid mayflies *Afronurus* sp. and *Cinygmina* sp. from field samples of populations showing asynchronous growth and used them to estimate production (see Table VII). Subsequent investigations by Salas and Dudgeon (2001) showed that these mayflies grew faster than was evident from the field data (and thus CPIs were shorter than had been assumed). If an appropriate CPI correction factor is applied, the adjusted figures for *Afronurus* sp. and *Cinygmina* sp. in Tai Po Kau Forest Steam calculated by Dudgeon (1996b), as reported in Table VII, would be increased by a factor of more than 4. This correction would make the estimates for *Afronurus* sp. more similar to the values recorded by Salas and Dudgeon (2003). It is tempting to suggest that such underestimation may be the rule rather than the exception in production studies of fast-growing tropical stream insects.

Given the scarcity of macroinvertebrate production estimates from the tropics and the difficulties of ensuring their accuracy or representativeness, there may be value in encouraging the use of models to determine secondary production (e.g. Morin and Bourassa, 1992; Benke, 1993; Morin and Dumont, 1994; Morin, 1997). Most models have been developed for temperate species, but some temperature versus growth rate relationships for tropical mayflies are given by Salas and Dudgeon (2002). While models are not perfect substitutes for actual measurements, they may indicate the approximate magnitude of secondary production in a particular habitat providing information that could be used to aid resource-management decisions (Benke, 1996). In addition, the outcome of modelling may help identify taxa worthy of special investigation and/or the environmental factors responsible for influencing productivity in the tropics. The results of subsequent investigations would yield data that could be used to refine or improve existing models by increasing the representation of tropical taxa.

## E. Trophic Basis of Macroinvertebrate Production

Widely-accepted notions of the factors that underlie production and community organization of stream macroinvertebrate communities have been derived mainly from research undertaken in temperate regions. The river continuum concept (RCC: Vannote *et al.*, 1980) is the most well known general model of stream ecosystems, and describes how community organization responds to longitudinal changes in the relative importance of different sources of allochthonous and autochthonous energy sources. The applicability of the RCC to tropical streams is not well established (and is discussed in Chapter 9 of this volume; see also Thorp and Delong, 2002), and the processing of allochthonous detritus by aquatic macroinvertebrates may differ in important ways in tropical and temperate latitudes (see Chapters 3 and 9) in part because shredding detritivores seem to be less abundant and diverse in many parts of the tropics. There are few insect shredders in streams in Brazil (Walker, 1987), Ecuador (Bojsen and Jacobsen, 2003) Puerto Rico (March *et al.*, 2001; Ramírez and Hernandez-Cruz, 2004; Wright and Covich, 2005) and Costa Rica (Pringle and Hamazaki, 1998) but their role may be partly substituted by shrimps, while freshwater crabs are reputed to play a similar role in Kenyan streams (Dobson *et al.*, 2002, 2004; Dobson, 2004) where insect shredders also appear to be scarce. In tropical Asia too, shredders occur at low densities (Dudgeon, 1994, 1999a, 2000; Dudgeon and Wu, 1999), although there are reports that they are relatively abundant in some tropical Australian streams (Cheshire *et al.*, 2005). This matter is addressed in more detail in Chapter 9. All that need be said here is that better elucidation of any latitudinal trends in shredder abundance must take account of both density and biomass of these animals (most published data refer to the former), and has to be based on sampling techniques that allow consideration of decapods as an integral part of benthic macroinvertebrate assemblages.

It is fundamental tenet of the RCC that small shaded streams have heterotrophic community metabolisms that depend upon allochthonous energy in the form of leaf litter, and benthic assemblages dominated by detritivores. Leaving aside the issue of the scarcity of shredders, there are other reasons to assume that tropical streams may not conform to the RCC (see also Greathouse and Pringle, 2006). Algal food may be more important than allochthonous detritus for invertebrate consumers in small tropical streams (see also Chapter 2 of this volume), and evidence from recent studies of food assimilation using stable isotopes (mainly of carbon) are confirming this. Here, we describe examples from Hong Kong streams.

By combining estimates of production and information on the stable isotope signatures of Salas and Dudgeon (2001, 2003) it was shown that more than 50% of total mayfly production in shaded Tai Po Kau Forest Stream was derived from autochthonous foods. In a second shaded stream nearby, 32% of production was supported by autochthonous foods during the wet season, when algal standing stocks were regularly depleted by spates, but the proportion rose to 72% during the dry season. However, to obtain an estimate of the importance of algal food to total mayfly production, these proportions need to be adjusted by weighting them for seasonal differences in mayfly production because more than 70% of the total production occurred in the wet season. Even with this adjustment, the trophic basis of mayfly production in Tai Po Kau was mostly autochthonous (66%), whereas in the second stream slightly less than half (42%) of annual production was depended on autochthonous energy sources. In a study of a wider range of sites, it appeared that despite the relative abundance of allochthonous foods, stable isotope analyses showed that mayflies preferred to feed on algae and cyanobacteria, and growth rates of mayflies were enhanced by algal food (Salas and Dudgeon, 2002), although consumption of detritus became more important during the wet season (Salas and Dudgeon, 2001).

These results do not apply solely to mayflies, as similar data are also available for *Caridina* shrimps in four Hong Kong streams (Yam and Dudgeon, 2005a, 2006). Dual-isotope multiple-source mixing models were used to analyse the relative contributions of leaf litter, fine detritus and periphyton to shrimps in shaded and unshaded streams. Leaf litter contributed <10% to the biomass of shrimps in unshaded streams and 10–20% in shaded streams. Periphytic algae contributed >60% to shrimp biomass in unshaded streams in the wet and (especially) the dry seasons, and 35–60% to *Caridina* biomass in the shaded stream. Autochthonous energy was the primary food source of shrimps regardless of shading conditions, and was supplemented by collection of fine detritus and limited direct consumption of allochthonous leaf litter.

Inclusion of more species (from Tai Po Kau Forest Stream only) was possible during a study of food webs in stream pools that used the complementary approach of gut content analyses and stable isotopes (Mantel *et al.*, 2004). The results showed a primary dependence on periphyton, and especially periphytic cyanobacteria, by virtually all of the non-predatory macroinvertebrate taxa; with the exception of Chironominae and Orthocladiinae leaf litter and fine organic matter made relatively minor contributions to consumer biomass (see Table 3 in Mantel *et al.*, 2004). There is no suggestion that the Hong Kong findings apply to all stream types in all parts of the tropics. Nonetheless, recent work in Brazil (e.g. Brito *et al.*, 2006, and references therein), Puerto Rico (March and Pringle, 2003) and tropical Australia (Douglas *et al.*, 2005: see also Chapter 2 of this volume) also demonstrate the importance of algal-based resources in small streams. The generality of such findings must await the results of research in a wider array of stream types and localities (especially Africa).

## IV. BIOASSESSMENT AND BIOMONITORING

The physical and chemical condition of many streams in tropical countries is deteriorating as a consequence of rapidly-growing human populations, land-use change, intensified agricultural practices and increased industrialization all of which tend to be accompanied by changes to natural flow regimes (e.g. Dudgeon, 1992, 2000; Pringle *et al.*, 2000; Wishart *et al.*, 2000; see Chapter 10 of this volume). Most streams draining inhabited areas are polluted, and quite often extremely polluted. Organic pollution from non-treated domestic and industrial sewage is by far the most common cause. This situation corresponds to what happened in many north-temperate countries decades ago, but recent investment in waste-water treatment and restoration projects combined with appropriate legislation and monitoring programs has led to long-term improvements in water quality. A prerequisite for success in these cases has been the access to tools capable of producing reliable assessments of stream health. Chemical, physical and biological parameters have all been used in such assessments, but biological indicators and, in particular, benthic macroinvertebrates, have proved particularly useful (e.g. Hellawell 1986, Rosenberg and Resh, 1993; Metcalfe-Smith 1994, Bonada *et al.* 2006 and references therein) and currently play a central role in the European Water Framework Directive (European Commission, 2000). The great advantage of biological assessment over chemical analyses is that macroinvertebrates integrate both short- and long-term changes in a range of environmental variables, regardless of whether they act separately or synergistically. Close to a century of experience in north-temperate streams has yielded information about the relative tolerances of different macroinvertebrate taxa thereby allowing use of macrobenthos assemblage structure for monitoring of water quality and environmental change.

Tropical countries have made only limited use of biomonitoring to assess stream conditions or water quality. This is despite the fact that, given trained manpower, bioassessment can

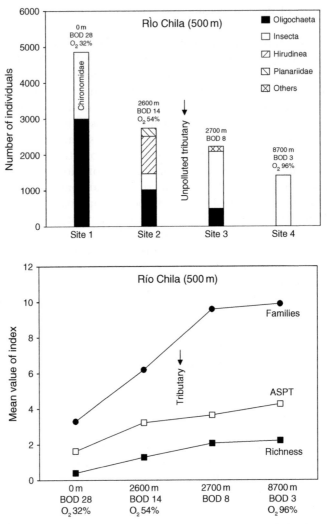

*FIGURE 4* Composition (above) and mean values of three different biological indices (below) for benthic macroin-vertebrate assemblages collected by semi-quantitative 'kick samples' at four sites at increasing distances from a point source of sewage in a coastal lowland stream (Río Chila) in Ecuador. Distances (m) from the source, mean biological oxygen demand (BOD mg $L^{-1}$), and minimum oxygen saturation (%) are given also. ASPT = Average Score Per Taxon biological index (Armitage *et al.*, 1983); Richness = Margaleff's richness index (D). Data are from Bang and Thomsen (1997).

be used to provide accurate information about the environment at a fraction of the cost and equipment required by other assessment approaches (Resh, 1995). There is certainly good reason to assume that biomonitoring and assessment using macroinvertebrates will work well in the tropics (Breen *et al.*, 1981). The characteristic longitudinal succession of macroinvertebrate assemblages downstream of a point-source of organic pollution described by Hynes (1960) applies just as well in tropical streams as in temperate ones. For example, the predominant macroinvertebrate groups immediately downstream of a source of domestic sewage in a small Ecuadorian lowland stream (Fig. 4) were Oligochaeta and chironomids (especially the pollution-tolerant genus *Chironomus*). Further downstream, they decreased in abundance and were replaced by Planariidae, leeches and insects aside from chironomids; further away still, a diverse assemblage dominated by insects but including crustaceans was re-established, although total densities were lower than close to the pollution source. Similarly, several numerical indices or metrics that have been used to describe assemblage condition such as the number of families, the ASPT index (see below) and Margaleff's richness index all showed a clear downstream transition with increased distance from the pollution source (Fig. 4).

Many researchers in tropical Latin America (e.g. Caicedo and Palacio 1998; Fenoglio *et al.*, 2002; Henne *et al.* 2002; Couceiro *et al.* 2007; Moya *et al.* 2007), Africa (e.g. Victor and Ogbeibu, 1985; Matagi, 1996; Victor and Onomivbori, 1996; Ndaruga *et al.* 2004; Kasangaki *et al.* 2006) and Asia (e.g. Augustine and Diwan, 1991; Khan and Kulshrestha, 1993; Trihadiningrum *et al.*, 1996; Mustow *et al.*, 1997; Mustow 2002) have used benthic macroinvertebrates for assessment of stream conditions. The approaches taken vary greatly: some have been simple descriptive studies using enumerations of individuals and taxa, or ecological indices of richness, evenness and diversity (e.g. Margaleff, Simpson or Shannon-Wiener indices); others have included multivariate statistical techniques (particularly ordination) or even artificial neural networks (Gutiérrez *et al.*, 2004) to relate assemblage structure to environmental variables. The diversity or richness indices used by ecologists to describe assemblages were not intended specifically for biomonitoring and assessment, and are therefore rarely good at detecting anything but the most severe cases of environmental degradation (Cao *et al.*, 1996). A better approach is the use of biotic indices in which specific taxa are given values corresponding to their presumed sensitivity/tolerance towards pollution. Unfortunately, few tropical countries have biotic indices designed for the regional stream fauna and validated through regressions against relevant physico-chemical parameters (e.g. Jacobsen, 1998; Henne *et al.*, 2002; Mustow, 2002) or by their ability to distinguish reference sites from impacted sites (e.g. Weigel *et al.*, 2002; Silveira *et al.*, 2005; Moya *et al.*, 2007).

Species identifications for invertebrates in topical countries is rarely possible and use of family-level biotic indices is more practical (Thorne and Williams, 1997). Three such indices that have been applied, with modifications, are the qualitative BMWP (Biological Monitoring Working Party) and ASPT (Average Score Per Taxon) indices, developed in the United Kingdom (see Armitage *et al.*, 1983), and the quantitative FBI (family biotic index) used in North America (Hilsenhoff, 1988). They have been employed with some success in Ecuador (Jacobsen, 1998), Thailand (Mustow, 2002), Mexico (Henne *et al.*, 2002), Brazil (Baptista *et al.*, 2007) and Bolivia (Jacobsen and Marín, 2007). One reason why it appears possible to obtain satisfactory results with these temperate indices in tropical streams is that most macroinvertebrate families are globally widespread (even though they are represented by different species) and the general tolerance of each family does not seem to vary much in different regions or on different continents. That said, the accuracy of individual biotic indices could be improved by minor adjustments to take account of known differences in family representation in tropical and temperate regions (Thorne and Williams, 1997), as has been done in Argentina (Domínguez and Fernández, 1998), Colombia (Zamora, 1999), Brazil (Junqueira and Campos, 1998) and Thailand (Mustow, 2002). This process reveals the constraint that tolerance levels of strictly-tropical taxa are imperfectly known. Ideally, tolerance scores for each taxon should be derived from knowledge of their occurrence in relation to water quality parameters (e.g. Hilsenhoff, 1988; Lenat, 1993), or from controlled laboratory experiments, but this is rarely the approach used and researchers often resort to 'scientific intuition'.

An additional problem with the use of biotic indices (or metrics in general) is that individual indices may show relatively weak relationships with some environmental variables. One solution is to integrate several indices, each one showing some degree of relationship with relevant environmental parameters, into a single multimetric index – an approach that is advocated for streams in Europe (e.g. Hering *et al.*, 2004). Examples of such composite indices have been tested using macroinvertebrate assemblage data from Thailand, Ghana and Brazil (Thorne and

Williams, 1997), and subsequent studies by Weigel *et al.* (2002: Mexico), Baptista *et al.* (2007: Brazil) and Moya *et al.* (2007: Bolivia) have applied a similar approach. Such multimetric indices include combinations of response variables such as family richness, BMWP and ASPT scores, and combined richness of Ephemeroptera, Plecoptera and Trichoptera (EPT taxa). Functional metrics such as percentages or abundances of various macroinvertebrate functional-feeding groups have occasionally been included as candidate metrics in bioassessments, but have rarely proved useful perhaps because of the scarcity of shredders in tropical streams (see Section III-E).

It should be stressed that most biotic indices currently in use have been formulated with the intent of detecting organic pollution. While they integrate overall environmental quality, they may not elucidate the precise cause of a change in macroinvertebrate assemblages. Moreover, not all perturbations are necessarily reflected in indices. For example, metal exposure from gold mining in Ecuadorian streams could not be unequivocally detected by BMWP or ASPT indices (Tarras-Wahlberg *et al.*, 2001). Water-quality parameters of importance to human health, such as concentrations of faecal bacteria and pathogens, are also unlikely to be reflected directly in biotic indices based on macroinvertebrates. Clearly, decisions about human use of stream water should not be based solely on such scores.

On the other hand, macroinvertebrate assemblage structure is not merely a function of the degree of human impact, and biotic indices vary in response to natural influences such as altitude (Jacobsen, 2003). Accurate bioassessment of stream water quality depends on a good knowledge of the extent and type of natural variation in assemblage structure, with environmental impact or stress being indicated by the deviation from the expected reference level (i.e. beyond the natural range of variation). For example, the general scarcity of EPT taxa (which are usually considered to be associated with good water quality) in streams on the Bolivian Altiplano (see Chapter 8 of this volume) complicates the use of conventional biotic indices because they yield low scores that can mislead the unaware (Jacobsen and Marín, 2007). A possible solution is application of existing predictive bioassessment models, such as the British RIVPACS (Clarke *et al.*, 2003) or the Australian AUSRIVAS (Wright, 1995; Davies, 2000), which use key environmental parameters unrelated to pollution (habitat stability, altitude and conductivity, etc.) to objectively predict the assemblage composition expected in a particular stream. The deviation of the observed composition from the predicted condition provides a measure of impact or stress. Existing models would need to be modified with respect to the taxa they include, and predicted environment-taxa relationships would require calibration for regional conditions; this has not yet been undertaken anywhere in the tropics.

Seasonal differences in stream discharge over much of the tropics can have marked effects on pollution loads, especially though dilution of effluents during the wet season. In Ecuadorian streams, Jacobsen (1998) recorded seasonal changes in BMWP and ASPT indices at upstream (unpolluted) and downstream (polluted) sites, with the upstream-downstream differences being most pronounced during the dry season when tubificid oligochaetes and *Chironomus* larvae became abundant at the lower sites. These findings, and the evident negative relationship between stream discharge and pollution loads, suggest that bioassessments using tropical macroinvertebrates should be carried out during the dry season.

## V. RESEARCH NEEDS

The fragmentary nature of the existing data on the streams throughout the tropics is a major challenge in the compilation of information on composition and biogeography of macroinvertebrates. Many stream ecologists find it difficult to stay abreast of research progress in tropical streams and rivers because, until quite recently, the literature was sparse and widely scattered (Jackson and Sweeney, 1995b) and comparatively few international publications are contributed by limnologists in tropical countries (Dudgeon, 2003). A lack of publications and limited collecting effort may have biased interpretations of the evolution of major taxa in tropical streams (Covich, 1988) and also their zoogeographical patterns. Development of an on-line Internet database on the composition and distribution of tropical stream macroinvertebrates (indeed, all tropical freshwater biodiversity) would be one way of enhancing the communication of information that has been collected but not yet disseminated adequately. Such a database might help to encourage fundamental systematic and taxonomic research on macroinvertebrates, and could be accompanied by establishment of regional reference collections (Davies and Hart, 1981). These initiatives would need to be complimented by the preparation of well-illustrated keys and field guides for different tropical regions.

A shortage of tools to aid macroinvertebrate identification has hampered ecological research on these animals: as a result, concepts and hypotheses developed in temperate regions have not been tested widely, nor has the potential application of macroinvertebrates in biomonitoring and assessment of stream condition been realized fully. In that regard, most tropical countries are far behind those in the north temperate region – the tropical portion of Australia is an exception in this regard – and the development of regionally-appropriate multimetric indices for biomonitoring and bioassessment should certainly be considered a research priority for ecologists and water scientists. Obtaining adequate funding for such efforts would be problematic but, given the importance of conserving steam ecosystems and the irreplaceable goods and ecosystem services that they provide to humans (for a discussion of these issues, see Dudgeon *et al.*, 2006), provision of financial support by nation or international agencies would be a sound investment that could enhance environmental security and sustainability.

Another result of our limited knowledge of tropical stream invertebrates is uncertainty over the patterns and of latitudinal trends in macroinvertebrate richness and diversity discussed earlier in this chapter. Research on the ecological and historical factors regulating local and regional species diversity of macroinvertebrates within the tropics is needed also. Influential factors might include the magnitude of climatic seasonality and the physical disturbance regime resulting from stream-flow fluctuations; these will also determine the extent of temporal variation in macroinvertebrate communities. The attributes responsible for the apparent high resilience of these communities is not fully understood, although it seems possible that the asynchronous growth and development that has been widely reported for many taxa may play some role.

Food webs and energy flow in tropical steams have not been extensively investigated. Understanding of stream ecosystem functioning and formulation of appropriate management strategies will depend upon better knowledge of the relative importance of autochthonous and allochthonous energy sources, combined with information on the functional organization of macroinvertebrate assemblages and degree of omnivory or generalism of major taxa. The apparent scarcity of shredders in some tropical streams, and the relative abundance of predators, is in need of explanation if – in fact – they are real patterns and not simply artefacts of insufficient sampling at a limited number of sites. Wider use of stable isotope techniques, stoichiometry and other techniques for tracing energy flow (e.g. essential fatty-acid biomarkers) might help to address some of these matters, but will probably also require a much more substantial data set of macroinvertebrate secondary production estimates – at present, both the quantity and quality of such estimates are inadequate to support robust generalizations that apply within or between tropical regions.

One obvious gap in studies of macroinvertebrate life histories is uncertainty about fate and behaviour of adult aquatic insects in the tropics. The fact that adults of many species do not show strong emergence and flight seasonality (although others certainly do) makes investigation of ecology more difficult than in temperate latitudes, as does the tendency for many tropical stream insects to fly at night. The present focus on aquatic stages of the life cycle presumes that this is where the main population constraints and regulating factors operate, but mortality or recruitment failure resulting from event in the land-water interface might be more significant. In addition, the emergence of adult insects could represent a considerable transfer of energy and nutrients from water to land (see land to water transfer of material in the form of plant litter) and may be an important food source for terrestrial consumers (bats, birds, spiders, etc.). To date, the potential or extent of reciprocal material transfers along tropical streams has received little attention, and it should be stressed that the water to land transfer is almost entirely mediated by macroinvertebrates.

In a review of this nature, decisions as to what is omitted are almost as consequential as those about what is included. We have said little about 'lower invertebrates' because knowledge of the ecology and taxonomy of most of them is poorer than for insects and decapods. Consideration of behaviour such as drift or interactions among macroinvertebrates, such as competition and predation, has also been given short shrift. This is both a matter of data scarcity and representativeness: the few manipulative studies of invertebrate interactions that have been carried out in the tropics have been undertaken at a very few locations (Puerto Rico and Costa Rica are notable in this regard), and we do not feel it is possible to make useful generalizations from them. More such studies are certainly needed across the tropics. It is especially evident that the amounts of research (of all types) devoted to macroinvertebrates in tropical Africa is much less than elsewhere in the tropics, yet the threats that streams in this region face (see Chapter 10 of this volume) are no less that those in the rest of the tropics. While there are manifest benefits to be gained from improving our knowledge of stream macroinvertebrates and their ecology throughout the tropics, in the prevailing context of limited research funds and resources, there is a compelling case for prioritizing research in Africa.

## REFERENCES

Abdallah, A.H., de Mazancourt, C., Elinge, M.M., Graw, B., Grzesiuk, M., Henson, K., Kamoga, M., Kolodziejska, I., Kristersson, M., Kuria, A., Leonhartsberger, P., Matemba, R.B., Merl, M., Moss, B., Minto, C., Murfitt, E., Musila, S.N., Ndayishiniye, J., Nuhu, D., Oduro, D.J., Provvedi, S., Rasoma, R.V., Ratsoavina, F., Trevelyan, R., Tumanye, N., Ujoh, V.N., van de Wiel, G., Wagner, T., Waylen, K., and Yonas, M. (2004). Comparative studies on the structure of an upland African stream ecosystem. *Freshw. Forum* **21**, 27–24.

Allan, J.D. (1995). "Stream Ecology. Structure and Function of Running Waters". Chapman and Hall, London, U.K.

Anderson, N.H., and Sedell, J.R. (1979). Detritus processing by macroinvertebrates in stream ecosystems. *Annu. Rev. Entomol.* **24**, 351–377.

Arkell, J.B.F. (1979). Aspects of the feeding and breeding biology of the Giant kingfisher. *Ostrich* **50**, 176–181.

Armitage, P.D., Moss, D., Wright, J.F., and Furse, M.T. (1983). The performance of a new biological water quality score system based on macroinvertebrates over a wide range of unpolluted running-water sites. *Water Res.* **17**, 333–347.

Arthington, A.H. (1990). Latitudinal gradients in insect species richness of Australian lotic systems: a selective review. *Trop. Freshw. Biol.* **2**, 179–196.

Augustine, S., and Diwan, A.P. (1991). Population dynamics of midge worms with relation to pollution in River Kshipra. *J. Environ. Biol.* **12**, 255–262.

Ball, S.L., and Baker, R.L. (1996). Predator-induced life history changes: antipredator behavior costs or facultative life history shifts? *Ecology* **77**, 1116–1124.

Bang, A.C., and Thomsen, F.B. (1997). 'Effekter af organisk forurening i tropisk og tempererede vandløb'. Masters dissertation. Freshwater Biological Laboratory, University of Copenhagen, Denmark.

Baptista, D.F., Buss, D.F., Egler, M., Giovanelli, A., Silveira, M.P., and Nessimian, J.L. (2007). A multimetric index based on benthic macroinvertebrates for evaluation of Atlantic forest streams at Rio de Janeiro State, Brazil. *Hydrobiologia* **575**, 83–94.

Benke, A.C. (1984). Secondary production of aquatic insects. *In* "The Ecology of Aquatic Insets" (V.H. Resh and D.M. Rosenberg, Eds), pp. 289–322. Praeger Scientific, New York, U.S.A.

Benke, A.C. (1993). Concepts and patterns of invertebrate production in running water. *Verh. Int. Verein. Limnol.* **25**, 15–38.

Benke, A.C. (1996). Secondary production of macroinvertebrates. *In* "Methods in Stream Ecology" (F.R. Hauer and G.A. Lamberti (Eds), pp. 557–578. Academic Press, New York, U.S.A.

Benke, A.C. (1998). Production dynamics of riverine chironomids: extremely high biomass turnover rates of primary consumers. *Ecology* 79, 899–910.

Benke, A.C., and Jacobi, D.I. (1986). Growth rates of mayflies in a subtropical river and their implications for secondary production. *J. N. Am. Benthol. Soc.* 5, 107–114.

Benke, A.C., Hauer, R.F., Stites, D.L., Meyer, J.L., and Edwards, D.L. (1992). Growth of snag-dwelling mayflies in a blackwater river: the influence of temperature and food. *Arch. Hydrobiol.* 125, 63–81.

Benzie, J.A.H., and de Silva, P.K. (1988). The distribution and ecology of the freshwater prawn *Caridina singhalensis* (Decapoda, Atyidae) endemic to Sri Lanka. *J. Trop. Ecol.* 4, 347–359.

Bishop, J.E. (1973). "Limnology of a Small Malayan River, Sungai Gombak". Dr W. Junk Publishers, The Hague, The Netherlands.

Blanco, J.F., and Scatena, F.N. (2006). Hierarchical contribution of river-ocean connectivity, water chemistry, hydraulics and substrate to the distribution of diadromous snails in Puerto Rico streams. *J. N. Am. Benthol. Soc.* 25, 82–98.

Bojsen, B., and Jacobsen, D. (2003). Effects of deforestation on macroinvertebrate diversity and assemblage structure in Ecuadorian Amazon streams. *Arch. Hydrobiol.* 158, 317–342.

Bonada, N., Prat, N., Resh, V.H., and Statzner, B. (2006). Development in aquatic insect biomonitoring: a comparative analysis of recent approaches. *Annu. Rev. Entomol.* 51, 495–523.

Boyero, L. (2002). Insect biodiversity in freshwater ecosystems: is there any latitudinal gradient? *Mar. Freshw. Res.* 53, 753–755.

Breen, C.M., Chutter, F.H., de Merona, B., and Saad, M.A.H. (1981). Rivers. *In* "The Ecology and Utilization of African Inland Waters" (J.J. Symoens, M. Burgis and J.J. Gaudet, Eds), pp. 83–92. UNEP Reports and Proceedings Series 1, United Nations Environment Programme, Nairobi, Kenya.

Brewin, P.A., Buckton, S.T., and Ormerod, S.J. (2000). The seasonal dynamics and persistence of stream macroinvertebrates in Nepal: do monsoon floods represent disturbance? *Freshw. Biol.* 44, 581–594.

Bright, G.R. (1982). Secondary benthic production in a tropical island stream. *Limnol. Oceanogr.* 27, 472–480.

Brito, E.F., Moulton, T.P., De Souza, M.L., and Bunn, S.E. (2006). Stable isotope analysis indicates microalgae as the predominant food source of fauna in a coastal forest stream, south-east Brazil. *Austral Ecol.* 31, 623–633.

Brown, D.S. (1980). "Freshwater Snails of Africa and their Medical Importance". Taylor & Francis, London, U.K.

Burger, J.F. (1987). Specialized habitat selection by black flies. *In* "Black Flies. Ecology, Population Management, and Annotated World List" (K.C. Kim and R.W. Merritt, Eds), pp. 129–145. Pennsylvania State University Press, University Park, U.S.A.

Butler, J.R.A., and du Toit, J.T. (1994). Diet and conservation status of Cape clawless otters in eastern Zimbabwe. *S. Afr. J. Wildl. Res.* 24, 41–47.

Butler, J.R.A., and Marshall, B.E. (1996). Resource use within the crab eating guild of the upper Kairezi River, Zimbabwe. *J. Trop. Ecol.* 12, 475–490.

Caicedo, O., and Palacio, J. (1998). Los macroinvertebrados bénticos y la contaminación orgánica en la quebrada La Mosca (Guarne, Antioquia, Colombia). *Actualidades Biol.* 69, 61–73.

Campbell, I.C. (1995). The life histories of three tropical species of *Jappa* Harker (Ephemeroptera: Leptophlebiidae) in the Mitchell River System, Queensland, Australia. *In* "Current Directions in Research on Ephemeroptera" (L. Corkum and J. Ciborowski, Eds), pp. 197–206. Canadian Scholars' Press, Toronto, Canada.

Cao, Y., Bark, A.W., and Williams, W.P. (1996). Measuring the responses of macroinvertebrate communities to water pollution: a comparison of multivariate approaches, biotic and diversity indices. *Hydrobiologia* 341, 1–19.

Cheshire, K., Boyero, L., and Pearson, R.G. (2005). Food webs in tropical Australian streams: shredders are not scarce. *Freshw. Biol.* 50, 748–769.

Clarke R.T., Wright, J.F., and Furse, M.T. (2003). RIVPACS models for predicting the expected macroinvertebrate fauna and assessing the ecological quality of rivers. *Ecological Modelling* 160, 219–233.

Collart, O.O., and Megalhães, C. (1994). Ecological constraints and life history strategies of palaemonid prawns in Amazonia. *Verh. Int. Verein. Limnol.* 25, 2460–2467.

Connell, J.H. (1978). Diversity in tropical rainforests and coral reefs. *Science* 199, 1302–1310.

Corbet, P.S. (1980). Biology of Odonata. *Annu. Rev. Entomol.* 25, 189–217.

Couceiro, S.R.M., Hamada, N., Luz, S.L.B., Forsberg, B.R., and Pimentel, T.P. (2007). Deforestation and sewage effects on aquatic macroinvertebrates in urban streams in Manaus, Amazonas, Brazil. *Hydrobiologia* 575, 271–284.

Covich, A.P. (1988). Geographical and historical comparisons of Neotropical streams: biotic diversity and detrital processing in highly variable habitats. *J. N. Am. Benthol. Soc.* 7, 361–386.

Covich, A.P., and MacDowell, W.H. (1996). The stream community. *In* "The Food Web of a Tropical Rain Forest" (D.P. Reagan and R.B. Waide, Eds), pp. 443–460. The University of Chicago Press, Chicago, U.S.A.

Covich, A.P., Crowl, T.A., and Scatena, F.N. (2003). Effects of extreme low flows on freshwater shrimps in a perennial tropical stream. *Freshw. Biol.* 48, 1199–1206.

Cressa, C. (1990). "Flujo de energía en una sección del Río Orituco Venezuela". Informe Final al Consejo de Desarrollo Científico y Humanístico, Caracas, Venezuela.

Cressa, C. (1994). Structural changes of the macroinvertebrate community in a tropical river. *Verh. Int. Verein. Limnol.* 25, 1853–1855.

Cressa, C. (1998). Community composition and structure of macroinvertebrates of the river Camurí Grande, Venezuela. *Verh. Int. Verein. Limnol.* 26, 1008–1011.

Cressa, C. (2003). "Taxonomía, Composición cualitativa y producción secundaria de la fauna béntica en el Parque Nacional El Avila". Informe final Consejo de Desarrollo Científico y Humanístico, Caracas, Venzuela.

Cressa, C., and Barrios, C. (2002). Larval growth rate and development time of egg, larvae and pupae of two species of Trichoptera from a stream in Venezuela. *Verh. Int. Verein. Limnol.* 28, 148–152.

Cressa, C., and Holzenthal, R. (2003). Trichoptera. *In* "Diversidad Biológica en Venezuela" (M. Aguilera, A. Azócar and E. González, Eds), pp. 412–425. Fundación Mendoza, Caracas, Venezuela.

Cressa, C., and Stark, B. (2003). Plecoptera. *In* "Diversidad Biológica en Venezuela" (M. Aguilera, A. Azócar and E. González, Eds), pp. 478–487. Fundación Mendoza, Caracas, Venezuela.

Crowl, T.A., McDowell, W.H., Covich, A.P., and Johnson, S.L. (2001). Freshwater shrimp effects on detrital processing and nutrients in a tropical headwater stream. *Ecology* 82, 775–783.

Currie, D.C., and Craig, D.A. (1987). Feeding strategies of larval black flies. *In* "Black Flies. Ecology, Population Management, and Annotated World List" (K.C. Kim and R.W. Merritt, Eds), pp. 155–170. Pennsylvania State University Press, University Park, U.S.A.

Darwall, W., Smith, K., Lowe, T., and Vié, J.-C. (2005). The Status and Distribution of Freshwater Biodiversity in Eastern Africa. IUCN SSC Freshwater Biodiversity Assessment Programme, IUCN, Gland, Switzerland and Cambridge, U.K.

Davies, B.R., and Hart, R.C. (1981). Invertebrates. *In* "The Ecology and Utilization of African Inland Waters" (J.J. Symoens, M. Burgis and J.J. Gaudet, Eds), pp. 51–68. UNEP Reports and Proceedings Series 1, United Nations Environment Programme, Nairobi, Kenya.

Davies, P.E. (2000). Development of a national river bioassessment system (AUSRIVAS) in Australia. *In* "Assessing the Biological Quality of Fresh waters: RIVPACS and Other Techniques" (J.F. Wright, D.W. Sutcliffe, and M.T. Furse, Eds), pp. 113–124. Freshwater Biological Association, Ambleside, U.K.

De Souza, M.L., and Moulton, T.P. (2005). The effects of shrimps on benthic material in a Brazilian island stream. *Freshw. Biol.* 50, 592–602.

Dobson, M. (2004). Freshwater crabs in Africa. *Freshw. Forum* 21, 3–26.

Dobson, M., Magana, A., Mathooko, J.M., and Ndegwa, F.K. (2002). Detritivores in Kenyan highland streams: more evidence for the paucity of shredders in the tropics? *Freshw. Biol.* 47, 909–919.

Dobson, M., Mathooko, J.M., Ndegwa, F.K., and M'Erimba, C. (2004). Leaf litter processing rates in a Kenyan highland stream, the Njoro River. *Hydrobiologia* 519, 207–210.

Dobson, M.K., Magana, A.M., Lancaster, J., and Mathooko, J.M. (2007). Aseasonality in the abundance and life history of an ecologically dominant freshwater crab in the Rift Valley, Kenya. *Freshw. Biol.* 52, 215–225.

Domínguez, E., and Fernández, H.R. (1998). Calidad de los ríos de la Cuenca del Salí (Tucumán, Argentina) medida por un índice biótico. *Fundación Miguel Lillo, Tucumán, Argentina, Serie Conservación de la Naturaleza* 12, 1–39.

Douglas, M.M., Bunn, S.E., and Davies, P.M. (2005). River and wetland food webs in Australia's wet-dry tropics: general principles and implications for management. *Mar. Freshw. Res.* 56, 329–342.

Dudgeon, D. (1982). The life history of *Brotia hainanensis* (Brot, 1872) (Gastropoda: Prosobranchia: Thiaridae) in a tropical forest stream. *Zool. J. Linn. Soc.* 76, 141–154.

Dudgeon, D. (1983). Spatial and temporal changes in the distribution of gastropods in the Lam Tsuen River, New Territories, Hong Kong, with notes on the occurrence of the exotic snail, *Biomphalaria straminea. Malacol. Rev.* 16, 91–92.

Dudgeon, D. (1985). The population dynamics of some freshwater carideans (Crustacea: Decapoda) in Hong Kong, with special reference of *Neocaridina serrata* (Atyidae). *Hydrobiologia* 120, 141–149.

Dudgeon, D. (1988a). The influence of riparian vegetation on macroinvertebrate community structure in four Hong Kong streams. *J. Zool., Lond.* 216, 609–627.

Dudgeon, D. (1988b). Flight periods of aquatic insects from a Hong Kong forest stream I. Macronematinae (Hydropsychidae) and Stenopsychidae (Trichoptera). *Aquat. Insects* 10, 61–68.

Dudgeon, D. (1989a). The influence of riparian vegetation on the functional organization of four Hong Kong stream communities. *Hydrobiologia* 179, 183–194.

Dudgeon, D. (1989b). Life cycle, production, microdistribution and diet of the damselfly *Euphaea decorata* (Odonata: Euphaeidae) in a Hong Kong forest stream. *J. Zool., Lond.* 217, 57–72.

Dudgeon, D. (1989c). Gomphid (Odonata: Anisoptera) life cycles and production in a Hong Kong forest stream. *Arch. Hydrobiol.* 114, 531–536.

Dudgeon, D. (1989d). Ecological strategies of Hong Kong Thiaridae (Gastropoda: Prosobranchia). *Malacol. Rev.* 22, 39–53.

Dudgeon, D. (1991). An experimental study of the effects of predatory fish on macroinvertebrates in a Hong Kong stream. *Freshw. Biol.* 25, 321–330.

Dudgeon, D. (1992). Endangered ecosystems: a review of the conservation status of tropical Asian rivers. *Hydrobiologia* **248**, 167–191.

Dudgeon, D. (1993). The effects of spate-induced disturbance, predation and environmental complexity on macroinvertebrates in a tropical stream. *Freshw. Biol.* **30**, 189–197.

Dudgeon, D. (1994). The influence of riparian vegetation on macroinvertebrate community structure and functional organization in six New Guinea streams. *Hydrobiologia* **264**, 65–85.

Dudgeon, D. (1995a). Life histories, secondary production and microdistribution of Psephenidae (Coleoptera: Insecta) from a tropical forest stream. *J. Zool., Lond.* **236**, 465–481.

Dudgeon, D. (1995b). Life history, secondary production and microdistribution of *Hydrocyphon* (Coleoptera: Scirtidae) in a tropical forest stream. *Arch. Hydrobiol.* **133**, 261–271.

Dudgeon, D. (1996a). The influence of refugia on predation impacts in a Hong Kong stream. *Arch. Hydrobiol.* **138**, 145–159.

Dudgeon, D. (1996b). Life histories, secondary production and microdistribution of heptageniid mayflies (Ephemeroptera) in a tropical forest stream. *J. Zool., Lond.* **240**, 341–361.

Dudgeon, D. (1996c). Life histories, secondary production and microdistribution of *Stenopsyche angustata* (Trichoptera: Stenopsychidae) in a tropical forest stream. *J. Zool., Lond.* **238**, 679–691.

Dudgeon, D. (1996d). The life history, secondary production and microdistribution of *Ephemera* spp. (Ephemeroptera: Ephemeridae) in a tropical forest stream. *Arch. Hydrobiol.* **135**, 473–483.

Dudgeon, D. (1997). Life histories, secondary production and microdistribution of hydropsychid caddisflies (Trichoptera) in a tropical forest stream. *J. Zool., Lond.* **243**, 191–210.

Dudgeon, D. (1999a). "Tropical Asian Streams: Zoobenthos, Ecology and Conservation". Hong Kong University Press, Hong Kong, China.

Dudgeon, D. (1999b). Patterns of variation in secondary production in a tropical forest stream. Arch. Hydrobiol. **144**, 271–281.

Dudgeon, D. (1999c). The population dynamics of three species of Calamoceratidae (Trichoptera) in a tropical forest stream. *In* "Proceedings of the 9th International Symposium on Trichoptera (H. Malicky and P. Chantaramongkol, Eds), pp. 83–91. University of Chiang Mai, Chiang Mai, Thailand.

Dudgeon, D. (2000). The ecology of tropical Asian rivers and streams in relation to biodiversity conservation. *Annu. Rev. Ecol. Syst.* **31**, 239–263.

Dudgeon, D. (2003). The contribution of scientific information to the conservation and management of freshwater biodiversity in tropical Asia. *Hydrobiologia* **500**, 295–314.

Dudgeon, D. (2006). The impacts of human disturbance on stream benthic invertebrates and their drift in North Sulawesi, Indonesia. *Freshw. Biol.* **51**, 1710–1729.

Dudgeon, D., and Wu, K.K.Y. (1999). Leaf litter in a tropical stream: food or substrate for macroinvertebrates? *Arch. Hydrobiol.* **146**, 65–82.

Dudgeon, D., Arthington, A.H., Gessner, M.O, Kawabata, Z., Knowler, D., Lévêque, C., Naiman, R.J., Prieur-Richard, A.-H., Soto, D., Stiassny, M.L.J., and Sullivan, C.A. (2006). Freshwater biodiversity: importance, threats, status and conservation challenges. *Biol. Rev.* **81**, 163–182.

European Commission (2000). Directive 2000/60/EC of the European Parliament and Council, establishing a framework for Community action in the field of water policy. *Official J. Eur. Community* L327, 1–72.

Fenoglio, S., Badino, G., and Bona, F. (2002). Benthic macroinvertebrate communities as indicators of river environmental quality: an experience in Nicaragua. Rev. Biol. Trop. 50, 1125–1131.

Flecker, A.S. (1992a). Fish trophic guilds and the structure of a tropical stream: weak direct versus strong indirect effects. *Ecology* **73**, 927–940.

Flecker, A.S. (1992b). Fish predation and the evolution of invertebrate drift periodicity: evidence from Neotropical streams.*Ecology* **73**, 438–448.

Flecker, A.S., and Feifarek, B.P. (1994). Disturbance and the temporal variability of insect assemblages in two Andean streams. *Freshw. Biol.* **31**, 131–142.

Flint, O.S., Jr, Holzenthal, R.W., and Harris, S.V. (1999). "Catalog of the Neotropical Caddisflies (Insecta: Trichoptera)". Special Publication of the Ohio Survey, Columbus, Ohio, U.S.A.

Flowers, R.W. (1991). Diversity of stream-living insects in northwestern Panama. *J. N. Am. Benthol. Soc.* **10**, 322–334.

Flowers, R.W., and Pringle, C.M. (1995). Yearly fluctuations in the mayfly community of a tropical stream draining lowland pasture in Costa Rica. *In* "Current Directions in Research on Ephemeroptera" (L.D. Corkum and J.H. Ciborowski, Eds), pp. 131–150. Canadian Scholars' Press, Toronto, Ontaria.

Fox, L.R. (1977). Species richness in streams – an alternative mechanism. *Am. Nat.* **111**, 1017–1021.

Freitag, H. (2004). Composition and longitudinal patterns of aquatic insect emergence in small rivers of Palawan Island, the Philippines. *Int. Rev. Hydrobiol.* **89**, 375–391.

Froehlich, C.G. (1969). *Caenis cuniana* sp.n., a parthenogenetic mayfly. *Beitr. Neotrop. Fauna* **6**, 103–108.

Gillies, M.T. (1988). Descriptions of the nymphs of some Afrotropical Baetidae (Ephemeroptera). I. *Cloeon* Leach and *Rhithrocloeon* Gillies. *Aquat. Insects* **10**, 49–59.

Gillies, M.T. (1994). Descriptions of some Afrotropical Baetidae (Ephemeroptera). II. *Baetis* Leach, *s.l.*, East African species. *Aquat. Insects* **16**, 105–118.

Greathouse, E.A., and Pringle, C.M. (2006). Does the river continuum concept apply on a tropical island? Longitudinal variation in a Puerto Rican stream. *Can. J. Fish. Aquat. Sci.* **63**, 134–152.

Greathouse, E.A., Pringle, C.M., McDowell, W.H., and Holmquist, J.G. (2006). Indirect upstream effects of dams: consequences of migratory consumer extirpation in Puerto Rico. *Ecol. Appl.* **16**, 339–352.

Guest, W.C. (1979). Laboratory life history of the palaemonid shrimp *Macrobrachium amazonicum* (Heller) (Decapoda, Palaemonidae). *Crustaceana* **37**, 141–152.

Gutiérrez, J.D., Riss, W., and Ospina, R. (2004). Bioindicación de la calidad del agua con macroinvertebrados acuáticos en la sabana de Bogotá, utilizando redes neuronales artificiales. *Caldasia* **26**, 151–160.

Hancock, M.A., and Bunn, S.E. (1997). Population dynamics and life history of *Paratya australiensis* (Decapoda: Atyidae) in upland rainforest streams, southeast Queensland, Australia. *Mar. Freshw. Res.* **48**, 361–369.

Harrison, A.D., and Rankin, J.J. (1975). Forest litter and stream fauna on a tropical island, St. Vincent, West Indies. *Verh. Int. Verein. Limnol.* **19**, 1736–1745.

Hellawell, J.M. (1986). "Biological Indicators of Freshwater Pollution and Environmental Management". Elsevier Applied Science, London.

Henne, L.J., Schneider, D.W., and Martinez, L.M. (2002). Rapid assessment of organic pollution in a West-Central Mexican river using a family-level biotic index. *J. Environ. Plan. Manag.* **45**, 613–632.

Hering, D., Moog, O., Sandin, L., and Verdonschot, P.F.M. (2004). Overview and application of the AQEM assessment system. *Hydrobiologia* **516**, 1–20.

Hill, M.P., and O'Keeffe, J.H. (1992). Some aspects of the ecology of the freshwater crab (*Potamonautes perlatus* Milne Edwards) in the upper reaches of the Buffalo River, eastern Cape Province, South Africa. *S. Afr. J. Aquat. Sci.* **18**, 42–50.

Hilsenhoff, W.L. (1988). Rapid field assessment of organic pollution with a family level biotic index. *J. N. Am. Benthol. Soc.* **7**, 65–68.

Hurlbert, S.H., Rodriguez, G., and Dias dos Santos, N. (1981). "Aquatic Biota of Tropical South America: Part 1, Arthropoda". San Diego State University, San Diego, U.S.A.

Huryn, A.D., and Wallace, J.B. (2000). Life history and production of stream insects. *Annu. Rev. Entomol.* **45**, 83–110.

Hynes, H.B.N. (1960). "The Biology of Polluted Waters". Liverpool University Press, Liverpool, U.K.

Hynes, H.B.N. (1970). "The Ecology of Running Waters". Liverpool University Press, Liverpool, U.K.

Hynes, J.D. (1975). Annual cycles of macroinvertebrates of a river in southern Ghana. *Freshw. Biol.* **5**, 71–83.

Illies, J. (1969). Biogeography and ecology of Neotropical freshwater insects, especially those from running waters. *In* "Biogeography and Ecology in South America" (E.J. Fittkau, J. Illies, H. Klinge, G.H. Schwabe and H. Sioli, Eds), pp. 685–708. Dr. W. Junk Publishers, The Hague, The Netherlands.

Illies, J. (1978). "Limnofauna Europaea. A Checklist of the Animals Inhabiting European Inland Waters, with Accounts of their Distribution and Ecology". Gustav Fischer Verlag, Stuttgart, Germany.

Jackson, J.K., and Sweeney, B.W. (1995a). Egg and larval development times for 35 species of tropical stream insects from Costa Rica. *J. N. Am. Benthol. Soc.* **14**, 115–130.

Jackson, J.K., and Sweeney, B.W. (1995b). Research in tropical streams and rivers: introduction to a series of papers. *J. N. Am. Benthol. Soc.* **14**, 2–4.

Jacobi, D.I., and Benke, A.C. (1991). Life histories and abundance patterns of snag-dwelling mayflies in a blackwater Coastal Plain river. *J. N. Am. Benthol. Soc.* **10**, 372–387.

Jacobsen, D. (1998). Influence of organic pollution on the macroinvertebrate fauna of Ecuadorian highland streams. *Arch. Hydrobiol.* **143**, 179–195.

Jacobsen, D. (2003). Altitudinal changes in diversity of macroinvertebrates from small streams in the Ecuadorian Andes. *Arch. Hydrobiol.* **158**, 145–167.

Jacobsen, D. (2004). Contrasting patterns in local and zonal family richness of stream invertebrates along an Andean altitudinal gradient. *Freshw. Biol.* **49**, 1293–1305.

Jacobsen, D., and Encalada, A. (1998). The macroinvertebrate fauna of Ecuadorian highland streams in the wet and dry season. *Arch. Hydrobiol.* **142**, 53–70.

Jacobsen, D., and Marín, R. (2007). Bolivian Altiplano streams with low richness of macroinvertebrates and large temporal fluctuations in temperature and dissolved oxygen. *Aquat. Ecol.*, DOI 10.1007/s10452-007-9127-x.

Jacobsen, D., Schultz, R., and Encalada, A. (1997). Structure and diversity of stream macroinvertebrates assemblages: the effect of temperature with altitude and latitude. *Freshw. Biol.* **38**, 247–261.

Jalihal, D.R., Sankoli, K.N., and Shenoy, S. (1993). Evolution of larval development patterns and process of freshwaterization in the prawn genus *Macrobrachium* Bate, 1868 (Decapoda, Palaemonidae). *Crustaceana* **65**, 365–376.

Johnson, S.L., Covich, A.P., Crowl, T.A., Estrada-Pinto, A., Bithorn, J., and Wurtsbagh, W.A. (1998). Do seasonality and disturbance influence reproduction in freshwater atyid shrimp in headwater streams, Puerto Rico? *Verh. Int. Verein. Limnol.* **26**, 2076–2081.

Junqueira, V.M., and Campos, S.C.M. (1998.) Adaptation of the "BMWP" method for water quality evaluation to Rio das Velhas watershed (Minas Gerais, Brazil). *Acta Limnol. Brasiliensia* **10**, 125–135.

Kasangaki, A., Babaasa, D., Efitre, J., McNeilage, A., and Bitariho, R. (2006). Links between anthropogenic perturbation and benthic macroinvertebrate assemblages in Afromontane forest streams in Uganda.*Hydrobiologia* 563, 231–245.

Khan, A.A., and Kulshrestha, S.K. (1993). Benthic fauna in relation to pollution: a case study at River Chambal near Kota in Central India. *Environ. Int.* 19, 597–610.

King, J.M. (1983). Abundance, biomass and density of benthic macro-invertebrates in a western Cape River, South Africa. *Trans. Roy. Soc. S. Afr.* 45, 11–34.

Kohler, S.L., and McPeek, M.A. (1989). Predation risk and the foraging behavior of competing stream insects. *Ecology* 70, 1811–1825.

Kottelat, M., and Whitten, T. (1996). Freshwater biodiversity in Asia with special reference to fish. *World Bank Tech. Pap.*, 343, 1–59.

Lake, P.S., Schreiber, E.S.G., Milne, B.J., and Pearson, R.G. (1994). Species richness over time, with stream size and with latitude. *Verh. Int. Verein. Limnol.* 25, 1822–1826.

Lenat, D.R. (1993). A biotic index for the southeastern United States: derivation and list of tolerance values, with criteria for assigning water quality ratings. *J. N. Am. Benthol. Soc.* 12, 279–290.

Mackey, A.P. (1977a). Growth and development of larval Chironomidae.*Oikos* 28, 270–275.

Mackey, A.P. (1977b). Quantitative studies on the Chironomidae (Diptera) of the rivers Thames and Kennet. IV. Production. *Arch. Hydrobiol.* 80, 327–348.

Maldonado, V., Perez, B., and Cressa, C. (2001). Seasonal variation on the ephemeropteran community of four tropical rivers. *In* "Trends in Research in Ephemeroptera and Plecoptera" (E. Domínguez, Ed.), pp. 125–134. Kluwer Academic/Plenum Academic, New York, U.S.A.

Mantel, S.K., and Dudgeon, D. (2004a) Growth and production of a tropical predatory shrimp, *Macrobrachium hainanense* (Palaemonidae), in two Hong Kong streams. *Freshw. Biol.* 49, 1320–1336.

Mantel, S.K., and Dudgeon, D. (2004b). Dietary variation in a predatory shrimp, *Macrobrachium hainanense* (Palaemonidae), in Hong Kong forest streams. *Arch. Hydrobiol.* 160, 305–328.

Mantel, S.K., and Dudgeon, D. (2004c). Effects of *Macrobrachium hainanense* predation on benthic community functioning in tropical Asian streams. *Freshw. Biol.* 49, 1306–1319.

Mantel, S.K., and Dudgeon, D. (2005). Reproduction and sexual dimorphism of the palaemonid shrimp *Macrobrachium hainanense* in Hong Kong streams. *J. Crust. Biol.* 25, 450–459.

Mantel, S.K., Salas, M., and Dudgeon, D. (2004). Food web structure in a tropical Asian forest stream. *J. N. Am. Benthol. Soc.* 23, 728–755.

March, J.G., and Pringle, C.M. (2003). Food web structure and basal resource utilization along a tropical island stream continuum, Puerto Rico. *Biotropica* 35, 84–93.

March, J.G., Benstead, J.P., Pringle, C.M., and Ruebel, M.W. (2001). Linking shrimp assemblages with rates of detrital processing along an elevational gradient in a tropical stream. *Can. J. Fish. Aquat. Sci.* 58, 470–478.

March, J.G., Pringle, C.M., Townsend, M.J., and Wilson, A.I. (2002). Effects of freshwater shrimp assemblages on benthic communities along an altitudinal gradient of a tropical island stream. *Freshw. Biol.* 47, 377–390.

March, J.G., Benstead, J.P., Pringle, C.M., and Scatena, F.N. (2003). Damming tropical island streams: problems, solutions, and alternatives. *Bioscience* 53, 1069–1078.

Marchant, R. (1982a). Life spans of two species of tropical mayfly nymph (Ephemeroptera) from Magela Creek, Northern Territory. *Aust. J. Mar. Freshw. Res.* 33, 173–179.

Marchant, R. (1982b). Seasonal variation in the macroinvertebrate fauna of billabongs along Magela Creek, Northern Territory. *Aust. J. Mar. Freshw. Res.* 33, 329–342.

Marchant, R., and Hehir, G. (1999). Growth, production and mortality of two species of *Agapetus* (Trichoptera: Glossosomatidae) in the Acheron Rivet, south-east Australia.*Freshw. Biol.* 42, 655–671.

Marchant, R., and Yule, C.M. (1996). A new method for estimating life-spans of aseasonal aquatic insects from streams on Bougainville Island, Papua New Guinea. *Freshw. Biol.* 35, 101–107.

Matagi, S.V. (1996). The effect of pollution on benthic macroinvertebrates in a Ugandan stream. *Arch. Hydrobiol.* 137, 537–549.

Mathooko, J.M. (1996). "Artificial Physical Disturbance at the Sediment Surface of a Kenyan Mountain Stream with Particular Reference to the Ephemeroptera Community". Unpublished Ph.D. thesis, University of Vienna, Vienna, Austria.

Mathooko, J.M. (1997). The influence of physical disturbance on recolonization patterns of a mayfly community in a tropical mountain stream. *In*: "Ephemeroptera and Plecoptera: Biology-Ecology-Systematics" (P. Landolt and M. Sartori, Eds), pp. 275–281. Mauron–Tinguely and Lachat, Fribourg, Switzerland.

Mathooko, J.M. (1998). Mayfly diversity in East Africa. *Afr. J. Ecol.* 36, 368–370.

Mathooko, J.M. (1999). Effects of differing inter-disturbance intervals on the diversity of mayflies recolonizing disturbed sites in a tropical stream. *Arch. Hydrobiol.* 146, 101–116.

Mathooko, J.M., and Mavuti, K.M. (1992). Composition and seasonality of benthic invertebrates, and drift in the Naro Moru River, Kenya. *Hydrobiologia* 232, 47–56.

McElravy, E.P., Wolda, H., and Resh, V.H. (1982). Seasonality and annual variability of caddisfly adults (Trichoptera) in a 'non-seasonal' tropical environment. *Arch. Hydrobiol.* **94**, 302–317.

Merritt, R.W., and Cummins, K.W. (1996). "An Introduction to the Aquatic Insects of North America. Third Edition". Kendall/Hunt Publishing Company, Dubuque, U.S.A.

Metcalfe-Smith, J.L. (1994). Biological water-quality assessment of rivers: use of macroinvertebrate communities. *In* "The Rivers Handbook. Volume 2" (P. Calow, and G.E. Petts, Eds), pp. 144–179. Blackwell Scientific Publications, Oxford, U.K.

Mizuno, T. (1982) Secondary production. Shrimps, crabs and mollusks. *In* "Tasek Bera: the Ecology of a Freshwater Swamp" (J.I. Furtado and S. Mori, Eds), pp. 307–314. Dr. W. Junk Publuishers, The Hague, The Netherlands.

Morin, A. (1997). Empirical models predicting population abundance and productivity in lotic systems. *J. N. Am. Benthol. Soc* **16**, 319–337.

Morin, A., and Bourassa, N. (1992). Modeles empiriques de la production annuelle et du rapport P/B d'invertebres benthiques d'eau courante. *Can. J. Fish. Aquat. Sci.* **49**, 532–539.

Morin, A., and Dumont, P. (1994). A simple model to estimate growth rate of lotic insect larvae and its value for estimating population and community production. *J. N. Am. Benthol. Soc.* **13**, 357–367.

Morse, J. (1997). Checklist of world Trichoptera. *In* "Proceedings of the 8th International Symposium on Trichoptera" (R.W. Holzenthal and O.S. Flint Jr, Eds), pp. 339–342. Ohio Biological Survey, Columbus, U.S.A.

Mossolin, E.C., and Bueno, S.L.S. (2002). Reproductive biology of *Macrobrachium olfersi* (Decapoda, Palaemonidae) in São Sebastião, Brazil.*J. Crust. Biol.* **22**, 367–376.

Moya, N. (2006). "Indice multimétrico de integridad biótica para la cuenca del Río Chiripiri, Cochabamba, Bolivia". Masters thesis, Universidad Mayor San Andrés, La Paz, Bolivia.

Moya, N., Tomanova, S., and Oberdorff, T. (2007). Initial development of a multi-metric index based on aquatic macroinvertebrates to assess stream condition in the upper Isidoro-Sécure Basin, Cochabamba, Bolivia. *Hydrobiologia*, **589**, in press.

Mustow, S.E. (2002). Biological monitoring of rivers in Thailand: use and adaptation of the BMWP score. *Hydrobiologia* **479**, 191–229.

Mustow, S.E., Wilson, R.S., and Sannarm, G. (1997). Chironomid assemblages in two Thai water courses in relation to water quality. *Nat. Hist. Bull. Siam Soc.* **45**, 53–64.

Mwangi, B.M. (2000). "Bedsediments, Organic Matter and Macroinvertebrate Responses to Changes in Catchment Land Use along a Low Order Tropical Stream: Sagana River, Kenya". Unpublished Ph.D. thesis, University of Vienna, Vienna, Austria.

Ndaruga, A.M., Ndiritu, G.G., Gichuki, N.N., and Wamicha,W.N. (2004). Impact of water quality on macroinvertebrate assemblages along a tropical stream in Kenya. *Afr. J. Ecol.* **42**, 208–216.

Ng, P.K.L. (1988) "Freshwater Crabs of Peninsular Malaysia and Singapore". Department of Zoology, National University of Singapore, Singapore.

Nolte, U., De Oliveira, M.J., and Sturs, E. (1997). Seasonal, discharge-driven patterns of mayfly assemblages in an intermittent Neotropical stream. *Freshw. Biol.* **37**, 333–343.

Ogada, M.O. (2006) "Effects of the Louisiana Crayfish Invasion and Other Human Impacts on the African Clawless otter in the Ewaso Ng'iro Ecosystem". Unpublished Ph.D. thesis, Kenyatta University, Nairobi, Kenya.

Oliver, D.R. (1979). Contribution of life history information to taxonomy of aquatic insects. *J. Fish. Res. Bd Can.* **36**, 318–321.

Patrick, R. (1964). A discussion of the results of the Catherwood Expedition to the Peruvian headwaters of the Amazon. *Verh. Int. Verein. Limnol.* **15**, 1084–1090.

Pearson, R.G., Benson, L.J., and Smith, R.E.W. (1986). Diversity and abundance of the fauna in Yuccabine Creek, a tropical rainforest stream. *In* "Limnology in Australia" (P. De Dekker and W.D. Williams, Eds.), pp. 329–342. Dr. Junk Publishers, Dordrecht, The Netherlands.

Peckarsky, B.L., Taylor, B.W., McIntosh, A.R., McPeek, M.A., and Lyttle, D.A. (2001). Variation in mayfly size at metamorphosis as a developmental response to risk of predation. *Ecology* **82**, 740–757.

Petr, T. (1970). The bottom fauna of the rapids of the Black Volta River in Ghana. *Hydrobiologia* **36**, 399–418.

Pringle, C.M. (1991). Geothermally modified waters surface at La Selva Biological Station, Costa Rica: volcanic processes introduce chemical discontinuities into lowland tropical streams. *Biotropica* **23**, 523–529.

Pringle, C.M. (1996) Atyid shrimps (Decapoda: Atyidae) influence the spatial heterogeneity of algal communities over different scales in tropical montane streams, Puerto Rico. *Freshw. Biol.* **35**, 125–140.

Pringle, C.M., and Hamazaki, T. (1998) The role of omnivory in a Neotropical stream: separating diurnal and nocturnal effects. *Ecology* **79**, 269–280.

Pringle, C.M., and Ramírez, A. (1998). Use of both benthic and drift sampling techniques to assess tropical stream invertebrate communities along an altitudinal gradient, Costa Rica. *Freshw. Biol.* **39**, 359–373.

Pringle, C.M., Blake, G.A., Covich, A.P., Buzby, K.M., and Finley, A. (1993). Effects of omnivorous shrimp in a montane tropical stream: sediment removal, disturbance of sessile invertebrates and enhancement of understorey algal biomass. *Oecologia* **93**, 1–11.

Pringle, C.M., Hemphill, N., McDowell, W.H., Bednarek, A., and March, J.G. (1999) Linking species and ecosystems: different biotic assemblages cause interstream differences in organic matter.*Ecology* 80, 1860–1872.

Pringle, C.M., Scatena, F.N., Paaby-Hansen, P., and Nuñez-Ferrera, M. (2000). River conservation in Latin America and the Caribbean. *In*: "Global Perspectives on River Conservation: Science, Policy and Practice" (P.J. Boon, B.R. Davies and G.E. Petts, Eds), pp. 41–77. John Wiley and Sons, Ltd, Chichester, U.K.

Purves, M.G., Kruuk, H., and Nel, J.A.J. (1994). Crabs *Potamonautes perlatus* in the diet of otter *Aonyx capensis* and water mongoose *Atilax paludinosus* in a freshwater habitat in South Africa. *Z. Säugetierk.* 59, 332–341.

Ramírez, A., and Hernandez-Cruz, L.R. (2004). Aquatic insect assemblages in shrimp-dominated tropical streams. *Biotropica* 36, 259–266.

Ramírez, A., and Pringle, C.M. (1998a). Invertebrate drift and benthic community dynamics in a lowland Neotropical stream, Costa Rica. *Hydrobiologia* 386, 19–26.

Ramírez, A., and Pringle, C.M. (1998b). Structure and production of a benthic insect assemblage in a Neotropical stream. *J. N. Am. Benthol. Soc.* 17, 443–463.

Ramírez, A., Paaby, P., Pringle, C.M., and Aguero, G. (1998) Effect of habitat type on benthic macroinvertebrates in two lowland tropical streams, Costa Rica. *Rev. Biol Trop.* 46 (**Suppl. 6**), 201–213.

Reice, S.R., Wissmar, R.C., and Naiman, R.J. (1990). Disturbance regimes, resilience, and recovery of animal communities and habitats in lotic ecosystems. *Environ. Manag.* 14, 647–659.

Resh, V.H. (1995). Freshwater benthic macroinvertebrates and rapid assessment procedures for water quality monitoring in developing and newly industrialized countries. *In* "Biological Assessment and Criteria – Tools for Water Resource Planning and Decision Making" (W.S. Davis and T.P. Simon, Eds), pp. 167–177. Lewis Publishers, Boca Raton, U.S.A.

Resh, V.H., Lévêque, C., and Statzner, B. (2004). Long-term, large-scale biomonitoring of the unknown: assessing the effects of insecticides to control river blindness (onchocerciasis) in West Africa. *Annu. Rev. Entomol.* 49, 115–139.

Rincón, J.A., and Cressa, C. (2000). Temporal variability of macroinvertebrates assemblages in a Neotropical intermittent stream, Venezuela. *Arch. Hydrobiol.* 148, 421–432.

Rohde, K. (1992). Latitudinal gradients in species diversity: the search for the primary cause.*Oikos* 65, 514–527.

Rosenberg, D.M., and Resh, V.H. (1993). "Freshwater Biomonitoring and Benthic Macroinvertebrates". Chapman and Hall, New York, U.S.A.

Ross, H.H. (1967). The evolution and past dispersal of the Trichoptera. *Annu. Rev. Entomol.* 12, 169–206.

Salas, M., and Dudgeon, D. (2001). Stable-isotope determination of mayfly (Insecta: Ephemeroptera) food sources in three tropical Asian streams. *Arch. Hydrobiol.* 151, 17–32.

Salas, M., and Dudgeon, D. (2002). Laboratory and field studies of mayfly growth in the tropics. *Arch. Hydrobiol.* 153, 75–90.

Salas, M., and Dudgeon, D. (2003). Life histories, production dynamics and resource utilisation of mayflies (Ephemeroptera) in two tropical Asian forest streams. *Freshw. Biol.* 48, 485–499.

Schultz, R. (1997). "Biologisk struktur i ecuadorianske lavlandsvandløb med forskellig grad af riparisk skygning". Masters thesis, Freshwater Biological Laboratory, University of Copenhagen, Denmark.

Shelley, A.J. (1988). Vector aspects of the epidemiology of onchocerciasis in Latin America. *Annu. Rev. Entomol.* 33, 337–366.

Shivoga, W.A. (1999). "Spatio-temporal Variations in Physicochemical Factors and Stream Fauna and Composition in the Zones of Transition Between Baharini Springbrook, Njoro River and Lake Nakuru, Kenya". Unpublished Ph.D. thesis, University of Vienna, Austria.

Silveira, M.P. Baptista, D.F. Buss, D.F., Nessimian, J.L., and Egler, M. (2005). Application of biological measures for stream integrity assessment in south-east Brazil. *Environ. Monit. Assess.* 101, 117–128.

Somers, M.J., and Nel, J.A.J. (1998). Dominance and population structure of freshwater crabs (*Potamonautes perlatus* Milne Edwards). *S. Afr. J. Zool.* 33, 31–36.

Stanford, J.A., and Ward, J.V. (1983). Insect diversity as a function of environmental variability and disturbance in stream systems. *In* "Stream Ecology – Application and Testing of General Theory" (J.R. Barnes and G.W. Minshall, Eds), pp. 265–278. Plenum Press, New York, U.S.A.

Stark, B.P. (2001). A synopsis of Neotropical Perlidae (Plecoptera). *In* "Trends in Research in Ephemeroptera and Plecoptera" (E. Domínguez, Ed.), pp. 405–422. Kluwer Academic/Plenum Academic, New York, U.S.A.

Starmühlner, F. (1984). Checklist and longitudinal distribution of the meso and macrofauna of mountain streams of Sri Lanka (Ceylon). *Arch. Hydrobiol.* 101, 303–325.

Statzner, B. (1975). Funktionsmorphologische Studien am Genitapparat von drei neuen *Cheumatopsche*-Arten (Trichoptera), Hydropsychidae. *Zool. Anz.* 193, 382–398.

Sternberg, R.V., and Cumberlidge, N. (2001). Notes on the position of the true freshwater crabs within the brachyrhynchan Eubrachyura (Crustacea: Decapoda: Brachyura). *Hydrobiologia* 449, 21–39.

Stout, J., and Vandermeer, J. (1975). Comparison of species richness for stream-inhabiting insects in tropical and mid-latitude streams. *Am. Nat.* 109, 263–280.

Sweeney, B.W. (1984). Factors influencing life-history patterns of aquatic insects. *In* "The Ecology of Aquatic Insets" (V.H. Resh and D.M. Rosenberg, Eds), pp. 56–100. Praeger Scientific, New York, U.S.A.

Sweeney, B.W., and Vannote, R.L. (1978). Size variation and the distribution of hemimetabolous aquatic insects: two thermal equilibrium hypotheses. *Science* 200, 444–446.

Sweeney, B.W., Vannote, R.L., and Dodds, P.J. (1986). Effects of temperature and food quality on growth and development of a mayfly, *Leptophlebia intermedia. Can. J. Fish. Aquat. Sci.* 43, 12–18.

Sweeney, B.W., Jackson, J.K., and Funk, D.H. (1995). Semivoltinism, seasonal emergence and adult size variation in a tropical stream mayfly (*Euthyplocia hecuba*). *J. N. Am. Benthol. Soc.* 14, 131–146.

Tarras-Wahlberg, N.H., Flachier, A., Lane, S.N., and Sangfors, O. (2001). Environmental impacts and metal exposure of aquatic ecosystems in rivers contaminated by small scale gold mining: the PuyangoRiver basin, southern Ecuador.*Sci. Tot. Environ.* 278, 239–261.

Thorne, St. T.R., and Williams, W.P. (1997). The response of benthic macroinvertebrates to pollution in developing countries: a multimetric system of bioassessment. *Freshw. Biol.* 37, 671–686.

Thorp, H.J., and Delong, D.M. (2002). Dominance of autochthonous autotrophic carbon in food webs of heterotrophic rivers. *Oikos* 96, 543–550.

Trihadiningrum, Y., De Pauw, N., Tjondronegoro, I., and Verheyen, R.F. (1996). Use of benthic macroinvertebrates for water quality assessment of the Blawi River (East Java, Indonesia). *In* 'Perspectives in Tropical Limnology' (F. Sciemer and K.T. Boland, Eds), pp. 199–221. SPB Academic Publishing bv, Amsterdam, The Netherlands.

Turcotte, P., and Harper, P.P. (1982). The macro-invertebrate fauna of a small Andean stream. *Freshw. Biol.* 12, 411–419.

Turnbull-Kemp, P.S.J. (1960). Quantitative estimations of populations of the river crab, *Potamon (Potamonautes) perlatus* (M. Edw.), in Rhodesian trout streams. *Nature* 185, 481.

Uieda, V.S., and Gajardo, I.C. (1996). Macroinvertebrados perifíticos encontrados em poções e corredeiras de um riacho. *Naturalia* 21, 31–47.

Valenti, W.C., de Mello, J.T.C., and Lobão, V.L. (1987). Growth of *Macrobrachium acanthurus* (Wiegmann, 1836) from Ribeira de Iguape River (São Paulo, Brazil) (Crustacea, Decapoda, Palaemonidae). *Rev. Bras. Biol.* 47, 349–355.

Vannote, R.L., and Sweeney, B.W. (1980). Geographical analysis of thermal equilibria: a conceptual model for evaluating the effect of natural and modified thermal regimes on aquatic insect communities. *Am. Nat.* 115, 667–695.

Vannote, R.L., Minshall, G.W., Cummins, K.W., Sedell, J.R., and Cushing, C.E. (1980). The River Continuum Concept. *Can. J. Fish. Aquat. Sci.* 37, 130–137.

Victor, R., and Ogbeibu, A.E. (1985). Macrobenthic invertebrates of a stream flowing through farmlands in southern Nigeria. *Environ. Poll. (Ser. A)* 39, 337–349.

Victor, R., and Onomivbori, O. (1996). The effects of urban perturbations on the benthioc macroinvertebrates of a southern Nigerian stream. *In* 'Perspectives in Tropical Limnology' (F. Sciemer and K.T. Boland, Eds), pp. 223–238. SPB Academic Publishing bv, Amsterdam, The Netherlands.

Walker, I. (1987). The biology of streams as part of Amazonian forest ecology. *Experientia* 43, 279–287.

Walker, I., and Ferreira, M.J.N. (1985). On the population dynamics and ecology of the shrimp species (Crustacea, Decapoda, Natantia) in the Central Amazonian river Tarumã-Mirim. *Oecologia* 66, 264–270.

Walsh, F. (1985). Onchocerciasis: river blindness. *In* "The Niger and its Neighbours" (A.T. Brove, Ed.), pp. 269–294. A.A. Balkema, Rotterdam, The Netherlands.

Walter, H., and Breckle, S.W. (1985). "Ecological Systems of the Geobiosphere. 1. Ecological Principles in Global Perspective". Springer-Verlag, Berlin, Germany.

Walter, H., and Medina, E. (1971). Caracterización climática de Venezuela sobre la base de climadiagramas de estaciones particulares. *Bol. Soc. Venez. Cienc. Nat.* 119/120, 211–240.

Ward, G.M., and Cummins, K.W. (1979). Effects of food quality on growth of a stream detritivore, *Paratendipes albimanus* (Meigen) (Diptera: Chironomidae). *Ecology* 60, 57–64.

Ward, J.V., and Stanford, J.A. (1982). Thermal responses in the evolutionary ecology of aquatic insects. *Annu. Rev. Entomol.* 27, 97–117.

Ward, J.V., and Stanford, J.A. (1983). The intermediate-disturbance hypothesis: an explanation for biotic diversity patterns in lotic ecosystems.*In* "Dynamics of Lotic Ecosystems" (T.D. Fontaine and S.M. Bartell, Eds), pp. 347–356. Ann Arbor Science Publishers, Ann Arbor, U.S.A.

Weigel, B.M., Henne, L.J., and Martínez-Rivera, L.M. (2002). Macroinvertebrate-based index of biotic integrity for protection of streams in west-central Mexico. *J. N. Am. Benthol. Soc.* 21, 686–700.

Williams, D.D., and Feltmate, B.W. (1992). "Aquatic Insects". C.A.B. International, Wallingford, U.K.

Williams, T.R., and Hynes, H.B.N. (1971). A survey of the fauna of streams on Mount Elgon, East Africa, with special reference to the Simuliidae (Diptera). *Freshw. Biol.* 1, 227–248.

Williams, T.R., Hynes, H.B.N., and Kershaw, W.E. (1964). Freshwater crabs and *Simulium neavei* in East Africa II. – Further observations made during a second visit to East Africa in February–April 1962 (the dry season). *Ann. Trop. Med. Parasitol.* 58, 159–167.

Wilson, K.D.P. (2003). 'Field Guide to the Dragonflies of Hong Kong'. Agriculture Fisheries and Conservation Department, Government of the Hong Kong SAR. Hong Kong.

Wishart, M.J., Davies, B.R., Boon, P.J., and Pringle, C.M. (2000). Global disparities in river conservation: "First World" values and "Third World" realities. *In*: "Global Perspectives on River Conservation: Science, Policy and Practice" (P.J. Boon, B.R. Davies and G.E. Petts, Eds), pp. 353–369. John Wiley and Sons, Ltd, Chichester, U.K.

Wolf, M.E., Matthias, U., and Roldán, G. (1988). Estudio del desarrollo de los insectos acuáticos, su emergencia y ecología en tres ecosistemas diferentes en el Departamento de Antioquia. *Actual. Biol.* **17**, 2–27.

Wowor, D. (1985). Population structure and spawning period of Udang Regang (*Macrobrachium sintangense*). *Berita Biologi* **3**, 116–120 (In Indonesian).

Wright, J.F. (1995). Development and use of a system for predicting the macroinvertebrate fauna in flowing waters. *Australian Journal of Ecology* **20**, 181–197.

Wright, M.S., and Covich, A.P. (2005). The effect of macroinvertebrate exclusion on leaf breakdown rates in a tropical headwater stream. *Biotropica* **37**, 403–408.

Yam, S.W.R., and Dudgeon, D. (2005a). Stable isotope investigation of food use by *Caridina* spp. (Decapoda: Atyidae) in Hong Kong streams. *J. N. Am. Benthol. Soc.* **24**, 68–81.

Yam, S.W.R., and Dudgeon, D. (2005b). Inter- and intraspecific differences in the life history and growth of *Caridina* spp. (Decapoda: Atyidae) in Hong Kong streams. *Freshw. Biol.* **50**, 2114–2128.

Yam, S.W.R., and Dudgeon, D. (2005c). Genetic differentiation of *Caridina* spp. (Decapoda: Atyidae) in Hong Kong streams. *J. N. Am. Benthol. Soc.* **24**, 845–857.

Yam, S.W.R., and Dudgeon, D. (2006). Production dynamics and growth of atyid shrimps (Decapoda: *Caridina* spp.) in 4 Hong Kong streams: the effects of site, season, and species. *J. N. Am. Benthol. Soc.* **25**, 406–416.

Yule, C.M. (1995). Benthic invertebrate fauna of an aseasonal tropical mountain stream on Bougainville Island, Papua New Guinea. *Mar. Freshw. Res.* **46**, 507–518.

Yule, C.M. (1996). The ecology of an aseasonal tropical river on Bougainville Island, Papua New Guinea. *In* "Perspectives in Tropical Limnology" (F. Schiemer and K.T. Boland, Eds), pp. 239–254. SPB Academic Publishing, Amsterdam, The Netherlands.

Yule, C., and Pearson, R.G. (1996). Aseasonality of benthic invertebrates in a tropical stream on Bougainville Island, Papua New Guinea. *Arch. Hydrobiol.* **137**, 95–117.

Yule, C.M., and Yong, H.S. (Eds). (2004). "Freshwater Invertebrates of the Malaysian Region". National Academy of Sciences, Kuala Lumpur, Malaysia.

Zamora, G.H. (1999). Adaptación del índice BMWP para la evaluación biológica de la calidad de las aguas epicontinentales en Colombia. *Rev. Unicauca Cienc.* **4**, 47–60.

Zwick, P. (1976). *Neoperla* (Plecoptera, Perlidae) emerging from a mountain stream in Central Africa. *Int. Rev. Hydrobiol.* **61**, 683–697.

Zwick, P. (1986). The Bornean species of the stonefly genus *Neoperla* (Plecoptera: Perlidae). *Aquat. Insects* **8**, 1–53.

# 5

# Fish Ecology in Tropical Streams

Kirk O. Winemiller, Angelo A. Agostinho, and Érica Pellegrini Caramaschi

This chapter emphasizes the ecological responses of fishes to spatial and temporal variation in tropical stream habitats. At the global scale, the Neotropics has the highest fish fauna richness, with estimates ranging as high as 8000 species. Larger drainage basins tend to be associated with greater local and regional species richness. Within longitudinal stream gradients, the number of species increases with declining elevation. Tropical stream fishes encompass highly diverse reproductive strategies ranging from egg scattering to mouth brooding and livebearing, with reproductive seasons ranging from a few days to the entire year. Relationships between life-history strategies and population dynamics in different environmental settings are reviewed briefly.

Fishes in tropical streams exhibit diverse feeding behaviors, including specialized niches, such as fin and scale feeding, not normally observed in temperate stream fishes. Many tropical stream fishes have greater diet breadth while exploiting abundant resources during the wet season, and lower diet breadth during the dry season as a consequence of specialized feeding on a subset of resources. Niche complementarity with high overlap in habitat use is usually accompanied by low dietary overlap. Ecological specializations and strong associations between form and function in tropical stream fishes provide clear examples of evolutionary convergence.

Several studies have revealed the major influence of fishes on benthic ecosystem dynamics in tropical streams including effects on primary production and nutrient cycling. Tropical stream fishes are important food resources for humans in many countries, and significant conservation challenges include drainage-basin degradation, pollution, dams, overfishing, and introductions of exotic species.

## I. INTRODUCTION

Freshwater fishes may comprise 25% of the vertebrate species on Earth (Stiassny, 1996). Like most major groups, they display greatest taxonomic diversity in tropical latitudes. Much of this diversity is associated with between-habitat (beta) and between-region (gamma) diversity among streams, and although species diversity within a particular habitat or stream (alpha diversity) can be impressive (Winemiller, 1996a), a broad survey of temperate and tropical streams reveals that, on average, tropical streams contain no greater, or only slightly greater, species richness than comparably sized temperate streams (Matthews, 1998). In this chapter, 'streams' are defined as fluvial systems with channel widths up to approximately 30 m. By applying this simple definition, streams of many different gradients and positions within stream-order networks are included (Fig. 1). Moreover, streams associated with drainage basins of

*FIGURE 1* Representative tropical stream habitats: top-left – upland stream, western Benin; top-right – creek draining uplands adjacent to Barotse floodplain, Upper Zambezi River, Zambia; middle-left – clearwater Río Aguaro in Venezuelan llanos; middle-right – blackwater 'morichal' in northeastern Venezuela; bottom-left – rainforest stream on Osa Peninsula, Costa Rica; bottom-right – lowland stream in coastal swamp-forest, Tortuguero, Costa Rica (see colour plate section).

different sizes and streams at different elevations will be discussed. Many of the issues and ecological mechanisms that are discussed in the context of stream fishes apply also to rivers (streams with channel widths > 30 m).

In this chapter, we describe the ecological strategies and responses of fishes to spatial and temporal variation in stream habitats at a range of scales. Streams are highly variable in both of these dimensions. Spatially, streams reveal large variation in local microhabitats, longitudinal patterns of zonation along elevation gradients, and inter-regional faunal differences. Temporal variation in many important environmental and ecological factors occurs on daily, seasonal, and inter-annual scales. Thus, it is impossible to make meaningful generalizations about stream fish ecology without a taking a scale-specific approach. This chapter is organised by first examining broad scales of variation in stream fish assemblages and subsequently describing ecological patterns and processes at successively smaller scales.

For many decades, research on tropical stream fishes lagged behind investigation of temperate stream fishes (e.g. Matthews and Heins, 1987), although there has been a significant increase in studies from several tropical regions in recent years, especially from Argentina, Brazil, French Guyana, Venezuela, and parts of West Africa. At present, however knowledge of tropical streams is less extensive than that of temperate areas, and this information shortfall is exacerbated by the fact that tropical regions contain a greater diversity of habitats and a much greater diversity of species. Moreover, our limited knowledge of tropical fish ecology limits our ability to seek broad ecological generalizations (Dudgeon, 2000a). Some of the conceptual models created to predict ecological patterns for temperate stream fishes probably apply well for tropical fishes, but others may not. Here our goal is to summarize current knowledge of stream fish ecology in the tropics, and to identify pertinent gaps in order to stimulate new research.

## II. STREAM HABITATS AND FISH FAUNAS IN THE TROPICS

### A. Landmasses and Watersheds

Perhaps the most obvious patterns of variation in stream fish assemblages occur at very broad spatial scales contrasting taxonomic composition of faunas from different continents and major river drainage basins. Historical contingencies in the evolution of fish lineages have resulted in biogeographic patterns. These patterns are particularly evident among freshwater-restricted lineages, because land and saltwater barriers impose a high degree of dispersal limitation. Thus, an interesting question is: what is the extent of ecological similarity (convergence) among stream fish assemblages inhabiting habitats with similar characteristics in different regions? We will attempt to address this question as we examine the ecological features of stream fishes, but first we compare and contrast the major geographic, landscape, and taxonomic characteristics within major regions of the Earth.

### B. Continental Basins

Freshwater fishes show strong patterns of phylogeography at the continental scale. Detailed discussion of the geological and evolutionary factors associated with present day continental distributions of tropical fishes is beyond the scope of this chapter, but the major ichthyofaunal patterns in each tropical region are reviewed briefly.

### 1. Neotropics

The South and Central American rivers and streams contain the greatest number of species on Earth, with recent estimates ranging as high as 8000 and 25% of global fish species richness

(Vari and Malabarba, 1998). The main Neotropical orders are the Characiformes (tetras, piranhas, and related forms), Siluriformes (catfishes), Gymnotiformes (electric knifefishes or 'eels'), Perciformes (represented mostly by the Cichlidae), and the Cyprinodontiformes, represented mostly by Poeciliidae (livebearers) and Rivulidae (killifishes). Freshwater families containing relatively few species include the Lepidosirenidae (lungfish), Osteoglossidae (bonytongues), and Synbranchidae (swamp eels). Freshwater fishes of marine origin include the Potamotrygonidae (stingrays), Clupeidae (shads), Engraulidae (anchovies), Atheriniformes (silversides), Gobiidae (gobies), Eleotridae (sleepers), and Achiridae (soles). South America contains great river basins that yield rivers with huge discharges, including the Amazon (ranked first in the world with a discharge of $212\,500\,m^3\,s^{-1}$), Paraná/La Plata (fifth; $18\,800\,m^3\,s^{-1}$), Orinoco (ninth; $17\,000\,m^3\,s^{-1}$), Magdalena (17th; $7500\,m^3\,s^{-1}$), and São Francisco (33rd; $2800\,m^3\,s^{-1}$). Although small by South American standards, the most ichthyologically significant river basins in Central America are the Usumacinta (southern Mexico) and the San Juan (Nicaragua and Costa Rica). Small independent drainage basins throughout Central America form a dual series of drainages running along east and west versants of the Cordillera Central.

High species richness of freshwater fishes in South America is derived from a combination of historical, geological, and ecological factors. After the breakup of Gondwanaland (ca. 110 million years ago), the continent's complex geological history, including the uplift of the Andes and its effects on drainage systems (including changing the outflow of the Amazon drainage from west, to north, to east), provided many opportunities for vicariant speciation and dispersal (Lundberg *et al.*, 1998). Like other tropical regions, the continent experienced alternating periods of mesic and dry conditions over geologic time (Prance, 1982) yet, because of the continent's relatively low elevation, overall conditions were probably not as dry those that are thought to have prevailed in Africa (see below). Moreover, because of the presence of ancient Guyana and Brazilian Shields, South America probably experienced less marine intrusion, and hence fewer extinctions of freshwater species, than Southeast Asia.

The highest fish species richness in the Neotropics (and in any river on Earth) is within the Amazon Basin. Around 2000 species are known from the region, although the actual number is certainly much higher as many new Neotropical taxa are described each year (Vari and Malabarba, 1998). Most fishes of large-lowland rivers are distributed widely throughout the Amazon Basin, and some species that typically inhabit peripheral streams are also broadly distributed, such as *Hoplias malabaricus* (Erythrinidae), *Rhamdia quelen* (Heptapteridae), *Helogenes marmoratus* (Cetopsidae) and *Acestrorhynchus microlepis* (Acestrorhynchidae). Others apparently have fairly restricted distributions: e.g. *Paracheirodon axelrodi* (Characidae) in the upper Rio Negro drainage; *Hyphessobrycon* spp. (Characidae) and *Corydoras* (Callichthyidae) catfishes in various tributary sub-basins.

The Amazon Basin is permanently or intermittently connected to the Orinoco via the Casiquiare River (Rio Negro tributary) in southern Venezuela, to the Essequibo River (Guyana) via the headwaters of the Rio Branco (Rio Negro tributary), and to the Paraná River via the Rio Paraguay headwaters in the Pantanal region of southwestern Brazil. Thus, some large species, such as the catfishes *Zungaro jahu* (=*Paulicea lutkeni*) and *Hypophthalmus edentatus*, are distributed across all three major river basins. Even certain small species, such as the knifefish *Eigenmannia virescens* and the tetra *Moenkhausia dichroura* appear to occur throughout all three river basins. However, given the rapid pace by which new phylogenetic research is changing our view of taxonomy and species distributions, it is possible that re-evaluation of the taxonomic status of wide-ranging species may reveal each of them to be two or more species. Small coastal river basins have depauperate freshwater fish faunas, although invasions by marine species can result in stream assemblages with species richness comparable to streams in large continental basins.

Several major groups of fishes are absent from Neotropical river basins beyond the Amazon. For example, osteoglossiformes are only present in the Amazon and Essequibo basins, and the lungfish (*Lepidosiren paradoxa*) is confined to the Amazon and Paraguay basins. Of relevance to stream ecology is the restriction of some fishes to streams having particular physicochemical characteristics. Three fairly distinctive water types have long been recognized in South America. Whitewaters drain uplands of relatively young geological formations, such as the Andes Mountains, and have comparatively high conductivity, nutrient loads, suspended sediment loads, and tend to have neutral or alkaline pH. Blackwater streams drain lowlands, especially those draining ancient geologic formations, such as the Guyana Shield in northern South America, and have low conductivity, dissolved nutrient concentrations, suspended sediment loads, and more acid pH. Humic acids leached from submerged vegetation produce the dark tea color and pH as low as 3–4. Clearwater streams can be found throughout the continent, but usually drain well-weathered uplands and carry relatively few suspended sediments. Depending on the landscape attributes and hydrology, clearwater streams may have neutral or acid pH, and low or high conductivity and nutrient concentrations. In general, streams with higher turbidity contain more siluriform and gymnotiform fishes since they can navigate and locate food under low light conditions using well-developed olfactory and electric sensory capabilities. Diurnal characiform and cichlid fishes comprise greater fractions of fish assemblages in clearwater and blackwater aquatic systems. Research by Menni *et al.* (1996, 2005, and references therein) in Argentina emphasizes the influence of local physicochemical conditions on stream fish distributions.

The Central American ichthyofauna is dominated by lineages derived from South American ancestors and freshwater species derived from marine lineages. Dispersal from South America into Central America occurred in two phases, with the earlier colonization by so-called secondary-division freshwater groups (Cichlidae, Poeciliidae) during the Late Cretaceous or Paleocene (Myers, 1966; Bussing, 1976). The second phase occurred after the final formation of the Isthmus of Panama during the Pliocene, and included taxa of the Characiformes, Siluriformes, Poeciliidae, and Cichlidae. Dominant marine-derived taxa include Eleotridae, Gobiidae, and mullets (Mugilidae). Several marine taxa (e.g. Centropomidae, Lutjanidae, Pomadasyidae, and Syngnathidae) enter streams and may occur dozens of kilometers inland, where they reside as integral parts of local fish assemblages.

## 2. Africa

The major rivers of Africa are the Congo (ranked second in the world with a discharge of $40\,480\,m^3\,s^{-1}$), Zambezi (19th; $7100\,m^3\,s^{-1}$), Niger (21st; $6100\,m^3\,s^{-1}$), and Nile (32nd; $2800\,m^3\,s^{-1}$); with the Volta ($1260\,m^3\,s^{-1}$) and Senegal ($690\,m^3\,s^{-1}$) also ranked within the top 50 rivers by discharge. Highest species richness (690) occurs in the Congo Basin (Hugueny, 1989), followed by the Niger (211), Volta (137), and Nile (127). The African ichthyofauna is dominated by Osteoglossiformes (weakly electric fishes of the Mormyridae), Characiformes (Alestidae, Hepsetidae, Citharinidae, and Distichodontidae), Cypriniformes (barbs), Siluriformes, Cyprinodontiformes (killifishes), and Perciformes (cichlids and Nile perches). Much of the African continent lies at higher elevation than South America or Southeast Asia, and consequently, fewer marine-derived species are represented in the African freshwater ichthyofauna. Marine-derived groups include anguillid eels and pufferfish (Tetraodontidae). The African ichthyofauna has traditionally been divided into faunal regions: Sudanian, Upper Guinea, Lower Guinea, Congo, East Coast, Zambezi, and South African (Lowe-McConnell, 1987). Distributions of individual species, however, do not necessarily adhere to these regional divisions. Lévêque (1997) provides examples of 10 distributional patterns for fishes of northern Africa.

Most of Africa has strongly seasonal rainfall. The wettest regions are the western and central Congo River Basin, Lower Guinea (Cameroon, Gabon, Central African Republic), and Upper Guinea in the region of the Niger River headwaters (Guinea) and along the coast (Sierra

Leone, Liberia, Cote d'Ivoire). Depending on soil composition and topography, these mesic areas usually contain mosaics of blackwater, clearwater, and whitewater streams. Blackwater streams mostly occur in rainforest areas that are poorly drained with oligotrophic soils. These mesic regions contain the richest and most ecologically-diverse fish faunas on the continent. As in South America and Asia, blackwater streams in Africa tend to support different fish species from whitewater streams; however some species such as *Hepsetus odoe* (Hepsetidae) and *Clarias gariepinus* (Clariidae) are fairly ubiquitous. Characiformes and cichlids tend to be relatively more abundant and diverse in blackwater streams than whitewater streams.

Streams of the Sahel, eastern Africa, and southern Africa lie within regions where seasonal precipitation has given rise to the development of extensive savanna and tropical dry forests. The Upper Niger River floodplains, Nile River, and mainstem Congo River have mostly whitewater characteristics. By contrast, the Upper Zambezi River, which drains sandy soils, has clearwater characteristics but after it accumulates dissolved and suspended solids from tributaries draining clay-rich soils, resembles a whitewater stream in its middle and downstream reaches. Relative to clearwater and blackwater streams, whitewater streams tend to be dominated by cyprinids, catfishes, and mormyrids. Relatively low fish species richness of the Sahel, eastern, and southern African river basins seems to be associated with major historic changes in rainfall as well as contemporary patterns of seasonal precipitation that frequently include extended periods of drought.

## 3. Asia–Australia

The tropical regions of Asia include southeastern China, the Indian subcontinent, Indo-China, the Malay Peninsula, and the East Indies. The major rivers of tropical Asia are the Brahmaputra (ranked fourth in the world with a discharge of $19\,800\,m^3\,s^{-1}$), Ganges (fifth; $18\,700\,m^3\,s^{-1}$), Mekong (13th; $11\,000\,m^3\,s^{-1}$), and Indus (22nd; $5600\,m^3\,s^{-1}$). The Asian fish fauna is dominated by Cypriniformes (barbs and loaches), Siluriformes, anabantoids (gouramis, climbing perches, snakeheads), and mastacembeloids (spiny eels), plus many freshwater species from the estuarine and marine groups including Beloniformes (halfbeaks and needlefish), Atheriniformes, Eleotridae, Gobiidae, and Tetraodontidae. The Cichlidae is represented by just three species (*Etroplus* spp.) in India and Sri Lanka that tolerate brackish water.

The rivers of continental Southeast Asia and the islands of Borneo and Sumatra have the highest fish species richness within Asia. However, despite the complex geologic history and great topographic diversity of Southeast Asia, some elements of the fish fauna reveal surprisingly high inter-basin similarity, with several species broadly distributed from India to western Borneo. Historic sea-level changes resulted in a series of marine intrusions and recessions over low-lying terrain of southeastern Asia that have created many opportunities for allopatric speciation and dispersal. In addition, the great river basins arising in the Himalayan Mountains (Ganges, Mekong, etc.) have provided high habitat diversity along elevation gradients and refuges for species throughout geologic and climatic changes.

Streams in the Australian-New Guinea region drain the land east of Wallace's Line, which divides Borneo and Sumatra in the west from Celebes (Sulawesi) and the Philippines in the east. The Fly is the largest river in this region. Stream fish assemblages contain fewer representatives of freshwater families and more marine-derived and marine taxa such as Mugilidae, Terapontidae, and Toxotidae. Rainbow fishes (Melanotaeniidae) are also an important and diverse group in this region. Thirty-five fish species occur in both New Guinea and the Cape York Peninsula of northeastern Australia, a reflection of land bridges that existed as recently as 6000–8000 years ago (Allen *et al.*, 2002). Northern Australia has over 100 freshwater fish species, with highest richness in the northeast or, and including many marine-derived or marine-dependent species (Allen *et al.*, 2002).

Habitats and physicochemical characteristics of streams vary considerably within the Asian-Australian tropical region depending on topography, geology, and soils. Blackwaters are present in lowland regions of the Malay Peninsula, Borneo, and New Guinea, while upland streams in the major river basins of India and Southeast Asia tend to have whitewater and clearwater characteristics. In particular, the many rivers that arise in the Himalayas and Tibetan Plateau carry heavy sediment and nutrient loads, and streams draining alluvial floodplains tend to be highly productive with high biomass of macrophytes, invertebrates, and fishes. Northern Australian streams tend to be of the clearwater type, but some blackwater streams can be found in the northeast where rainfall is higher and landscapes more forested.

## 4. Coastal Drainages and Islands

Streams that drain directly into the sea have fish assemblages that are minor subsets of regional faunas and, with some exceptions, tend to have low endemism. Due to dispersal limitations, these coastal streams can be particularly depauperate. The species that are present are often those with some degree of salinity tolerance, such as the classification of certain groups a secondary freshwater groups proposed by Myers (1966) for Central America and the Caribbean. Salinity tolerance experiments with characid (low tolerance) and poeciliid (higher tolerance) fishes have supported this classification (Winemiller and Morales, 1989). Small coastal drainages in West Africa have fish assemblages that are subsets of the Niger-Chad-Sudan faunal province, although, some drainages contain endemic taxa (Hugueny and Lévêque, 1994). Fish species richness is a function of catchment surface area and discharge African rivers of West, with some areas containing more species and others containing fewer species than predicted by chance (Hugueny, 1989; see also Fig. 2). Patterns of species extinction and colonization associated with climatic variation (extended periods of drought conditions) during the Quaternary Period could account for the latter observation. A strong linear relationship between species richness and catchment area has also been obtained for coastal rivers of the Guyana region of northern South America (Mérigoux *et al.*, 1998). Coastal streams along the Brazilian Atlantic coast can be classified into three groups according to their endemic fish faunas (Buckup, 1999). Fish diversification in these drainages depends on the history of headwater drainage captures and especially vicariance and dispersion caused by sea level changes during the last 300 000 years (Weitzman *et al.*, 1988).

Marine fishes can dominate on islands and peninsular areas that have few species from freshwater-affiliated groups. For example, small streams on the Osa Peninsula of Costa Rica that drain directly into the Pacific Ocean contain only four freshwater species, but support large populations of eleotrids (three species), gobies (two species), mugilids, syngnathids, pomadasyids,

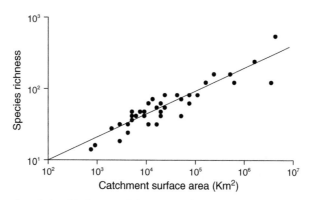

*FIGURE 2* Plot showing the relationship between fish species richness and river drainage area in West Africa (from Hugueny, 1989).

lutjanids, gerreids, and tetraodontids (one species each: Winemiller, 1983). Fish assemblages on the Caribbean island of Puerto Rico are comprised entirely of marine and marine-dependent fish species (Covich and McDowell, 1996). *Sicydium plumieri* (Gobiidae) penetrates far inland (to 600 m asl) by climbing over waterfalls using its mouth and suctorial disk (fused pelvic fins). *Sicydium* species from islands (e.g. Hawaii) and coastal regions (e.g. Osa Peninsula) in other regions of the world show similar behavior and distribution.

## C. Latitudinal Gradients

A strong correlate of latitudinal gradients in both the Northern and Southern hemispheres is annual variation in photoperiod and solar insolation yielding temperate–subtropical–tropical transitions in environmental conditions. How do patterns of fish species richness in small streams change with latitude in each region? Do any important taxonomic or functional groups of fishes drop out at subtropical or temperate latitudes within major biotic regions?

In the Americas, the northern transition zone occurs in Mexico, with mild temperate conditions in northern Tamaulipas yielding to subtropical conditions in southern Tamaulipas, and a tropical climate prevailing southward from Veracruz. All of Central America, the Caribbean, and South America down to the tropic of Capricorn is tropical, and within these regions, elevation gradients (see below) dominate climatic conditions more than latitude, particularly along the Andean Cordillera.

In northern Central America, there is a replacement of South American families (Characidae, Poeciliidae, Cichlidae) with North American families (Cyprinidae, Catostomidae, and Ictaluridae) (Bussing, 1976), and fish species seem to display less ecomorphological specialization northward across the temperate-tropical transition zone. For example, the species-rich stream-fish assemblages of the Usumacinta Basin of southern Mexico contain cichlids and poeciliids that exhibit a wider range of body sizes and diverse feeding strategies than assemblages from streams in San Luis Potosí and Tamaulipas. These more northern streams typically support one or two native cichlid species (*Herichthys* spp.), all of which are trophic generalists, and a few poeciliids that feed either on periphyton (*Xiphophorus* spp.) or small arthropods (*Gambusia* spp.).

Disregarding the high elevations of the Andes (but see Chapter 8 of this volume), the southern zone of climatic transition in the Western Hemisphere occurs from southern and middle Argentina (temperate) to northern Argentina, Paraguay and the Brazilian states of Paraná and São Paulo (subtropical); areas to the north of this are tropical. There is a decline in the number of families and genera in the ichthyofauna between La Plata-Paraná-Paraguay Basin and the smaller river basins in northern Argentina that is probably associated with historical biogeography more than contemporary climatic factors associated with the latitudinal gradient, although the latter undoubtedly plays a role in the distributions of individual species. Many groups display less ecological specialization at higher latitudes to the south. For example, the geophagine cichlids show large interspecific variation in body size and snout length, and feeding behavior ranging from specialized substrate sifters to epibenthic gleaners in the La Plata-Paraná-Paraguay Basin, whereas the few geophagine species in southern drainages tend to be generalists: i.e. medium-sized, epibenthic omnivores.

In Africa, the northern climatic transition zone occurs mostly in the Saharan region where there are few perennial streams. Subtropical conditions are encountered in the vicinity of the lower Nile in Egypt, where the only streams present are those associated the floodplain drainage. The entire continent south of the Sahara to the Okavango and Zambezi rivers in Botswana and Zambia/Mozambique (respectively) is tropical, the only exceptions being the mountain peaks of east Africa. Despite their situation within the Tropic of Capricorn, Botswana, Zambia and southern Mozambique experience conditions that are more typical of subtropical climates. The region is dominated by flat terrain and floodplains, but the elevation is sufficiently high to provide temperatures as low as 8°C during June and July. Further southward through South Africa, the climate becomes increasingly temperate.

Many higher taxa that are species-rich in the Congo Basin have few representatives in the drainages of Angola, the Okavango-Zambezi Basin, and the region to the south. For example, the siluriform families Schilbeidae, Bagridae, and Amphiliidae have relatively fewer species in the Zambezi Basin, where cyprinids (especially species of *Barbus* and *Labeo*) and cichlids (particularly *Serranochromis* spp.) assume relatively greater functional importance in stream communities. A few genera are endemic to temperate regions of southern Africa, including *Pseudobarbus* (Cyprinidae) and *Sandelia* (Anabantidae) (Skelton, 1993), but overall species richness decreases as latitude increases. As in the southern Neotropics, there is a decline in morphological and trophic diversity of stream fishes from tropical through to temperate parts of southern Africa. The latitudinal trend in species attributes is less apparent in West Africa, where causal mechanisms may have more to do with biogeography than climatic gradients.

In eastern Asia, the southward transition from temperate to subtropical and to tropical occurs in southern China. In the continental interior, elevation changes in the Himalayas drive climatic transitions more than latitude, and fish assemblages change accordingly (see below). Lowland regions throughout Indochina and the East Indies have tropical climates and high fish species richness. As in southern Africa, the latitudinal transition from tropical to temperate is associated with a decline in the number of higher taxa plus the addition a smaller number of new ones, but this may reflect habitat changes associated with elevation rather than climatic changes associated with latitude. For example, ambassids, ariids, and other marine-derived species are limited to southern, lowland streams, whereas certain hillstream taxa (*Coraglanis* and *Glyptosternon*: Sisoridae) are confined to northern high-elevation regions. Within this group of upland stream fishes, there is also some latitudinal differentiation, in *Glyptothorax* (Sisoridae), for example, there is a group of seven species confined to southern India and another 25 species limited to northern India (Jayaram, 1999).

The southern tropical–temperate transition occurs over northern Australia, with tropical climates largely restricted to the northernmost coastal regions, especially eastern portions of the continent. Ichthyofaunal change along the southern latitudinal gradient appears to be weak, with greater faunal variation associated with the longitudinal axis that is more strongly associated with precipitation and geology (Allen *et al.*, 2002).

## D. Elevation Gradients

Topography and climate at different elevations influence stream geomorphology, and discharge dynamics. High elevation streams in the tropics have fast currents, and are characterized by plunge pools, riffles, and rapids dominated by boulders and cobbles, with few or no aquatic macrophytes (see also Chapter 8 of this volume). Strong currents select for species that have adaptations for maintaining position in rapid current and foraging on substrates (Fig. 3). Benthic fishes associated with swift currents often have dorsoventrally compressed bodies, reduced swim bladders, and large, broad pectoral and pelvic fins; these provide a surface for pressure from the water current to press against the substrate. Examples include genera from a number of families including *Amphilius* (Amphilidae), *Euchiloglanis* (Sisoridae), *Balitora*, *Homaloptera* (Balitoridae), *Psilorhynchoides* (Psilorynchidae) and *Parodon* (Parodontidae). In some cases, the body and fins form a suction cup that enhances adhesion to the substrate, as in *Sicydium* (Gobiidae), *Gobiesox* (Gobisocdiae), *Gastromyzon* (and other Balitoridae), and certain *Euchiloglanis* (Sisoridae). *Pseudecheneis sulcatus* (Sisoridae), which lives in high elevation streams from Nepal to northern Burma, has an oval patch of 13–14 transverse muscular folds on the ventral surface of its thoracic region that are used for adherence to surfaces. The same mechanism is observed in the dorsal suction pad (modified dorsal spines) of marine remoras

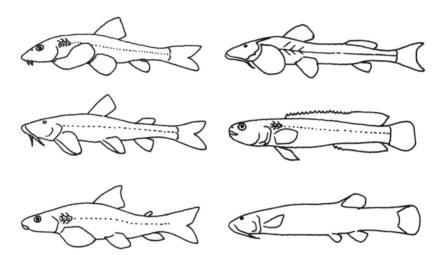

FIGURE 3   Illustration of fishes that inhabit high-gradient streams highlighting the relationship between body shape and position normally occupied in water column: top-left – Asian balitorid loach; top-right – Asian sisorid catfish; middle-left – African amphiliid catfish; middle-right – African cichlid; bottom-left – South American parodontid characin; bottom-right – South American trichomycterid catfish.

(Echeneidae). Other fishes use their fleshy mouths and oral suction to adhere to solid surfaces, notably the Loricariidae, Astroblepidae, as well as some Mochokidae (*Chiloglanis*) and the sisorid *Oreoglanis siamensis*. Another ecomorphological syndrome observed in fishes inhabiting high-elevation streams is body elongation and fossorial behavior. This syndrome is observed in South America (trichomycterid catfishes), Africa (clariid catfishes) and Asia (cobitid loaches).

Allochthonous resources, including drifting terrestrial invertebrates and detritus, are important in high-elevation streams. Periphyton and associated invertebrates provide the other major food resource for mountain and hillstream fishes. Spates and periodic scouring floods displace aquatic organisms, increase mortality (especially on juveniles), move sediments and restructure habitats, decrease the availability of benthic food resources and influence on the intensity of biotic interactions involving fishes (Dudgeon, 1993). As a result, fish populations are subject to large fluctuations and ecological interactions may be only weakly density dependent. However, because virtually no research has been conducted on the ecology of fishes in tropical high-elevation streams, this proposition is largely inferred from work conducted on temperate stream fishes (e.g. Grossman *et al.*, 1998).

At middle elevations (piedmont and other upland regions), streams have well-defined pool-riffle geomorphology, high substrate diversity (bedrock, cobble, gravel, sand, and terrestrial litter), and seasonal discharge variation, including periodic spates and flash floods during the wet season. Fish species tend to segregate according to pool-glide-riffle mesohabitats. For example, *Ancistrus triradiatus* (Loricariidae), *Lebiasina erythrinoides* (Lebiasinidae) and *Synbranchus marmoratus* (Synbranchidae) were essentially restricted to shallow riffle and glide habitats in Caño Volcán, a Venezuelan piedmont stream (K. O. Winemiller, unpublished observations). Even within pools, species used different areas of the habitat: *Aequidens pulcher, Crenicichla geayi* (Cichlidae) *Hypostomus argus* (Loricariidae), *Rhamdia* sp., *Hoplias malabaricus,* and *Prochilodus mariae* (Prochilodontidae) occurred on or close to the bottom in mid-pool; the characids *Astyanax* spp., *Bryconamericus beta, Brycon whitei, Roeboides dayi* were in mid-water at mid-pool; *Corynopoma riisei* occurred close to the surface in mid-pool; and *Poecilia reticulata* (Poeciliidae), *Bryconamericus deuteronoides,* and *Creagrutus melasma* (Characidae) were found in shallow water near pool margins. Juveniles of many larger species also were confined largely to shallow riffle, glide, and pool edge habitats with high structural complexity.

Similar patterns of habitat segregation, which may be associated with morphological specialization, have been reported from upland and lowland streams in other parts of the Neotropics (Angermeier and Karr, 1983; Arratia, 1983; Winemiller, 1983; Sabino and Castro, 1990; Flecker, 1997; Aranha *et al.*, 1998; Sabino and Zuanon, 1998; Mazzoni and Lobón-Cerviá, 2000; Lemes and Garutti, 2002) and other tropical regions (Inger and Chin, 1962; Moyle and Senanayake, 1984; Schut *et al.*, 1984; Bhat, 2005).

At low elevations, streams have slow currents, pools, runs, sloughs, organic-rich sediments, abundant and diverse aquatic macrophytes, allochthonous and autochthonous resources, and seasonal patterns of discharge. In lowland streams of large continental basins, water level varies predictably in response to seasonal precipitation. During the wet season, high discharge often maintains stream channels at bank-full levels for several months. Depending on local topography, lateral flooding of riparian zones may be extensive. Many fishes migrate upstream from larger, deeper channels to spawn in these shallow flooded areas that serve as productive nursery habitats (Welcomme, 1969; Winemiller, 1996a). Thus, reproductive activity is synchronized with the flood season with one spawning peak per year. In some parts of Asian that experience two monsoonal floods each year, stream fishes generally spawn twice (Dudgeon, 2000a).

Rapid remineralization of nutrients associated with organic matter accumulated in sediments during the dry season can lead to releases of inorganic nutrients into the water column during the wet season. In savanna regions with few trees to intercept solar radiation, this increased nutrient availability promotes production rates by aquatic macrophytes and periphyton, resulting in high rates of production of aquatic invertebrates and juvenile fishes. Some lowland streams that drain the floodplains of large rivers provide important deepwater habitats during the dry season, as well as conduits for fish lateral migrations in response to seasonal changes in water level (see also Chapter 7 of this volume). Fishes inhabiting broad lowland marshes that are often associated with floodplains partition their use of habitat based on water depth and vegetation structure (Winemiller, 1987, 1990). Small fishes vulnerable to predation occupy the shallowest marginal areas and dense stands of aquatic vegetation, including root masses of Water hyacinth (*Eichhornia crassipes*: Pontederiaceae) and other floating plants. Larger fishes, including piscivores and those feeding lower in the food web but protected from predation by virtue of size or possession of spines or armor, occupy deeper open-water areas. In the Neotropics, Nocturnal knifefishes (Gymnotiformes) and catfishes from various families hide in vegetation or woody debris during daytime, but swim in open waters at night.

Lowland streams in coastal regions are strongly influenced by proximity to the sea, and the hydrology and physicochemistry of those near sea level may be affected by tidal fluxes. Coastal stream faunas usually contain marine elements, some of which (e.g. eleotrids, mullets and certain gobies) have life cycles that require a period of residence in estuarine or marine habitats (Winemiller, 1983; Winemiller and Morales, 1989; Winemiller and Leslie, 1992). Streams on coastal alluvial plains can show faunal transitions over distances of only a few kilometers (Winemiller and Leslie, 1992): in French Guiana, for example, fish assemblages at locations 3–10 km from the coast contained only two marine-derived species, *Megalops atlanticus* (Megalopidae) and *Eleotris amblyopsis* (Eleotridae) (Mérigoux *et al.*, 1998). These transitions are driven by biotic factors (giving rise to functional species replacements) in addition to differential responses to physicochemical factors. In eastern Costa Rica, for example, the large omnivorous/detritivorous cichlid, *Vieja maculicauda*, is common in coastal lagoons and streams, but is replaced in upland streams by *Tomocichla tuba*, another cichlid of similar size and feeding habits. In the coastal lowlands, these species never co-occur although they may be present in habitats less than a kilometer apart. Similar patterns of spatial segregation are observed among other functional 'equivalents' in the coastal zone (e.g. the poeciliids *Brachyrhaphis parismina* on the coast and *B. holdridgei* inland).

Several studies have examined variation in fish assemblages over broad elevation gradients in the tropics. Bistoni and Hued (2002) found that fish species richness in Argentinean streams was correlated with altitude, distance from source, and stream order. Other studies of lotic fish assemblages in Chile (Campos, 1982), Ecuador (Ibarra and Stewart, 1989), Colombia (Jiménez *et al.*, 1998), and southern Brazil (Abes and Agostinho, 2001) identified significant patterns of species distribution and assemblage structure in relation to physicochemical conditions and elevation gradients. One of the best illustrations of the strong effect of a large-scale elevation gradient on fish assemblages comes from a study of the Gandaki River, Nepal (Edds, 1993). Assemblage structure at 81 sites distributed between 50 and 3100 m asl was strongly correlated with stream bed gradient and other physical and environmental features associated with elevation.

## E. Stream Size and Habitat Gradients

Stream size generally co-varies with order within drainage networks. First- and second-order streams in drainage system headwaters are narrow, but pools vary and depth depends on topographic gradients and substrates. As stream order increases, stream mesohabitats become larger and more diverse with well-defined pool-riffle structure, especially in areas with high topographic relief (Schlosser, 1987). As small tributaries coalesce into higher order streams, mesohabitats become larger but often less differentiated, particularly in low-gradient landscapes with deep alluvial substrates. Stream depth and width do not necessarily co-vary, and broad lowland streams can be quite shallow when compared to high gradient streams of the same drainage system.

The number of species generally increases in a downstream direction. Several factors contribute to this trend: larger mesohabitats; shorter distance to downstream source populations; reduced barriers to migration from downstream locations; and more stable habitats (fewer local extinction events). In addition to accumulation of additional species, species turnover also occurs along streams especially over long gradients involving major elevation changes (Edds, 1993). Among lowland streams, patterns of species turnover may be stronger within coastal streams due to the influence of marine-dependent species, such as eleotrids and some gobies, as well as invasions by marine lutjanids and carangids (Winemiller and Leslie, 1992). A similar trend can be observed in small streams that drain directly to large rivers, with more species and genera present relative to similar-sized streams affluent to lower-order streams (Penczak *et al.*, 1994; Agostinho and Penczak, 1995; Pavanelli and Caramaschi, 1997, 2003).

In the Bandama Basin of West Africa, fish assemblages show a pattern of longitudinal zonation into three sections: headwaters subject to seasonal desiccation with relatively few species of small fishes (characids, rivulids, and *Barbus* spp.); a long middle reach with high species diversity; and a short lower reach where brackish water sometimes penetrates several kilometers inland and marine and marine-dependent fishes are present (de Mérona, 1981). The middle reach contains high habitat diversity with a succession of deep pools and shallow swift riffles. Genera characteristic of riffles include *Nannocharax* (Citharinidae), *Amphilius* (Amphiliidae), and *Aethiomastacembelus* (Mastacembelidae). Species common in slower pool areas of the middle reach include the alestiids *Alestes baremoze*, *Brycinus macrolepidotus*, *B. nurse* and *Hydrocynus forskalii*, and the catfish *Schilbe mandibularis* (Schilbeidae).

In upland streams of Sri Lanka, fish species richness increases downstream from waterfalls in the headwaters (Moyle and Senanayake, 1984). Stream discharge increases, slope decreases, and new fish species are added at locations successively lower in the longitudinal gradient, so that richness increases from one or two species at upstream sites to as many as 14 species further downstream. Upstream sites support benthic fishes adapted to life in swift currents, such as the benthopelagic cyprinids *Garra lamta* and *Noemacheilus notostigma* (Balitoridae) and small pelagic species with rapid life cycles (e.g. *Rasbora daniconius*: Cyprinidae). Carnivorous *Ophiocephalus gaucha* (Channidae) and herbivorous *Awaous* (=*Gobius*) *grammepomus* (Gobiidae) were confined to the most downstream sites.

The Luongo River in northern Zambia has an unusual pattern of longitudinal zonation, with swift high-gradient sections in both the upper and lower reaches separated by a long middle section with low gradients and sluggish flow (Balon and Stewart, 1983). The three sections of the river are separated from the middle reach by waterfalls that appear to serve as effective barriers to dispersal and, as a result, the two high-gradient reaches contain mostly fishes associated with turbulent waters but with almost no species in common. The Luongo also has a fairly high number of endemic fish species. Rather different habitat associations occur in the Limpopo River Basin of Mozambique and the Transvaal region of South Africa.

Gaigher (1973) identified five groups: widespread species with non-specific habitat affinities (e.g. *Barbus paludinosus, Clarias gariepinus*), species that are widespread but absent from coldwater, high-elevation streams (e.g. *Micralestes acutidens*: Characidae; *Labeo cylindricus*: Cyprinidae), pool-dwelling species from low-elevation warm-water streams (e.g. the mormyrid *Petrocephalus catostoma* and the mochokid catfish *Synodontis zambezensis*), species confined to middle- and high-elevation streams with perennial flow (e.g. *Barbus lineomaculatus* and the cichlid *Chetia flaviventris*), and species with restricted distributions and/or habitat requirements, such as lungfish and specialized catfish (e.g. *Protopterus annectens* and *Amphilius platychir*), that are most at risk of local extirpation by human impacts.

Research from the temperate zone indicates that small headwater streams have more stochastic environments (in terms of discharge, depth, and water quality) and tend to have more randomly-assembled local communities made up of species with good colonizing ability (Schlosser, 1987; Grossman *et al.*, 1998). In tropical systems, headwater streams, especially those at low elevations, often show lower assemblage variability than larger downstream reaches as seen, for instance in small headwater streams of central Amazonian rainforests (Schwassman, 1978). The composition of these assemblages is fairly constant and fluctuations in population densities are low (Bührnheim and Fernandes, 2001, 2003). Likewise, monthly variations in the fish assemblage of Caño Volcán, a small headwater stream in the Andean piedmont, was much lower than that seen in Caño Maraca, a lowland creek in the Venezuelan llanos (Winemiller, 1987). In contrast to these Neotropical systems, fish assemblages in headwaters of the upper Ogun River in Nigeria exhibited large seasonal changes (Adebisi, 1988). These streams are reduced to series of isolated pools during the dry season, when samples were dominated by omnivorous species, such as *Brycinus* spp. and *Tilapia zillii* (Cichlidae), and insectivores (e.g. smaller mormyrid species and the bagrid catfish *Chrysichthys auratus*). During the early stages of the wet season, large piscivorous fishes, including some mormyrids (*Mormyrops deliciosus*), tigerfish (*Hydrocynus forskalii*) and bagrids (*Bagrus docmak*), begin to appear in gillnet catches.

The patch-dynamics model of species richness in streams predicts that highest species richness occurs in habitats with intermediate spatio-temporal scales of disturbance. This prediction was tested by Mérigoux *et al.* (1999) who examined the relationship between variation in habitat characteristics and species richness of juvenile fishes at 20 sites in tributaries of the Sinnamary River, French Guiana. Habitat variability accounted for only 36% of variation in species richness in multiple-regression models, but this may reflect the influence of rare species on total community variation, and analyses restricted to common species might show stronger support for the patch dynamics model. Furthermore, the Sinnamary River is bisected by a major impoundment that is likely to influence population changes of highly-mobile species. Seasonal and inter-annual environmental variation probably influences species with different life history strategies in dissimilar but predictable ways (as discussed below) so that, even if there are winners and losers in a given habitat patch during any particular period, the fish assemblage would appear relatively stable when viewed over longer temporal scales.

Much of the environmental variation in low-gradient tropical streams is relatively predictable in accordance with seasonal precipitation. Rising water levels create new aquatic habitats, increase ecosystem productivity and total production, and stimulate fish reproduction (Lowe-McConnell, 1964, 1987; Winemiller, 1996a). During the dry season, aquatic habitats contract, production decreases, and water quality often declines (with low dissolved oxygen and high hydrogen sulfide), so that fish mortality increases. Fish samples from isolated dry-season pools tend to contain many species with special respiratory adaptations (Lowe-McConnell, 1964; Winemiller, 1996a), including fishes capable of efficiently skimming the oxygen-rich surface layer (Kramer *et al.*, 1978, Winemiller 1989a). This annual wet-dry cycle is probably associated with essentially density-independent population dynamics during the early part of the wet season, followed by increasing importance of density-dependent regulation of populations (e.g. by food limitation and predation) as water levels drop during the dry season. It is not uncommon for lowland streams in the tropics, particularly those in savanna regions, to be reduced to a series of isolated pools during the dry season. Chapman and Kramer (1991) concluded that populations of *Poecilia gilli* in isolated pools in small, high-gradient streams in Costa Rica are regulated more by density-independent factors linked to extreme hydrologic changes (spates or drying), compared to fishes in low-gradient streams that experience more gradual and predictable hydrological changes (Chapman *et al.*, 1991).

Fish species richness was correlated with water depth in a survey of habitats in the Niandan River in eastern Guinea (Hugueny, 1990) and in three streams in western Guinea (Pouilly, 1993), and this trend seems to be typical of most tropical drainages. Table I summarizes information on stream width, dominant fish species, and taxonomic richness at the species and genus level for stream surveys from tropical regions worldwide. Species richness ranges from 4 to 83, with lowland streams tending to support more species than upland and mid-elevation streams. Species richness is positively correlated with stream width ($r = 0.46$, df $= 43$, $p = 0.0015$) as is generic richness, although the latter relationship is weaker ($r = 0.31$, df $= 43$, $p = 0.047$). The average number of species in lowland streams of around 15 m width was similar in the Neotropics (23) and Asia (25) but slightly lower in Africa (20.5), and may be due to the absence of data for streams from the Congo – the most species-rich basin in Africa for fishes. The ratio of number of species to number of genera was lower for Neotropical streams (1 : 20) than those in Asia (1 : 37) and Africa (1 : 50) and, although more surveys were included for the Neotropics, this result is unlikely to be an artifact. The number of coexisting congeneric species in African and Asian streams (e.g. cyprinids such as *Barbus* in Africa versus *Puntius* and *Rasbora* in Asia) does indeed appear to be generally higher than in the Neotropics, although the effects of locality and habitat are also influential.

Angermeier and Karr (1983) studied fish assemblage-habitat relationships in a series of low-gradient streams entering Lake Gatún, Panama. The number of species of algivores and omnivores increased with stream width; insectivore richness increased also but declined in the widest streams. Average total biomass of algivores and fishes feeding on terrestrial plants (seeds, fruits) both increased linearly with stream width, whereas insectivore biomass showed the opposite trend (Fig. 4). Such data illustrate the general pattern that algal food is more available in relatively open and unshaded downstream reaches within densely forested landscapes. The abundance of small fishes (<40 mm standard length), by contrast, decreased with stream width but tended to be higher in pools than riffles and runs (Angermeier and Karr, 1983). Many small fishes in tropical streams use shallow areas, including pool margins, as refugia from piscivorous fishes; for example, 21 of 29 species inhabiting an Amazonian stream were located along the margins (Sabino and Zuanon, 1998). In Trinidadian streams, the killifish *Rivulus hartii* is abundant in small tributaries where piscivores are absent, but also uses shallow marginal

TABLE I   Taxonomic Richness in Tropical Stream Fish Assemblages at Different Elevations and Regions: Location, Mean Stream Width (m), Number of Species, Number of Genera, Dominant Species, and Literature Source

| Site | Width | Species | Genera | Dominant species | Source |
|------|-------|---------|--------|------------------|--------|
| High elevation (>1000 m above regional lowlands) | | | | | |
| Gandaki, Nepal | – | 4 | 3 | *Schizothorax richardsoni* | Edds, 1993 |
| Mergulhão, Brazil | 5 | 26 | 8 | – | Aranha *et al.*, 1998 |
| San Pablo, Colombia | 7 | 19 | 18 | *Brycon henni* | Cardona *et al.*, 1998 |
| Mid elevation (100–1000 m above regional lowlands) | | | | | |
| Córrego Capivara, Brazil | 20 | 26 | 20 | *Phalloceros caudimaculatus* | Uieda and Barretto, 1999 |
| R. Pardo, Brazil | 3 | 19 | 15 | *Astyanax fasciatus* | Castro and Casatti, 1987 |
| Córrego Cedro, Brazil | – | 21 | 20 | *Poecilia reticulata* | Lemes and Garutti, 2002 |
| C. Gameleira, Brazil | – | 8 | 7 | *Astyanax bimaculatus* | Alves and Vono, 1997 |
| C. Tabajara, Brazil | 2.5 | 18 | 15 | – | Uieda, 1984 |
| Morro do Diablo, Brazil | 9 | 22 | 19 | – | Casatti *et al.*, 2001 |
| C. Acaba Saco, Brazil | 4.5 | 28 | 19 | *Loricaria* sp. | Miranda and Mazzoni, 2003 |
| C. Água Boa, Brazil | 2.5 | 35 | 32 | *Ancistrus aguaboensis* | Miranda and Mazzoni, 2003 |
| C. Cavalo, Brazil | 3 | 36 | 23 | *Ancistrus minutus* | Miranda and Mazzoni, 2003 |
| Chajeradó, Colombia | <10 | 19 | 14 | – | Sánchez-Botero *et al.*, 2002 |
| Chajeradó, Colombia | <30 | 18 | 16 | – | Sánchez-Botero *et al.*, 2002 |
| Caño Volcán, Venezuela | 6 | 20 | 16 | *Bryconamericus beta* | Winemiller, 1990 |
| R. Todasana, Venezuela | 5.5 | 9 | 9 | *Sicydium plumieri* | Penczak and Lasso, 1991 |
| Tosso, Benin | 7 | 16 | 11 | *Brycinus longipinnis* | K. O. Winemiller, unpublished observations |
| Parakou, Benin | 15 | 11 | 10 | *Barbus boboi* | K. O. Winemiller, unpublished observations |
| Sina Sinarou, Benin | 12 | 11 | 7 | *Barbus callipterus* | K. O. Winemiller, unpublished observations |
| Kouande, Benin | 3 | 13 | 5 | *Barbus macrops* | K. O. Winemiller, unpublished observations |
| Perma, Benin | 15 | 12 | 7 | *Sarotherodon galilaeus* | K. O. Winemiller, unpublished observations |
| Tchan Duga, Benin | 2 | 14 | 9 | *Epiplatys* sp. | K. O. Winemiller, unpublished observations |
| Low elevation (<100 m above regional lowlands) | | | | | |
| Sábalo, Costa Rica | 5 | 21 | 19 | *Poecilia gilli* | Burcham, 1988 |
| Quebrada, Costa Rica | 2.5 | 21 | 20 | *Eleotris amblyopsis* | Winemiller and Leslie, 1992 |

(continued)

*TABLE I   (continued)*

| Site | Width | Species | Genera | Dominant species | Source |
|------|-------|---------|--------|------------------|--------|
| Agua Fría, Costa Rica | 26 | 56 | 42 | *Astyanax fasciatus* | Winemiller and Leslie, 1990 |
| Camaronal, Costa Rica | 5.5 | 16 | 16 | *Astyanax fasciatus* | Winemiller, 1983 |
| Pedro Miguel, Panama | – | 12 | 12 | *Gephyrocharax atricaudata* | Zaret and Rand, 1971 |
| Tarumã, Brazil | 2.5 | 6 | 6 | *Aequidens tetramerus* | Knöppel, 1970 |
| Barro Branco, Brazil | 1 | 17 | 15 | *Aequidens tetramerus* | Knöppel, 1970 |
| Stream-41, Brazil | <3 | 11 | 10 | *Hemigrammus* sp. | Bührnheim and Fernandes, 2003 |
| Gavião, Brazil | <3 | 19 | 18 | *Hyphessobrycon heterorhabdus* | Bührnheim and Fernandes, 2003 |
| Porto Alegre, Brazil | <3 | 12 | 12 | *Hemigrammus* sp. | Bührnheim and Fernandes, 2003 |
| R. Indaiá, Brazil | – | 8 | 8 | *Deuterodon iguape* | Sabino and Castro, 1990 |
| R. Ubatiba, Brazil | 13 | 22 | 20 | *Astyanax* cf *hastatus* | Costa, 1984 |
| R. Mato Grosso, Brazil | 3 | 17 | 16 | – | Costa, 1987 |
| Z. Campus, Brazil | 2.8 | 27 | 21 | *Astyanax bimaculatus* | Lobón-Cerviá *et al.*, 1994 |
| Z. Barbará, Brazil | 2.6 | 24 | 22 | *Astyanax bimaculatus* | Lobón-Cerviá *et al.*, 1994 |
| R. Fazenda, Brazil | 20 | 21 | 21 | *Deuterodon* cf *pedri* | Uieda and Uieda, 2001 |
| I. Guaraná. Brazil | – | 29 | 27 | – | Sabino and Zuanon, 1998 |
| R. Ubatiba, Brazil | 2.5 | 22 | 19 | *Deuterodon* cf *hastatus* | Mazzoni and Lobón-Cerviá, 2000 |
| Nareuda, Bolivia | 10 | 38 | – | *Corydoras loretoensis* | Chernoff and Willink, 1999 |
| Maraca, Venezuela | 22.5 | 83 | 72 | *Cheirodon pulcher* | Winemiller, 1990 |
| Siapa trib, Venezuela | 5 | 34 | 27 | *Bryconops giancopinii* | K. O. Winemiller, unpublished observations |
| Siapa trib, Venezuela | 4 | 36 | 28 | *Moenkhausia copei* | K. O. Winemiller, unpublished observations |
| Siapa trib, Venezuela | 5 | 26 | 23 | *Characidium* sp. | K. O. Winemiller, unpublished observations |
| Siapa trib, Venezuela | 8 | 20 | 20 | *Apistogramma uaupesi* | K. O. Winemiller, unpublished observations |
| Kambo, Guinea | – | 23 | 21 | *Chiloglanis occidentalis* | Pouilly, 1993 |
| Balisso, Guinea | – | 29 | 26 | *Epiplatys* sp. | Pouilly, 1993 |
| Kilissi, Guinea | – | 16 | 16 | *Brienomyrus brachyistius* | Pouilly, 1993 |
| Bugungu, Kenya | – | 14 | 9 | *Barbus kerstenii* | Welcomme, 1969 |
| Kataba, Zambia | 5 | 12 | 7 | *Aplocheilichthys* sp. | Winemiller, 1993 |
| Kapuas-6, Borneo | 5 | 21 | 16 | – | Roberts, 1989 |
| Kapuas-37, Borneo | 5 | 33 | 22 | – | Roberts, 1989 |
| Bulu, Borneo | 2 | 16 | 12 | *Noemacheilus spiniferus* | Watson and Balon, 1984 |
| Payau, Borneo | 2 | 23 | 18 | *Osteochilus kahajenensis* | Watson and Balon, 1984 |
| Kaha, Borneo | 4 | 32 | 22 | *Glaniopsis hanitschi* | Watson and Balon, 1984 |

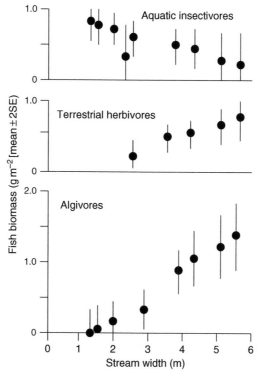

FIGURE 4 Relationship between stream width and fish biomass (density per unit area) for three feeding guilds in lowland streams of Panaman (from Angermeier and Karr, 1983).

habitats of rivers (where predaceous fishes occur) for dispersal, foraging, and reproduction (Fraser *et al.*, 1999).

At a smaller spatial scale, a few studies have demonstrated how tropical stream fishes partition different areas (microhabitats) in pool habitats. Zaret and Rand (1971) showed how fishes use different vertical and lateral regions of a stream pool in Panama (Fig. 5). Comparable patterns of spatial segregation occur in small streams of the Andean piedmont in Venezuela (K. O. Winemiller, unpublished observations), and similar findings have been reported for fishes inhabiting pools of rainforest streams in Borneo (Inger and Chin, 1962), Sri Lanka (Moyle and Senanayake, 1984; Schut *et al.*, 1984), the Amazon (Sabino and Zuanon, 1998), and southeastern Brazil (Sabino and Castro, 1990; Mazzoni and Lobón-Cerviá, 2000). All such studies note that interspecific variation in body shape and fin placement indicates the manner in which fishes use microhabitats. In a study of three upland streams in coastal Guinea, West Africa, during the dry season, Pouilly (1993) found consistent differences in assemblages in low-velocity pool habitats and high-velocity riffles and runs, with a strong association between morphology and habitat. Fishes in shallow high-velocity habitats further partitioned space based on substrate type, but those in pools had weaker associations between morphology and interspecific segregation of space. Martin-Smith (1998) recorded lower species richness but higher fish abundance in riffles of a Borneo stream where the assemblage was dominated by seven benthic balitorid species; 18 species of cyprinids occurred in pools, whereas 13 species were ubiquitous occurring in both habitats. In a Brazilian rainforest stream, algivorous loricariid catfishes with high dietary overlap showed evidence of spatial segregation by microhabitat and were active during different periods (Buck and Sazima, 1995).

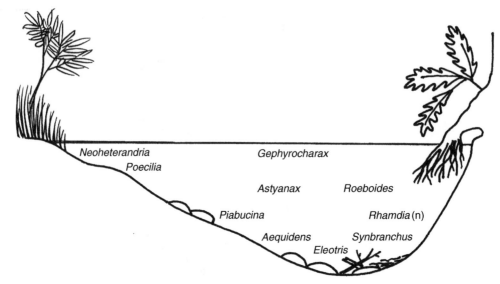

*FIGURE 5*   Relative vertical and horizontal positions occupied by fishes within a cross-section view of a dry-season pool in lowland stream in Panama where (n) designates nocturnal fishes; based on Zaret and Rand (1971).

## III. REPRODUCTIVE STRATEGIES AND POPULATION DYNAMICS

Given the taxonomic diversity of tropical stream fishes and their distribution among various habitats and regions, it is not surprising that they also display diverse strategies and modes of reproduction. Virtually any attribute associated with reproduction reveals high interspecific variation. The challenge is to understand how these different modes of reproduction allow fishes to be adapted to patterns of environmental variation in streams. To organize thinking along these lines, two frameworks for interpreting modes of reproduction as adaptive strategies will be examined. The first is the reproductive guild concept of Balon (1975) in which physiological, morphological, and behavioral features of adult and early life stages of fishes are understood in terms of the physicochemical and ecological challenges imposed by aquatic habitats. The second is Winemiller's (1989a, 1992) trilateral life-history continuum that proposes optimal demographic and reproductive strategies for different patterns of environmental variation.

There is marked interspecific variation in age and size of sexual maturity and, indeed, the smallest of all vertebrates are fishes from tropical marine and freshwater habitats. The trichomycterid *Ammoglanis pulex* (de Pinna and Winemiller, 2000) is a miniature, worm-like catfish that does not exceed 15 mm length and lives among the interstices of coarse sand substrates in clearwater rivers of the Guyana Shield region of South America. The cyprinid *Danionella translucida* from Burma is slightly smaller (<12 mm standard length: Roberts, 1986), and probably feeds on meiofauna. Tropical fresh waters also contain some of the largest bony fishes on Earth, including the Mekong catfish (*Pangasianodon gigas*: Pangasiidae) and the Amazonian Pirarucu (*Arapaima gigas*: Arapaimidae), but these massive species do not normally inhabit streams less than 30 m wide. Other notably large tropical stream fishes are certain Asian carps (Cyprinidae: *Catlocarpio siamensis*, for example), South American electric 'eels' (actually Gymnotiformes: Electrophoridae), detritivorous prochilodontids (South America), herbivorous and piscivorous characids (South America), schilbeid and clariid catfishes (Africa and Asia), pimelodid catfishes (South America), ariid catfishes (Asia-Australia), barramundi (Asia-Australia), and lungfishes (South America, Africa, and Australia). Fish growth can be quite rapid in many lowland tropical ecosystems, so some of these large tropical species (exceeding 5 kg) probably mature in their second year. Studies of age and growth in tropical stream fishes are still scarce (Agostinho *et al.*, 1991; Amaral *et al.*, 1999), mostly due to the difficulty of age determination by standard methods.

Tropical stream fishes display a wide range of values for fecundity and egg size – traits that tend to vary inversely in fishes. Some species spawn only a few eggs or give birth to one or a few well-developed offspring per clutch. Many annual killifishes spawn a few eggs at intervals of a day to a few days throughout their adult lifespan, whereas, the unusual poeciliid *Tomeurus gracilis* releases a single fertilized egg per spawning event, and South American freshwater stingrays give birth to a few fully-developed offspring each time they reproduce. In contrast, some tropical stream fishes have batch fecundities well over 100 000 (e.g. *Brycon, Labeo*). The relative investment in reproduction depends on fish size, fecundity, egg size, and spawning interval. Eggs diameters in African teleosts vary from 0.6 mm (*Micralestes acutidens*) to 3.2 mm (*Chrysichthys maurus*) (Lévêque, 1997); egg diameter ranged from 0.45 mm (*Steindachnerina argentea*: Curimatidae) to 4.0 mm (*Ancistrus triradiatus*) in fishes among telecosts from a stream in the Venezuelan llanos (Winemiller, 1989b). Spawning may occur daily in certain killifishes but only once per year in many species of high-fecundity batch spawners. In many cases, the interval between spawning events seems to be correlated with the periodicity of major environmental variation in streams: in the case of partial spawners, short intervals between reproductive bouts are associated either with rather constant habitats or with habitats that undergo stochastic variation over short time scales in the order of weeks. Most fishes in small central Amazonian rainforest streams, relatively constant habitats, are partial spawners (Schwassman, 1978). Livebearing poeciliid fishes of rainforest streams in Costa Rica breed continuously, but also display seasonal patterns of reproductive effort (Winemiller, 1993).

The ultimate environmental driver for variations in habitat quality and quantity, and hence breeding events, is probably stream hydrology (Schwassman, 1978; Lowe-McConnell, 1979). Total spawners respond to large-scale environmental variation that is largely predictable, usually on an annual scale in relation to the flood regime. Most stream fishes spawn just before, but more often just after, the onset of flooding to take advantage of expanded, productive aquatic habitats. In floodplain areas, many species migrate up tributary creeks to spawn in seasonally flooded marshes (Welcomme, 1969; Winemiller, 1989b, 1996a). In upland streams, spates and scouring floods during wet seasons favor spawning in small tributary streams or small embayments created when rain-swollen tributaries join the mainstem. Some fish species in upland streams have extended spawning periods, including Characiformes, such as *Corynopoma riisei* (Winemiller, 1989b), *Astyanax bimaculatus* (Garutti, 1989), and *Characidium* sp. (Mazzoni *et al.*, 2002); the loricariid *Hypostomus punctatus* (Menezes and Caramaschi, 1994); and the pimelodid *Pimelodella pappenheimi* (Amaral *et al.*, 1998). Some fishes, such as *Roeboides dayi* and the cichlid *pulcher* Aequidens even have dry-season spawning peaks (Winemiller, 1989b). In Panamanian lowland streams, several characiform species reproduce throughout the dry season, but others breed during the wet season (Kramer, 1978a).

One of the most fascinating aspects of tropical stream fish reproduction is the variation observed in parental care. Parental care can be organized according Balon's (1975) reproductive guilds framework (Table II). Open substrate spawners scatter eggs and milt in the water column or over substrates in a seemingly non-selective manner. Most of these species have relatively high fecundity and small eggs, with high larval mortality. Brood hiders may spawn in the water column above or in contact with specific substrates, often gravel or aquatic vegetation, that provide cover for developing eggs and larvae. Brood hiders may have low or high fecundity and breed at frequent intervals (e.g. killifishes that burrow into loose, humic-rich substrates to spawn) or seasonally (e.g. many medium-sized Neotropical and African characids). Substrate guarders usually deposit fertilized eggs onto vegetation, root wads, or in holes. These species usually do not construct a nest, but defend the area around their eggs for one to a few

*TABLE II*  Genera from Each Tropical Region Providing Examples of the Reproductive Guilds
Described by Balon (1975)

| Region | Neotropics | Africa | Asia/Indo-Pacific/Australia |
|---|---|---|---|
| Open substrate spawners | Brycon | Alestes | |
| | Creagrutus | Opsaridium | Brachydanio |
| | Prochilodus | Labeo | Chela |
| | Pimelodella | Schilbe | Pangasius |
| | Rhamdia | Ctenopoma | |
| Brood hiders | Hyphessobrycon | Arnoldichthys | Osteochilus |
| | Moenkhausia | Neolebias | Puntius |
| | Copella | Barbus | Rasbora |
| | Pterolebias | Synodontis | Melanotaenia |
| | Rachovia | Nothobranchius | Parambassis |
| Substrate guarders | Pygocentrus | Heterotis | Notopterus |
| | Serrasalmus | Polypterus | Mogurnda |
| | Gymnotus | Notopterus | Oxyeleotris |
| | Sternopygus | Citharinus | |
| Nest guarders | Aequidens | Protopterus | Betta |
| | Crenicichla | Gymnarchus | Osphronemus |
| | Ancistrus | Hepsetus | Trichogaster |
| | Hoplosternum | Clarias | Channa |
| | Synbranchus | Hemichromis | Heteropneustes |
| External bearers | Osteoglossum | Arius | Scleropages |
| | Loricariichthys | Oreochromis | Betta |
| | Loricaria | Sarotherodon | Channa |
| | Geophagus | Haplochromis | Sphaerichthys |
| | Oostesthus | Serranochromis | Luciocephalus |
| Internal bearers | Potamotrygon | Dasyatis | Himantura |
| | Belonesox | | Dermogenys |
| | Brachyrhaphis | | Hemirhamphodon |
| | Poecilia | | Nomorhamphus |
| | Xiphophorus | | |

days before leaving the area. Nest guarding takes many forms among tropical stream fishes, including creation of depressions on the substrate, cavities within dense stands of aquatic macrophytes, and bubble nests (a froth formed at the surface by 'blowing' mucus-coated bubbles). Nest guarders may remain with their broods for several weeks after hatching. They typically have relatively low batch fecundity and large eggs, but the bubblenest-building Asian belontiids, some of which lay tiny eggs and have high fecundity, are an exception to this rule. The external bearers include the mouth brooders and fishes that carry adhesive eggs and larvae on their body surface, usually the ventrum as in *Loricaria* (Loricariidae) and *Oostethus* (Syngnathidae). Mouth brooding has evolved multiple times in many highly divergent taxa (Table II). Internal bearing (ovoviparity and viviparity) has likewise evolved in a range of taxa; however, it appears to be uncommon among African stream fishes.

Balon (1975) has explained the relationships between environmental conditions, parental care, and the anatomical and physiological adaptations of eggs, larvae and reproductive adults. He viewed reproductive guilds as adaptive strategies for maximizing larval survival in spatially and temporally heterogeneous habitats. Behaviors such as nest construction and brood guarding (i.e. frequent 'mouthing' of eggs and larvae, fanning, etc.) can decrease silt accumulation and increase water circulation and delivery of dissolved oxygen to developing early life stages. Placing eggs in bubble nests at the water surface increases oxygen availability under anoxic conditions in seasonal wetlands. Egg characteristics are correlated with the modes of spawning and parental care. For instance, eggs of external bearing loricariid catfishes (*Loricaria* and *Loricariichthys* spp.) have thick zona radiata apparently to protect against abrasion, and the zona granulosa produces secretions that probably contributes to egg adhesion (Suzuki *et al.*, 2000). Among Paraná River loricariids, the cavity nesting loricariids (*Hypostomus* and *Megalancistrus* spp.) have the largest eggs with the thickest zona granulosa.

Migration to spawning habitats favorable for egg and larval development can enhance survival of early life stages. During the wet season in Panama, *Brycon petrosus* migrate to the headwaters of Panamanian rainforest streams to spawn, and spawning aggregations have been observed on partially submerged leaf litter (Kramer, 1978b). Headwater migrations also have been observed in affluents of the Upper Paraná River. Characiformes such as *Salminus maxillosus*, *Brycon orbignianus,* and *Prochilodus lineatus*, among others, migrate to the headwaters during the wet season to spawn in shallow water (<3 m), and eggs drift downstream while developing. Drifting larvae reach the nursery areas (lagoons) when the river overflows its banks (Agostinho *et al.*, 2003b). In piedmont streams of Venezuela, the small characid *Bryconamericus dueterodonoides* spawns during the dry season, and larvae drift downstream at night (Flecker *et al.*, 1991). Drifting larvae may be transported to more productive habitats in lower stream reaches, with young fishes migrating back upstream during the ensuing wet season. Other larger Characiformes inhabiting the same streams (e.g. *Brycon whitei*, *Salminus hilarii,* and *Prochilodus mariae*) have evolved a different strategy and migrate to lowland floodplains for spawning during the early wet season. Similar breeding migrations are observed among some of the large cyprinids of Himalayan piedmont streams (e.g. *Catla catla, Tor* spp.).

Fish spawning migrations in lowland creeks tend to be fairly local, with fishes swimming up or down creeks or laterally across flooded zones in search of productive marshes. These seasonal wetlands contain abundant food (microcrustaceans) and cover (aquatic macrophytes) for early life stages. At the onset of the flood period, the area of these shallow marshes increases rapidly, and per-unit-area densities of larval predators are low. A quarter of the most common species captured from a seasonal marsh associated with a small lowland creek in the Venezuelan llanos only entered the habitat for reproduction and feeding during the wet season (Winemiller, 1989b, 1996a). Most of the fish species encountered in a creek draining marginal wetlands around Lake Victoria in Uganda likewise used the habitat for reproduction and feeding only during the wet season, with downstream migrations during the dry season associated with the onset of anoxic conditions (Welcomme, 1969).

How do reproductive strategies relate to habitat variation and dynamics, and are there consistent patterns of fish reproductive strategies across the tropics? Based on multivariate analysis of 10 life-history and demographic traits of fishes inhabiting a lowland creek in the Venezuelan llanos, Winemiller (1989a) found a continuum that identified three endpoint life-history strategies. This continuum describes essential life history tradeoffs among fishes (Winemiller and Rose, 1992) and other groups of organisms (Winemiller, 1992). Of these three endpoints, the opportunistic strategy is marked by rapid maturation at small size and sustained high reproductive effort. It is associated with small size, low fecundity, frequent reproductive intervals, and extended breeding seasons as exemplified by annual killifishes, guppies (*Poecilia reticulata*), and small characids. The opportunistic strategy most efficiently maximizes the intrinsic rate of population increase, and should be a superior strategy among the three identified for increasing fitness under density-independent environmental settings, for example, when population density is reduced by habitat disturbance or predation. Reznick and Endler (1982) have demonstrated that increases in predation intensity on adult guppies in Trinidadian streams results in the evolution of earlier ages and smaller sizes of maturation.

The key attribute of the second endpoint – the equilibrium strategy – is high parental investment for individual offspring, either by egg provisioning, parental care, or usually both. Fishes

*TABLE III*   Genera from Each Tropical Region Exemplifying the Trilateral Life-history Strategies of Winemiller (1989a, 1992): Opportunistic (O), Periodic (P), and Equilibrium (E)

| Region | Neotropics | Africa | Asia/India/Australia |
|---|---|---|---|
| Opportunistic | Hyphessobrycon | Barbus (small) | Brachydanio |
| | Nannostomus | Neolebias | Danionella |
| (early maturation) | Pterolebias | Leptoglanis | Rasbora |
| (high reproductive effort) | Rachovia | Aplocheilichthys | Pseudomugil |
| | Brachyrhaphis | Nothobranchius | Dermogenys |
| | Poecilia | Ethmalosa | |
| Periodic | Acestrorhynchus | Marcusenius | Cyclocheilichthys |
| | Brycon | Alestes | Osteochilus |
| (high fecundity) | Bryconops | Distichodus | Tor |
| (seasonal spawning) | Myleus | Citharinus | Botia |
| | Leporinus | Barbus (large) | Bagrichthys |
| | Curimata | Labeo | Leiocassis |
| | Prochilodus | Schilbe | Mystus (large) |
| | Eigenmannia | Bagrus | Bagarius |
| | Pimelodella | Chrysichthys | Pangasius |
| | Rhamdia | Clarias | Kryptopterus |
| | Amblydoras | Synodontis | Osphronemus |
| Equilibrium | Ancistrus | Polypterus | Heteropneustes |
| | Loricaria | Protopterus | Channa |
| (parental care) | Loricariichthys | Gymnarchus | Mogurnda |
| | Hypostomus | Hepsetus | Oxyeleotris |
| | Hoplosternum | Pelvicachromis | Scleropages |
| | Aequidens | Hemichromis | Ctenops |
| | Biotodoma | Oreochromis | Arius |
| | Crenicichla | Sarotherodon | |
| | Geophagus | Tilapia (large) | |
| | Satanoperca | Serranochromis | |
| Intermediate – E/P | Pygocentrus | Notopterus | Chaca |
| | Serrasalmus | Pollimyrus | Mystus (small) |
| | Gymnotus | Clarias | Sphaerichthys |
| | Synbranchus | Parauchenoglanis | Trichogaster |
| Intermediate – E/O | Corydoras | Ctenopoma | Betta |
| | Belonesox | Pseudocrenilabrus | Hemirhamphodon |
| | Archocentrus | Tilapia (small) | Luciocephalus |
| Intermediate – O/P | Astyanax | Micralestes | Puntius |
| | Moenkhausia | Hemigrammocharax | Melanotaenia |
| | Roeboides | Chiloglanis | Parambassis |

exhibiting the equilibrium strategy are mostly of intermediate body size, with low fecundity, large eggs, and well-developed parental care as exemplified by brood-guarding cichlids and catfishes (such as *Hoplosternum littorale* and loricariids). This strategy should maximize parental fitness under conditions where density-dependent mortality is important, especially where food is limiting or the threat from predation is high. The third endpoint – the periodic strategy – is associated with high fecundity, small eggs, a contracted and synchronized spawning period, and little or no parental care, and it was the most common strategy in the assemblage studied by Winemiller (1989a). Most of these periodic strategists were characiformes, gymnotiformes, and siluriformes with intermediate body sizes. In its most extreme manifestation, periodic strategists mature at large sizes, have high fecundities, pulse spawning, and migratory behavior (e.g. *Prochilodus mariae*). This strategy appears to maximize fitness in habitats with strong seasonal variation in environmental quality and food availability (Winemiller, 1989b; Winemiller and Rose, 1992). Relatively stable habitats, such as low-order streams draining lowland rainforests in central Amazonia, seem to favor opportunistic life-history strategists (e.g. small characiformes and dwarf cichlids such as *Apistogramma* spp.) and equilibrium strategists such as *Hoplias*, *Gymnotus*, and the cichlid genus *Aequidens*.

Small fishes with high reproductive efforts are represented in all tropical regions (Table III), particularly in headwater streams that experience frequent hydrological disturbances, as well as seasonal lowland streams associated with higher species richness. Opportunistic strategists in Africa include the small barbs (*Barbus* spp.), characiformes (*Neolebias* and *Hemigrammocharax* spp.), and killifishes (*Aphiosemion*, *Aplocheilichthys*, and *Nothobranchus* spp.). In the Neotropics, small species of killifishes (*Cynolebias*, *Rachovia*, and *Pterolebias* spp.), poeciliids (*Fluviphylax*, *Neoheterandria*, and *Poecilia* spp.), and characiforms (*Characidium*, *Deuterodon*, and *Hemigrammus* spp.) are typical opportunistic strategists (Winemiller, 1989b; Mazzoni and Petito, 1999; Mazzoni *et al.*, 2002). In Asia, the opportunistic strategy is observed among small cyprinids (e.g. *Brachydanio*, *Rasbora*, *Microrasbora*, and some *Puntius* spp.) and halfbeaks (e.g. *Dermogenys* spp.). Within the Asian cyprinid genus *Puntius*, continuously breeding opportunistic strategists and seasonally-spawning periodic strategists can be found in the same stream (de Silva *et al.*, 1985), paralleling the situation seen in Neotropical characids (Winemiller, 1989b). Equilibrium species with well-developed parental care, including mouth brooders, occur in multiple families in the streams of all tropical regions (Table III). Migratory periodic-type fishes with large body sizes, high fecundities and synchronized spawning periods are also found throughout the tropics (Table III). The triangular life history continuum also describes fundamental patterns of variation in reproduction and population dynamics of fish assemblages in temperate floodplain rivers (Winemiller, 1996b; Humphries *et al.*, 1999).

## IV. FEEDING STRATEGIES AND FOOD-WEB STRUCTURE

Fishes in tropical streams display diverse feeding behaviors, including specialized trophic niches not normally observed in temperate stream fishes (e.g. seed, fruit, scale, fin, and mucus feeding). Many tropical freshwater fishes are trophic generalists (Knöppel, 1970), sometimes accompanied by a contraction of the diet during periods of reduced resource availability. An increase in dietary specialization accompanied by a decrease in interspecific dietary overlap has been documented during the dry season for stream fishes in Panama, Costa Rica, and Venezuela (Zaret and Rand, 1971; Winemiller, 1987, 1989c; Winemiller and Pianka, 1990). In Sri Lankan rainforest streams, fishes revealed patterns of niche complementarity in which high overlap in habitat use was accompanied by low dietary overlap (Moyle and Senanayake, 1984; Fig. 6). Dietary specializations in such streams tended to be associated with consumption of allochthonous foods and morphological specializations. Other studies of tropical stream fishes have also documented significant patterns of association between diet and morphology (Watson and Balon, 1984; Winemiller, 1991; Winemiller *et al.*, 1995; Winemiller and Adite, 1997; Hugueny and Pouilly, 1999; Ward-Campbell *et al.*, 2005). The impression that emerges from these studies is that many tropical stream fishes increase dietary breadth to take advantage of abundant resources during the wet season, then resort to more specialized feeding during the dry season when interspecific competition for limited resources favors consumption of foods for which each species has greatest relative foraging efficiency based on morphology. For example, the scale-feeding glass characid, *Roeboides dayi*, eats large amounts of seasonally-abundant aquatic insects during the wet season, but shifts to a diet of mostly fish scales during the dry season when densities of a diverse array of invertebrate-feeding fishes tend to increase as the habitat diminishes (Peterson and Winemiller, 1997). Like several other scale-feeding

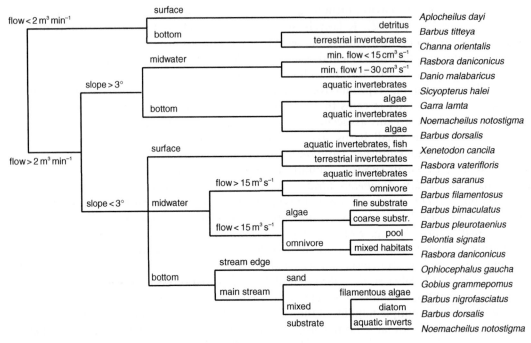

*FIGURE 6* Ecological 'key' illustrating the high degree of microhabitat and food resource partitioning reducing ecological overlap among fishes in a Sri Lankan stream (from Moyle and Senanayake (1984)).

Neotropical characids, *R. dayi* has external teeth on the snout used to dislodge scales from the flanks of fishes, but this morphological specialization is of little service during the wet season when aquatic insects and other invertebrates are plentiful.

Tropical stream fishes occupy almost the entire spectrum of trophic niches that can occur in aquatic communities (Fig. 7). Periphyton grazers are present in stream habitats in almost every region and elevation. Most of them possess inferior mouths, often with fleshy lips, and numerous spatulate teeth for rasping; cyprinids, which lack jaw teeth, often possess horny oral ridges used for rasping. Periphyton is abundant in both high- and low-gradient streams, but can be limited by availability of light and solid substrates. For this reason, specialist grazers may be uncommon in streams with shifting sand substrates, especially where there is dense shading by riparian forest. In lowland streams with muddy beds, the surfaces of aquatic macrophytes or woody debris often support sufficient periphyton to support grazing fishes. In these and other streams, many grazing fishes shift to feeding on detritus and sediments rich in organic matter when periphyton stocks are reduced (as may happen on a seasonal basis). However, detritus is a less nutritious resource for grazers, such as loricariid catfishes (Power, 1984a), and it is probably only consumed when periphyton is scarce.

Phytoplankton is rarely a major component of the diet of tropical stream fishes, perhaps reflecting its relative scarcity in streams compared to lakes and wetlands, but it is a significant component of the diet of the tiny Asian cyprinid *Pectenocypris balaena* (Roberts, 1989). Zooplankton tend to be rare in the water column of upland streams, but many fishes in lowland streams consume large amounts. Most zooplankton feeders, including larvae of nearly all species, consume individual zooplankton, but some African and Neotropical catfishes, such as *Hemisynodontis membranaceus* and *Hypophthalmus edentatus*, and the Neotropical cichlid *Chaetostoma flavicans* have morphological specializations allowing them to consume zooplankton by filter feeding.

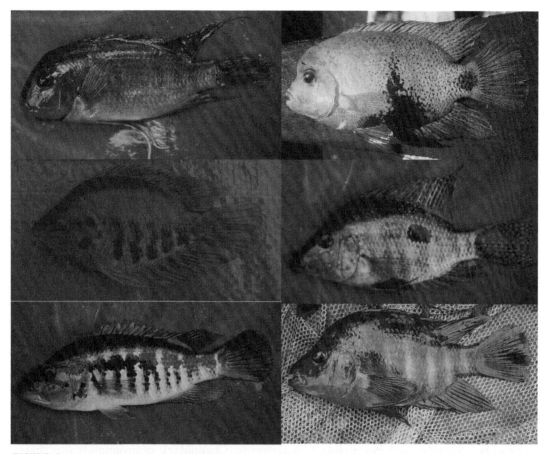

FIGURE 7 Ecologically divergent cichlid fishes from the Tortuguero region of the Costa Rican Caribbean coast: top-left – *Neetroplus nematopus* (algae grazer); top-right – *Vieja maculicauda* (detritivore); middle-left – *Archocentrus centrarchus* (invertebrate picker); middle-right – *Astatheros rostratum* (invertebrate feeding substrate sifter); bottom-left – *Parachromis dovii* (piscivore); bottom-right – *Amphilophus citrinellus* (generalized carnivore). Similar patterns of interspecific dietary and ecomophological diversity are observed among cichlid faunas throughout the Neotropics and tropical Africa (see colour plate section).

Aquatic invertebrates are important food sources for stream fishes throughout the tropics. Taxa consumed depend on stream elevation, topography, and other landscape features that influence aquatic habitat characteristics, but insects of the major aquatic orders are common in the diets of many fishes, and small crustaceans are important dietary items in low-gradient streams. Larger decapod crustaceans and mollusks are also important food, although mollusks and some other taxa can be virtually absent in extreme blackwater conditions of low pH and hardness. Invertebrate meiofauna are selected by specialist feeders, usually tiny benthic fishes such as the burrowing trichomycterid catfishes of the Neotropics, or may be consumed in bulk by sediment feeders (e.g. Neotropical prochilodontids and curimatids).

Terrestrial arthropods are consumed by many tropical fishes, and these allochthonous resources assume greater importance in small forest streams and along the land–water interface of larger streams and wetlands. Their importance in fish diets has been shown in all tropical regions, including Amazonian rainforest (Knöppel, 1970), Sri Lanka (Moyle and Senanayake, 1984) and West Africa (Paugy and Bénech, 1989). Species with surface-oriented morphologies seem to specialize on terrestrial insects (e.g. the characid *Corynopoma riisei* and poeciliid *Alfaro cultratus* in the Neotropics), and the terrestrial invertebrates that live on floating mats of aquatic vegetation can be of great importance in the diets of fishes associated with such habitats: examples from the Neotropics include *Astronotus ocellatus* (Cichlidae) and the driftwood catfish *Parauchenipterus galeatus* (Auchenipteridae).

Although herbivory upon macrophytes is common in tropical fish assemblages, relatively few tropical fishes consume the non-reproductive tissues of living plants (Agostinho *et al.*, 2003a). Exceptions are Neotropical *Pterodoras* spp. (Doradidae), that sometimes consume large amounts of aquatic macrophytes. *Schizodon* spp. (Anostomidae) specialize on aquatic vegetation, and have morphological adaptations (oral and pharyngeal teeth, gill rakers, alimentary canal) for this diet (Ferretti *et al.*, 1996). Elsewhere, *Tilapia rendalli* in the Upper Zambezi consume large amounts of emergent and floating grasses (Winemiller and Kelso-Winemiller, 2003). More often, are consumed by fishes the non-reproductive parts of macrophytes in the form of detritus: examples include curimatids and prochilodontids in South America, distichodontids and tilapiine cichlids in Africa, and cyprinids and pangasiid catfishes in Asia. These and other detritivores have long guts and other morphological and physiological adaptations for extracting energy and nutrients from refractory organic material (Fugi and Hahn, 1991; Delariva and Agostinho, 2001).

Terrestrial plants tissues, especially flowers, fruits and seeds, are very important resources for fishes in tropical streams. Many fruit- and seed-eating fishes have dentition specialized for crushing (e.g. *Brycon*, *Colossoma*, *Myleus*, *Metynnis*, and other bryconine and serrasalmine Characidae), and when seeds are not destroyed, fishes can be significant seed dispersal agents for riparian and floodplain trees in the Neotropics (Goulding, 1980; Souza-Stevaux *et al.*, 1994; Horn, 1997; Mannheimer *et al.*, 2003). Most fishes that feed on fruits and seeds also consume invertebrates, but the proportion depends on season and habitat.

Piscivorous fishes exhibit variable degrees of feeding specialization: some feeding non-selectively while others are highly specialized for the pursuit and capture of particular types of prey in terms of morphology and/or behavior. Piscivores are well represented among the bonytongues (Osteoglossiformes), cichlids, catfishes (especially the Ariidae, Bagridae, Clariidae, Schilbeidae, Siluridae and Pimelodidae), snakeheads (Channidae), centropomids (e.g. barramundi and Nile perches), and characiformes (certain groups within the Atestidae and Characidae). Erythrinids (e.g. *Hoplias*, *Hoplerythrinus*) are the most common and widespread piscivores in small Neotropical streams (Araújo-Lima *et al.*, 1995). Other groups contain relatively few piscivorous genera or species (e.g. the Mormyridae, Cyprinidae, and Gymnotiformes). Piscivores tend to be represented by fewer species at lower population abundance than other trophic groups. Most consume their prey whole, but a few piscivores bite pieces of flesh or fins from prey that may be as large or larger than themselves (e.g. the South American *Serrasalmus* and *Cetopsis* spp., African *Hydrocynus brevis* and *Ichthyoborus* spp., and Asian *Channa* spp.). Scale-feeding fishes are found in streams of the Neotropical (*Roeboides* spp. and *Exodon paradoxus*: Characidae) and Indian-Asian-Australian regions (*Chanda nama*: Ambassidae). Some members of the Neotropical catfish family Trichomycteridae feed on mucus (*Ochmacanthus* spp.: Winemiller and Yan, 1989) or blood (e.g. species of *Vandellia*, *Paravandellia*, *Stegophilus*, and *Acanthopoma*: Machado and Sazima, 1982) from the gill filaments from other fishes. In contrast to this extreme specialization, virtually all piscivores eat aquatic invertebrates during their larval and juvenile stages, and adults of some species continue to include quantities of decapod crustaceans and other large invertebrates in their diets. A few Neotropical piscivores even consume tetrapod vertebrates. Perhaps most remarkable among these is the South American bonytongue *Osteoglossum bicirrhosum*, which can leap more than a meter above the water surface to capture snakes, birds and bats (Goulding, 1980).

The feeding-guild structure of several tropical stream fish assemblages has been examined by analysing stomach contents. In a Panamanian stream studied by Angermeier and Karr (1983), the assemblage comprised seven guilds: insectivores consuming aquatic insects (11 species/size classes), general insectivores (six species), grazers of algae (five species), omnivores (two species), and one species each of a terrestrial herbivore, a piscivore, and a scale eater. A Sri Lankan stream had a similar guild structure: six species eating aquatic insects, six grazer species, four omnivores, three species feeding on terrestrial insects, two piscivores (that also ate invertebrates), and a detritivore (Moyle and Senanayake, 1984). A small coastal creek in Costa Rica supported seven species feeding on algae and detritus, five piscivores (that also ate invertebrates), and three omnivores; the equivalent species totals for each guild in a larger creek in the same area were six, eleven, and two, respectively, plus eleven species that ate invertebrates and two piscivores (Winemiller, 1987). In an Andean piedmont stream in Venezuela, the fish assemblage was made up of four species feeding on algae and detritus, eight species that ate invertebrates, six omnivores, and a piscivore, whereas a lowland creek in the Venezuelan llanos had equivalent species totals of 9, 21, 12, 6 plus a further 6 species that fed on fish and invertebrates. The dietary data of these four stream assemblages in Costa Rica and Venezuela were analyzed by Winemiller and Pianka (1990) using null-model algorithms. There was statistically-significant guild structure within all assemblages, and niche partitioning within guilds, with the guild structure being more developed during the dry season when, as discussed above, resources are more limited and density-dependent factors influence populations.

Several studies have revealed the important influence of fishes on ecosystem dynamics in tropical streams. Strong effects of a migratory detritivore, *Prochilodus mariae*, on sediments – as well as algal and invertebrate community structure – in a Venezuelan piedmont stream (Río Las Marías) have been experimentally demonstrated by Flecker (1996). Reduced discharge during the dry season results in sedimentation of suspended clay particles. By ingesting and resuspending fine sediments, *P. mariae* shift the periphyton assemblage from dominance by relatively inedible cyanobacteria to dominance by diatoms. Changes in sediments and algal stocks also influences nutrient dynamics in the ecosystem (Flecker *et al.*, 2002). For reasons not yet understood, *P. mariae* migrations into the stream are low in the dry season of some years and high in others; consequently, the stream ecosystem shifts between two alternative states depending on the abundance of *P. mariae*. When *P. mariae* are rare, the stream has clear water, a thick layer of fine sediments on the stream bed, and dominance of the periphyton by cyanobacteria which are responsible for high rates of nitrogen fixation. During years when large numbers of *P. mariae* migrate into the stream, it has turbid water, a thin layer of sediments on the stream bed, diatom-dominated periphyton, and lower nitrogen fixation rates. *Prochilodus* inhabit deep pools and runs, but other benthic herbivorous and detritivorous fishes in Río Las Marías have similar ecosystem 'engineering' effects in shallow riffles (Flecker, 1997). *Parodon apolinari* (Parodontidae), *Ancistrus triradiatus,* and *Chaetostoma milesi* (Loricariidae) prefer to graze algae from stone surfaces, but they will ingest overlying sediments in order to access periphyton; this sediment removal has potential implications for primary production. When Power (1990) experimentally manipulated densities of the *Ancistrus spinosus* in a Panamanian stream, she found that benthic algal stocks and rates of photosynthesis were greatest under light grazing pressure that removed accumulated fine sediments.

Recent work by McIntyre *et al.* (2007) has shown strong effects of fishes on nutrient cycling in a Neotropical stream (Río Las Marías, Venezuela). Different species varied significantly in the rate at which they excreted nitrogen and phosphorus, with excretion of one or other nutrient – and hence nutrient cycling – being dominated by a relatively small subset of the 69 species in the steam. Simulations showed that elimination of one or more of these species would cause significant reductions in the rate of nutrient cycling, with the greatest changes being associated with loss of fishes targeted by fishermen.

Benthivorous fishes have been shown to undergo seasonal shifts in diet, apparently in response to changes in relative availability of algae and organic-rich sediments. Virtually all benthic algivorous fishes inhabiting a lowland creek in the Venezuelan llanos (Caño Maraca) had guts containing mostly algae during the wet season, but diets were dominated by detritus during the dry season (Winemiller, 1990, 1996b). Omnivorous characid fishes in the same system remove periphyton from the roots of floating aquatic macrophytes, and showed less extreme seasonal shifts in the amounts of algae that they ingested. Other consequences of fish consumption of algae have been shown by Power (1983, 1984b) who demonstrated that the distribution of loricariid catfishes – and hence periphyton – in a Panamanian stream was influenced by the threat of predation by wading birds. Because grazing by loricariids reduces algal standing stocks, fish avoidance of shallow-water areas where the predation threat was high led to a 'bath-tub ring' of algae in the shallow marginal areas of deep pools. A small Venezuelan piedmont stream that contained a similar loricariid fauna did not show the same pattern of algal distribution, which has been attributed to the relative rarity of piscivorous birds and/or additional smaller species of grazers in shallow pool margins (Winemiller and Jepsen, 1998).

Pringle and Hamazaki (1997, 1998) experimentally manipulated fish access to benthic periphyton in a Costa Rican lowland stream. In the presence of seven species of algae-gleaning fishes, the periphyton assemblage was dominated by cyanobacteria and chironomid (Diptera) larvae. In the absence of these fishes, diatoms dominated the periphyton, and aquatic insects were more diverse and abundant. The fish effects were modified by the occurrence of periodic scouring flash floods that tended to cause relatively greater reductions of periphyton stocks and insect abundance in fish-exclusion areas. Additional complexity in this system arises from the fact that the effect of fishes, which are diurnal feeders, is modified by decapod crustaceans (prawns) that feed at night. Experiments manipulating access to patches by fishes, prawns, or both revealed an additive effect of diurnal and nocturnal grazers, but a greater effect of diurnal fishes.

Food chains in tropical streams are consistently short, usually only three or four trophic levels (Winemiller, 1990). Food webs are comprised of dozens, if not hundreds of food chains,

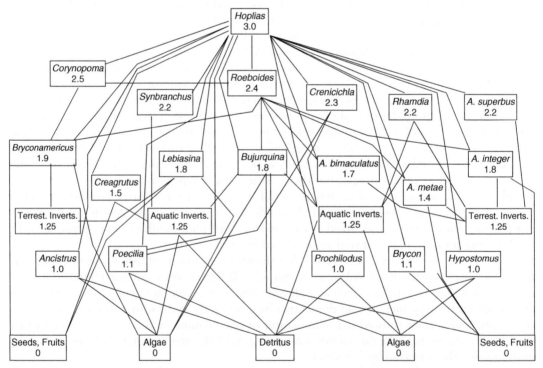

*FIGURE 8*  Caño Volcán food web (piedmont stream, Venezuela) illustrating the dominant functional groups and trophic links (based on Winemiller, 1990). Algae, detritus, seeds/fruit, aquatic macroinvertebrates, and terrestrial invertebrates have been aggregated. Numbers are estimated trophic level values.

FIGURE 9 Convergent evolution in piscivorous fishes with specialized morphology and feeding behavior: *Macrochirichthys macrochirus* (Cypriniformes: Cyprinidae) from Southeast Asia and Indonesia (top); *Rhaphiodon vulpinus* (Characiformes: Cynodontidae) from South America (bottom) (see colour plate section).

originating from aquatic and terrestrial primary producers and detritus to fishes (Fig. 8). In the tropics, stream fishes seem to have claimed, in evolutionary terms, some of the niche space occupied by aquatic invertebrates in temperate regions. Fishes that live as periphyton grazers and detritivores are more common and diverse in the tropics. The abundances of fishes species feeding at lower trophic levels as primary consumers results in direct and relatively efficient conversion of primary production into fish biomass, implying that food chains leading to fish biomass are more efficient in tropical streams than their temperate equivalents. Unfortunately this hypothesis is difficult to test because of a number of potential factors that confound latitudinal comparisons (e.g. phylogenetic history/constraints, latitudinal differences in photoperiod and net annual production, etc.), although it is apparent that tropical freshwater systems frequently support impressively high fish production and harvest (Welcomme, 1985). In surprising contrast to the high fish production in large tropical rivers, however, small streams in Venezuela (Penczak and Lasso, 1991), Borneo (Watson and Balon, 1984), and Brazil (Agostinho and Penczak, 1995; Mazzoni and Lobón-Cerviá, 2000) are less productive then similar-sized streams at temperate latitudes.

As mentioned above, tropical stream fishes reveal consistent patterns of association between feeding behavior and morphology (Moyle and Senanayake, 1984; Watson and Balon, 1984; Wikramanayake, 1990; Winemiller, 1991; Mérigoux and Ponton, 1998). Herbivores and detritivores have long alimentary canals and often possess specialized dentition for scraping or raking materials from substrates. Seed and fruit eaters have intermediate gut lengths and dentition that allows efficient mastication. Zooplanktivorous filter feeders have long, comb-like gill rakers. Piscivores have large, often upturned, mouths, sharp conical or triangular teeth, oro-pharyngeal tooth plates, and short guts. Body shape, fin dimensions and placement, and the relative position of the eyes and mouth combine to indicate swimming behavior and habitat affinities. Because these general patterns are robust, tropical freshwater fishes reveal extensive convergent evolution in ecomorphology (Fig. 9). Convergent morphologies and associated ecological attributes have been demonstrated by statistical comparisons among the assemblages of weakly electric African mormyrids and South American gymnotiformes (Winemiller and Adite, 1997), cichlid fishes from assemblages in Africa, and South America (Winemiller et al., 1995), and entire fish assemblages from lowland habitats in Africa, Central America, South America, and two temperate regions (Winemiller, 1991). One inference arising from comparative studies is that the highly-diverse Central American cichlids have undergone more recent adaptive radiation than the other fluvial cichlid faunas.

## V. CONSERVATION OF FISH BIODIVERSITY

The high taxonomic and ecological diversity of stream fishes in the tropics provides a unique and extensive record of evolutionary biological diversification for scientific study. Tropical stream fishes provide striking examples of ecological convergence, and the limited dispersal ability of many freshwater fish clades makes them particularly amenable to evolutionary investigations. However, this diverse fauna is not merely of scientific or aesthetic interest. Tropical stream fishes are important food resources for humans and, in large areas of Africa and Asia in particular, freshwater fishes are the primary source of animal protein. They are also the genetic reservoirs of current and future aquaculture stocks. Across much of the tropics, stream fishes are collected for export via the aquarium and ornamental fish trade, and can be of considerable economic importance. Sport fishing in some tropical rivers supports ecotourism that may have potential for expansion. Leaving aside discussion of the complex and unpredictable effects of global climate change on stream fishes (reviewed by Poff *et al.*, 2001), the major threats to stream fishes and the ecological integrity of their habitats, are outlined below. This is followed by a brief account of some possible management responses to alleviate the worst effects of human impacts on stream fishes. A more detailed treatment of conservation issues relevant to tropical streams is given in Chapter 10 of this volume.

### A. Drainage-basin Degradation and Land-use Change

The primary threat to the ecological integrity of tropical streams and the long-term survival of their fish faunas is degradation of watersheds by a variety of human activities. The largest impact is from deforestation and conversion of land to agriculture, which modifies light regimes, increases mortality of eggs and larvae due to more ultraviolet radiation reduces, inputs of allochthonous energy and woody debris, and increases sedimentation, nutrient loading, and agrochemical inputs (see review by Pusey and Arthington, 2003). Removal of trees is poorly regulated or practiced illegally in some tropical countries, and the impacts on drainage basins and riparian zones can be devastating. Complete deforestation can lead to streams that are waterless during the dry season, and the wet season run-off from spates that result in soil erosion and transport of sediments, in addition to flash floods of higher amplitude and shorter duration than are experienced under forested conditions. Together, these changes result in the elimination of all but the most resistant and resilient fish species. Complete deforestation of local drainage basins in the Andean piedmont of Venezuela during the last 30 years have reduced fish assemblages in many streams from approximately 25 species to fewer than 10. Resistant species, including *Hoplias malabaricus*, *Poecilia reticulata*, and *Bujurquina pulcher*, are tolerant of habitat disturbances and can reproduce year-round. Migratory species, such as *Prochilodus mariae*, *Brycon whitei*, and *Salminus hilarii*, are eliminated from streams that they formerly occupied during the dry season (Lilyestrom and Taphorn, 1978; Winemiller *et al.*, 1996). In areas of Ecuador with fragmented forests, beta diversity is higher among stream fishes in forested areas relative to sites that have been deforested, indicating greater heterogeneity in species composition (Bojsen and Barriga, 2002). In addition, the percentage of rare species making up assemblages in Ecuadorian streams is positively correlated with canopy cover.

Deforestation of riparian areas in savanna regions can be particularly devastating. Poff *et al.* (2001) report that the number of fish species in a stream draining the Guyana Shield region of southern Venezuela declined from 80 to 5 following riparian deforestation (associated with

*PLATE 1* Representative tropical stream habitats: top-left – upland stream, western Benin; top-right – creek draining uplands adjacent to Barotse floodplain, Upper Zambezi River, Zambia; middle-left – clearwater Río Aguaro in Venezuelan llanos; middle-right – blackwater 'morichal' in northeastern Venezuela; bottom-left – rainforest stream on Osa Peninsula, Costa Rica; bottom-right – lowland stream in coastal swamp-forest, Tortuguero, Costa Rica (See Fig. 1 in Chapter 5 on page 108).

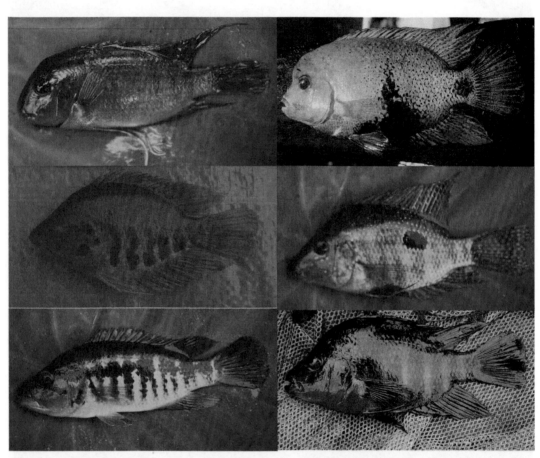

*PLATE 2* Ecologically divergent cichlid fishes from the Tortuguero region of the Costa Rican Caribbean coast: top-left – *Neetroplus nematopus* (algae grazer); top-right – *Vieja maculicauda* (detritivore); middle-left – *Archocentrus centrarchus* (invertebrate picker); middle-right – *Astatheros rostratum* (invertebrate feeding substrate sifter); bottom-left – *Parachromis dovii* (piscivore); bottom-right – *Amphilophus citrinellus* (generalized carnivore). Similar patterns of interspecific dietary and ecomophological diversity are observed among cichlid faunas throughout the Neotropics and tropical Africa (See Fig. 7 in Chapter 5 on page 131).

*PLATE 3* Convergent evolution in piscivorous fishes with specialized morphology and feeding behavior: *Macrochirichthys macrochirus* (Cypriniformes: Cyprinidae) from Southeast Asia and Indonesia (top); *Rhaphiodon vulpinus* (Characiformes: Cynodontidae) from South America (bottom) (See Fig. 9 in Chapter 5 on page 135).

(a) (b) (c)

*PLATE 4*   (a) A step-cascade stream running through high Andean forest at 3200 m asl elevation in southern Ecuador. (b) A deep, narrow moorland stream channel with submerged macrophytes draining páramo at 4000 m asl elevation in Central Ecuador. (c) A braided stream on the Bolivian Altiplano (4000 m asl elevation) draining the Cordillera Real. Huayna Potosí (6091 m asl) is seen in the background (Photos by D. Jacobsen) (See Fig. 1 in Chapter 8 on page 220).

*PLATE 5* (a) A stream surrounded by *Espletia* shrubs on the El Angel páramo at 3800 m asl in northern Ecuador. (b) A stream enclosed by *Polylepis* woodlands at 3600 m asl in Central Ecuador covered by dense *Polylepis* woodlands. (c) Low flow in a wide stream channel on the Bolivian Altiplano (3800 m asl). (d) A stream at 3000 m asl in a Dry Inter-Andean valley close to La Paz, Bolivia (Photos by D. Jacobsen) (See Fig. 2 in Chapter 8 on page 223).

PLATE 6 Thick mats of *Microspora* (Chlorophyta) covering cobbles in a milky, kryal stream 4240 m asl in Central Ecuador. Additional data from this stream are given in Table IV (Photo by D. Jacobsen) (See Fig. 9 in Chapter 8 on page 234).

(a)                                    (b)                                    (c)

PLATE 7 (a) A springbrook in Ecuador (4100 m asl) dominated by *Myriophyllum* and *Callitriche*. Mt Antisana (5758 m asl) in the background. (b) A lake outlet stream in Central Ecuador (3600 m asl) dominated by *Lilaeopsis*, *Myriophyllum*, and *Potamogeton*. (c) An unstable, braided stream on the Bolivian Altiplano (3900 m asl) with dense patches of *Hydrocotyle* and *Elodea* along the banks (Photos by D. Jacobsen) (See Fig. 10 in Chapter 8 on page 235).

(a)          (b)                              (c)

PLATE 8   (a) *Isoetes* dominating a small rhithral stream at 4000 m asl in Ecuador. (b) A springbrook at 4100 m asl in Ecuador with patches of *Callitriche* and *Myriophyllum*. (c) Dense growth of *Potamogeton* in a turbid, sluggish Altiplano stream in Bolivia at 3900 m asl (Photos by D. Jacobsen) (See Fig. 11 in Chapter 8 on page 236).

PLATE 9   Left: *Astroblepus* sp. (length 9 cm) from a large, swiftly flowing boulder-filled stream at 2300 m, Ecuador. Right: *Trichomycterus* sp. (length 4 cm) from a stream at 3900 m asl on the Bolivian Altiplano (Photos by D. Jacobsen) (See Fig. 12 in Chapter 8 on page 238).

PLATE 10  Changes along the upper course of a small, milky kryal draining a glacier on Volcán Antisana (5758 m asl), Ecuador. (a) Close to the glacier snout at 4850 m asl; $T_{max} = 0.5°C$, and there are no benthic macrofauna. (b) Descending the moraine at 4770 m asl and 0.5 km from the glacier snout; $T_{max} = 6°C$, with benthic macrofauna dominated by Podonominae; only two other Diptera taxa (Diamesinae and Muscidae) present. (c) At 4590 m asl and 1.2 km from the glacier snout; $T_{max} \approx 10°C$, and two taxa (Simuliidae and Hydrobiosidae) have been added to the macrofauna. (d) At 4430 m asl and 2.2 km from the glacier; $T_{max} = 13.4°C$, and eight taxa are present due to the addition of Baetidae, Elmidae and Turbellaria (Photos by D. Jacobsen) (See Fig. 16 in Chapter 8 on page 246).

highway construction) that resulted in high turbidity, reduced flow, loss of shade, and increased water temperatures. In tropical Asia, where the loss rate of forest cover is estimated from 0.9% to 2.1% per year, deforestation is a major cause of declining fish biodiversity, because the allochthonous materials from riparian zones and seasonally-inundated forest provide important food resources for numerous fishes (Dudgeon, 2000a).

## B. Pollution

Many tropical countries lack resources for adequate sewage treatment, and considerable organic matter and other substances are released directly into streams. Throughout tropical Africa, and in parts of tropical Asia, untreated waste from cities, towns and villages are discharged into streams with negative influences on ecosystems and human health. Certain agricultural practices, especially sugar-cane processing, also increase organic loads in streams, and in South America, the use of alcohol fuel for vehicles has greatly increased land conversion. Signs of impacts on streams are widespread: pollution has eliminated fish from 5% of the total length of China's major rivers (Dudgeon (2000a), and pollution impacts are reported to be apparent in most Indian rivers (Natarajan, 1989). Many cities in tropical Latin America still have grossly inadequate sewage treatment, and release industrial wastes into surface waters. Cities in São Paulo State, Brazil, treat only about 8% of their effluents (Agostinho *et al.*, 1995). Excessive loads of dissolved and particulate organic matter result in increased biological oxygen demand and acute and chronic reductions in dissolved oxygen in the water column of streams, with dire consequences for most fish species.

Some forms of agriculture (e.g. rice, cotton), as practiced in developing tropical countries, result in large inputs of pesticides and herbicides into streams. In the Venezuelan llanos, for instance, applications of pesticides in rice fields have resulted in mass mortality of stream fishes. Large floodplain ecosystems may dilute agricultural chemicals (Nico *et al.*, 1994). Excessive loading of dissolved organic matter from sugarcane processing plants causes dry-season fish kills in some streams of the South American llanos that are important seasonal refugia for regional fish populations (Winemiller *et al.*, 1996). Urban and industrial pollution in the Lake Valencia Basin of Venezuela has resulted in depauperate stream fish assemblages and the possible extirpation of at least one endemic species (Provenzano *et al.*, 2003). Mercury contamination from gold-mining operations, which is often combined with degradation of watersheds, has impacted stream fishes throughout the Neotropics (Winemiller and Morales, 1989; Nico and Taphorn, 1994). More subtle effects on fishes have been observed: a pollution gradient in a southern Brazilian river was associated with reductions in oocyte diameter and gonadal indices as well as altered gonadosomatic relationships in *Astyanax fasciatus* relative to conspecifics from unpolluted sites (Schulz and Martins-Junior, 2001).

## C. Dams and Impoundments

The impacts of dams on river hydrology and biodiversity are well recognized (Allan and Flecker, 1993; Dudgeon, 2000b). Streams are often impounded to create reservoirs for local water supplies, for aquaculture or sport fishing, or to produce hydroelectric power. Construction of small impoundments for hydroelectric power is increasing in Southern Brazil, because most of the large rivers already have been fully exploited for this purpose. Moreover, it is relatively easy and inexpensive to dam low-order streams, and small impoundments built for various purposes have proliferated throughout the tropics. In arid regions, these impoundments may enhance survival of fishes that would otherwise suffer high mortality when streams dry out periodically. Notwithstanding, the primary concern arising from construction of dams and impoundments is the loss of habitat connectivity and migration corridors, both of which have serious consequences for fish populations (Winemiller *et al.*, 1996; Benstead *et al.*, 1999).

## D. Overfishing

Rural people in many tropical countries rely extensively on fishery resources in small streams. Most higher-order streams contain significant populations of large fishes, even but the smallest lowland streams support stocks of large fishes, such as clariid catfishes (Africa), mastacembelid eels (Asia), and erythrinid characiforms and callichthyid catfishes (South America). Many stream fish populations are greatly overexploited, especially in densely settled areas. The placement of gillnets or barrage-type fish traps across stream channels to intercept migrating fishes usually have significant negative consequences for local stocks. Depletion of particular species that are responsible for high rates of phosphorus or nitrogen excretion can also have implications for nutrient recycling and hence local stream productivity (McIntyre *et al.*, 2007). Catches from heavily-fished streams consist mainly of smaller opportunistic species with high reproductive effort and rapid population turnover. Examples include the small cyprinids and catfishes that are dietary staples of impoverished rural people throughout Southeast Asia, India, and Africa. Improved fishery management could increase productivity if fish populations were allowed to contain greater numbers of larger-bodied adults with higher fecundity. Unfortunately, degradation of aquatic habitats and drainage basins usually accompanies overexploitation, which further reduces stream fish production.

## E. Exotic Species

Introductions of exotic (non-indigenous) species are a threat to native fishes and the ecological integrity of streams worldwide. The problem has been better documented in temperate regions, but Lévêque (1997) listed 27 documented exotic fish introductions into African countries. Most other reported cases of inter-continental introductions in the tropics involve African tilapiine cichlids ('tilapias'), originally imported for aquaculture. These fishes thrive in lacustrine habitats in many parts of the Neotropics, but their establishment in streams seems to be limited to low-diversity coastal drainages and degraded habitats where native species were reduced or eliminated prior to tilapia introductions. However, invasions are still proceeding. Many small lakes have been created to support pay-to-fish businesses in Brazil, most of them based on African tilapia. In Maringá municipality (Paraná state), at least 40 of these small businesses use tilapia, and a recent survey of streams in the area found tilapia in all samples, sometimes as the only species present (Fernandes *et al.*, 2003). Tilapia populations also have become established in India and parts of Southeast Asia and northern Australia. Another widespread exotic is the mosquitofish (*Gambusia affinis*), a North American poeciliid that has been introduced throughout the tropics and subtropics to control mosquito larvae. Like tilapia, mosquitofish populations seem to flourish in degraded streams and low-diversity streams of coastal drainages.

A prevalent problem in the tropics is inter-basin transfers of non-indigenous fishes within regions. Such transfers have increased in southern Brazil over the past two decades. For the most part, species native to the Amazon Basin (e.g. *Cichla* spp., *Astronotus ocellatus* and the sciaenid *Plagioscion squamosissimus*) have become established in the Upper Paraná and Paraguay basins. Many introductions were initially to small impoundments for sport fishing. In addition to being transferred between basins within South America, peacock cichlids (*Cichla* spp.), which are voracious piscivores and popular sportfish, have been introduced to Florida, Hawaii, Puerto Rico, Panama, and Malaysia. As well as deliberate introductions, exotic ornamental fishes frequently are released by accident. The illegal development of fish farms for exotic species along stream banks, and other violations of guidelines designed to prevent escape, have been responsible for the release of an estimated 1.3 billion non-native fishes into streams of a single sub-basin of the Paraná River during floods in January, 1997 (Orsi and Agostinho, 1999). A subsequent increase in lerniosis (a disease caused by the parasitic copepod crustacean *Lernaea cyprinacea*) among native fishes increased sharply (Gabrielli and Orsi, 2000). A striking example of the extent of exotic fish introductions has been recorded from the Paraiba do Sul River Basin, southeastern Brazil, which is the main center of ornamental fish farming in South America. Magalhães *et al.* (2002) reported 22 non-indigenous ornamental species in a single stream and a reservoir in the region, and there was evidence of reproduction by several species.

## VI. MANAGEMENT TO ALLEVIATE HUMAN IMPACTS AND RESTORE DEGRADED STREAMS

Appropriate drainage-basin management is essential to avoid stream sedimentation, altered hydrology, and increased loads of nutrients and organic matter. Protection of vegetation cover, maintenance of the integrity of riparian zones, and reductions in point- and non-point-source pollution are essential components of conservation and management strategies for stream fishes. Appropriate regulations exist in many tropical countries, but, for a variety of reasons, are not enforced or are enforced weakly. In some regions land-use practices are unregulated, or follow traditions that have existed for centuries. Sociological and economic tools and incentives will be an essential part of any solution to the persistent environmental problems associated with poverty in developing tropical countries.

Dam construction should, ideally, be limited to streams where fishes do not migrate or where such migrations are not an essential part of the life cycle or breeding. In regions where fish migrations occur (as in most of the tropics), dams should be limited to one or few tributaries within drainage basins thereby preserving some connectivity within system. In cases where this approach is impractical due to economic or social imperatives, research will be needed on the construction of appropriate fish passageways combined with monitoring of their effectiveness.

Relative to temperate streams, very little research has been conducted on tropical stream fishes in support of management. More research is needed on many aspects of the biology numerous poorly-known species, such as fish-habitat relationships and factors influencing population dynamics and production (for further information, see below). Existing fisheries regulations are weakly enforced in many tropical regions, and community-based approaches to regulate fishing effort that involve local people need to be developed and implemented. Promotion of existing examples of self-regulation (e.g. the Tonlé Sap fishery in Cambodia), including descriptions of possible pitfalls and shortcomings, could offer models for the management of tropical stream resources. Inter-basin and inter-regional transfers of fishes must be halted by legislation and all other means. While some exotic introductions are well intended, most still occur in a context of ecological naivety. The potential benefits of exotics pose unacceptable risks for native biota, and eradication of established exotics is virtually impossible in most instances. As an alternative, increased efforts to explore the suitability of native fishes for aquaculture and sport fishing are needed.

## VII. RESEARCH NEEDS

Despite the many threats to tropical stream fishes, and the societal and economic constraints upon measures to address them, the steps needed to improve both the scientific foundations for management of tropical stream fishes and the effectiveness of management can be described succinctly (Stiassny, 1996). One of the most pressing needs is additional surveys of habitats and biotic diversity in regions of the tropics, such as the Congo Basin, where the ichthyofauna is still poorly known. Existing survey data need to be taxonomically verified and updated, and regional databases should be compiled and maintained for analysis of new impacts, long-term trends, and management needs. Regional databases could be combined to facilitate analysis of broader-scale biogeographic patterns, regional variation in species richness, and invasions by exotic species.

Knowledge of critical fish-habitat relationships in tropical streams lags far behind that available for temperate streams. This information is essential for assessment of human impacts and management of fish stocks. Incredibly, the habitat affiliations of most tropical stream fishes remain undocumented, and habitat use at critical stages of their life cycles (e.g. breeding) is often not well understood. This information will be essential to broader understanding of the influence of environmental variation at different scales on fish-habitat relationships. Population and assemblage dynamics, both seasonally and inter-annually, as well as the reproductive biology and life-history strategies of the great majority of tropical stream fishes remain undocumented. In addition to such fundamental research, studies are needed to assess the negative impacts of human activities (e.g. channel obstructions, deforestation, nutrient enrichment) on fish populations and the stream ecosystems that support them. Such studies could pave the way for development of new approaches or technologies to mitigate impacts (e.g. fish passageways, riparian management, wetland restoration).

Much greater understanding of species interactions (especially food webs) and their influence on population dynamics in species-rich communities is required for effective fisheries management, and this information gap is particularly large for tropical freshwater systems. Continued and improved assessment of past and current management practices on fish populations will be needed if we are to achieve sustainable harvest of fish stocks.

## REFERENCES

Adebisi, A.A. (1988). Changes in the structural and functional components of the fish community of the upper Ogun River, Nigeria. *Arch. Hydrobiol.* 113, 457–463.

Abes, S.S., and Agostinho, A.A. (2001). Spatial patterns in fish distributions and structure of the ichthyocenosis in the Água Nanci stream, upper Paraná River Basin, Brazil. *Hydrobiologia* 445, 217–227.

Agostinho, A.A., and Penczak, T. (1995). Populations and production of fish in two small tributaries of the Paraná River, Paraná, Brazil. *Hydrobiologia* 312, 153–166.

Agostinho, A.A., Barbieri, G., and Verani, J.R. (1991). Idade e crescimento do cascudo preto *Rhinelepis áspera* (Siluriformes, Loricariidae) no rio Paranapanema, bacia do rio Paraná. *UNIMAR, Maringá* 13, 249–258.

Agostinho, A.A., Vazzoler, A.E.A. de M., and Thomaz, S.M. (1995). The high Paraná River Basin: limnological and ichthyological aspects. In "Limnology in Brazil". (J.G. Tundisi, C.E.M. Bicudo, and T. Matsumura-Tundisi, Eds.), pp. 59–103. ABC/SBL, Rio de Janeiro, Brazil.

Agostinho, A.A., Gomes, L.C., and Julio, H.F., Jr (2003a). Relações entre macrófitas e fauna de peixes: implicações no controle de macrófitas. In "Ecologia e Manejo de Macrófitas Aquáticas" (S.M. Thomas and L.M. Bini, Eds.), pp. 261–279. Editora da Universidade Estadual de Maringá, Maringá, Brazil.

Agostinho, A.A., Gomes, L.C., Suzuki, H.I., and Júlio, H.F., Jr (2003b). Migratory fishes of the upper Paraná River Basin, Brazil. In "Migratory Fishes of South America: Biology, Fisheries and Conservation Status" (J. Carolsfeld, B. Harvey, C. Ross, and A. Baer, Eds.), pp. 19–99. World Fisheries Trust, Victoria, Canada.

Allan, J.D., and Flecker, A.S. (1993). Biodiversity conservation in running waters. *BioScience* 43, 32–43.

Allen, G.R., Midgley, S.H., and Allen, M. (2002). "Field Guide to the Freshwater Fishes of Australia". Western Australian Museum, Perth.

Alves, C.B.M., and Vono, V. (1997). A ictiofauna do córrego Gameleira, afluente do rio Grande, Uberaba (MG). *Acta Limnologica Brasiliensia* 9, 23–31.

Amaral, M.F., Aranha, J.M.R., and Menezes, M.S. (1998). Reproduction of the freshwater catfish *Pimelodella pappenheimi* in Southern Brazil. *Stud. Neotrop. Fauna Environ.* 33, 106–110.

Amaral, M.F., Aranha, J.M.R., and Menezes, M.S. (1999). Age and growth of *Pimelodella pappenheimi* (Siluriformes, Pimelodidae) from an Atlantic Forest stream in Southern Brazil. *Braz. Arch. Biol. Tec.* 42, 449–453.

Angermeier, P.L., and Karr, J.R. (1983). Fish communities along environmental gradients in a system of tropical streams. *Environ. Biol. Fish.* 9, 117–135.

Aranha, J.M.R., Takeuti, D.F., and Yoshimura, T.M. (1998). Habitat use and food partitioning of the fishes in a coastal stream of Atlantic forest, Brazil. *Rev. Biol. Trop.* 46, 955–963.

Araújo-Lima, C.A.R.M., Agostinho, A.A., and Fabré, N.N. (1995). Trophic aspect of fish communities in Brazilian rivers and reservoirs. In "Limnology in Brazil" (J.G. Tundisi, C.E.M. Bicudo, and T. Matsumura-Tundisi, Eds.), pp. 105–136. ABC/SBL, Rio de Janeiro, Brazil.

Arratia, G.F. (1983). Preferencias de habitat de peces siluriformes de águas continentales de Chile (Fam. Diplomystidae y Trichomycteridae). *Stud. Neotrop. Fauna Environ.* 18, 217–237.

Balon, E.K. (1975). Reproductive guilds of fishes: a proposal and definition. *J. Fish. Res. Bd. Can.* 32, 821–864.

Balon, E.K., and Stewart, D.J. (1983). Fish assemblages in a river with unusual gradient (Luongo, Africa Zaïre system), reflections on river zonation, and description of another new species. *Environ. Biol. Fish.* 9, 225–252.

Benstead, J.P., March, J.G., Pringle, C.M., and Scatena, F.N. (1999). Effects of a low-head dam and water abstraction on migratory tropical stream biota. *Ecol. Appl.* 9, 656–668.

Bhat, A. (2005). Ecomorphological correlates in tropical stream fishes of southern India. *Environ. Biol. Fishes*, 73, 211–225.

Bistoni, M.A., and Hued, A.C. (2002). Patterns of fish species richness in rivers of the central region of Argentina. *Braz. J. Biol.* 62, 753–764.

Bojsen, B.H., and Barriga, R. (2002). Effects of deforestation on fish community structure in Ecuadorian Amazon streams. *Freshw. Biol.* 47, 2246–2260.

Buck, S., and Sazima, I. (1995). An assemblage of mailed catfishes (Loricariidae) in southeastern Brazil: distribution, activity, and feeding. *Ichthyol. Explor. Freshw.* 6, 325–332.

Buckup, P.A. (1999). Sistemática e Biogeografia de Peixes de Riachos. In "Ecologia de Peixes de Riachos" (E.P. Caramaschi, R. Mazzoni, and P.R. Peres-Neto, Eds.), Série Oecologia Brasiliensis, Vol. 6, pp. 91–138. PPGE-UFRJ, Rio de Janeiro.

Bührnheim, C.M., and Fernandes, C.C. (2001). Low seasonal variation of fish assemblages in Amazonian rain forest streams. *Ichthyol. Explor. Freshw.* 12, 65–78.

Bührnheim, C.M., and Fernandes, C.C. (2003). Structure of fish assemblages in Amazonian rain-forest streams: effects of habitat and locality. *Copeia*, 2002, 255–262.

Burcham, J. (1988). Fish communities and environmental characteristics of two lowland streams in Costa Rica. Revista Biologia Tropical 36, 273–285.

Bussing, W.A. (1976). Geographical distribution of the San Juan Ichthyofauna of Central America with remarks on its origin and ecology. In "Investigations of the Ichthyofauna of Nicaraguan Great Lakes" (T.B. Thorson, Ed.), pp. 157–175. University Nebraska Press, Lincoln.

Campos, H.C. (1982). Zonacion de los peces em los rios del sur de Chile. Actas del VIII Congreso Latinoamericano de Zoologia. (P.J. Salinas, Ed.), pp. 1417–1431. Mérida, Venezuela.

Cardona, M., Roman-Valencia, J.L.L., and Hurtado, T.H. (1998). Composicion y diversidad de los peces de la quebrada San Pablo en el alto Cauca, Colombia. *Bol. Ecotrop. Ecossistemas Trop.* 32, 11–24.

Casatti, L., Langeani, F., and Castro, R.M.C. (2001). Peixes de riacho do Parque Estadual Morro do Diabo, bacia do alto rio Paraná, SP. Biota Neotropica, Vol. 1, No. 1, http://www.biotaneotropica.org.br/v1n12/pt/abstract?inventory + BN 1122001.

Castro, R.M., and Casatti, L. (1987). The fish fauna from a small forest stream of the upper Paraná River Basin, Southeastern Brazil. *Ichthyol. Explor. Freshw.* 7, 337–352.

Chapman, L.J., and Kramer, D.L. (1991). The consequences of flooding for the dispersal and fate of poeciliid fish in an intermittent tropical stream. *Oecologia* 87, 299–306.

Chapman, L.J., Kramer, D.L., and Chapman, C.A. (1991). Population dynamics of the fish *Poecilia gilli* (Poeciliidae) in pools of an intermittent tropical stream. *J. Anim. Ecol.* 60, 441–453.

Chernoff, B., and Willink, P.W., Eds. (1999). A Biological Assessment of the Aquatic Ecosystems of the Upper Rio Orthon Basin, Pando, Bolivia. Bulletin of Biological Assessment 15, Conservation International, Washington, DC.

Costa, W.J.E.M. (1984). Peixes fluviais do sistema lagunar de Maricá, Rio de Janeiro, Brazil. *Atlântica* (Rio Grande, Brazil) 7, 65–72.

Costa, W.J.E.M. (1987). Feeding habits of a fish community in a tropical coastal stream, Rio Mato Grosso, Brazil. *Stud. Neotrop. Fauna Environ.* 22, 145–153.

Covich, A.P., and McDowell, W.H. (1996). The stream community. In "The Food Web of a Tropical Rain Forest" (D.P. Reagan and R.B. Waide, Eds.), pp. 434–459. University of Chicago Press, Chicago.

Delariva, R.L., and Agostinho, A.A. (2001). Relationship between morphology and diets of six Neotropical loricariids. *J. Fish Biol.* 58, 832–847.

de Mérona, B. (1981). Zonation ichthyologique du bassin du Bandama (Côte d'Ivoire). *Rev. Hydrobiol. Trop.* 16, 103–113.

de Pinna, M.C.C., and Winemiller, K.O. (2000). A new species of *Ammoglanis* (Siluriformes, Trichomycteridae) from Venezuela. *Ichthyol. Explor. Freshw.* 11, 255–264.

de Silva, S.S., Schut, J., and Kortmulder, K. (1985). Reproductive biology of six *Barbus* species indigenous to Sri Lanka. *Environ. Biol. Fish.* 3, 201–218.

Dudgeon, D. (1993). The effects of spate-induced disturbance, predation and environmental complexity on macroinvertebrates in a tropical stream. *Freshw. Biol.* 30, 189–197.

Dudgeon, D. (2000a). The Ecology of tropical Asian rivers and streams in relation to biodiversity conservation. *Annu. Rev. Ecol. Syst.* **31**, 239–263.

Dudgeon, D. (2000b). Large-scale hydrological changes in tropical Asia: prospects for riverine biodiversity. *BioScience* **50**, 793–806.

Edds, D.R. (1993). Fish assemblage structure and environmental correlates in Nepal's Gandaki River. *Copeia*, 1993, 48–60.

Fernandes, R., Gomes, L.C., and Agostinho, A.A. (2003). Pesque e pague: negócio ou fonte de dispersão de espécies exóticas? *Acta Sci. (Brazil)* **25**, 115–120.

Ferretti, C.M.L., Andrian, I.F., and Torrente, G. (1996). Dieta de duas espécies de *Schizodon* (Characiformes, Anostomidae), na planície de inundação do Alto Rio Paraná e sua relação com aspectos morfológicos. *Bol. Inst. Pesca (Brazil)* **23**, 171–186.

Flecker, A.S. (1996). Ecological engineering by a dominant detritivore in a diverse tropical stream. *Ecology* **77**, 1845–1854.

Flecker, A.S. (1997). Habitat modification by tropical fishes: environmental heterogeneity and the variability of interaction strength. *J. N. Am. Benthol. Soc.* **16**, 286–295.

Flecker, A.S., Taphorn, D.C., Lovell, J.A., and Feifarek, B.P. (1991). Drift of characin larvae, *Bryconamericus deuterodonoides*, during the dry season from Andean piedmont streams. *Environ. Biol. Fish.* **31**, 197–202.

Flecker, A.S., Taylor, B.W., Berhardt, E.S., Hood, J.M., Cornwell, W.K., Cassatt, S.R., Vanni, M.J., and Altman, N.S. (2002). Interactions between herbivorous fishes and limiting nutrients in a tropical stream ecosystem. *Ecology* **83**, 1831–1844.

Fraser, D.F., Gilliam, J.F., MacGowan, M.P., Arcaro, C.M., and Guillozet, P.H. (1999). Habitat quality in a hostile river corridor. *Ecology* **80**, 597–607.

Fugi, R., and Hahn, N.S. (1991). Espectro alimentar e relações morfológicas com o aparelho digestivo de três espécies de peixes comedores de fundo do Rio Paraná, Brasil. *Rev. Bras. Biol.* **51**, 873–879.

Gabrielli, M.A., and Orsi, M.L. (2000). Dispersão de *Lernaea cyprinacea* (Linnaeus) (Crustacea, Copepoda) na região norte do estado do Paraná, Brasil. *Rev. Bras. Zool.* **17**, 395–399.

Gaigher, I.G. (1973). The habitat preferences of fishes from the Limpopo River System, Transvaal and Mozambique. *Koedoe* **16**, 103–116.

Garutti, V. (1989). Contribuição ao conhecimento reprodutivo de *Astyanax bimaculatus* (Ostariophysi, Characidae), em cursos de água da bacia do Paraná. *Rev. Bras. Biol.* **49**, 489–495.

Goulding, M. (1980). "The Fishes and the Forest". University of California Press, Berkeley.

Grossman, G.D., Ratajczak, R.E., Jr, Crawford, M., and Freeman, M.C. (1998). Assemblage organization in stream fishes: effects of environmental variation and interspecific interactions. *Ecol. Monogr.* **68**, 395–420.

Horn, M.H. (1997). Evidence for dispersal of fig seeds by the fruit-eating characid fish *Brycon guatemalensis* Regan in a Costa Rican tropical rain forest. *Oecologia* **109**, 259–264.

Hugueny, B. (1989). West African rivers as biogeographic islands: species richness of fish communities. *Oecologia* **79**, 236–243.

Hugueny, B. (1990). Richesse des peuplements de poissons dans le Niandan (Haut Niger, Afrique) en fonction de la taille de la riviére et de la diversité du milieu. *Rev. Hydrobiol. Trop.* **23**, 351–364.

Hugueny, B., and Lévêque, C. (1994). Freshwater fish zoogeography in west Africa: faunal similarities between river basins. *Environ. Biol. Fish.* **39**, 365–380.

Hugueny, B., and Pouilly, M. (1999). Morphological correlates of diet in an assemblage of West African freshwater fishes. *J. Fish Biol.* **54**, 1310–1325.

Humphries, P., King, A.J., and Koehn, J.D. (1999). Fish, flows and flood plains: links between freshwater fishes and their environment in the Murray-Darling River system, Australia. *Environ. Biol. Fish.* **56**, 129–151.

Ibarra, M., and Stewart, D.J. (1989). Longitudinal zonation of sandy beach fishes in the Napo River Basin, Eastern Ecuador. *Copeia*, 1989, 364–381.

Inger, R.F., and Chin, P.K. (1962). The freshwater fishes of North Borneo. *Fieldiana, Zool.* **45**, 1–268.

Jayaram, K.C. (1999). "The Freshwater Fishes of the Indian Region". Narendra Publishing, Delhi.

Jiménez, J.L., Román-Valencia, C., and Cardona, M. (1998). Distribución y constancia de las comunidades de peces em la Quebrada San Pablo, Cuenca del Rio La Paila, Alto Cauca, Colombia. *Actual. Biol.* **20**, 21–27.

Knöppel, H.-A. (1970). Food of central Amazonian fishes: contribution to the nutrient-ecology of Amazonian rain-forest streams. *Amazoniana* **2**, 257–352.

Kramer, D.L. (1978a). Reproductive seasonality in the fishes of a tropical stream. *Ecology* **59**, 976–985.

Kramer, D.L. (1978b). Terrestrial group spawning of *Brycon petrosus* (Pisces: Characidae) in Panama. *Copeia*, 1978, 536–537.

Kramer, D.L., Lindsey, C.C., Moodie, G.E.E., and Stevens, E.D. (1978). The fishes and aquatic environment of the central Amazon Basin, with particular reference to respiratory patterns. *Can. J. Zool.* **56**, 717–729.

Lemes, E.M., and Garutti, V. (2002a). Ecologia da ictiofauna de um córrego de cabeceira da bacia do alto rio Paraná, Brasil. *Iheringia* **92**, 69–78.

Lemes, E.M., and Garutti, V. (2002b). Ictiofauna de poção e rápido em um córrego de cabeceira da bacia do alto rio Paraná. *Com. Mus. Ciênc. Tecnol. PUCRS, Sér. Zool., Porto Alegre*, 15, 175–199.

Lévêque, C. (1997). "Biodiversity Dynamics and Conservation : The Freshwater Fish of Tropical Africa". Cambridge University Press, Cambridge.

Lilyestrom, C.G., and Taphorn, D.C. (1978). Aspectos sobre la biología y conservación de la palambra (Brycon whitei) Myers y Weitzman, 1960. *Rev. UNELLEZ Ciénc. Tecnol., Guanare, Venezuela* 1, 53–59.

Lobón-Cerviá, J., Utrilla, C.G., and Querol, E. (1994). An evaluation of the 3-removal method with electrofishing techniques to estimate fish numbers in streams of the Brazilian Pampa. *Arch. Hydrobiol.* 130, 371–381.

Lowe-McConnell, R.H. (1964). The fishes of the Rupununi Savanna District of British Guiana, South America. Pt I. Ecological groupings of fish species and effects of the seasonal cycle on the fish. *J. Linn. Soc. (Zool.)* 45, 103–144.

Lowe-McConnell, R.H. (1979). Ecological aspects of seasonality in fishes in tropical waters. In "Fish Phenology" (P.J. Miller, Ed.), pp. 219–241. Symposium of the Zoological Society, No. 44, London Academic Press.

Lowe-McConnell, R.H. (1987). "Ecological Studies in Tropical Fish Communities". Cambridge University Press, Cambridge.

Luiz, E.A., Agostinho, A.A., Gomes, L.C., and Hahn, N.S. (1998). Ecologia trófica de peixes em dois riachos da bacia do rio Paraná. *Rev. Bras. Biol.* 58, 273–285.

Lundberg, J.G., Marshall, L.G., Guerrero, J., Horton, B., Malabarba, M.C.S.L., and Wesselingh, F. (1998). The stage for Neotropical fish diversification: a history of tropical South American rivers. In "Phylogeny and Classification of Neotropical Fishes" (L.R. Malabarba, R.E. Reis, R.P. Vari, Z.M.S. Lucena, and C.A.S. Lucena, Eds.), pp. 13–68. EDIPUCRS, Porto Alegre, Brasil.

Machado, F.A., and Sazima, I. (1982). Comportamento alimentar do peixe hematófago *Branchioca bertonii* (Siluriformes, Trichomycteridae). *Ciênc. Cult.* 35, 344–348.

Magalhães, A.L.B., Amaral, I.B., Ratton, T.F., and Brito, M.F.G. (2002). Ornamental exotic fishes in the Gloria reservoir and Boa Vista stream, Paraíba do Sul River Basin, state of Minas Gerais, southeastern Brazil. *Comun. Mus. Cienc. Tecnol. PUCRS, Ser. Zool., Porto Alegre* 15, 265–278.

Mannheimer, S., Bevilacqua, G., Caramaschi, E.P., and Scarano, F.R. (2003) Evidence for seed dispersal by the catfish *Auchenipterichthys longimanus* in an Amazonian lake. *J. Trop. Ecol.* 19, 215–218.

Martin-Smith, K.M. (1998). Relationships between fishes and habitat in rainforest streams in Sabah, Malaysia. *J. Fish Biol.* 52, 458–482.

Matthews, W.J. (1998). "Patterns in Freshwater Fish Ecology". Chapman and Hall, New York.

Matthews, W.J., and Heins, D.C. (1987). "Community and Evolutionary Ecology of North American Stream Fishes". University Oklahoma Press, Norman.

Mazzoni, R., and Lobon-Cerviá, J. (2000). Longitudinal structure, density and rates of a Neotropical stream fish assemblage: River Ubatiba in the Serra do Mar, Southeast Brazil. *Ecography* 23, 588–602.

Mazzoni, R., and Petito, J. (1999). Reproductive biology of a tetragonopterinae (Osteichthyes, Characidae) of the Ubatiba fluvial system, Maricá, RJ. *Braz. Arch. Biol. Technol.* 42, 455–461.

Mazzoni, R., Caramaschi, E.P., and Fenerich-Verani, N. (2002). Reproductive biology of a Characidiinae (Osteichtyes, Characidae) from the Ubatiba River, Maricá, RJ. *Braz. J. Biol.* 62, 487–494.

McIntyre, P., Jones, L.E., Flecker, A., and Vanni, M.J. (2007). Fish extinctions alter nutrient recycling in freshwaters. *PNAS* 104, 4461–4466.

Menezes, M.S., and Caramaschi, E.P. (1994). Características reprodutivas de *Hypostomus* grupo *H. punctatus* no rio Ubatiba, Marica, RJ (Osteichthyes, Siluriformes). *Rev. Bras. Biol.* 54, 503–513.

Menni, R.C., Gómez, S.E., and López Armengol, F. (1996). Subtle relationships: freshwater fishes and water chemistry in southern South America. *Hydrobiologia* 328, 173–197.

Menni, R.C., Miquelarena, A.M., and Volpedo, A.V. (2005). Fishes and environment in northwestern Argentina: from lowland to Puna. *Hydrobiologia* 544, 33–49.

Mérigoux, S., and Ponton, D. (1998). Body shape, diet and ontogenetic diet shifts in young fish of the Sinnamary River, French Guiana, South America. *J. Fish Biol.* 52, 556–569.

Mérigoux, S., Ponton, D., and de Merona, B. (1998). Fish richness and species-habitat relationships in two coastal streams of French Guiana, South America. *Environ. Biol. Fish.* 51, 25–39.

Mérigoux, S., Hugueny, B., Ponton, D., Statzner, B., and Vauchel, P. (1999). Predicting diversity of juvenile Neotropical fish communities: patch dynamics versus habitat state in floodplain creeks. *Oecologia* 118, 503–516.

Miranda, J.C., and Mazzoni, R. (2003). Composição da ictiofauna de três riachos do alto rio Tocantins, GO. Biota Neotropica v3 (n1) http://www.biotaneotropica.org.br/v3n1/pt/abstract?article + BN 00603012003.

Moyle, P.B., and Senanayake, F.R. (1984). Resource partitioning among the fishes of rainforest streams in Sri Lanka. *J. Zool. Lond.* 202, 195–223.

Myers, G.S. (1966). Derivation of the freshwater fish fauna of Central America. *Copeia*, 1966, 766–773.

Natarajan, A.V. (1989). Environmental impacts of Ganja basin development on Renepool and fisheries of the Ganga river system. *Can. Spec. Publ. Aquat. Sci.* 106, 545–560.

Nico, L.G., and Taphorn, D.C. (1994). Mercury in fish from gold-mining regions in the upper Cuyuni River system, Venezuela. *Fresenius Environ. Bull.* 3, 287–292.

Nico, L.G., Schaeffer, D.J., Taphorn, D.C., and Barbarino-Duque, A. (1994). Agricultural chemical screening and detection of chlorpyrifos in fishes from the Apure drainage, Venezuela. *Fresenius Environ. Bull.* 3, 685–690.

Orsi, M.L., and Agostinho, A.A. (1999). Introdução de espécies de peixes por escapes acidentais de tanques de cultivo em rios da bacia do Rio Paraná, Brasil. *Rev. Bras. Zool.* 16, 557–560.

Paugy, D., and Bénech, V. (1989). Poissons d'eau douce des bassins côtiers du Togo (Afrique de l'Ouest). *Rev. Hydrobiol. Trop.* 22, 295–316.

Pavanelli, C.S., and Caramaschi, E.P. (1997). Composition of the ichthyofauna of two small tributaries of the Paraná River, Porto Rico, Paraná State, Brazil. *Ichthyol. Explor. Freshw.* 8, 23–31.

Pavanelli, C.S., and Caramaschi, E.P. (2003). Temporal and spatial distribution of the ichthyofauna in two streams of the Upper Rio Paraná Basin. *Braz. Arch. Biol. Technol.* 46, 271–280.

Penczak, T., and Lasso, C. (1991). Problems of estimating populations parameters and production of fish in a tropical rain forest stream in northern Venezuela. *Hydrobiologia* 215, 121–133.

Penczak, T., Agostinho, A.A., and Okada, E.K. (1994). Fish diversity and community structure in two small tributaries of the Paraná River, Paraná state, Brazil. *Hydrobiologia* 294, 243–251.

Peterson, C.C., and Winemiller, K.O. (1997). Ontogenetic diet shifts and scale-eating in *Roeboides dayi*, a Neotropical characid. *Environ. Biol. Fish.* 49, 111–118.

Poff, L., Angermeier, P., Cooper, S., Lake, P., Fausch, K., Winemiller, K., Mertes, L., Rahel, F., Oswood, M., and Reynolds, J. (2001). Climate change and stream fish diversity In "Future Scenarios of Global Biodiversity" (O.E. Sala, T. Chapin, and E. Huber-Sannwald, Eds.), p. 315–349. Springer–Verlag, New York.

Pouilly, M. (1993). Habitat, écomorphologie et structure des peuplements de poisons dans trios petits cours d'eau tropicaux de Guinée. *Rev. Hydrobiol. Trop.* 26, 313–325.

Power, M.E. (1983). Grazing responses of tropical freshwater fishes to different scales of variation in their food. *Environ. Biol. Fish.* 9, 103–115.

Power, M.E. (1984a). The importance of sediment in the grazing ecology and size class interactions of an armored catfish, *Ancistrus spinosus*. *Environ. Biol. Fish.* 10, 173–181.

Power, M.E. (1984b). Depth distribution of armored catfish: predator-induced resource avoidance? *Ecology* 65, 523–528.

Power, M.E. (1990). Resource enhancement by indirect effects of grazers: armored catfish, algae, and sediment. *Ecology* 71, 897–904.

Prance, G.T. (1982). "Biological Diversification in the Tropics". Columbia University Press, New York.

Pringle, C.M., and Hamazaki, T. (1997). Effects of fishes on algal response to storms in a tropical stream. *Ecology* 78, 2432–2442.

Pringle, C.M., and Hamazaki, T. (1998). The role of omnivory in a Neotropical stream: separating diurnal and nocturnal effects. *Ecology* 79, 269–280.

Provenzano, F.R., Schaefer, S.A., Baskin, J.N., and Royer-Leon, R. (2003). New, possibly extinct lithogenine loricariid (Siluriformes, Loricariidae) from Northern Venezuela. *Copeia*, 2003, 562–575.

Pusey, B.J., and Arthington, A.H. (2003). Importance of the riparian zone to the conservation and management of freshwater fish: a review. *Mar. Freshw. Res.* 54, 1–16.

Reznick, D., and Endler, J.A. (1982). The impact of predation on life history evolution in Trinidadian guppies (*Poecilia reticulata*). *Evolution* 36, 160–177.

Roberts, T.R. (1986). *Danionella translucida*, a new genus and species of cyprinid fish from Burma, one of the smallest living vertebrates. *Environ. Biol. Fish.* 16, 231–241.

Roberts, T.R. (1989). The freshwater fishes of western Borneo (Kalimantan Barat, Indonesia). *Mem. California Acad. Sci.* 14, 1–210.

Sabino, J., and Castro, R.M.C. (1990). Alimentação, período de atividade e distribuição espacial dos peixes de um riacho da floresta atlântica (Sudeste do Brazil). *Rev. Bras. Biol.* 50, 23–36.

Sabino, J., and Zuanon, J. (1998). A stream fish assemblage in Central Amazônia: distribution, activity patterns and feeding behavior. *Ichthyol. Explor. Freshw.* 8, 201–210.

Sánchez-Botero, J.I., Garcez, D.S., and Palacio, B.J. (2002). Distribuón de la ictiofauna y actividad pesquera en la microcuenca del Rio Chajeradó Atrato Medio, Antioquia Colombia. *Actualidades Biológicas, Medellín, Colômbia* 24, 157–161.

Schlosser, I. (1987). A conceptual framework for fish communities in small warmwater streams. In "Community and Evolutionary Ecology of North American Stream Fishes" (W.J. Matthews and D.C. Heins, Eds.), pp. 17–24. University Oklahoma Press, Norman.

Schulz, U.H., and Martins-Junior, H. (2001). *Astyanax fasciatus* as bioindicator of water pollution of Rio dos Sinos, RS, Brazil. *Braz. J. Biol.* 6, 615–622.

Schut, J., de Silva, S.S., and Kortmulder, K. (1984). Habitat associations and competition of eight *Barbus* (=*Puntius*) species (Pisces, Cyprinidae) indigenous to Sri Lanka. *Netherlands J. Zool.* 34, 159–181.

Schwassman, H.O. (1978). Times of annual spawning and reproductive strategies in Amazonian fishes. In "Rhythmic Activity of Fishes" (J.E. Thorpe, Ed.), pp. 187–200. Academic Press, London.

Skelton, P.H. (1993). "An Illustrated Guide to the Freshwater Fishes of Southern Africa". Southern Book Publishers, Halfway House, South Africa.

Souza-Stevaux, M.C., Negrelle, R.R.B., and Citadini-Zanette, V. (1994). Seed dispersal by the fish *Pterodoras granulosus* in the Paraná River Basin, Brazil. *J. Trop. Ecol.* 10, 621–626.

Stiassny, M.L.J. (1996). An overview of freshwater biodiversity: with some lessons from African fishes. *Fisheries* 21, 7–13.

Suzuki, H.I., Agostinho, A.A., and Winemiller, K.O. (2000). Relationship between oocyte morphology and reproductive strategy in loricariid catfishes of the Paraná River, Brazil. *J. Fish Biol.* 56, 791–807.

Uieda, V.S. (1984). Ocorrência e distribuição de peixes em um riacho de água doce. *Rev. Bras. Biol.* 44, 203–213.

Uieda, V.S., and Barretto, M.G. (1999). Composição da ictiofauna de quatro trechos de diferentes ordens do rio Capivara, Bacia do Tietê, Botucatu, SP. *Rev. Bras. Zoociênc.* 1, 55–67.

Uieda, V.S., and Uieda, W. (2001). Species composition and spatial distribution of a stream fish assemblage in the east coast of Brazil; comparison of two field study methodologies. *Braz. J. Biol.* 61, 377–388.

Vari, R.P., and Malabarba, L.R. (1998). Neotropical ichthyology: an overview., eds. In "Phylogeny and Classification of Neotropical Fishes" (L.R. Malabarba, R.E. Reis, R.P. Vari, Z.M.S. Lucena, and C.A.S. Lucena, Eds.), pp 1–11. EDIPUCRS, Porto Alegre.

Ward-Campbell, B.M.S., Beamish, F.W.H., and Kongchaiya, C. (2005). Morphological characteristic in relation to diet in five coexisting Thai fish species. *J. Fish Biol.* 67, 1266–1279.

Watson, D.J., and Balon, E.K. (1984). Structure and production of fish communities in tropical rain forest streams of northern Borneo. *Can. J. Zool.* 62, 927–940.

Weitzman, S.H., Menezes, N.A., and Weitzman, M.J. (1988). Phylogenetic biogeography of the Glandulocaudini (Teleostei: Characiformes, Characidae) with comments on the distributions of other freshwater fishes in Eastern and Southeastern Brazil. In "Proceedings of a Workshop on Neotropical Distribution Patterns" (P.E. Vanzolini and W.R. Heyer, Eds.), pp. 379–427. Academia Brasileira de Ciências, Rio de Janeiro.

Welcomme, R.L. (1969). The biology and ecology of the fishes of a small tropical stream. *Zool. J. Linn. Soc.* 158, 485–529.

Welcomme, R.L. (1985). River fisheries. *FAO Fish. Tech. Pap.* 262, 1–330.

Wikramanayake, D. (1990). Ecomorphology and biogeography of a tropical stream fish assemblage: evolution of assemblage structure. *Ecology* 71, 1756–1764.

Winemiller, K.O. (1983). An introduction to the freshwater fish communities of Corcovado National Park, Costa Rica. *Brenesia* 21, 47–66.

Winemiller, K.O. (1987). "Tests of Ecomorphological and Community Level Convergence among Neotropical Fish Assemblages". Unpublished Ph.D. dissertation, University of Texas, Austin.

Winemiller, K.O. (1989a). Development of dermal lip protuberances for aquatic surface respiration in South American characid fishes. *Copeia*, 1989, 382–390.

Winemiller, K.O. (1989b). Patterns of variation in life history among South American fishes in seasonal environments. *Oecologia* 81, 225–241.

Winemiller, K.O. (1989c). Ontogenetic diet shifts and resource partitioning among piscivorous fishes in the Venezuelan llanos. *Environ. Biol. Fish.* 26, 177–199.

Winemiller, K.O. (1990). Spatial and temporal variation in tropical fish trophic networks. *Ecol. Monogr.* 60, 331–367.

Winemiller, K.O. (1991). Ecomorphological diversification of freshwater fish assemblages from five biotic regions. *Ecol. Monogr.* 61, 343–365.

Winemiller, K.O. (1992). Life history strategies and the effectiveness of sexual selection. *Oikos* 62, 318–327.

Winemiller, K.O. (1993). Reproductive seasonality in livebearing fishes inhabiting rainforest streams. *Oecologia* 95, 266–276.

Winemiller, K.O. (1996a). Dynamic diversity: fish communities of tropical rivers. In "Long-term Studies of Vertebrate Communities" (M.L. Cody and J.A. Smallwood, Eds.), pp. 99–134. Academic Press, San Diego.

Winemiller, K.O. (1996b). Factors driving spatial and temporal variation in aquatic floodplain food webs. In "Food Webs: Integration of Patterns and Dynamics" (G.A. Polis and K.O. Winemiller, Eds.), pp. 298–312. Chapman and Hall, New York.

Winemiller, K.O., and Adite, A. (1997). Convergent evolution of weakly-electric fishes from floodplain habitats in Africa and South America. *Environ. Biol. Fish.* 49, 175–186.

Winemiller, K.O., and Jepsen, D.B. (1998). Effects of seasonality and fish movement on tropical river food webs. *J. Fish Biol.* 53 (Suppl. A), 267–296.

Winemiller, K.O., and Kelso-Winemiller, L.C. (2003). Food habits of tilapiine cichlids of the Upper Zambezi River and floodplain during the descending phase of the hydrologic cycle. *J. Fish Biol.* 63, 120–128.

Winemiller, K.O., and Leslie, M.A. (1992). Fish communities across a complex freshwater-marine ecotone. *Environ. Biol. Fish.* 34, 29–50.

Winemiller, K.O., and Morales, N.E. (1989). Comunidades de peces del Parque Nacional Corcovado luego del cese de las actividades mineras. *Brenesia* 31, 75–91.

Winemiller, K.O., and Pianka, E.R. (1990). Organization in natural assemblages of desert lizards and tropical fishes. *Ecol. Monogr.* **60**, 27–55.

Winemiller, K.O., and Rose, K.A. (1992). Patterns of life-history diversification in North American fishes: implications for population regulation. *Can. J. Fish. Aquat. Sci.* **49**, 2196–2218.

Winemiller, K.O., and Yan, H.Y. (1989). Obligate mucus feeding in a South American trichomycterid catfish. Copeia 1989, pp. 511–514.

Winemiller, K.O., Kelso-Winemiller, L.C., and Brenkert, A.L. (1995). Ecological and morphological diversification in fluvial cichlid fishes. *Environ. Biol. Fish.* **44**, 235–261.

Winemiller, K.O., Marrero, C., and Taphorn, D.C. (1996). Perturbaciones causadas por el hombre a las poblaciones de peces de los llanos y del piedemonte Andino de Venezuela. *Biollania Guanare, Venezuela* **12**, 13–48.

Zaret, T.M., and Rand, A.S. (1971). Competition in tropical stream fishes: support for the competitive exclusion principle. *Ecology* **52**, 336–342.

# 6

# Aquatic, Semi-Aquatic and Riparian Vertebrates

Nic Pacini and David M. Harper

The course of aquatic vertebrate evolution has been moderated by extreme climatic events such as sea-level rise and flooding, with areas such as floodplains providing productive feeding grounds and refuges in times of drought. Aquatic and semi-aquatic vertebrates in the tropics have achieved a remarkable range of anatomical and physiological adaptations, in contrast to temperate latitudes where rivers froze during the last glaciations. Convergent adaptations across different families has been the norm, dictated by the need to swim and to efficiently exploit aquatic niches inhabited by invertebrates and fish. Most taxa are adapted to the lowland sections of large rivers, while few have evolved within fast streams. The current geographic distribution of aquatic, semi-aquatic and riparian species is a reflection of the abundance and variability of water supply in different regions, the obstacles presented to migration by mountain ranges, and the presence of large rivers which may serve as barriers or biocorridors. These vertebrates include a wide range of taxa, some that are more dependent on water than others, and a survey of the main taxa of herpetofauna and mammals is given here.

Tropical streams are characterised by seasonal flows and inundation of the floodplain. Many aquatic and semi-aquatic vertebrates have become highly specialised and dependent upon these seasonal flood pulses conditions. They are vulnerable to habitat degradation or alteration associated with flow regulation, irrigation schemes and the conversion of floodplains to agriculture. Because of their specialism, occupation of more upland habitats is not possible, and aquatic and semi-aquatic vertebrates now include some of the most highly threatened species on Earth. Loss of these species, or large declines in their densities, will impact the ecology of tropical river ecosystems because of the important roles that aquatic, semi-aquatic and terrestrial vertebrates play in energy flow, nutrient cycling and riparian landscape engineering.

## I. INTRODUCTION

Freshwater habitats provide food and shelter to a large number of vertebrates besides fish, but only few are wholly adapted to inland waters. They belong to the mammalian orders Cetacea and Sirenia and are both represented by exclusively tropical species. Others can be considered *semi-aquatic* on the principle that, although they could not survive without close proximity to aquatic habitats, they are incompletely adapted (Dunstone, 1998) and need frequent contact with land especially during crucial phases of their life cycle, such as during reproduction.

Semi-aquatic reptiles are represented by some of the most ancient living vertebrate species; these include the 23 species of crocodilians, two thirds of the 250 species of modern testudinids and a small number of freshwater snakes; nearly all of these are tropical species. Amphibians and birds contain even more semi-aquatic species; these are globally distributed, but characterised by a greater diversity in the tropics than elsewhere. Between 100 and around 200 species of mammals are closely associated with aquatic habitats. Thirty-four of these are marine carnivores including seals, sea lions, walruses and the sea Otter (*Enhydra lutris*). Most of the remaining species live closely associated with tropical inland aquatic habitats; only few are present in the temperate regions.

Semi-aquatic mammals such as otters and hippopotamuses display remarkable anatomical and physiological adaptations; riparian species exhibit characteristic behavioural traits (e.g. cats, monkeys and antelopes that swim) and have life cycles that are dependent on the seasonal inundation of floodplains and consequential changes in the availability of resources and habitat.

Most semi-aquatic vertebrates swim at the water surface where drag forces are maximal and locomotion is slowed and may be energetically expensive. Predators such as crocodilians, birds, mustelids, and some rodents and otter shrews must move quickly in water in order to attack their prey, and have therefore evolved streamlined bodies and efficient propulsive organs such as webbed feet and flattened heavy muscular tails. Skeletal adaptations include a relatively short ileum and femur in relation to tibia, which reduces resistance to the forward movement of the limb in the highly viscous medium, and long manual and pedal phalanges for increased propulsion. Other morphological adaptations to aquatic life are related to the tendency of endotherms to suffer significant heat loss in water. To reduce this, aquatic and semi-aquatic mammals have developed large body mass compared to their nearest land relatives and thick, impermeable, water-repellent furry coats. This feature has given rise to significant commercial exploitation of some species (especially those in temperate latitudes), pushing them to the brink of extinction.

This chapter provides a synthetic overview of the distribution and ecology of aquatic, semi-aquatic and riparian vertebrates throughout four broadly defined intertropical biomes: the Afrotropics, the Neotropics, the Indo-Malayan Region and Australo-Papuasia. Semiaquatic vertebrates are treated in some detail while riparian species are illustrated by citing a number of case studies as examples, with no attempt to make a complete listing. Birds are not included herein, while fishes are treated in Chapter 5 of this volume.

## A. The Distribution of Tropical Vertebrate Assemblages

Within different continents, organisms possess convergent ecophysiological adaptations that can often be explained by a common phylogenetic history (true vicariants) or by a convergent adaptation to environmental conditions (ecological vicariants). Tropical aquatic and semi-aquatic vertebrates provide good examples of both. Continental drift and past connections between southern land masses explain why many tropical species are more closely related to one another than any forms in northern temperate regions. Contact between South America and Africa during the Cretaceous, gave rise to related lineages of animals and plants such as caecilians, marsupials and the large flightless ratites. Physiological constraints imposed by living in aquatic habitats contributed to the development of closely convergent and often competing life forms among phylogenetically unrelated taxa such as between otters, mice and tenrecs (otter shrews). Among vertebrates, ecological vicariants are more frequent in the tropics than in parts of the world where glaciation events disturbed the continuity of vegetation belts (Newton, 2003).

Past climatic history has preserved ancient and diversified lineages in the tropics while, in temperate regions, extinctions prevailed over speciation. This pattern, which has been demonstrated in birds by molecular genetic techniques (Newton, 2003), is widely accepted for other vertebrate groups. Throughout the Quaternary, climate forcing caused irregular 'warm-wet' and 'cold-dry' cycles, while higher latitudes experienced long-term glaciations and inter-glacials. Periodic formation of extensive savannahs led to the isolation of marginal aquatic habitats and to the allopatric speciation of animals and plants adapted to wet environments. This process would also have been influenced by significant changes in sea levels associated with some glaciations, with recurrent island exposures and estuarine shifts. These factors combined with warm or temperate climate with no ice cover, and the presence of floodplain pulsing, are likely to have been the evolutionary drivers for a high diversity of riverine vertebrates in the tropics. They provide an interpretational key for understanding the natural history of mammals and reptiles in tropical streams, with an intensity that varies between continents.

In inter-tropical Africa, the shrinking/expansion of vegetation and desert belts greatly influenced the connections between humid ecosystems, causing repeated waves of animal and plant colonisation of marginal wetlands and then fragmentation of former distribution ranges. This repeated displacement partly mimicked the more extensive latitudinal vegetation shifts that occurred in the lowlands of the northern hemisphere as the ice expanded during glaciations and retreated during interglacials promoting the evolution of vicariant species. The interpretation of present animal distributions needs to take into account also the former shape of the hydrological network and past connections between river basins and lakes. Relating river network and animal distributions is confounded by the fact that rivers may constitute either impassable barriers or dispersal routes depending on the adaptations of the taxon concerned (see below). While tropical lowlands were affected by recurrent floods, droughts and sea-level changes, zones of higher relief offered a permanent refuge to humid forest species throughout the Quaternary glaciations and made a large contribution to shaping current species distribution ranges. Tropical mountains, as well as a limited number of lowlands enjoying higher rainfall due to their particular geographical location (e.g. large river deltas and coastal forests in heavy rainfall areas), represent centres of endemism from where animals and plants recolonised surrounding habitats at the onset of favourable conditions (Fig. 1). While this 'refuge theory' has been defended by Haffer and Prance (2001), better knowledge of palaeoenvironments in the tropics (especially the Amazon) will be needed to confirm it.

In addition to long-term climatic and geomorphological events, short-term climatic cycles appear to have been equally important for generating biodiversity, as they produce a medium-intensity disturbance which favours the continuous rejuvenation of habitats and of the communities living within them. The Flood Pulse Concept proposes that all large tropical catchments are naturally and predictably pulsing systems, driven by the seasonal alternation between dry and wet seasons (Junk, 2001; see also Chapter 7 of this volume). In the Neotropics, for example, migration, habitat transformation and the seasonal establishment/displacement of vertebrate communities is controlled by the flood regime and a transition from the 'limnophase' to the 'potamophase' (Neiff, 2001). In South-eastern Brazil, many species of rodents and marsupials, such as *Didelphis* opossums, give birth at the time of the fruiting season (Bergallo and Magnusson, 1999; Caceres, 2003). This close coupling of biological cycles implies that exceptional changes in the frequency and/or intensity of the flood regime may cause disruptions and damage to well established communities (Neiff, 2001), and indicates the potential consequences of human-induced changes on riverine ecosystems. The latter are evident from studies of large river basins in India and in China: these densely-populated systems with highly-regulated flows now sustain aquatic and riparian biodiversity that has been greatly depleted as a result of human activities (Dudgeon, 2000a).

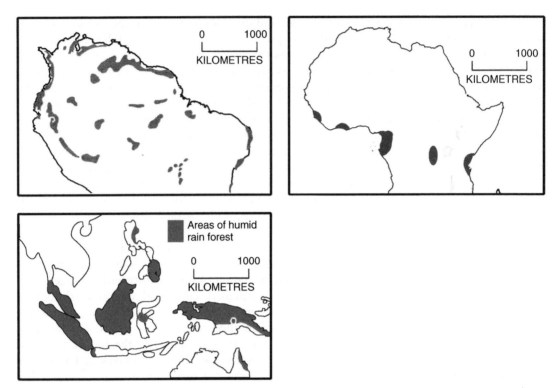

*FIGURE 1* Main lowland humid forest refuges at the height of the most recent dry period in the three southern continents [Source: Figure 11.1 in Newton (2003)].

## B. Rivers as Barriers, Filters and Biocorridors

Large river channels tend to coincide with the edges of the distribution ranges of terrestrial animals, suggesting that rivers are barriers to dispersal. By the end of the 19th century, Wallace (1852, 1876) had noted that the distribution ranges of related species of primates and birds tended to be delimited by major Amazonian rivers such as the Negro, the Madeira and the Amazon. Amazon tributaries act as barriers to the dispersal of Saddle-back tamarins (*Saguinus fuscicollis* and *S. tripartitus*: Cebidae) (Peres *et al.*, 1996; Heymann *et al.*, 2002), while the main river limits the range of a number of other primates – an effect that shows an increasing trend from headwaters to mouth as the channel widens (Ayers and Clutton-Brock, 1992). The lower, but not the upper and the mid Amazon, and possibly some blackwater rivers of the Guyana highlands, prevent the dispersal of *Ateles* spider monkeys (Collins and Dubach, 2000). The distribution of the Dark-handed agile gibbon (*Hylobates agilis*: Hylobatidae) in Asia is limited by the Perak and the Muda rivers in peninsular Malaysia (Earl of Cranbrook, 1988). In Africa, closely-related monkeys evolved into separate species on either bank of the Congo (Table I), and a major biogeographical subdivision of the central African forest is defined by the course of the Congo and its major tributaries (Grubb, 2001).

While rivers may be a barrier for a large number of riparian and terrestrial taxa, they serve as bio-corridors for aquatic and semi-aquatic vertebrates such as Amazon manatees (*Trichechus inunguis*: Trichechidae) that use them during long-distance migrations. In the Afrotropics, river channels and gallery forests play an important role as biodiversity refuges as well as serving as biocorridors for the dispersal of animals and plants across inhospitable dry savannahs. Colyn (1991) proposed the existence of a fluvial lowland super-refuge located in the central Congolese

TABLE 1   Primate Distributions Across the Congo River, Africa. IUCN Status is included also where EN = Endangered; VU = Vulnerable; LR nt = Lower Risk (Near Threatened); and DD = Data Deficient

| Left bank | | | Right bank | | |
| --- | --- | --- | --- | --- | --- |
| Name | Species | Status | Name | Species | Status |
| Angolan colobus | Colobus angolensis | VU | Crowned guenon | Cercopithecus pogonias | EN |
| | | DD | | | |
| Wolf's guenon | Cercopithecus wolfi | DD | L'hoest guenon | Cercopithecus lhoesti | LR nt |
| Bonobo | Pan paniscus | EN | Chimpanzee | Pan troglodytes | EN |
| Golden-bellied mangabey | Cercocebus galeritus chrysogaster | DD | Agile mangabey | Cercocebus galeritus agilis | LR nt |
| Black mangabey | Lophocebus aterrimus | LR nt | Grey-cheeked mangabey | Lophocebus albigena | LR nt |
| Demidoff's galago | Galagoides demidovii phasma | | Demidoff's galago | Galagoides demidovii anomurus | |
| Dryad guenon | Cercopithecus dryas | | Demidoff's galago | Galagoides demidovii murinus | |
| Southern talapoin | Miopithecus talapoin | | Northern talapoin | Miopithecus ogouensis | |

Source: Modified from IUCN (2000)

forest zone, which would have maintained species diversity during the last severe Quaternary arid climatic phase between 15 000 and 20 000 years ago (Maley, 2001), although there are alternative views about the nature of such refuges (Grubb, 2001). Their importance in the Amazon, which was moister than Africa throughout the Quaternary, may have been smaller. Conversely however, river barrier effects are likely to have been greater due to the higher discharge of Neotropical rivers.

Gascon et al. (2000) observed that a consistency between species distributions and the hydrographical network implicitly assumes a parallel evolution between the lineages supposedly separated by riverine barriers and the historical development of drainage basins. Nonetheless, modern ranges are not always related to present-day drainage patterns. Along the slow-flowing Juruá River (western Amazonia), west-east discontinuities in the distribution of amphibian populations were thought to correlate with ancient ridges that originated 5–10 million years ago, a date that is consistent with the genetic distances between distinct populations of the Dart-poison frog (Epipedobates femoralis) (Lougheed et al., 1999). However, further studies showed that the Juruá does not influence the gene flow of the arboreal spiny rat Mesomys (Patton et al., 1994) while riparian habitat differences have significant effects irrespective of bank affiliation (Gascon et al., 2000). Furthermore, a west-east trend in vertebrate species richness was apparent, with distribution ranges ending perpendicularly to the direction of the main channel, in contrast to the pattern that would be predicted by the river-barrier hypothesis.

The effectiveness of river barriers is taxon specific and also depends upon channel characteristics, including width, discharge and current velocity (Colwell, 2000). Rivers therefore tend to act rather as 'filters' rather than simple barriers. Several processes can contribute to faunistic continuity between river banks. Extensive lateral inundation of lowland floodplains can transfer islands of floating vegetation which serve as migration rafts for small terrestrial vertebrates. Lateral migration of the main channel, including meander cut-offs and frequent changes of direction, provide additional processes which enhance population continuity between banks. Broadening of the floodplain and the occurrence of deposition zones close to the delta are conducive to the formation of islands, which provide stepping stones for passage.

## II. AQUATIC MAMMALS

Few vertebrates (apart from fish) are entirely adapted to life in water. Only freshwater dolphins (Cetacea) and manatees (Sirenia) live permanently in rivers and estuaries, although some may visit coastal areas for short periods.

### A. Cetaceans

Two families and two genera of dolphins inhabit rivers and wetlands in South America. Both belong to an ancient lineage (the Platanistoidea) that has a common Miocene ancestor with species present in South East Asia (Grabert, 1984). The Boto (*Inia geoffrensis*: Iniidae) is endemic to the Neotropics where it comprises two well-differentiated evolutionarily-significant units: one is confined to Bolivian river basins, and the other is distributed across the Amazon and Orinoco (Banguera-Hinestroza *et al.*, 2002). Boto live predominantly in slow moving, turbid 'white waters' (see Chapter 1 of this volume) where they fish using echolocation and by probing the substrate with their sensitive snout. The platanistid *Pontoporia blainvillei* lives in salt water in the estuary of the Rio de la Plata and rarely travels upstream. Tucuxi (*Sotalia fluviatilis*: Delphinidae) are representatives of a marine dolphin family that have become adapted to life in estuaries and freshwaters. Their range comprises most of the Amazon basin, including Ecuador, Surinam, Guyana, and northeastern Nicaragua (Carr and Bonde, 2000), and a number of putative subspecies are recognised (Rosas and Monteiro-Filho, 2002).

Niche segregation between Boto and Tucuxi is reflected by their stomach contents (Best, 1984). Boto can reach 2.6 m and 160 kg and are solitary, preying upon benthic fish, crustaceans and even small riverine turtles. They are strictly confined to fresh water and avoid waterfalls and estuaries. The Tucuxi is one of the smallest dolphins, with a maximum length of 1.5 m and 53 kg. It lives in small family groups hunting pelagic fish (Best, 1984), but may congregate into larger schools at times of displacement or for communal fishing (Monteiro-Filho, 2000). It sometimes associates with parties of Giant otter (*Pteronura brasiliensis*: Mustelidae) taking advantage of the fish schools driven by them. Boto are able to exploit a variety of freshwater aquatic habitats including smaller rivers and riparian wetlands (shallow várzeas and igapós: see also Chapter 7 of this volume), whereas Tucuxi are found in the pelagic waters of floodplain lakes, deep channels, estuaries and coastal waters up to 5 m deep (Edwards and Schnell, 2001; de Oliveira Santos *et al.*, 2002). Both dolphin species were observed to congregate at the confluence of rivers during low water levels but moved onto the inundated floodplain during spates (Leatherwood *et al.*, 2000; McGuire, 2002). In contrast to their marine relatives, they have poor sight and use echolocation as a primary means of prey detection. Neotropical river dolphins typically give birth at times of seasonally low river flows, when receding water levels force fish into narrow channels and they become easier to catch. Such increased food availability reduces the stress imposed upon lactating dolphins (Best, 1984).

Dolphins are not hunted for food and rarely for their hide or fat. However, they may often become entangled in fishing gear (Monteiro-Neto *et al.*, 2003) and are poisoned by fishermen to reduce damage to nets and limit their impact on the fish stocks. Other potential threats include boat strikes, oil spills, water and noise pollution, and prey depletion by commercial fisheries (McGuire, 2002). On a larger scale, deforestation and aquatic habitat degradation may significantly reduce dolphin prey species, such as the many fish species feeding on forest fruits (Best and da Silva, 1989; see also Chapter 5 of this volume). Dam construction prevents fish migration, thus reducing resources for dolphins, and causes fragmentation of populations into smaller, more vulnerable groups (Vidal, 1993; Pringle *et al.*, 2000). As top predators, riverine and estuarine dolphins are at risk from pollutants due to bioaccumulation and biomagnification effects of metals such as lead, cadmium and mercury (Monteiro-Neto *et al.*, 2003).

Coastal areas, estuaries and the lower portion of river basins throughout tropical Asia and northern Australia are visited by the estuarine Irrawaddy dolphin *Orcaella brevirostris* (Delphinidae) which represents the oriental equivalent of the Amazonian Tucuxi. It ascends the Irrawady, the Ganga, the Mekong and other rivers, and can live permanently in freshwater. Little is known about its habits but there are reports that hunting, by-catch and the reduction of fish prey by commercial fisheries are causes of population decline (Dudgeon *et al.*, 2000). Similar threats are faced by the poorly-known finless porpoise *Neophocaena phocaenoides* (Phocoenidae), a small-toothed cetacean common in estuaries throughout Asia (Nowak, 1999; Dudgeon, 2000b) and which may have separated into an endemic subspecies (*Neophocaena phocaenoides phocaenoides*) in the Yangtze. Three species of true river dolphins are known from Asia. The Baiji or Yangtze dolphin (*Lipotes vexillifer*: Lipotidae) is probably extinct; recent systematic population estimates reported some 23 individuals in a surveyed river length of 1687 km (Zhang *et al.*, 2003), but repeated surveys in 2006 yielded no Baiji. Like the Amazonian Boto, Baiji tended to congregate at confluences and sand bars with large eddies, and to migrate onto floodplain wetlands during the high water period. The causes of its decline include habitat modification, pollution, fisheries by-catch and collisions with boat traffic.

In the Indus and in the Ganga-Brahmaputra-Meghna floodplains, the survival of two strictly riverine dolphin species (or sub-species), *Platanista minor* and *P. gangetica* (Platanistidae), is seriously compromised by pervasive hydrological modifications. Both have very small eyes and poor vision but are not completely blind. Reproduction in *P. gangetica* responds to flood increases during the monsoon and may be accompanied by lateral migrations. It tends to migrate upstream towards the incoming flood and onto the outer floodplain during the rising flood and return during recession periods. Similar behaviour is known in the Baiji and the Amazonian Boto. Major threats to *Platanista* spp. are population fragmentation and stranding on drying floodplains as a result of barriers to movement caused by numerous barrages and irrigation networks (Dudgeon *et al.*, 2000). Indeed, disruption of riverine connectivity and habitat modification are significant threats for all freshwater cetaceans.

## B. Sirenians

Sirenians evolved from protosirenian ancestors that colonised the southern American continent some 35 million years ago after dispersing from the Old World (Best, 1984; O'Shea, 1994). Up to the late Miocene, sirenian trichechids – or manatees – lived in fresh and in marine waters along coastal rivers and estuaries in South America while their sister family, the Dugongidae, evolved (and have since remained) in marine habitats. During the Pliocene, trichechids spread throughout the Caribbean giving origin to the West Indian coastal manatee (*Trichechus manatus*). Repeated sea introgression and flood pulsing resulted in the colonisation of the Amazon basin where river captures caused a progressive isolation of manatee populations and induced freshwater adaptations that contributed to the evolution of the Amazon manatee (*T. inunguis*). Manatees have acute hearing and use acoustic signals as a basis for communication and mutual individual recognition (Sousa-Lima *et al.*, 2002). They have very secretive habits, and thus information concerning their ecology and distribution is rather scant.

One characteristic that has contributed to the success of manatees is a highly specialised dentition of supernumerary molars which are replaced throughout life. They enabled *Trichechus inunguis* to exploit the plentiful plant resources of Amazon floating meadows and allow it to thrive even in places where submerged aquatic plants are limited by river turbidity. The estuarine *T. manatus* instead has access to sea grasses (Hydrocharitaceae and Potamogetonaceae) and exploits a wider habitat range than *T. inunguis* (Domning, 1982). Amazon manatees feed on a wide range of aquatic, semiaquatic and grassland species, and maintain grazed 'lawns' along slow flowing channels and within inundated floodplain meadows (Best, 1981; Colares and

Colares, 2002). They also eat highly-invasive aquatic weeds such as Water hyacinth (*Eichhornia crassipes*: Pontederiaceae) and the floating fern *Salvinia* spp. (Salviniaceae); *T. inunguis* have been used to control infestations of aquatic weeds in canals in Guyana and elsewhere. At Curua-Una Dam (Santarém, Brazil), for example, where 42 manatee were introduced to control weeds, Best (1984) calculated that their total consumption approached 20 t of plants per month.

The behaviour of *Trichechus inunguis* is linked to the seasonal flood cycles, which induce migration onto the inundated floodplain during high water and a return to deep channels or lakes where they rest during the dry season. The retention of a low metabolic rate, which is a primitive evolutionary trait that enhances diving capabilities, allows manatee to persist without feeding during extended periods (up to 7 months) during the dry season when they survive on their fat reserves (O'Shea, 1994). As with most other Amazonian vertebrates, manatee have become vulnerable to hunters during low-water periods (Best, 1984) when they are intensively sought for their fat and skins (O'Shea, 1994).

The West African manatee (*Trichechus senegalensis*) originated 3–4 million years ago from American trichechids (O'Shea, 1994). They occur in both freshwater and in shallow coastal areas but prefer large estuaries and coastal brackish wetlands where they feed on mangroves and overhanging bank growth, and may also consume large quantities of rice plants (Domning, 1982). Their habits are less well known than those of Neotropical manatees but all three extant species are threatened by human activities – especially hunting. The present distribution of *T. senegalensis* ranges from the Senegal River in the North to the Cuanza River (Angola) in the South. Populations in 1980–81 were estimated at between 9 000 and 15 000 individuals (Nishiwaki, 1984), but recent surveys of *T. senegalensis* indicate that illegal hunting has caused very significant population declines (L. Luiselli, unpublished observations).

## III. REPTILES

The early origin and the apparently 'primitive' character of reptiles contrast with their highly developed physiological adaptations that optimise locomotion, feeding and reproduction in aquatic habitats, and survival under anoxic stress (Storey, 1996). The South American continent hosts some 1100 reptiles, 300 of which are considered endemic to the region (Harcourt and Sayer, 1996). However, semi-aquatic reptiles, such as caimans and turtles, are represented by a relatively small number of species in comparison with the large number of lizards and especially snakes. Few of these can be considered semi-aquatic or amphibious, while most are terrestrial and adapted to dry habitats. By contrast, in the mosaic of marshes, streams, lakes, rivers and ponds constituting the Niger River Delta – probably the largest and herpetologically best known wetland in Africa – snakes reach a remarkable diversity with up to 43 sympatric species (Akani *et al.*, 1999, Luiselli and Akani, 2002, 2003; Luiselli *et al.*, 1998, 2005). Australia has an extremely rich reptile fauna with 15 families out of 38 for the entire globe; many species are associated with aquatic habitats and several families are endemic to the continent and/or are the sole remaining representatives of taxa that were formerly widely distributed (Archer *et al.*, 1994).

### A. Crocodilians

Crocodilians first appeared some 200 million years ago. They spend most of their time in water and have highly complex behaviour with social interactions encompassing dominance hierarchies, coordinated feeding, vocalisation, and well developed maternal care (Ross, 1998). They swim relatively slowly in relation to body mass and strength; the main means of propulsion is the tail, while limbs are used for paddling gently to optimise balance when moving at

TABLE II   Conservation Status and Global Distribution of Crocodylidae (Ross, 1998; IUCN, 2006).
EN = Endangered; VU = Vulnerable; LR nt = Lower Risk (Near Threatened); and DD = Data Deficient

| Region | Common name | Species | IUCN status | Distribution | Habitat |
|---|---|---|---|---|---|
| Afrotropics | African gavial | Crocodylus cataphractus | DD | Western and central Africa | Swamp forest, streams |
| | Nile crocodile | Crocodylus niloticus | LR | Egypt, sub-Saharan and southern Africa | Large rivers in savannah rivers, Savannah |
| | Dwarf crocodile | Osteolaemus tetraspis | LR | Western and central Africa | Swamp forest, streams |
| Neotropics | American crocodile | Crocodylus acutus | VU | Central America | Fresh and brackish wetlands |
| | Orinoco crocodile | Crocodylus intermedius | CR | Lower Orinoco Basin | Open river banks |
| | Morelet's crocodile | Crocodylus moreletii | DD | Atlantic Mexican coast | Fresh and brackish wetlands |
| | Cuban crocodile | Crocodylus rhombifer | EN | Zapata swamp, Cuba | Swamps |
| Asian tropics | Philippine crocodile | Crocodylus mindorensis | CR | The Philippines | Marshes and ponds |
| | Mugger | Crocodylus palustris | VU | India, Pakistan, Nepal, Sri Lanka | Rivers |
| | Siamese crocodile | Crocodylus siamensis | CR | Southeastern Asia | Swamps and rivers |
| | False gavial | Tomistoma schlegelii | DD | Sundaic Region | Swamps, rivers |
| Australo-Papuasia | Australian freshwater crocodile | Crocodylus johnstoni | LR | Northern Australia | Wetlands and rivers |
| | New Guinea crocodile | Crocodylus novaeguineae | LR | New Guinea | Wetlands |
| | Saltwater crocodile | Crocodylus porosus | LR | Coastal India through Southeast Asia to northern Australia | Lowland aquatic habitats, coastal waters |

lower speed (Seebacher *et al.*, 2003). Crocodilians include 23 species subdivided into three families of unequal importance and distribution (Table II). The Crocodylidae (14 species) occur throughout the tropics and include *Crocodylus niloticus* and *C. porosus*, which are the dominant crocodilians in the Afrotropics and in Australo-Papuasia respectively. The Alligatoridae (eight species) represent a central and southern American crocodilian radiation, with only one member, the rare *Alligator sinensis*, present in Asia (Table III). The Gavialidae are represented by the Indian gharial, *Gavialis gangeticus*. All species are predators on fish, crustaceans, aquatic birds, small and larger mammals, although juveniles may also take insects. Unfortunately, there is little evidence in the literature of any ecological or structuring effect of crocodilians on prey populations. Adult crocodiles have been widely hunted by humans while hatchlings and juveniles are eaten by many other semi-aquatic and riparian vertebrates.

Most crocodilians prefer shallow and slow-flowing freshwater habitats at low altitudes, including floodplain wetlands, river channels and man-made habitats. Some *Crocodylus* spp. (*C. palustris* and *C. porosus*) live in brackish water and access estuaries and mangroves, but *C. niloticus* can dehydrate if exposed to high salinity for long periods (Leslie and Spotila, 2000). All crocodilians are highly sensitive to temperature, which has particular effects on

*TABLE III*   Conservation Status and Global Distribution of Tropical and Subtropical Alligatoridae and Gavialidae (Ross 1998; IUCN, 2006). EN = Endangered; VU = Vulnerable; LR nt = Lower Risk (Near Threatened); and DD = Data Deficient

| Region | Common name | Species | IUCN status | Distribution | Habitat |
|---|---|---|---|---|---|
| Asia | Chinese alligator | *Alligator sinensis* | CR EN | Lower Yangtze basin Indus, Ganges, Mahanadi, and Brahmaputra Rivers | Riverine wetlands, swamps, farms |
| | Gharial (Gavialidae) | *Gavialis gangeticus* | | | Slow sections of large rivers |
| Americas | American alligator | *Alligator mississippiensis* | LR | Southeastern United States | Swamp, marsh, streams, estuaries |
| | Caiman | *Caiman crocodilus* | LR | Central America and Amazon basin | Lowland aquatic habitats |
| | Broad-snouted caiman | *Caiman latirostris* | LR | Paraná and Saõ Francisco Rivers | Floodplain wetlands |
| | Yacaré | *Caiman yacare* | LR | Paraná and Paraguay Rivers (Pantanal) | Marsh, swamp, wetland |
| | Black caiman | *Melanosuchus niger* | EN | Amazon River basin and beyond | Rivers and floodplain wetlands |
| | Dwarf caiman | *Paleosuchus palpebrosus* | LR | Amazon and Orinoco Rivers and beyond | Forest streams |
| | Smooth-fronted caiman | *Paleosuchus trigonatus* | LR | Upper Amazon basin and beyond | Forest streams |

reproduction and sex ratios of offspring. Changes to habitat structure, such as alterations in riparian vegetation may change the thermal environment and can have significant consequences for their populations. In the Greater Saint Lucia Wetland Park (South Africa), for example, riverbank shading by the exotic grass *Chromolaena odorata* cools riparian habitats and is responsible for a female-biased sex ratio and reduced survival of embryos in *C. niloticus* nests (Leslie and Spotila, 2001). Reproductive behaviour of crocodilians is complex and involves protection of the young and nests by females. Some species (e.g. *C. palustris* and *Alligator sinensis*) dig burrows in the river bank that are occupied during droughts or colder periods of the year.

The Nile crocodile (*Crocodylus niloticus*) is the best studied crocodilian species with a distribution that formerly included the entire African continent plus Israel and Jordan. It is the only extant crocodile in the Nile and Zambezi River systems and throughout southern Africa, but is absent from the north and much of the west of the continent (Ross, 1998; see also Table II). Hunting for skins and meat has depleted some populations despite the fact that crocodile harvesting is illegal in most African countries; while the exploitation of captive populations is becoming common. The Slender-snouted African gavial (*C. cataphractus*) used to be broadly distributed across western and central Africa, where it inhabited densely-vegetated riverine corridors (Waitkuwait, 1989). This poorly-known species has been depleted over much of the remainder of its former range and is now seriously threatened (Table II); the largest remaining population is in the Ogoue River (Gabon), but survey data are scant (Ross, 1996, 1998). The Dwarf crocodile (*Osteolaemus tetraspis*: Fig. 2) is another poorly-known species restricted to small streams and swamps of western and central African lowland equatorial

*FIGURE 2*   The African Dwarf crocodile, *Osteolaemus tetraspis*, caught for the bush meat market (Source: L. Luiselli, unpublished).

forests, and is often sympatric with *C. cataphractus* (Kofron, 1992; Ross, 1996). During rainy periods it moves about on land, mainly at night, and colonises small floodplain and forest pools in the riparian zone (Riley and Huchzermeyer, 1999). Breeding takes place at the onset of the wet season, when Dwarf crocodiles build nests out of mounds of vegetation among tree-trunks and roots 1–2 m above the water level. They are common in some areas such as the Niger delta where *C. niloticus* is relatively depleted. They feed mainly on crabs and molluscs and do not take larger prey (Luiselli *et al.*, 1999). Habitat partitioning among the three sympatric African crocodilians has been reported in Liberia and southern Nigeria (Kofron, 1992; Akani *et al.*, 1999), with *C. niloticus* abundant in mangrove swamps and river mouths, *C. cataphractus* along forested river corridors, and Dwarf crocodiles in smaller streams and river tributaries.

Like their African relatives, the caimans of South America (Table III) have long been exploited by humans for meat and skin. The Black caiman (*Melanosuchus niger*) the largest crocodilian on the continent, was formerly hunted so intensively that its fat was mixed with petrol to run electricity generators (Best, 1984), and populations are now significantly depleted (Brazaitis *et al.*, 1996; da Silveira and Thorbjarnarson, 1999). Decreases in *M. niger* prompted the exploitation of smaller species, such as the Spectacled caiman (*Caiman crocodilus*), known to have poorer quality skin but tasty meat. In consequence, this caiman is threatened in parts of its range and is considered locally endangered in some areas (Rebelo and Magnusson, 1983; Peres and Carkeek, 1993; Brazaitis *et al.*, 1996). The Yacaré caiman (*C. yacare*) and the Broad-snouted caiman (*C. latirostris*) are subject to similar pressures from humans. Besides direct hunting, the most critical impact upon caiman populations is from extensive traditional gold mining activities with consequences that include deforestation, habitat degradation and water pollution by mercury and lead (Brazaitis *et al.*, 1996). A ban on caiman hunting and trade in skins was put in place by the Brazilian authorities in 1967, but enforcement and control has been weak (Peres and Carkeek, 1993). Most *C. crocodylus* are caught at sizes which correspond with sexual maturity, but this is not so for the larger *M. niger* and its populations are thus more vulnerable to hunting impacts (Rebelo and Magnusson, 1983; Herron, 1991). Black caiman do not breed until they are 10–12 years of age and more than 2 m long (Brazaitis *et al.*, 1996).

The natural habitat of *Melanosuchus niger* includes floodplain lakes, slow-flowing streams and riparian wetlands where it feeds on small fishes, amphibians, crustaceans and snails, occasionally on birds and small mammals (Best, 1984). Breeding coincides with the low-water period, and eggs are deposited in a nest of rotting leaves that is guarded for 5–6 weeks. The habitat use of *Caiman crocodylus* is more varied; they are common in floodplain lakes and streams and abound in the Pantanal where recent estimates suggest there may be as many as 3.5 million individuals (da Silva *et al.*, 2001). Nests are built of vegetation in the riparian zone during the high-water period, and this may be an adaptation to reduce the chances of inundation by further water-level increases (Best, 1984). Caimans may undertake seasonal migrations: for example, *C. yacare* disperses from the riparian habitats in the middle Rio Negro during the wet season, but returns at the beginning of the dry season (Wang *et al.*, 2004).

Schneider's smooth-fronted caiman (*Paleosuchus trigonatus*) and the Dwarf caiman (*P. palpebrosus*) are typical of small forest streams, fast currents and rapids, and considered to be ecological equivalents of the African *Osteolaemus tetraspis* (Ross, 1998). *Paleosuchus palpebrosus* is relatively rare but may be present in a wider variety of habitats. *Paleosuchus* spp. may be the most terrestrial crocodilians of the Amazon; adult *P. trigonatus* catches other reptiles and mammals significantly more often than fish, and probably represents the largest biomass of vertebrate predators in terra firme habitats (Magnusson *et al.*, 1987; Magnusson and Lima 1991). *Paleosuchus* spp. are generally solitary and less numerous than other caimans, which occasionally prey on them or with which they may compete. Investigations in the Amazonian Jau National Park (Brazil) showed evidence of significant negative correlation between the distributions of *P. trigonatus* and *Caiman crocodylus*, which are of a similar body size, and of mutual exclusion between *P. trigonatus* and the larger *Melanosuchus niger*. However, the nature of these interspecific interactions has yet to be uncovered (Rebelo and Lugli, 2001).

Six species, including crocodiles, gharials and one alligator (Table III), demonstrate the great phylogenetic diversity of the 'crocodilian niche' in Asia. They are all globally endangered, apart from the Saltwater crocodile (*Crocodylus porosus*), which remains common in Australia and in Papua New Guinea. For instance, the subtropical Chinese alligator (*Alligator sinensis*) is confined to a single reserve along the lower Yangtze River where its survival depends on captive breeding. The northernmost crocodilian in the world, *A. sinensis*, spends the cool months hibernating underground in burrows.

The Gharial (*Gavialis gangeticus*: Gavialidae), was once common in many of rivers across Pakistan, northern India and Burma (Table III). It is one of the largest and most endangered crocodilian species, and the few remaining populations are protected in reserves such as the National Chambal (Gharial) Wildlife Sanctuary on the Ganga River. The Sundaic equivalent of the Gharial, the False-gharial (*Tomistoma schlegelii*) barely survives in some parts of its range, and its habits and ecology have received little scientific study. Even more rare and critically endangered crocodiles persist on few Philippine islands (*Crocodylus mindorensis*) and in Cambodia (*C. siamensis*). Their distribution in the wild is highly fragmented and populations are small (see Dudgeon, 2000b); the majority of individuals may now be represented by individuals in captive-breeding programmes (IUCN, 2006). In contrast, the Marsh crocodile or Mugger (*C. palustris*) survives well in modified habitats and under brackish water conditions. Its range spans a large part of the Indian subcontinent, where it is vulnerable but not seriously threatened (IUCN, 2006). The main current threat to this and other crocodilians is probably habitat degradation. Direct exploitation alone probably does not threaten any crocodilian with extinction (Ross, 1998); however when associated to habitat loss, its impact is greatly increased.

Within Australo-Papuasia, crocodiles are less threatened than in other regions of the world (Table II). *Crocodylus novaeguineae* has the most restricted distribution and is endemic to New Guinea where it occurs in two disjunct populations in the north and the south of the island. The Australian freshwater crocodile (*C. johnstoni*) is well distributed across the Northern Territory in tropical Australia. It is a small- to medium-sized, narrow-snouted crocodilian, restricted to freshwaters but adapted to a broad range of prey. Its range appears to be limited by the distribution of the larger *C. porosus*. This latter species is the largest and the most widely-distributed crocodilian on earth, occurring from India across to southern China and southwards to Australia (Ross, 1998). *Crocodylus porosus* is an estuarine species that breeds in floodplain wetlands and lowland rivers. Populations have been substantially reduced in Asia due to intensive hunting for its valuable hide, but populations in Australia and New Guinea are protected and secure. Salt-water crocodiles sometime attack humans, and conservation initiatives have to combine breeding programmes with release plans, commercial exploitation, and crocodile control in problem areas.

## B. Lizards

Monitor lizards (Varanidae), originated over 90 million years ago in northern Asia (Bennett, 1995) and expanded during the Miocene to reach Europe, Africa and Australia. They include 46 species and are the largest lizards in Africa, Asia and Australia, but are absent from the Neotropics. Some monitors, such as the African monitor (*Varanus niloticus*) spend a considerable amount of time in and around shallow water (Lenz, 1995; de Buffrénil and Hémery, 2002). They dive readily and can swim strongly when disturbed (Bennett, 1995; Branch, 1998). Nineteen species of varanids are known from Indonesia, including the relatively-common Asian or Malayan water monitor (*V. salvator*) that also occurs in India, southern China, the Philippines and New Guinea. This varanid is rarely found far from streams but also occurs in mangroves and may even swim far out to sea. The nostrils located towards the tip of nose indicate the aquatic habits of *V. salvator*. Like other varanids, *V. salvator* is a catholic feeder with a diet that includes crustaceans, fish, insects, small mammals, birds, amphibians and other reptiles (Rashid, 2004). Varanids achieve their greatest diversity in the Northern Territory of Australia, where most of the 26 species present on the continent are found. Here, semi-aquatic varanids include the Large water monitor (*V. mertensi*), and the Mangrove monitor (*V. indicus*). The larger *V. mertensi* is confined to freshwater and shows significant adaptations to the aquatic habitat: it has pronounced lateral compression of the elongated tail, nostrils situated on top of the snout that become sealed when the animal submerges, and an ability to remain active at low body temperatures (Bennett, 1995). Its distribution range is limited to northern Australia while *V. indicus* is more widespread occurring in New Guinea and parts of Southeast Asia. In parts of Africa and Asia, varanids are hunted for their skins or as medicinal ingredients, but mainly for their meat.

Other semi-aquatic lizards include the agamid water dragons *Physignathus cocincinus* (Green or Chinese water dragon), *P. lesueurii* (Australian water dragon) and *P. temporalis* (Asian water dragon). Like some varanids, water dragons have a laterally-compressed tail and an opportunistic diet. They bask on overhanging branches, and dive readily into water when disturbed. *Physignathus lesueurii* can remain submerged for as much as one hour (Cogger, 1996). Among the Skinks (Scincidae), a small number of species have distinctive aquatic habits. These include the Eastern water skink (*Eulamprus quoyii*) in Australia and the Chinese waterside skink (*Tropidophorus sinicus*) in Southeast Asia. Both species have a diet that includes shrimps and aquatic insects.

## C. Snakes

Most snakes are able to swim well and some families (Laticaudidae and Hydrophiidae) have become well adapted to the marine environment. Despite this, relatively few snakes can be considered restricted to streams or freshwaters and the great majority of them live in tropical or subtropical latitudes where water temperatures are relatively warm (Greene, 1997). Semi-aquatic snakes tend to have small eyes positioned in a dorsal-anterior position compared to their terrestrial relatives and nostrils situated relatively high on the head (Fig. 3); their bodies are covered with small scales and the ventral scutes tend to be narrow. In addition, the lungs of some semi-aquatic snakes extend along much of the length of the body contributing to buoyancy

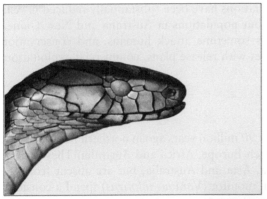
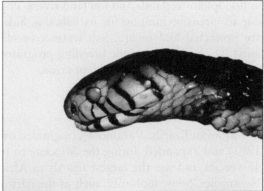
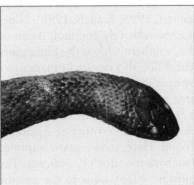

*FIGURE 3* Head morphology in relation to habitat use of snakes. From left to right: semiaquatic *Boulengerina annulata* (Elapidae) and *Afronatrix anoscopus* (Colubridae), arboricolous *Dendroaspis jamesoni* and ground-dwelling *Naja melanoleuca* (both Elapidae) (Illustrations by L. Luiselli and N. Pacini).

and helping them to stay afloat. Species of *Hydraethiops* and *Grayia* (Colubridae) have nostrils situated near the top of the head that have valves which can close when the animal submerges (Akani and Luiselli, 2001). The Water cobra *Boulengerina annulata* (Elapidae) is a superb swimmer that hunts fish and is widely distributed across the Congo basin and in parts of central Africa; individuals dive down to 25 m and may stay submerged for 10 min (Spawls and Branch, 1995). The possibility of efficient locomotion in water is associated with higher energetic costs of locomotion on land, this may account for the observation that some semi-aquatic taxa tend to have a lower reproductive investment than terrestrial snakes (Shine, 1988). Swimming speed in snakes is related to species and sex but not to body weight (Shine *et al.*, 2003) – a fact that is exploited by the large Boidae which are far more at ease while moving in water than on land. The threatened *Python sebae*, the largest snake in Africa continent, forages along large water courses hunting for riparian vertebrates including water birds and cattle (Luiselli, 2001). In Nigeria, they use rivers as dispersal corridors to penetrate towns where they feed on dogs, poultry and other domestic animals (Luiselli *et al.*, 2001). They are known to occasionally catch humans (Spawls and Branch, 1995).

Many semi-aquatic snakes are also semi-arboreal, and this may be attributed to the fact that they forage among submerged trees in inundated riparian forests and wetlands, and along rainforest streams. Frogs and tadpoles are a favoured prey, and some species also eat fish. There may be a seasonal switch from amphibians during the wet season to fish during dry periods when amphibians are less abundant, or a seasonal shift from terrestrial to aquatic prey as reported for colubrids in the genus *Natriciteres* (Luiselli, 2003). This may be accompanied by a change

*TABLE IV*   Semiaquatic Snakes of the Afrotropics: Diets and Habitats

|  | Common name | Genus | Species | Diet | Habitat |
|---|---|---|---|---|---|
| Colubridae | Black-bellied snakes | *Hydraethiops* | *laevis melanogaster* | Fishes | Rainforests, marshes |
|  | African water snakes | *Grayia* | *caesar furcata ornata smythii tholloni* | Fishes, anurans | Swamps, riparian forests |
|  | Marsh snakes | *Natriciteres* | *fuliginoides olivacea variegata* | Invertebrates, tadpoles | Rainforest ponds and marshes |
|  | Striped swamp snake | *Limnophis* | *bicolor bangweolicus* | Fishes | Streams |
|  | African brown water snakes | *Afronatrix* | *anoscopus aequifasciata* | Anurans, fishes | Rainforest streams |
| Elapidae | Water cobras | *Boulengerina* | *annulata christyi* | Anurans, fishes | Rainforest rivers and streams |
|  | Tree cobras | *Pseudohaje* | *goldii nigra* | Birds, frogs, fishes, rodents | Marshes, rainforest streams |
|  | Forest cobra | *Naja* | *melanoleuca* | Generalist; frequently eats fishes | Swamps, rivers, streams |
|  | Spitting cobra | *Naja* | *nigricollis* | Generalist; feeds also on fishes | Ponds and marshes |
| Boidae | African rock python | *Python* | *sebae* | Riparian vertebrates and cattle | Large rivers near forest |

in habitat use by some semi-aquatic snakes, which visit mangroves during the dry season when frog abundance decreases (L. Luiselli, personal communication). Further information on the diet and habitat use of semi-aquatic snakes in the Afrotropics are given in Table IV; note that this region has no snake guilds specialised for life in brackish habitats.

At least 24 species of tropical Asian snakes are semi-aquatic and associated permanently with freshwater (Kottelat and Whitten, 1996). Most of them are colubrids – mainly semi-nocturnal, viviparous, rear-fanged Homalopsinae (Gyi, 1970; Murphy and Voris, 2005) some of which also occur in mangroves (Karns *et al.*, 2002). Other more-or-less semi-aquatic snakes in the region include *Bungarus* banded kraits (12 riparian species: Elapidae) and *Python reticulatus* and *P. molurus*, which are similar to African Boidae. Adaptations to aquatic life in Homalopsinae include slit-like valvular nostrils, dorsally oriented eyes, a glottis that can be extended to fit into the internal nares, and a shallowly notched rostral that permits tight closure of the mouth (Gyi, 1970). The centre of radiation of Homalopsinae is Indochina and the Sundaic region from where they expanded into India, southern China, the Philippines and Australia. *Enhydris* includes 23 species (Murphy and Voris, 2005) primarily associated with vegetated freshwater ecotones. Some *Enhydris* (e.g. *Enhydris bennettii*) and several other Homalopsinae (species of *Bitia, Cantoria, Cerberus, Fordonia* and *Gerarda*) possess a gland to excrete excess salt (Murphy *et al.*, 2005). The degree of snake adaptation to aquatic habitats can also be shown by trophic specialisation. The Crab-eating water snake (*Fordonia leucobalia*), ranging from India, through Indonesia into New Guinea and northern Australia, possesses sturdy teeth adapted for crushing crustacean exoskeletons. The Tentacled fishing snake (*Erpeton tentaculatus*: Colubridae) of

Southeast Asia is an ambush predator of fish and the only snake known to also consume aquatic plant matter. Given their potential ecological role in wetlands, information about homalopsine ecology and population status is poor. This is despite that fact that in Tonlé Sap (Cambodia) large numbers of Homalopsinae, including the endemic *Enhydris longicauda,* are captured and traded for skin and as food for humans and farmed crocodiles representing what is probably the world's largest harvest of a single snake assemblage (Campbell *et al.*, 2006).

The wart snakes (Acrochordidae) are a small family of three strictly aquatic species inhabiting the Indo-Australian region (Greene, 1997). They are characterised by a rugose skin texture with bristle-tipped tubercles and, as is the case in some other water snakes, have dorsally-positioned eyes, valvular nostrils, and a flap for closing the lingual opening of the mouth. The File snake (*Acrochordus arafurae*), which lives in Australia and New Guinea, is active at night and ambushes large fish under water; the 2-m-long Javan wart snake (*A. javanicus*) specialises on eels and catfish (Greene, 1997). Rough scales on the body surface and tubercles that contain sensory organs aid in the detection and capture of fish, when the snake slams the coils together around the prey and holds it firmly using mouth and tail (Shine, 1991b). *Acrochordus arafurae* migrates seasonally into flooded grasslands to return to backwater pools during dry periods.

Australia hosts as many as 142 snake species; more than 90 of these are elapids that possess venom glands and fangs. Most elapid snakes found in the inland waters of Australo-Papuasia and South-eastern Asia are derived from marine-adapted clades (Slowinski and Keogh, 2000). Among them, the Hydrophiinae (sometimes considered to be an independent family from the rest of the Elapidae) comprise some 50 species worldwide and possess distinctive aquatic adaptations. *Hydrophis semperi* and *Laticauda crockeri* from the Philippines and Solomon Islands are freshwater species which belong to a primarily marine lineage (Shine and Shetty, 2001). Australian Grey swamp snakes *Hemiaspis* spp. take refuge in billabongs during the dry season, but migrate towards the floodplain to feed on fish during the wet season (Shea, 1999). The poisonous *Notechis* tiger snakes (*N. scutatus* and *N. ater*) that dwell in riparian areas within southern Australia are habitat generalists but have semi-aquatic adaptations such as high-burst swimming speed and can hold their breath for extended periods (Aubret, 2004). Other elapids (e.g. *Pseudechis porphyriacus*) are abundant in marshes, swamps and riparian habitats, and frequently enter water to forage and escape from predators (Shine and Shetty, 2001).

Other semi-aquatic snakes in Australia include the Water python (*Liasis fuscus*: Boidae) which swim readily, but spend time mainly in riparian habitats. They move to more elevated parts of the floodplain during the wet season following the migration of Dusky rats (*Rattus colletti*: Muridae) forced onto floodplain levées by rising water levels. Dusky rats represent the principal prey of Water pythons during the wet season, and python reproductive success is lined to the abundance of these rats (Madsen and Shine, 1996; Shine and Madsen, 1997). *Tropidonophis mairii* (Colubridae: Natricinae) is a semi-aquatic keelback common along much of the coastline of eastern and northern Australia (Webb *et al.*, 2001). It has glossy water-repellent scales, provided with a strong central ridge which assists locomotion on muddy substrates (Ehmann, 1992) and is the only Australian snake able to consume the highly-toxic invasive Cane toad, *Bufo marinus* (Shine, 1991a).

The Neotropics, and in particular the Amazon, has an exceptional diversity of semi-aquatic snakes. Most notable are four species of anacondas (*Eunectes* spp.: Boidae) provided with the characteristic anatomical adaptations for aquatic life of dorsally-situated eyes and nostrils (Greene, 1997). The Green anaconda *Eunectes murinus*, up to 7 m long – the largest snake in the world – is distinctively semi-aquatic. Like other anacondas, it feeds primarily on mammals (mainly peccaries), frogs and occasionally lizards. False water cobras (Colubridae: 17 species) such as the Queen snake (*Hydrodynastes gigas*) can reach 2 m in length and hunt fish by probing with their tails among giant shoreline bromeliads (Greene, 1997). Other semi-aquatic colubrids include some members of the genera *Helicops, Liophis* and *Tretanorhinus,* as well as *Pseudoeryx plicatilis* and *Hydrops martii*; the latter is a capable diver and can remain submerged for some time while hunting or escaping predators. The Common garter snake (*Thamnophis sirtalis*) is also semi-aquatic, and enters a hypometabolic state when conditions are unfavourable (Storey, 1996). Semi-aquatic elapids include also more than 50 species of *Micrurus* and *Micruroides*, and some representatives more closely associated with marine habitats. Finally, the Xenodontidae represent a large and poorly-known group of New World snakes comprising as many as 600 species among which several can be considered semi-aquatic.

## D. Turtles

Freshwater turtles are semi-aquatic specialists. Their phylogeny is ancient, complex and relatively poorly known, a factor which frustrates biodiversity assessments and conservation.

Several are able to combine breathing atmospheric air and oxygen exchange from the water, through the lining of the nasal passage and/or through the cloaca. They are also the most anoxia-tolerant, air-breathing vertebrates (Jackson, 2000). Freshwater turtles live in a variety of aquatic habitats and several taxa withstand brackish conditions. Moll and Moll (2004) have given an excellent overview of their ecology and habits. Most species forage in wetlands and slow-moving rivers and streams, and lay eggs within riparian zones and in sandbanks; only few are adapted to fast-flowing streams. Floodplain wetlands tend to support the greatest turtle density, especially sites that are frequently inundated (Bodie *et al.*, 2000), and migrations between floodplain wetlands and river channels may occur during the breeding period. For instance, *Podocnemis expansa* (Pelomedusidae) reside in Amazonian floodplain lakes when water levels are high but return to the river to nest when levels fall (Alho and Padua, 1982). Similar behaviour has been observed in Australasian *Carettochelys insculpta* (Carettochelyidae) (Doody *et al.*, 2002), whereas the New Guinea snake-necked turtle *Chelodina novaeguineae* (Chelidae) nests during the dry season and its small eggs have a long incubation period (Kennett *et al.*, 1992). A number of studies have documented downstream dispersal during periods of high water level by many river turtles (Pluto and Bellis, 1988; Moll and Moll, 2004) and although most species are highly mobile, *Hydromedusa maximiliani* (Chelidae), endemic to the Atlantic forests of coastal Brazil, is highly sedentary with a mean daily displacement of only 2 m; this has consequences for genetic differentiation among populations (Souza *et al.*, 2002). Turtles often require different habitats for feeding, nesting and aestivation, and different life stages may require different conditions. Turtle conservation is a challenge as they may use much of the riparian zone (Bodie, 2001) and many tropical species also undertake migrations that are related to the annual flood pulse. Detailed studies on the ecological significance of most tropical turtles are lacking; they consume invertebrates, fish, carrion, seeds and fruit as well as other plant material, and are likely to influence food webs and energy flow (Moll and Moll, 2004) – both as consumers and as a source of prey for other animals. Adults have few enemies (aside from humans), but hatchlings are predated by monitor lizards, crocodilians and large birds such as corvids and certain raptors, and predation risk may influence nest-site selection (Spencer, 2002).

Asian tortoises and freshwater turtles number around 100 species (van Dijk, 2000) – approximately one third of the global total – and are part of the most diverse testudinid assemblage in the world (Thirakhupt and van Dijk, 1994). At least three quarters of Asian turtles are endangered, 28% of them are critically endangered, and several are considered 'data-deficient' by the World Conservation Union (IUCN, 2006). At least one (the batagurid, *Cuora yunnanensis*) is considered extinct (van Dijk *et al.*, 2000) and another five *Cuora* species are critically endangered (IUCN, 2006). There are at least 18 turtle species in India, with a major turtle radiation centered within the Indo-Gangetic Plain and the riverine wetlands of Assam (van Dijk, 2000; Gupta, 2002). Five species of *Kachuga* (Bataguridae) from this region are considered to be at risk, among a number of other turtles. During the monsoon, two species of endangered giant freshwater turtles, *Aspideretes gangeticus* and *A. hurum* (Trionychidae), migrate upstream to breed in sandbanks along the Brahmaputra (Boruah and Biswas, 2002). Their survival is threatened as the migration coincides with the peak fishing season, and human activities have degraded nesting sites. A congeneric species, *A. leithii*, is classified as vulnerable while *A. nigricans* is extinct in the wild (IUCN, 2006). Another Asian trionychid, *Amyda cartilaginea*, was formerly widely distributed across lowland freshwaters, including peat swamps throughout Southern Asia, but its range and numbers have been depleted by overexploitation. Indonesia supports numerous freshwater turtles, perhaps as many as 29 species (Samedi and Iskandar, 2000), as does Burma (23 species, six endemics: Platt *et al.*, 2000), but virtually all species through Asia have been impacted by the demand for wild turtles as food in China and as ingredients in traditional Chinese medicine (Shepherd, 2000; Tana *et al.*, 2000; Cheung and Dudgeon, 2006). This serious problem is a reflection of a major overall global decline in turtles (Gibbons *et al.*, 2000; Moll and Moll, 2004; see also Chapter 10 of this volume), but is nowhere more serious than in Asia.

Although there are only 26 freshwater turtle species in Australia, this diversity can be considered high relative to the scarcity of surface water in the arid continent. Few taxa found in Asia extend beyond Wallace's Line aside from the circumtropical Trionychidae. These include *Pelochelys bibroni* which is endemic to Papua New Guinea and has been much reduced by hunting to supply markets in China; *P. cantorii,* present in New Guinea and Australia, is likewise endangered (IUCN, 2006). Most Australian turtles have a restricted distribution and only rather remote distant relationships with turtles living on other continents and, even within Australia, differentiation among taxa seems to have occurred in the remote past (Burbidge *et al.*, 1974). Studies of the carnivorous snake-necked *Chelodina rugosa* (Chelidae) in northern Australia shows that they have been able to sustain populations despite human impacts (Fordham *et al.*, 2004) because juveniles grow rapidly and mature early in highly-productive seasonal floodplain environments where selection for rapid growth is intense (Kennett, 1996). During the dry season *C. rugosa* aestivates and does not feed (Grigg *et al.*, 1986), but it becomes active in the wet season and nests under water while wetlands are inundated. Embryonic development begins as floodwaters recede during dry season, and hatchlings emerge at the start of the following wet season. Laboratory experiments confirm that prolonged immersion has no effect on egg mortality (Kennett *et al.*, 1998). Other endemic Australasian chelids include the endangered *C. mccordi* (endemic to Roti Island, Indonesia), three species of snapping turtles (*Elseya*), six species of short-necked *Emydura*, and the Fitzroy River turtle (*Rheodytes leukops*) which is unusual in that it is adapted to live in fast-flowing riffles (Gordos *et al.*, 2004). Adsorption of dissolved oxygen obtained by hyperventilation of the cloaca allows this species to survive in places where fast current would prevent turtles surfacing to breathe (Legler and Georges, 1993). *Rheodytes leukops* is threatened by flow regulation in the Fitzroy catchment which will alter habitat conditions (Tucker *et al.*, 2001). The vulnerable Pig-nosed turtle, *C. insculpta*, found only in New Guinea and northern Australia, is the sole Australasian representative of the Carettochelyidae. It inhabits streams as well as freshwater and brackish wetlands, and is the only turtle with paddle-like limbs and two claws. Two clutches of eggs are laid every second year, with the intervening 'rest year' being spent accumulating energy for breeding the following year (Doody *et al.*, 2003).

African turtles and tortoises number 45 species; the Pelomedusidae (African side-necked mud turtles) are particularly diverse. African turtles have not yet been subject to the intense hunting pressure evident in Asia (Moll and Moll, 2004; and see above), but populations of some species have been reduced by habitat degradation and others, such as the Madagascan pelomedusid *Erymnochelys madagascariensis*, are increasingly exploited for food (Kuchling, 1988).

Oil pollution of streams in the Niger Delta has reduced the number of turtles and caused dietary shifts among the survivors (Luiselli and Akani, 2003; Luiselli *et al.*, 2004). Organic pollution of streams also sustains leeches; these cause anaemia, favour bacterial and fungal infections, and the transmission of reptilian haemoparasites (Brites and Rantin, 2004; Luiselli *et al.*, 2004).

South America has 61 species turtles and tortoises, including 23 species of endemic mud and musk turtles (Kinosternidae). Some are of considerable significance to humans: the Yellow-spotted side-necked turtle (*Podocnemis unifilis*: Pelomedusidae) is widely distributed throughout the Orinoco and Amazon basins, and constitutes staple food during the dry season when animals are easier to capture (Thorbjarnarson *et al.*, 1993). Several species of freshwater turtles coexist in lowland stretches of major channels of the Amazon. In a tributary of the Tocantins-Araguaia, the Amazon turtle (*Podocnemys expansa*) and *P. unifilis* come ashore during flood recession period in search of pointbars within the interior meanders for that represent favoured sites for egg deposition (Ferreira Jr and Castro, 2003). The Amazon turtle builds 60-cm deep nests 2.5–3.3 m from the water; *P. unifilis* nests are shallower (15 cm) and dug only 1.5 m from the water. Depth and distance from the water edge affect rates of predation on nests (Kolbe and Janzen, 2002), and incubation period varies significantly according to sediment grain size, nest temperature and humidity. Differences in habitat range, extent of seasonal movements, and choice of habitat among sympatric Amazonian turtle species may indicate the evolution of habitat partitioning to minimise competition. Such habitat dependence makes turtles more vulnerable to the alteration of riparian habitats resulting from human activities and stream management practices (Bodie, 2001; see also Table V). Not all species are impacted equally by habitat change: Geoffroy's side-necked turtle (*Phrynops geoffroanus*) has been reported to be able to sustain high populations even in mildly-polluted streams in Brazil provided that substrate conditions were suitable and they were not subject to intense predation or hunting (Souza and Abe, 2000).

*TABLE V* The Impact of Stream Management Practices upon Freshwater Turtles (from Bodie, 2001)

| Practice | Direct effects | Impacts on turtles |
|---|---|---|
| Woody debris removal | Loss of riparian and channel habitat diversity | Loss of nesting sites; diversity and biomass reduction of prey |
| Riparian drainage | Loss of floodplain wetlands | Loss of nursery sites |
| Channelisation | Loss of habitat diversity; creation of fast-flowing streams | Loss of nesting sites; impeded hibernation in sediments; barriers to turtle migration |
| Reservoir development | Loss of habitat; creation of homogeneous lentic conditions | Loss of turtle diversity; greater density of predators; diseases |
| Flow regulation | Loss of natural flow regime; artificial floods and low flows | Impaired synchronisation of reproductive cycles; reduced nesting sites quality and hatchling survival |
| Reduction in sandbars/beaches | Habitat destruction | Nesting impairment |
| Agriculture, cattle grazing, urbanisation | Habitat destruction | Nesting impairment; pollution; prey reduction |
| Pollution and siltation | Habitat change; direct toxicity | Nesting impairment; sex reversal; morbidity; mortality; prey reduction |
| Monotypic landscape management | Loss of the necessary habitat diversity mosaic | Change in distribution and abundance |
| Unsustainable use (killing) | Direct mortality | Reduced populations and species richness |

## IV. AMPHIBIANS

A large majority of anurans (frogs and toads) develop from tadpoles and can therefore be considered semi-aquatic. Among caecilians (167 species), described as limbless tropical fossorial organisms, an unknown proportion of species includes an aquatic larval stage (Funk *et al.*, 2004); only few of them, such as the neotropical Typhlonectidae can be considered semi-aquatic. Lungless salamanders (Plethodontidae) reverted back to semiaquatic modes of life in the course of recent evolution (Chippindale *et al.*, 2004), but this taxon has a primarily temperate distribution.

As of early 2007, the number of amphibian species totalled 6158 (see www.AmphibiaWeb.org); nearly one third of them are considered as threatened. Populations of 43% of amphibian species are in decline, 165 may already be extinct, and more than 20% of the global species total are so insufficiently known as to be classified as 'data deficient' (IUCN, CI, and SN, 2004; AmphibiaWeb, 2007; see also Chapter 10). In Australia alone, for example, 40 out of 213 species are considered as threatened, a further 32 have experienced marked population reductions, and 15 are extinct in the wild (Hero and Morrison, 2004). Documentation of such declines and the searches for their causes has dominated most ecological research on amphibians in recent years. Collins and Storfer (2003) divided hypotheses about the causes of amphibian decline into 'well-known mechanisms' such as introductions of alien species, land-use change, and over-exploitation, and 'poorly-understood causes' such as global change (including changes in UV radiation and climate), environmental contamination, and infections. These factors may interact, creating complex cause and effect relationships. While amphibians have declined over the past 20 years, there has been a dramatic increase in the description of new amphibian species over the same period. A single survey conducted between 2000 and 2001 in Cambodia doubled the number of known species records for the country and described three species new to science (Ohler *et al.*, 2002). Increases such as this reflect surveys conducted in less accessible tropical forests and the availability of increasingly sophisticated techniques for measuring genetic variation (Köhler *et al.*, 2005). Key challenges to enhancing our knowledge of the global amphibian fauna (summarised by Banguera-Beebee and Griffiths, 2005) include a lack of taxonomic expertise to adequately survey the large number of species present in some parts of the tropics, the need for surveys of a significant proportion of the distributional range of many species, the difficulty of distinguishing short-term fluctuation in abundance from longer-term population declines, and the need to assess population changes over several generations.

The vulnerability of amphibians to human impacts and environmental change is exacerbated by the fact that many of them are confined to specific habitats and unable to disperse over significant distances. As a consequence, some workers believe that habitat destruction, alteration and fragmentation may be the most serious cause of current and future amphibian population declines and species extinctions (Dodd and Smith, 2003). A recent detailed herpetological survey in southern Madagascar (Lehtinen, *et al.*, 2003) demonstrated that the proximity of altered microclimates along the edges of fragmented dispersal ranges affects the distribution of amphibian and reptile populations, and that species that tended to avoid edges were relatively extinction-prone. Amphibians, which tend to be more aquatic in their habits (e.g. stream-dwelling mantellid frogs in Madagascar) have greater resilience to vegetation fragmentation than relatively terrestrial species (Vallan, 2000) and research on Mexican frogs has shown that semi-aquatic species are less sensitive to canopy loss and habitat change when the anthropogenic impact did not directly affect the presence or quality of water bodies (Pineda and Halffter, 2004). Nonetheless, there is clear evidence that many amphibians seem very responsive to habitat parameters, such as elevation, microclimates, vegetation diversity and structural complexity, as well as to the characteristics of freshwater habitats (e.g. Parris and McCarthy, 1999).

Over 80% of the 32 amphibian species found in the Ambohitantely Forest, Madagascar, lived within 10 m of permanent running water (Vallan, 2000); 7% occurred near intermittent streams and only 4% were associated with standing water. The six remaining species occurred far from aquatic habitats. Although lowland sites in the tropics seem to be more affected by anthropogenic impacts than uplands, declines in montane and stream-associated amphibian populations have been confirmed in Australia and the Neotropics (Hero and Morrison, 2004; Eterovick *et al.*, 2005). Morrison and Hero (2003) attribute this to intrinsic biological characteristics common to many high-altitude taxa such as shorter activity periods, longer larval periods, larger body size, later maturity, lower frequency of breeding, and so on. They stress that explanation of amphibian declines requires understanding of how species vulnerability interacts with concurrent threat factors (Hero and Morrison, 2004). In the wet tropics of Australia, low clutch size and high habitat specialisation – especially in association with stream habitats – were the primary ecological characteristics that distinguished declining species (Williams and Hero (2001).

The sensitivity of amphibians to environmental changes is often attributed to their permeable glandular skin and gelatinous eggs that lack a protective barrier to prevent the uptake of contaminants. Atrazine, organophosphates, carbamates as well as acid rain have undoubtedly had significant impacts on amphibian populations (Banguera-Beebee and Griffiths, 2005). Amphibian infections have also been correlated with a number of cases of decline. The chytrid fungus *Batrachochytrium dendrobatidis* is highly effective in impeding cutaneous respiration and osmoregulation in salamanders and frogs (Berger *et al.*, 1998). Chytridomycosis was recorded for the first time in southern Africa recently (Weldon *et al.*, 2004), but it may well have a cosmopolitan incidence among the Anura and Caudata and has been reported from 14 families and 93 species worldwide (Berger, 2000).

Little is known about the status of afrotropical amphibians, but they have certainly been affected by human activities (such as water pollution) in some areas (Akani *et al.*, 2004). Assemblages are certainly diverse: a survey of the rainforest of central Gabon yielded 41 amphibian species, only four were considered aquatic while 17 species lived within riparian zones close to river channels (Blanc and Fretey, 2004). There are perhaps up to 1000 amphibian species in Asia, a large number in Indonesia (87 anurans are known from Sumatra alone) where new species are discovered frequently (Kottelat and Whitten, 1996). Knowledge of Amphibian distribution Asia is distorted by the effects of logging, and over parts of Indonesia, the original distribution of species can hardly be defined (Inger and Iskandar, 2005). Large amphibians in Asia are threatened because they are heavily exploited for food, and even tadpoles are eaten in some areas (Lao PDR; Baird, 1999).

More than half of the global amphibian fauna occurs in the Neotropics (some 90% endemic), and virtually all temporary and permanent fresh waters are used as breeding sites by some species (Bertoluci and Rodrigues, 2002). Seasonally-inundated wetlands are preferred by many frogs as they are less likely to support large populations of tadpole predators. Eggs are often laid in shallow pools as an adaptation to reduce predation by fish. Another anti-predator adaptation is seen in leptodactylid females that build foam nests for egg deposition suspended on various substrates above the water surface. Interestingly, a similar habit was recorded in the Afrotropical racophorid *Chiromantis rufescens* (Fig. 4), which nests above streams in the Nigerian delta; shortly after birth, the young tadpoles fall directly into the water and swim away. Predation pressure seems to be an important determinant of the distribution of riparian frog species in the Amazon (e.g. Hero *et al.*, 2001) and has driven the evolution of a variety of anti-predator traits such as unpalatability and poisons in Anura throughout the Neotropics. There is evidence that amphibian assemblages in this region may be structured according to reproductive strategies. In the Atlantic forest of Boracéia, Brazil, where information on breeding by 28 anurans was collected, there was segregation in seasonal timing of breeding, and

(a)

(b)

FIGURE 4    Foam nests of *Chiromantis rufescens* (Rhacophoridae) in the Niger Delta (Photograph by L. Luiselli).

species-specific differences in activity periods and in the degree of reproductive opportunism (Bertoluci and Rodrigues, 2002.

## V. SEMI-AQUATIC MAMMALS

Unlike Cetacea and Sirenia, other non-marine mammal orders are only partially adapted to life in aquatic habitats. Despite this, the 'aquatic experiment' has been attempted during the evolution of virtually all mammal groups, leading to a remarkable array of anatomical, physiological and behavioural adaptations. Some, like the Platypus (*Ornithorhynchus anatinus*: Ornithorhynchidae), cannot be considered as tropical but many others are associated with the riparian zones, gallery forests and floodplains of tropical streams and rivers.

### A. Otters

The otters (Mustelidae: Lutrinae) are fully amphibious carnivores with adaptations to aquatic life that include palmed hindpaws, thick waterproof fur, and a particular conformation of the surface of the retina shared by hippos and whales. They consume a wide range of prey. Niche overlap between sympatric otters is minimised by a spatial segregation of hunting grounds, but there is evidence of dietary segregation especially during times of food scarcity (Rowe-Rowe and Somers, 1998). Despite eating more than other carnivores in relation to their body weight, otters exist close to their energetic limits (Kruuk, 1995). They are thus vulnerable to competition from fisheries and to pollution and habitat perturbations which affect their prey. The preferred habitat of most otters is secluded and well-vegetated stream or river banks, particularly those shaded by riparian trees (Perrin and Carugati, 2000; Perrin *et al.*, 2000). Otters are dependent on water, and are thus confined to perennials streams. This limits their distribution in parts of the tropics (East Africa for example) where streams dry up seasonally. The African Spotted-neck otter (*Lutra maculicollis)* has fully-webbed feet and is an agile swimmer. Prey is captured in the mouth (Rowe-Rowe and Somers, 1998), and preferred items include freshwater crabs, amphibians, slow-moving benthic fishes and *Tilapia* (Cichlidae) between 10 and 20 cm, but occasionally up to 60 cm (Kingdon, 1977). Parties of *L. maculicollis* sometimes cooperate in driving schools of fish to increase individual foraging success but, in parts of their range, these animals tend to be nocturnal and solitary (Kruuk, 1995), perhaps as a predator-avoidance strategy. The Cape clawless otter (*Aonyx capensis*) is relatively large (up to 16 kg) and has an extensive distribution from the Senegal Delta to the Upper Nile and down to South Africa. It prefers swamps, floodplains and alluvial wetlands. Only the hind feet are webbed; it has well-developed molars and strong cranial muscles that allow it to feed on unionid bivalves, turtles, crustaceans and *Potamonautes* crabs (Potamidae) as well as fishes (Kingdon, 1977; Somers and Purves, 1996). Prey items are captured with the forefeet. *Aonyx capensis* will undertake seasonal migrations in places where rivers and wetlands dry out for part of the year, returning at the onset of the wet season (Kruuk, 1995). Analysis of the pesticide and toxin concentrations of African otter scats has shown that *A. capensis* concentrates three times more pollutants than *L. maculicollis* due to the importance of crustaceans and other sediment-associated invertebrates in its diet (Mason and Rowe-Rowe, 1992). In central and western Africa, the range of *A. capensis* is partially restricted by the presence of the Congo clawless otter (*A. congica*), which inhabits rivers, streams and wetlands surrounded by gallery forest. It has relatively thin and short fur and it is the least adapted of the otters to aquatic life. The clawless forepaws (Fig. 5) are hairless and adapted to comb soft substrates and mud searching for invertebrates. It will also take amphibians and fish and, like other otters, it will attack small mammals and ravage bird nests. Crocodiles prey on otters and are the main cause for their scarcity in African rivers where crocodiles are abundant (Kruuk, 1995).

Two otter species inhabit the tropical latitudes of South America. They have overlapping ranges, but differ greatly in size. As its name suggests, the Giant otter (*Pteronura brasiliensis*) is the largest otter in the world. The distribution of the Giant otter has been substantially reduced due to hunting for its pelt, and some populations are seriously endangered. They are mainly confined to remote upstream localities in the Amazon, Orinoco and La Plata basins (Carter and Rosas, 1997), and remain widespread only in Surinam and in Guyana (Gutleb *et al.*, 1997). The preferred habitat of Giant otters is large slow-flowing rivers, oxbow lakes and floodplain swamps; they avoid strong currents (Schenck and Staib, 1998). Dens are built on higher grounds above the level of seasonal inundations; hunting occurs in open water and among marginal vegetation. Spraint analysis suggests high prey selectivity. Habitat surveys indicate that Giant otters are excluded from high-quality habitat by human disturbance, pollution and diseases spread by domestic animals – especially dogs (Schenck and Staib, 1998). Gold mining in the Amazon has resulted in mercury contamination of rivers and fishes to levels that render them unsafe for human consumption (Gutleb *et al.*, 1997). This pollutant is taken up efficiently in the intestine of otters and is subsequently transferred to otter pups and may be a contributing factor in recent declines of Giant otter populations in the Amazon.

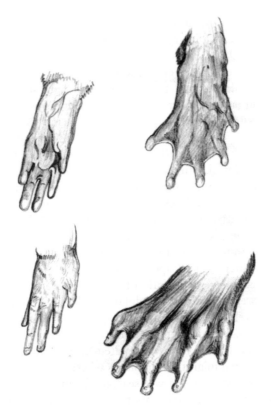

FIGURE 5     Fore- and hindpaws of the Congo clawless otter, *Aonyx congica* [Source: Kingdon (1974)].

The Long-tailed otter [*Lontra* (= *Lutra*) *longicaudis*] is poorly known and classified as data deficient by the IUCN (2006). Its range extends from Mexico to northern Argentina and Uruguay but is highly fragmented. These otters make dens in natural cavities along riverbanks above the high-water level (Pardini and Trajano, 1999). Long-tailed otters are more tolerant of human disturbance than Giant otters, and may inhabit degraded forest where humans are present, but they are not tolerant of pollution or substantial habitat alteration.

Another mustelid adapted to swimming and hunting in Neotropical freshwaters is the rare and poorly-known Grison (*Galictis vittata*), distributed from southern Mexico southwards across the Amazon Basin. It is a solitary, nocturnal riparian carnivore, feeding on amphibians, fish and other small vertebrates (Emmons, 1997).

Otters of tropical Asia include the Small-clawed otter (*A. cinereus*), the Smooth-coated otter (*Lutrogale perspicillata*), and the Hairy-nosed otter (*L. sumatrana*). They occur mostly in freshwater, but also inhabit mangroves, all three can occur sympatrically. The Smooth-coated otter and Small-clawed otter can also coexist with the Eurasian otter (*Lutra lutra*) in parts of Thailand, where spraint markings are used to delimit foraging grounds (Kruuk, 1995). *Lutrogale perspicillata* The Smooth-coated otter has the widest range of the three tropical Asian otters and is distributed from the Himalaya throughout the Indian subcontinent eastwards to western China and southwards to Indonesia. (An isolated subspecies is endemic to the southern Iraqi marshes.) It feeds mainly on fishes, and has strong preferences for prey type despite seasonal change in fish species composition (Hussain and Choudhury, 1998); it may also take birds such as ducks when rearing pups. Asian *Lutra* spp. are also specialist piscivores, but *Aonyx cinereus* feeds on crustaceans, especially crabs, as do African *Aonyx* spp. Human overexploitation of river fisheries are a potential threat for otter populations in Asia, although mutual associations between otters and fishermen are reported from China and India where otters have been used to drive fish into nets (Kingdon, 1977).

## B. Mongooses and Civets

The African Marsh mongoose (*Atilax paludinosus*: Herpestidae) shares several adaptations for life in freshwater with otters. The fur is thicker than other mongooses and, while it lacks webbed feet, it has soft long-fingered forepaws reminiscent of *Aonyx* spp.; this characteristic is not present in other herpestids. The Marsh mongoose feeds mainly by 'dredging' for food in shallow water, combing through the bottom substrate of shallow swamps and wetlands. It is highly opportunistic and takes a variety of prey, using stones to crack hard-shelled snails or unionid bivalves; it may also consume seeds and fruit (Rowe-Rowe and Somers, 1998). The Marsh mongoose is less dependent on water than otters are; this enables it to distribute widely across areas with few permanent streams and to survive severe dry seasons. Its diet and habitat use can be influenced by competition with otters, especially during periods of resource scarcity and drought (Somers and Purves, 1996). Other African mongooses that utilise riparian habitats are four species of cusimanse (*Crossarchus* spp.). They eat a range of foods and often forage in mall bands, perlustrating shallow water in search of crabs and other aquatic prey. In the Asian tropics, two species of *Herpestes* mongooses occupy a variety of habitats and their diet includes crustaceans, frogs, fish, and other small vertebrates. As its name suggests, the Crab-eating mongoose (*H. urva*) is often associated with water and is a good swimmer.

Some civets (Viverridae) have also evolved adaptations to aquatic life. The rare Aquatic genet (*Osbornictis piscivora*) inhabits the right bank of the Congo/Lualaba River and the Semliki Valley in eastern Congo. It spots fish and frogs from the water surface and although it may dive in to pursue prey, it avoids getting wet. The palms and soles of the fore- and hindpaws lack hair, which may be an adaptation for catching aquatic prey. The Aquatic genet is restricted to forested streams and localities where competition from fish-eating birds is limited (Kingdon, 1990). The endangered *Cynogale* otter civets comprise two species: the Sunda otter civet (*C. bennettii*) in peninsular Malaysia and possibly parts of Indonesia, and the enigmatic Tonkin otter civet (*C. lowei*) known only from northern Vietnam. *Cynogale* spp. display several striking adaptations to semi-aquatic life including a broad muzzle and expanded upper lip that allows the rhinarium to remain above the water line while the animal is submerged. Both the nostrils and ears can be closed by flaps, and the feet are broad and webbed to facilitate swimming. *Cynogale bennettii* lives in close proximity to streams and wetlands, and may ambush prey from a submerged position. Prey items include crustaceans, birds, and small mammals coming to drink.

## C. Otter Shrews

The otter shrews (Tenrecidae: Potamogalinae) are confined to mainland Africa, and are stenotopic habitat specialist that live in burrows along stream- and riverbanks. They are nocturnal, and burrow entrances are submerged. Giant otter shrews (*Potamogale velox*; Fig. 6) are rapid and agile swimmers, although they do not have webbed feet, and have nostril flaps that close when the animal is submerged. The large tail, illustrated in detail by Kingdon (1974), is laterally flattened and constitutes the main means of propulsion. Swimming involves muscular side-to-side movements of the lumbus that cause sinusoidal motion of the tail. This locomotory adaptation is exceptionally well developed in Giant otter shrews in which the last sacral vertebra is freed for increased mobility (Gingerich, 2003). By contrast, and despite superficial similarities in their locomotion, movement in otters is dominated by the hind limbs. The locomotory adaptations of the Giant otter shrew, combined with the loss of the clavicle, indicate the extreme

FIGURE 6    The Giant otter shrew, *Potamogale velox*, of Africa [Source: Kingdon (1974)].

specialism of *P. velox* and its deviation from the ancestral tenrecid body plan (Kingdon, 1974). Giant otter shrews range from the Cross River in Nigeria through central and eastern Africa down to the Zambezi (Timberlake, 1998). Preferred preys are crabs, amphibians and fish that are detected by scent or touch, involving the well-developed whiskers, because their eye-sight is weak.

The rare and endangered Dwarf otter shrew (*Micropotamogale lamottei*) occurs within a 5000-km² area within the Guinea highlands (especially Mount Nimba) and is one of the few semi-aquatic vertebrates associated with mountain streams and rivers. Unlike the Giant otter shrew, its tail is not conspicuously flattened. The Ruwenzori otter shrew (*Mesopotamogale ruwenzorii*) is also endangered, surviving in small streams and rivers in central Africa (Zaire and western Uganda). It is the only Potamogalinae with webbed feet, it digs tunnels and builds grass retreats along stream banks. Both of these otter shrews are opportunist predators that eat crustaceans and other aquatic invertebrates, frogs and some fish. They are sometimes caught in fish traps (Nowak, 1999).

Close relatives to otter shrews are the tenrecs of Madagascar, a lineage dating back 50 million years ago when tenrecids were widespread in Africa (Kingdon, 1990). The Web-footed tenrec (*Limnogale mergulus*) is a rare predator of small frogs, fish, and shrimps confined to eastern Madagascar. Once thought to be extinct, it is now considered endangered (IUCN, 2006). Related taxa are the Rice tenrecs (three species of *Oryzorictes*) and the Long-tailed tenrecs (at least 13 *Microgale* species) that appear to be associated to aquatic habitats although their ecology has received little study.

## D. Hippos

Both extant species of Hippopotamidae are confined to Africa: the Hippo (*Hippopotamus amphibius*), which was formerly widely distributed, and the forest-adapted Pigmy hippo [*Hexaprotodon* (*Choeropsis*) *liberiensis*]; their ranges do not overlap. Recent research implies a phylogenetic relationship to whales dating back over 65 million years (MacDonald, 2001), and ancestral hippopotamids were more diverse and widely distributed in Africa, Madagascar and Europe than is presently the case (Grubb, 1993; Manlius, 2000). Hunting by early human populations has caused extinction of three hippo species and caused great reductions in the distribution and abundance of *H. amphibius*.

*Hippopotamus amphibius* inhabits the lower reaches of rivers and spends the day in water emerging at night to feed on short grass swards. They seldom move more than 3 km from water and visit previously-grazed floodplain sites where their feeding activities maintain 'hippo lawns' (O'Connor and Campbell, 1986). Hippos have stumpy legs, splayed toes and no protective hooves reflecting the amount of time spent in water where their movement is facilitated by long supple back and kicking movements of the hind limbs (Fig. 7; see also Kingdon, 1974). Water is used as a daytime refuge, and the ears and nostrils have muscular valves allowing submergence. Hippos have a marked preference for deep backwaters rather than for fast-flowing reaches.

FIGURE 7   Hippo (*Hippopotamus amphibius*) movement in water [Source: Kingdon (1974)].

They are gregarious during the day, and form unstable groups of females and bachelors (Klingel, 1991); mating takes place in water, and males fight fiercely over access to females. Hippos have thick skin devoid of sweat glands, and thus they depend on water to cool their bodies (MacDonald, 2001). They appear physiologically adapted to minimise energy expenditure and accumulate large amounts of fat (Kingdon, 1974).

Principal threats to hippopotamus survival include the loss of grazing lands to cultivation as river banks and floodplains are converted to agriculture and large-scale irrigation schemes are developed. In some instances, however, they may benefit from artificial weirs and irrigation dams that retain deep water during droughts (Jacobsen and Kleynhans, 1993; Viljoen and Biggs, 1998). Their habit of concentrating in one or a few water bodies during the day makes hippos vulnerable to hunting. They are usually killed because of crop damage and attacks on fishermen and, more occasionally, for ivory from their large teeth or for meat (Eltringham, 1993).

The smaller Pygmy hippopotamus has relatively longer limbs in relation to body shape, a proportionally smaller and narrower head, and its orbits are not raised above the skull roof as in the Hippo. The Pigmy hippopotamus could be confined to the forests of Guinea and Liberia, with a total population of not more than 3000 individuals; it is classified as endangered (IUCN, 2006). They are solitary or occur in small groups, and live in riverine forest. Unlike the Hippo, they are browsers rather than grazers, and feed nocturnally on roots, water plants and fallen fruit, spending much of the day in the water. The feet appear more adapted to terrestrial life than those of its larger relative, although in other respects they share similar adaptations to aquatic life. In terms of ecology and social behaviour, Pygmy hippos are similar to tapirs (see below); mating occurs both in water and on land (MacDonald, 2001). Conflicts with humans are rarer than in the case of the Hippo, and threats to the Pygmy hippopotamus from hunting are less.

## VI. RIPARIAN MAMMALS

Many terrestrial mammals have a close relationship with the riparian zones of tropical streams. For some rodents and shrews, this has resulted in anatomical adaptations to feeding and moving in water. Others, such as monkeys, buffaloes, antelopes and wild pigs developed migration patterns to make use of riparian sites for feeding and access to water especially during seasonal dry periods. Several mammals are attracted to rivers and streams because of the potential prey that can be found there, and others take advantage of river corridors to migrate through otherwise inhospitable habitats. As a result, riparian zones (including gallery forest and floodplains) can be host to a diverse array of mammal species. The Pantanal of tropical South America, for example, is home to 122 species of mammals among which 15 are carnivores (Fonseca *et al.*, 1996).

### A. Felids

Tigers (*Panthera tigris*) are the largest extant carnivores in Asia. Despite their adaptability to a wide range of habitats and altitudes, tigers are able swimmers and their activities are often associated with wetlands, including swamp forest and mangroves. The riparian zones of large rivers and grassy floodplains constitute the preferred habitat and offer a sufficient density of large-bodied ungulates (Sunquist *et al.*, 1999). Prey are usually ambushed from dense vegetation, but swamp deer such as the Barasingha (*Cervus duvauceli*: Cervidae) that take refuge from predation by submerging are not safe from swimming tigers. Other Asiatic riparian felids include the Fishing cat and Flat-headed cat (*Prionailurus viverrinus* and *P. planiceps*), the Marbled cat and Clouded leopard (*Pardofelis marmorata* and *P. nebulosa*), and the Bay cat (*Catopuma badia*). Information on their habitat use and conservation status in the region is given by Dudgeon (2000b). African felids are not particularly associated to riparian habitats. In South America, Jaguars (*Panthera onca*) are the largest riparian predators after caimans. Peruvian jaguars catch fish, aquatic reptiles such as Spectacled caiman and freshwater turtles (*Podocnemis* spp.), as well as Capybara (see below) and riparian ungulates.

### B. Racoons and Marsupials

The Crab-eating racoon (*Procyon cancrivorous*: Procyonidae) is confined to waterside habitats throughout inter-tropical South America, from Costa Rica to northern Argentina. Its toes are not webbed but the paws are large and it is an agile climber and swimmer. Crab-eating racoons are nocturnal and find refuge in tree hollows during the day; they forage at night on fish, frogs, crabs and other invertebrates, as well as fruit and nuts. Among the Neotropical marsupials, The Water opossum (*Chironectes minimus*: Didelphidae) is the only species significantly adapted to streams and wetlands where it excavates burrows along the banks. It extends throughout the Neotropics from Mexico to Argentina, but this range is now highly fragmented. The Water opossum has a streamlined, otter-like body covered by water-repellent fur, and swims using its webbed hind feet. The pouch is watertight and opens at the rear, enabling the female to swim while carrying young (Nowak, 1999). Prey are detected by touch using sensitive naked forepaws; the diet consists mainly of fish, shrimp and other crustaceans although aquatic vegetation and fruit are eaten occasionally (Nowak, 1999). Other didelphids, such as the Thick-tailed opossum (*Lutreolina crassicaudata*) occur in riparian forests subject to seasonal inundation and swamps. They swim and feed mainly on fish and aquatic invertebrates, but they lack the adaptations to aquatic life seen in the Water opossum.

## C. Rodents

The rodents include a host of stenotopic and opportunistic species, distributed across the widest range of habitats of any mammalian order. Some are associated with wetlands and running waters, but knowledge of the ecology of most of them is fragmentary, and only a few examples are given here.

### 1. Mice and Rats

The Muridae is the largest family of mammals (numbering over 1300 species), with a great variety of adaptations to life in and around water. Oddly, however, there are no water rats in the Asian tropics. The Neotropical web-footed marsh rats (*Holochilus* spp.) are distinctive in that they feed almost exclusively on grass and herbs on floodplains, along stream banks or in marshes, occasionally they may take molluscs. Fish-eating rats (*Ichthyomys* spp.) are widely distributed in tropical America (Table VI). They have large, partially-webbed hind feet, stout whiskers, small eyes and ears, and a tail with a bristly underface. They are excellent swimmers. Venezuelan *Ichthyomys pittieri* sometimes capture small fish but more often feed on crustaceans and other aquatic invertebrates. The Central American mice of the genus *Rheomys* include *R. underwoodi*, which is well adapted to life in water and has dorsally-situated nostrils that have posterior valves that exclude water (Starret and Fisler, 1970). The hind feet are long and very large, with compressed lateral digits; these water mice often catch fish (Nowak, 1999). The South American water mice, placed in the genus *Neusticomys*, are typically found along fast-flowing streams in the eastern Andes (Table VI). They have sharp incisors with surfaces inclined towards each other allowing them to catch and hold slippery aquatic prey (Nowak, 1999). The Ecuador fish-eating rat (*Anotomys leander*) is an endangered rodent (IUCN, 2006) confined to high altitude streams and wetlands. It has well-developed vibrissae, velvety fur, and the ears are sealed during immersion by a muscular membrane. The broad hind feet are not fully webbed but have stiff hairs which aid in swimming (Nowak, 1999). These and other Neotropical mice and rats listed in Table VI are sometimes treated as distinct from the Muridae, and most are placed in the Sigmodontinae within the Cricetidae which contains all endemic South American rodents. The genus *Ichthyomys* is placed in the Ichthyomyinae.

*TABLE VI* Semiaquatic Murid Rodents of the Neotropics: Adaptations to Semiaquatic Life, Diet and Habitats

| Common name | Genus (No. of species) | Adaptations | Diet | Habitat |
|---|---|---|---|---|
| Rice rat | *Oryzomys* (36 spp.) | Long tail | Shrimps, crabs, fruit, seeds | Rainforest, marshes |
| Water rat | *Nectomys* (3 spp.) | Partially webbed feet; broad palm | Crabs, fruit, fungi | Swamps, riparian forest |
| Crab-eating rat | *Ichthyomys* (4 spp.) | Large, partially webbed feet | Crabs, small vertebrates | Streams |
| Central American water mouse | *Rheomys* (5 spp.) | Nostrils with valves; webbed feet | Small vertebrates | Streams |
| South American water mouse | *Neusticomys* (4 spp.) | Inclined incisors; narrow feet | Some species eat crabs | Rainforest |
| Marsh rat | *Holochilus* (4 spp.) | Partially webbed feet | Grass, vegetation | Marshland |
| Chibchan water mouse | *Chibchanomys trichotis* | Feet not webbed; behavioural | Insects, crustaceans | Streams |
| Fish-eating rat | *Anotomys leander* | Stiff hair; ear membranes | Insects, crustaceans | Streams |
| Water rat | *Scapteromys tumidus* | Elongated, with long claws | Insects, grass, seeds | Marshes, wetlands |

Many African rats and mice are sometimes treated as members of the family Nesomyidae (and its various subfamilies) while others are still referred to the Old World Muridae along with all of the Asian species. In the Afrotropics, an herbivore-frugivore semi-aquatic rodent niche can be defined, including species in the genera *Otomys, Pelomys, Dendromus, Dasymys* as well as *Delanymys brooksi, Thryonomys swinderianus* and *Mus bufo* (Kingdon, 1974). Although most of them do not show marked anatomical adaptations to aquatic life, they occur only in close proximity to permanent water and swim readily. The dendromurine genus *Dendromus* (four semiaquatic species) and *Delanymys brooksi* (Delanymyinae) have prehensile tails adapting them to climb and nest in riparian forest. A second Afrotropical niche includes carnivorous semi-aquatic rodents that display significant adaptations to water. Among them are species of the genera *Lophuromys, Malacomys* and *Deomys, Nilopegamys plumbeus* and *Colomys goslingi*. According to Kingdon (1971, 1974), the genera *Colomys* and *Malacomys* illustrate two alternative evolutionary strategies in the carnivorous murid adaptation to the aquatic medium with the former having a rather primitive tooth structure and modified anatomy while the latter has specialised dentition and a narrow skull but anatomical adaptations are otherwise limited. Both feed on molluscs, worms, and crustaceans, and occur in forested parts of central Africa with *C. goslingi* having a more eastern distribution, whereas *Malacomys* is commoner in the west. The former replaces the latter in wetter habitats where ranges overlap, and is found along muddy rivers and in swamps that flood regularly (Kingdon, 1974). *Colomys goslingi* hunts in water where it wades and sifts sediment with the front paws. The rare Ethiopian water mouse (*Nilopegamys plumbeus*) is known only from a mountain tributary of the Blue Nile, and is critically endangered (IUCN, 2006); it may even be extinct. It is the only African rodent to show the degree of adaptation to aquatic life and swimming ability evident in the Neotropical Ichthyomyinae and in the Australian Hydromyinae (Peterhans and Patterson, 1995). All three taxa share have relatively large brains that may have evolved in response to the demands of hunting in water, large snouts and with specialised vibrissae that may help detect movement under water, a pronounced cranial deflection allowing the nostrils to remain above water while the animals is in deep water, markedly reduced ears, dense soft fur, hairy tails, and elongated hairy feet used for swimming. The enlarged feet of *N. plumbeus* are different from those of *C. goslingi* or *Malacomys* spp., which belong to the 'wading murid' niche and have feet that are elongated and thin. They serve the purpose of raising the body so that these animals can hunt by wading in shallow water (Kingdon, 1974).

In Australo-Papuasia, a lineage of semi-aquatic rodents evolved within the mountain regions of New Guinea to produce 10 genera and 13 species of Hydromyini (Muridae: Hydromyinae) (Watts and Baverstock, 1994). They show convergence with the anatomy of the Neotropical Ichthyomyinae and are the Australasian murids most adapted to aquatic life. They are distinctive in that they lack hind molars and have strong incisors used to crack the mollusc shell and the carapaces of Crustacea. All Hydromyini have feet that are at least partially webbed, even species of *Pseudohydromys, Neohydromys* and *Meyermys* that have recently evolved a predominant terrestrial habit. *Hydromys* includes five species of large beaver rats that can attain 1.3 kg; one – *H. chrysogaster* – is widespread in Australia and New Guinea (Nowak, 1999). It lives in almost all types of permanent freshwater habitat and mangroves and hunts fish, crustaceans, frogs, waterbirds and small mammals carrying them to a preferred resting site before eating in a manner similar to otters. *Hydromys chrysogaster* has an enlarged, flattened and massive skull that might facilitate in foraging for food under stones and among riverbed substrates. They swim with eyes, nostrils and the dorsum of the head above the water surface; details of their swimming and locomotion are given by Fish and Baudinette (1999). Other Hydromyinae, such as the Mountain water rat (*Parahydromys asper*) and the Earless water rat (*Crossomys moncktoni*) live along fast streams in the mountains of New Guinea at altitudes above 600 m.

They have webbed feet and extensive vibrissae for prey detection; *C. moncktoni* has water-proof fur and a crest of hair along the underside of the tail that may serve as a rudder.

## 2. Coypu and Capybara

The Coypu or Nutria (*Myocastor coypus*: Myocastoridae) originated from wetlands in southern South America, West of the Andes. It is a semi-aquatic rodent, occurring on floodplains, backwaters, slow-flowing rivers, and wetlands, with thick water-repellent fur and feet armed by long claws. Coypu eat the stems, roots, tubers and leaves of a number of aquatic and shoreline plants, and are usually associated with well-vegetated open habitat. Coypu have been widely introduced outside their range where they were farmed for their fur, and feral populations showing evidence of adaptations to new conditions have become established (Guichón *et al.*, 2003).

The Capybara (*Hydrochaeris hydrochaeris*: Hydrochoeridae) is the largest extant rodent reaching approximately 50 kg. It is widely distributed across South America where it frequents dense vegetation on floodplains, wetlands and along river channels. These animals are also hunted and bred commercially for their meat. Capybara feed mainly on aquatic plants and grass, and breeding is synchronised with the annual flood cycle and the maximum productivity of floodplain grasses (Best, 1984). Capybara are amphibious and submerge up to the nostrils in water, where they are able to swim with agility due to the splayed webbed digits of their 'hooves'. They make use of this ability to hide among aquatic vegetation from predators such as Jaguar and caimans and anacondas. Capybara live in groups; there is evidence of some interspecific mutualism with Wattled jacana (*Jacana jacana*: Jacanidae) that groom them of ticks (Marcus, 1985).

## D. Ungulates

Ungulates or hoofed mammals (orders Artiodactyla and Perissodactyla) occur in Asia, Africa and the Neotropics, but their diversity varies greatly among these regions. For example, South America has only 21 species of ungulates (and only 16 in tropical latitudes); whereas there are 91 species in Africa most of which are bovids and the majority of which are antelopes. A number of species use riparian forest and shallow water as refuge from predators, or are associated with the productive grasslands of seasonally-inundated floodplains. Native ungulates are not present in Australia or Papua New Guinea.

## 1. Antelopes

Antelopes occur in Africa and Eurasia and a few African species can be considered semi-aquatic. Sitatunga (*Tragelaphus spekei*) are amphibious antelopes associated with wetlands and are represented by three subspecies (*T. spekei spekei, T. spekei gratus* and *T. spekei selousi*) that occur (respectively) in the Congo basin, around Lake Victoria, and in the Bangweulu, Zambesi and Okavango basins. They have small home ranges close to water and, when threatened by land-based predators, will submerge in water up to the level of their nostrils. Sitatunga have splayed hooves that improve swimming ability and enable them to walk on floating islands of vegetation (Kingdon, 1977). The coat is shaggy, and is covered with an oily, water-repellent secretion. Sitatunga browse leaves from riparian trees but also graze on wetland grasses and sedges. The Lechwe (*Kobus leche*) has similar adaptations to the Sitatunga, and is distributed across parts of central and southern Africa; subspecies associated with different river basins and parts of the Zambezi are recognised (Williamson, 1990, 1994; Nefdt, 1996; Timberlake, 1998). Lechwe population densities have been reduced greatly during the past century: for example, declines in the Kafue lechwe (*K. leche kafuensis*), which is endemic to the Kafue flats, were due to intensive hunting and wetland degradation following construction of a large hydroelectric scheme along the Zambezi (Jeffery, 1993). The Nile lechwe *(Kobus megaceros)* is endemic to the Sudd swamps, and frequently enters water (Kingdon, 1990).

Riparian swamps and river floodplains in Africa are also home to Bohor reedbuck (*Redunca redunca*), the Bushbuck (*Tragelaphus scriptus*), the Puku (*K. vardonii*) and the Waterbuck (*K. ellipsiprymnus*). The Waterbuck is the larger of the two *Kobus* spp. and has a waterproof coat that is impregnated with grease; it is a much larger (up to 2000 kg) and more abundant than Puku. It consumes reeds such as *Phragmites* and *Typha* as well as riparian grasses. *Kobus* spp. are highly territorial, and territories are arranged in a linear fashion along the banks of rivers and wetlands allowing easy access to the water (Goldspink *et al.*, 1998). The Water chevrotain (*Hyemoschus aquaticus*) is a member of the Tragulidae, an artiodactyl family closely related to Eurasian cervids. It has a highly fragmented range from Sierra Leone through the Congo to western Uganda (Istituto Ecologia Applicata, 1999) where it is confined to lowland gallery forests. Chevrotains frequently visit marshes and wetlands, and live in close proximity of water.

## 2. Deer

While large Bovidae tend to dominate riverine areas in the African tropics, especially in savannah, dense forest and rugged terrain favoured evolution of smaller, secretive, less gregarious Artiodactyla in the families Cervidae, Tragulidae, and Moschidae. Most are adapted to concealment in forests but make seasonal use of floodplains as a source of fresh grasses and forage. Examples of riparian deer include the endangered Bawean deer (*Axis kuhlii*) endemic to Bawean Island off Java, and the Javan rusa (*Cervus timorensis*), a floodplain species surviving in disjointed populations on several Indonesian islands. The endangered Spotted deer (*C. alfredi*) is a predominantly frugivorous forest species endemic to Visayan Islands in the central Philippines. The Brown-antlered deer (*C. eldi eldi*) surviving in a wildlife park in Manipur, India, is critically endangered as a result of conversion of its floodplain habitat to agriculture (IUCN, 1998). The same factor combined with hunting led to the extinction of Schomburgk's deer (*C. schomburgki*) in Thailand during the 20th century (Dudgeon, 2000b). The Barasingha swamp deer (*C. duvauceli*) is less specialised (classified as vulnerable: IUCN, 2006) occurring in a range of lowland habitats in northern India and Nepal. They live in tall flooded grasslands, frequently visit shallow wetlands and have been reported to feed upon aquatic plants. As is the case for many other floodplain ungulates, their reproduction, migrations and social behaviour (congregation into herds during drought and dispersion during the wet season) are linked with the annual river flood cycle. Other Asian floodplain deers include the Sambar (*C. unicolor*) distributed across much of mainland Asia into Indonesia, and (*Elaphurus davidianus*) a primitive swamp-adapted form thought to be the lone survival of a Pliocene deer lineage (Geist, 1999). It swims readily and has wide hooves that are sensitive to hard ground but would be adaptive on marshy ground. Limited habitat flexibility lead to the extinction of Père David's deer in the wild in China as floodplain habitats were progressively settled and converted to agriculture. However, a population of captive-bred animals has been used to establish a self-sustaining population in a nature reserve in the middle Yangtze River (Dudgeon, 2000b). The Chinese water deer (*Hydropotes inermis*) still thrives among the tall grasslands which border the lower reaches of the Yangtze and its floodplain wetlands. It rises to higher ground only during floods.

Due to low population numbers and to their distribution in remote, seasonally-inundated habitats, little information is available on some Asian deer species. The least known among all are perhaps the Asian mouse deer (Tragulidae), which include *Tragulus javanicus*, *T. napu* and *Moschiola meminna* and are further subdivided into a complex of insular sub-species. *Tragulus napu* is highly adapted to swamp forests and is closely dependent upon aquatic habitats (Dudgeon, 2000b). Also poorly known is the Hog deer (*Axis porcinus*) which consists of two subspecies with wide but fragmented distribution ranges. *Axis porcinus porcinus* survives in open floodplain grasslands along the Ganga, Brahmaputra and Indus rivers put formerly extended east into China; *Axis porcinus annamiticus* is endemic to the Chao Phraya basin in Thailand where populations have been greatly reduced as a result of habitat loss through conversion of floodplain to agriculture. In India and Nepal, Hog deer have benefited indirectly from wet grassland habitat protection measures taken in favour of more endangered swamp deer and other large mammals (IUCN, 1998).

The New World Cervidae have been separated from their Old World relatives since the late Miocene, some 12.4 to 9.3 million years ago (Miyamoto *et al.*, 1990). The Marsh deer (*Blastocerus dichotomus*) is the largest South American cervid with thin legs, membranous hooves and a long shaggy coat similar to the African Sitatunga (Geist, 1999). It is adapted to floodplain wetlands and seasonally-inundated grasslands in central and southern America, east of the Andes. It feeds on grasses, reeds, shrubs and aquatic plants; more than one third of the diet is made up of hydrophytes such as *Nymphaea* spp. (Nymphaeaceae). Preferred aquatic habitats are less than 70 cm deep; however marsh deer may move to deeper waters in search for aquatic plants, and again like the Sitatunga, may occur on floating mats of vegetation (Geist, 1999). The relatively small Red brocket deer (*Mazama americana*) also lives on floodplains in the Neotropics. Their ecology is not well known, but they seem to prefer forested sites and are excellent swimmers.

### 3. Buffaloes and Rhinos

Southern Asia, from India and Nepal to Malaysia and Vietnam, is the origin of the world-wide distributed Asian water buffalo *Bubalus bubalis* (Bovidae, synonymous with *B. arnee*) native to southern Asia, from India and Nepal to Malaysia and Vietnam, but domesticated and introduced widely elsewhere. It is a floodplain species adapted to move in swamps and wet ground that does not favour other cattle. Wild populations of *B. bubalis* are now restricted to few floodplain areas in India, although northern Australia supports a population of feral animals. Cross-breeding with domestic cattle poses a threat to the wild genotype of *B. bubalis*, and few Indian populations represent the true wild type. Other Asian buffaloes associated with swamps and water include the rare and endangered forest species *B. mindorensis* living in the Philippines, and the Lowland anoa (*Bubalis depressicornis*) restricted to Sulawesi. A specific problem characterising buffalo conservation is due to inbreeding with domestic cattle.

Asia hosts three species of rhinoceros, relics of the diverse Rhinocerontidae (Perissodactyla) which formerly occurred widely in Eurasia, Africa and North America (Cerdeño, 1998). Lineages adapted to swampy habitats evolved several times among rhinos, during their past evolutionary history. Unlike African rhinos, the extant Asian species are adapted to live in floodplain wetlands and tropical peat swamps. The Javan rhino (*Rhinoceros sondaicus*) is one of the most critically endangered large mammals in the world, with a population reduced to less than 100 individuals (Dudgeon, 2000b; IUCN, 2006). It survives in Java and probably Vietnam and occurs in swamps along water courses. The Sumatran rhino (*Dicerorhinus sumatrensis*) prefers humid riparian forests and wallows in shallow water. During the wet season, it migrates to higher grounds. The range of the Indian rhino (*Rhinoceros unicornis*) has been reduced greatly from its former wide distribution to floodplains in Thailand and along the Indus, the Ganga and Brahmaputra (Dudgeon, 2000b). This rhino, and other floodplain grazers in Asia, have been greatly reduced in their abundance and distribution through conversion of floodplains to agriculture, and alteration of natural river flow and inundation patterns (Dudgeon, 2000a).

### 4. Tapirs

Tapirs (Perissodactyla: Tapiridae) were once widely distributed across the globe, but Pleistocene glaciations forced them into climatic refuges in Southeast Asia (including Sumatra) and South America. They consume a great variety of leaves and fruit but are selective about what they eat. They are dependent on the proximity of water and often occur in seasonally-inundated sites although they move to higher ground during the wet season. Tapirs are nocturnal, and hide in dense forest during the day. They often enter water, and also use it as a refuge from predators (Brooks *et al.*, 1997). The Lowland tapir (*Tapirus terrestris*) is among the largest herbivores in South America East of the Andes and plays an important role as seed disperser in riparian and humid forest communities (Brooks *et al.*, 1997). Their diet is varied: they take fruit, leaves, grasses and aquatic plants in the Amazonian floodplain and in the Brazilian Pantanal (Bodmer, 1990), and bamboo leaves and twigs in the Atlantic forest of eastern Brazil (Rodrigues *et al.*, 1993; Brooks *et al.*, 1997). A related species, the endangered *T. pinchaque*, survives in high-rainfall regions of the Andes, but is less strongly associated with water. *Tapirus bairdii* is likewise endangered although its original range includes Central and South America. Like the Lowland tapir, its habitats include marshes, floodplains and forests, and it has a preference for secondary successional communities rather than mature primary forest. The Malayan Tapir (*T. indicus*) also uses a wide range of habitats: they hide in dense forest during the day and emerge to feed in open floodplain areas at night (Brooks *et al.*, 1997); they may move upland during the wet season. Forest edges along streams are favoured sites, and because Malayan tapirs sometimes remain submerged for long periods, proximity to water is an important component of suitable habitat (Dudgeon, 2000b).

## 5. Peccaries and Suids

Peccaries are placed in the Dicotylidae, a Neotropical family related to the Eurasian Suidae and African Hippopotamidae. It comprises three species among which the White-lipped peccary (*T. tayacu*) occurs in riparian habitats and floodplains. It is widely distributed and undertakes herd migrations through inundated forest of various depths and even across wide rivers (Bodmer, 1991; March, 1993). Their behaviour does not seem to be disrupted by floods and herds typically move within extensive ranges. White-lipped peccary are mainly frugivorous (Altrichter *et al.*, 2000) and the search for fruit may stimulate their migrations. These peccaries play an important role as seed dispersers and predators in the Neotropics analogous to that of bearded pigs (*Sus barbatus*) in the forests of Southeast Asia (Caldecott *et al.*, 1993). Hunting is a major threat, and conservation plans need to include protection of substantial tracts of forest to take into account the broad spectrum of habitats utilised during their migrations (March, 1993).

Tropical Asian Suidae include a large number of highly adaptable species, exploiting seasonally inundated floodplains. The composition of feral assemblages is influenced by widespread husbandry of pigs that may have begun as early as 40 000 years ago, and human impacts on tropical floodplain forests that have given a competitive advantage to open-habitat suids. The matter is complicated by uncertain relationships among lineages due to successive interbreeding between native, introduced, feral, domestic and hybrid populations (e.g. Groves, 1981). The Pigmy hog (*Sus salvanius*) is one of the most critically endangered mammals in Asia. Two disjunct populations of a few hundred individuals remain in tall grassland reserves in northern India (Oliver and Deb Roy, 1993). Pigmy hogs build nests within riverine and wetland riparian areas and reproduce at the onset of the monsoon, and their life histories are closely linked to the seasonal development of tall grasses during the annual inundation cycle of river floodplains. The Bearded pig (*Sus barbatus*) thrives in the Indonesian Archipelago where a number of subspecies have evolved on different islands. Bearded pigs are highly omnivorous and adaptable to a wide variety of habitats, but prefer evergreen dipterocarp and swamp forest. Herds of up to 3000 individuals make periodic migrations (with a periodicity of 2–4 years), when they swim across rivers and sometimes out to sea. Lowland populations make seasonal movements out of inundated forests to higher ground during the rainy season returning to swamp forest as waters recede in the dry season (Blouch, 1984). The Babirusa (*Babyrousa babyrussa*) is a primitive phylogenetically-isolated suid surviving mainly on Sulawesi. They are good swimmers and often wallow in mud (Nowak, 1999). Preferred habitats include lowland riparian zones covered with cane thickets and ponds with abundant plants although human activities have forced them to make use of marginal upland sites.

African Suids diversified during the Tertiary into three endemic genera comprising five species. Of these, the Red river hog (*Potamochoerus porcus*) and the Bush pig (*P. larvatus*) are associated with rain forest and swamps with distribution ranges divided by the Congo River. The western *P. porcus* is smaller, lives along the Senegal River and in lowland forests of central Africa. Both species are good swimmers and enjoy wallowing in water, but their habits are not well known and the extent of association with aquatic or riparian habitats is uncertain. A third Afrotropical suid, *Hylochoerus meinertzhageni*, is even less known and has a highly fragmented range throughout the formerly continuous intertropical forest belt. This nocturnal pig occurs in forest, floodplain forests and marshes. They do not stray far from water and tend to favour ecotones that offer a diversity of successional habitats (d'Huart, 1993).

## E. Primates

River channels and floodplain forests have played a major role in primate biogeography. While Old World primates evolved quadrupedalism as a main means of locomotion, Neotropical monkeys comprise mainly arboreal forms, many of which have a fully prehensile tail. Riparian habitats offer abundant and high-quality food resources for primates, especially for highly-arboreal species with sufficient habitat flexibility to forage above the inundated landscape during floods. Examples include the *Alouatta* howler monkeys: in flooded forest at the confluence of Paraná and the Paraguay Rivers (northeastern Argentina), Black howlers (*A. caraya*) consume leaves, flowers and fruit from a variety of riparian and aquatic plants, including *Ipomoea* spp. (Convolvulaceae) and Water hyacinth (Bravo and Sallenave, 2003). They prefer forest on islands or along river banks because they are fertilised by deposited sediments and contain more food resources than forests on the mainland. However, riparian troops are part of larger metapopulations, which include conspecifics in mainland forest where riparian monkeys retreat during severe floods (Bravo and Sallenave, 2003). The Capuchin (*Cebus apella*) is a similarly floodplain forest species (Heymann *et al.*, 2002), which combines arboreal life with swimming and wading. The Mangrove capuchin (*C. apella*) feed on oysters that are opened with sharp tools (Fernandes, 1991). Uakaris (*Cacajao calvus* and *C. melanocephalus*) inhabit forests along small streams and wetlands throughout the northern region of South America, but avoid the margins of large rivers.

In the Afrotropics, persistent centres of humid gallery forest in the centre of the continent have sustained a number of primate species during harsh climatic periods. These forests also provided a route for recolonisation of surrounding habitats during more favourable climes. For large ground-dwelling primates, however, the river network can represent a significant barrier to dispersal (Table I). *Cercopithecus* guenons originated from ground-dwelling savannah primates some 2.9 million years ago, evolving arboreal habits and a smaller body size as they became confined to humid forest fragments along streams (Kingdon, 1990). Most guenon lineages are thought to have evolved within the Congo basin during repeated glacial/interglacial cycles that fragmented the forest into separate wooded habitats along major rivers (Hamilton, 2001; Maley, 2001). A similar speciation occurred in *Procolobus*; this genus comprises 14 morphs most of which are dependent upon riparian forest (e.g. the Tana River red colobus: *P. rufomitratus*). Allen swamp monkeys (*Allenopithecus nigroviridis*) live seasonally-inundated in palm and swamp forests, and are one of the few monkeys to be present on both sides of the Congo River. Several other species are separated by the Congo (Table I) as in the case of the talapoins (*Miopithecus* spp.) that comprise two species, one on each side of the river (Oates, 1996; Grubb *et al.*, 2003).

Most species of mangabeys (*Cercocebus* spp.) and vervet monkeys (*Cercopithecus* spp. such as *C. lhoesti*) remain in close proximity to streams and swamps where they forage on fruit and riparian herbs. Several monkeys consume aquatic plants; one subspecies of the Blue monkey (*Cercopithecus mitis doggetti*), living in the swamps of South-western Uganda, feed on the soft stem bases of Papyrus (*Cyperus papyrus*: Cyperaceae). Savannah primates such as the Yellow baboon (*Papio cynocephalus*) converge on riparian forests and wetlands during dry periods of the year and may compete for food and feeding sites with more stenotopic humid forest monkeys (Wahungu, 2001). Populations of primates that live close to rivers climb into the canopy during inundations and then return to the ground as the flood recedes. In Likouala Swamp (northwestern Congo), for example, the Lowland western gorilla (*Gorilla gorilla*) spends most of the year living within the swamp (Fay *et al.*, 1989); they are poor swimmers, but wade in search of fruit, sedges and other aquatic and terrestrial herbs (Nishihara, 1995). During the flood season, gorilla build nests in trees or seek refuge in forests located at a higher levels (Doran and McNeilage, 1998), but they do not abandon riparian zones as they offer more fruit than the flooded swamp (Poulsen and Clark, 2004). The availability of large riparian trees providing sleeping sites may be a limiting factor for primate activities in these habitats (Anderson, 1984).

In Asia, riparian primates include Proboscis monkeys (*Nasalis larvatus*) endemic to Borneo, and inhabiting swamp forest and mangroves. They have partially webbed fingers and toes and bathe frequently (Boonratana, 2000); some consider them the best swimmers among non-human primates (Kern, 1964). Proboscis monkeys are highly mobile and do not defend home ranges; in the lower Kinabatangan River floodplain (northern Borneo), they were reported to cross river sections up to 150 m wide (Boonratana, 2000). Peat swamp forests, riparian forests and mangroves are also inhabited by the Long-tailed or Crab-eating macaque (*Macaca fascicularis*), distributed across much of tropical continental Asia and Indonesia and the Philippines archipelagos (Dudgeon, 2000b) and, less frequently, by the Southern Pig-tailed macaque (*Macaca nemestrina*).

Riparian, floodplain and peat swamp forests are among the essential habitats of Orang-utan (*Pongo pygmaeus*) living within two distinct populations in Sumatra and in Borneo (Meijaard, 1997). The Kapuas River in Borneo and the Alas River in Sumatra effectively separate further the island populations into recognisably distinct groups. Orang-utans are the only great apes outside Africa, and are perhaps the least understood ape (Delgado and van Schaik, 2000). Several other primates not closely associated with aquatic habitats, depend however, on the preservation of riparian forests. Among these are the gibbons (Hylobatidae), such as the Siamang or Black gibbon (*Symphalangus syndactylus*), the Dark-handed agile gibbon (*Hylobalis agilis*), and certain colobine leaf monkeys such as *Presbytis rubicunda* and *P. cristata* (see Dudgeon, 2000b).

## F. Shrews and Bats

Many opportunistic insectivores such as shrews are attracted to riparian habitats due to the relative abundance of insects, even in the dry season. The Hero shrew (*Scutisorex somereni*: Soricidae) is common in swampy or waterlogged habitats of Uganda, western Congo and forested central Africa (Kingdon, 1971). It can swim but does not hunt in water, and consumes frogs and invertebrates. Asia is home to the only truly semi-aquatic shrews in tropical latitudes. *Chimarrogale hantu*, *C. phaeura* and *C. sumatrana* occur in small mountain streams of the Peninsular Malaysia, Borneo and Sumatra, respectively. They are believed to forage under water, but little information about their habits or their diving ability is available, as these shrews are rarely encountered and probably occur at low population densities (Churchfield, 1998). The Elegant water shrew (*Nectogale elegans*) occurs in stony upland streams on mainland Asia (from China to the Himalaya) and is strongly aquatic, living in burrows along the banks. It lacks external ears, the nostrils are closed by valves when under water, and the eyes are covered by skin so that the animal is functionally blind. The diet consists of aquatic insect larvae, crustaceans and small fish. Elegant water shrews swim well, using their webbed feet fringed by hairs, and the tail. The palms of the feet bear hairy pads that may facilitate locomotion on slippery surfaces. Given the diversity of semi-aquatic shrews in tropical Asia, it is tempting to conclude that the absence of water rats from this region (see above) is due to their pre-emption by shrews. Interestingly, other tropical regions that support water rats lack semi-aquatic shrews.

On a global scale, bats (order Chiroptera) are highly diverse comprising some 175 genera and over 800 species. They are not commonly numbered among the vertebrates that live in close contact with streams. Nonetheless, aquatic and riparian habitats provide bats with shelters, roosting and hunting sites; several insectivorous bats, such as species of *Pipistrellus* and *Eptesicus* as well as other vespertilionids, hunt by flying over water. Most bats need to drink regularly and many can do so while in flight over the surface of streams; the vespertilionid genus *Scotophilus* drinks while flying at speed. The roosts of African *Eidolon* fruit bats (Pteropodidae) are often in close proximity of waterfalls and fast running streams that seem to cover the noise produced by the colonies, reducing predation risk from raptorial birds (Kingdon, 1974). Other riparian Afrotropical bats include the Hammer-headed fruit bat (*Hypsignathus monstrosus*: Pteropodidae), and the Yellow-winged bat (*Lavia frons*: Megadermatidae). The most conspicuous adaptations to an association with streams are provided by those bats that feed on aquatic animals. Among them is the Neotropical Fringe-lipped bat (*Trachops cirrhosus*: Phyllostomidae) that has strong jaws and a wide gape and is particularly well suited for preying on frogs (Findley and Wilson, 1982), being able to discriminate between poisonous and non-poisonous species (French, 1997) and to search for them by following frog calls as well as by using echolocation (Barclay *et al.*, 1981).

Few bats specialise in catching fish. In the Neotropics, the bulldog bats *Noctilio leporinus* and *N. albiventris* (Noctilionidae) are adapted to catch fish by dipping their feet armed with sharp claws and skimming the water as they fly over the surface (Fig. 8; Emmons, 1997). Their

*FIGURE 8* The Neotropical bat *Noctilio leporinus* (Noctilionidae) catching fish (Source: www.morcegolivre.vet.br).

echolocation detects tiny movements at the water surface precipitating an attack. The catch is stuffed into large mouth pouches. If the bat falls into water, it swims to the nearest shore before it takes flight. The only other New World bat specialising on catching fish is the Mexican vespertilionid *Myotis vivesi.*

In Southeast Asia bats are plentiful, diverse and very poorly known. As in other tropical biomes, they comprise frugivores and insectivores and a large number of them are found preferentially at river margins and in the proximity of aquatic habitats. Rickett's big-footed bat (*Myotis ricketti:* Vespertilionidae) is a piscivore, catching prey at the surface using its large, clawed feet (in a manner similar to Neotropical bulldog bats). Asian false vampire bats *Megaderma spasma* and *M. lyra* (Megadermatidae) are closely associated with aquatic habitats where they catch frogs, fish and even small bats by use of echolocation, but they lack the specialised adaptations of other fishing bats.

## VII. ECOLOGICAL ROLES OF VERTEBRATES IN TROPICAL STREAMS

At one time, many limnologists might have wandered if vertebrates other than fish had a role to play in the ecology of rivers and streams. Definitions such as 'autochthonous' and 'allochthonous' served the purpose to separate and simplify our understanding of processes, habitats and even species within river channels and riparian areas. Specific attention began to be paid to land-water ecotones and their roles in processing and transport of nutrients and energy as well as sites of species interaction; this could occur only after earlier, more-reductionist approaches, which viewed the ecotone as a frontier rather than a range of habitat had been abandoned. From what is known about the natural history and the ecology of semi-aquatic and riparian mammals (as outlined above), it is evident that ecotone definitions are highly species-specific and that there is little to be gained from a rigid separation of species into categories based on their habitat use or degree of dependence on water. In part this reflects the fact that streams and their ecotones are highly dynamic systems, and vertebrates will use different parts of these environments at different times of their lives or on a seasonal basis.

Great advancements have been achieved in the formalisation of our perception of river systems, with the progressive introduction of the now well-established River Continuum Concept (Vannote *et al.*, 1980), the Flood Pulse Concept (Junk, 2001; see Chapter 7 of this volume) and the multi-dimensional approach to characterising rivers and their floodplains (Boon, 1998; Ward *et al.*, 1998). This progression towards a better understanding of the integration of aquatic and riparian habitats has led to greater consideration being given to 'riverine' species of non-fish vertebrates. Nonetheless, the present compilation of vertebrates associated with tropical streams and rivers indicates the limited scope of our current knowledge; it is easier to compile a list of taxa, and speculate or question their ecological roles, than to list current achievements in understanding their interaction with the habitat and their overall ecological significance.

### A. Trophic Interactions and Nutrient Cycling

Semi-aquatic vertebrate predators potentially contribute significantly to the structuring of tropical stream fish communities (see also Chapter 5 of this volume). Studies on temperate otter populations indicate that their densities are directly dependent on fish productivity: in some Scottish streams, otters remove 53–67% of the annual salmonid production (Kruuk *et al.*, 1993). Any top-down effect of otter predation would be magnified by their localised foraging activity and selective removal of preferred prey species (Kruuk, 1995). In the tropics, otters often specialise on fish, but some species take also molluscs, crustaceans, amphibians and insects. However, their generally low density (at least in recent decades) is likely to limit their importance as community structuring agents. Potentially more relevant is the impact of crocodilians, which act both as otter competitors and predators. In most tropical riparian ecosystems, crocodiles and caimans represent the greatest portion of the vertebrate predator biomass and are likely to be 'keystone species' influencing ecosystem structure and function (Craighead, 1968; King, 1978). On the other hand, crocodile eggs and hatchlings are preyed upon by many larger vertebrates. Our current understanding of the ecological significance of crocodilians in tropical streams and rivers is poor, but it is likely to include elective predation on fish species, nutrient recycling, and maintenance of wet refugia during droughts by digging (i.e. 'landscaping'). For example, consumption of large fish by crocodiles may help unblock a significant nutrient bottleneck in rivers and streams, where long-lived and large-bodied fishes tend to retain a significant proportion of the nutrients available within the ecosystem which can depress overall fish productivity. Consumption of terrestrial prey by crocodiles must also contribute nutrients to aquatic habitats. Classical studies by Fittkau (1970, 1973) showed that nutrient release by caimans amounts to 0.2–0.3% of their body weight each day. By consuming small mammals and other terrestrial fauna, they realise a net import of nutrients into Amazonian fresh waters that has significant effects on primary and secondary production in streams and floodplain lakes (Fittkau, 1973; Best, 1984). Declines in populations of large river fishes (see Chapter 10 of this volume), and reductions in crocodile range and abundance throughout the tropics (see above), are likely to be causing changes in stream ecology in ways that are not fully appreciated; understanding and reconstructing these interactions (e.g. Rebelo and Lugli, 2001) will be challenging.

There are other aspects of the role of vertebrates in the mediation of nutrient transfers within tropical river basins. Ungulate migrations across the savannahs of Africa seem likely to impact floodplain nutrient cycles through their seasonal grazing activities, but there has been little quantitative analysis of this process. In parts of the Senegal floodplain, manuring by cattle has been estimated to amount to $350 \text{kg ha}^{-1}$ (Reizer, 1984), which is considered to represent a potentially-significant contribution to riverine fish productivity (Welcomme, 1985). Dams and irrigation schemes tend to lead to the loss on natural floodplain pastures, causing the overexploitation of managed grazing areas, the loss of natural fertilisation in traditionally-managed plots, and nutrient impoverishment of riparian habitats with concomitant impacts on stenotopic fauna.

Hippos transfer nutrients from riparian pastures ('hippo lawns') where they feed at night, directly into aquatic systems where they defecate. Depending on local conditions, hippo densities can be substantial: in Kruger National Park (South Africa), they account for around 6% of large herbivore biomass, while in Queen Elizabeth National Park (Uganda) this proportion formerly exceeded 50% (Lock, 1972). Hunting, conflict with humans, habitat alteration, and culling operations have greatly reduced hippo densities; in South Africa regular culls are undertaken to optimise hippo densities (Viljoen and Biggs, 1998). Despite significant data-collection efforts, the consequences of changed hippo densities for nutrient transfers are not well quantified. In South Africa, populations of 0.1 hippo $\text{h}^{-1}$ water can produce as much as 1 t dung $\text{h}^{-1}$ (Naiman and Rogers, 1997), and it is seems obvious that changes in hippo densities are likely to have significant consequences for aquatic productivity.

## B. Riparian Landscape Engineering

The presence of hippo gives a distinctive appearance to Afrotropical wetlands and river systems compared to other tropical floodplains, and the riverine wetlands of Africa would look very different without their presence. Hippo wallowing deepens and expands floodplain pools; they construct complex networks of trails through papyrus swamps and in doing so enhance the connectivity between adjacent wetland habitats; at the same time they maintain grazed 'lawns' close to river margins enhancing habitat diversity (Naiman and Rogers, 1997; McCarthy *et al.*, 1998). Other, more terrestrial grazers, such as Impala (*Aepyceros melampus*) graze hippo lawns during the dry season. Hippo trails through riparian papyrus swamp enhance water circulation and oxygen availability thereby opening new habitat for fishes and invertebrates (Harper and Mavuti, 1996). In the Okavango Delta (Northern Botswana), hippos have a catalytic effect on geomorphological change and their movements through the swamp prevent channel closure by encroaching vegetation; their pathways to backswamps can lead to development of new drainage pathways during channel avulsion (McCarthy *et al.*, 1998). Networks of hippo paths may also be stabilised by the development of a luxuriant vegetation fringe fertilised by hippo dung. Further evidence of the role of hippos as landscape engineers is evident from declines in their number (due to hunting) along the lower Zambezi River: between 1997 and 1990, the Marromeu hippo population fell from 2820 to 260; as a consequence, Water hyacinth overgrowth developed, clogging channels and reducing natural floodplain inundation (Timberlake, 1998).

Some semi-aquatic mammals make very particular contributions to riparian geomorphology, often acting in a mutualistic fashion. Sitatunga antelope trample small areas of swamp vegetation within their territories, creating rained microhabitats that can be colonised by small animals and by secondary vegetation. Crocodiles, caimans, coypu, otters, semi-aquatic mice and tenrecs excavate holes, cavities or burrows in the riverbank that can be colonised by other semi-aquatic mammals. During the dry season, various mammals seeking water will enlarge small pools within the drying floodplain thereby providing refuges for fish and other aquatic animals. Selective grazers and browsers such as Bushbuck, Puku and Waterbuck contribute to structuring the riparian vegetation assemblage. In the Tana River gallery forest (Kenya), the reduction of Elephants (*Loxodonta africana*) and the local extinction of Black rhino (*Diceros bicornis*) altered grazing patterns allowing overgrowth of the forest understorey changing the microhabitat of the riparian zone (Marsh, 1985). Such alterations are important because the composition and the density of the riparian vegetation have important consequences for crocodilians and other reptiles nesting along riverbanks (Leslie and Spotila, 2001).

Riparian zones include a number of vertebrates that eat fruit and disperse seeds and, besides plant propagules, semi-aquatic mammals are known to inadvertently disperse aquatic invertebrates by carrying them attached to their fur (Peck, 1975). Megachiropteran pteropodid fruit bats are involved in the pollination and especially the seed dispersal of several riparian tree species in the Old World tropics. Primates associated with riparian forest are also likely to be significant seed dispersers in places where their populations have not been depleted greatly, as are also Malayan tapirs in Asia where this role may once have been supplemented by the activities of Javan and Sumatran rhinos. In South America, Lowland tapirs are key seed dispersers because they do not masticate ingested seeds thoroughly and they tend to defecate in water, which is of particular benefit to waterside or riparian species and trees that are characteristic of seasonally-inundated forest. As tapirs are large unspecialised consumers of fruit, it is possible that they are responsible for seed dispersal once undertaken by large Pleistocene vertebrates now extinct. Forest fragmentation and direct pressures on the forest fauna over much of the tropics have reduced the efficacy of plant-vertebrate mutualisms with visible consequences on the structure of the vegetation mainly due to human-induced declines in frugivorous species and in the large mammals that played a role in structuring habitats (Dirzo and Miranda, 1991; Wright *et al.*, 2000; Cordeiro and Howe, 2001). Similar changes are affecting tropical floodplains and riparian forests and peat swamps (Dudgeon, 2000b), and combine with the direct habitat conversion that is typical of floodplains where human populations are either already large or growing rapidly.

## VIII. FINAL REMARKS

Rivers are not just for fish. They are globally important for a large number of semiaquatic and riparian amphibians, reptiles, mammals and birds. In tropical regions, aquatic and semi-aquatic vertebrates include lineages, which have survived many millennia through climate changes, by offering wildlife refugia characterised by permanent moisture. The present-day functioning of tropical floodplains and riparian habitats is conducive to the persistence of rich animal and plant communities which are permanently or seasonally dependent on river dynamics and habitat availability. Because tropical regions are characterised by seasonal peaks in discharge causing extensive and persistent floods, which last for several months before returning to baseflow conditions, tropical animals and plants have evolved distinct anatomical, physiological and behavioural adaptations to flooding. Such adaptations are not generally present at higher latitudes. Behavioural adaptations include swimming, horizontal migration across the basin and vertical migrations into the canopy. While arboreality is prevalent among Amazon vertebrates, in the Afrotropics – and to some extent Asia – large mammals conduct extensive migrations.

The examples given in this chapter illustrate the diversity of semiaquatic and riparian vertebrates, and indicate that it is likely that they interact with all major processes structuring geomorphology and vegetation processes in the land-water ecotone and riparian zones of the tropics. In so doing, they may contribute to the resistance to change and to the resilience of the river and stream ecosystems (Naiman and Rogers, 1997). The biotic integrity of a riverine corridor depends on the presence of a number of key semi-aquatic and riparian vertebrates; defaunated river floodplains are unlikely to provide the range of habitats or ecological functions needed to sustain other animals and plants. This conclusion is particularly relevant for tropical river basins, where semi-aquatic mammals and riparian vertebrates are relatively more diverse and abundant than in temperate latitudes.

The extensive use of the riparian fringe by floodplain vertebrates and the role of browsers in determining the structure of the riparian vegetation pin-points the shortcomings of theoretical studies which attempt to estimate the optimal width of protective buffer zones established around streams and wetlands. For a fully-aquatic organism, the riparian fringe can appear as a linear frontier but for amphibians, most aquatic insects, reptiles, birds and semi-aquatic mammals, it should be considered as a core area or one that is essential for part of the year. The floodplain itself may be the sole habitat of some vertebrates, but used facultatively on a seasonal basis by others. For some species distinguishing precise habitat occupancy will be crucial to inform conservation or management efforts: for example, the dry-season habitats of the Congolese Dwarf crocodile and the Amazonian Smooth-fronted caiman are shallow pools located within riparian ecotones, rather than within rivers or among backswamps and associated wetlands (Riley and Huchzermeyer, 1999). The status and accessibility of different riparian habitats must be taken into consideration in designing conservation measures for the benefit of semi-aquatic and riparian vertebrates. This is a research area that needs more attention in the tropics. A review of data from North America (Table VII) indicates that a buffer width of around 300 m will be needed to ensure the survival of amphibians and semi-aquatic reptiles (Semlitsch and Bodie, 2003). The information needed to make equivalent estimates for the tropics is lacking. Nonetheless, it seems likely that because tropical riparian zones experience seasonal flood pulses, they are likely to require far more extensive buffer zones than equivalent temperate habitats. Matters are complicated by the fact that in many cases, the survival of semiaquatic and riparian species in the tropics requires the preservation of habitat mosaics, including wetter and drier habitats. Ideally, protection of intact aquatic ecosystems and entire drainage basins, or at least some examples of them in each tropical region, is needed to ensure effective conservation of riverine vertebrates. In practice, human activities such as

*TABLE VII*   Mean Minimum and Maximum Width of Core Terrestrial Habitats Required by Amphibians and Reptiles in North America; These are Indicative of the Widths of Riparian Buffer Strips Needed to Protect Their Habitats. Data are from Semlitsch and Bodie (2003)

|  | *Mean minimum (m)* | *Mean maximum (m)* |
|---|---|---|
| Frogs | 205 | 368 |
| Salamanders | 117 | 218 |
| Amphibians | 159 | 290 |
| Snakes | 168 | 304 |
| Turtles | 123 | 287 |
| Reptiles | 127 | 289 |
| Herpetofauna | 142 | 289 |

water engineering, land-use change and deforestation have already modified hydrographs and the characteristics of tropical floodplains. Climate change will further move these ecosystems from their initial conditions and seasonal discharge patterns to which riverine organisms are adapted and upon which they depend.

## ACKNOWLEDGEMENTS

Many comments and discussions have been provided by Luca Luiselli from the Environmental Studies Centre 'Demetra' in Rome and by Spartaco Gippoliti from the Pistoia Zoological Garden. Franco Iozzoli from the Italian Environment Protection Agency designed the graphics. Constructive comments by two anonymous referees and David Dudgeon of the University of Hong Kong were highly appreciated.

## REFERENCES

Akani, G.C., Barieenee, I.F., Capizzi, D., and Luiselli, L. (1999). Snake communities of moist forest and derived savanna sites of Nigeria: biodiversity patterns and conservation priorities. *Biodiversity and Conservation* 8, 629–642.

Akani, G.C., and Luiselli, L. (2001). Ecological studies on a population of the water snake *Grayia smythii* in a rainforest swamp of the Niger Delta, Nigeria. *Contributions to Zoology* 70, 39–146.

Akani, G.C., Politano, E., and Luiselli, L. (2004). Amphibians recorded in forest swamp areas of the River Niger Delta (southeastern Nigeria), and the effects of habitat alteration from oil industry development on species richness and diversity. *Applied Herpetology* 2, 1–22.

Alho, C.J.R., and Padua, L.F.M. (1982). Reproductive parameters and nesting behaviour of the Amazon turtle *Podocnemis expansa* (Testudinata: Pelomedusidae) in Brazil. *Canadian Journal of Zoology* 60, 97–103.

Altrichter, M., Saenz, J.C., Carrillo, E., and Fuller, T.K. (2000). Dieta estacional del *Tayassu pecari* (Artiodactyla: Tayassuidae) en el Parque Nacional Corcovado, Costa Rica. *Revista de Biologia Tropical* 48, 689–701.

AmphibiaWeb (2007). Information on amphibian biology and conservation. (web application). Berkeley, CA, USA. http://www.amphibiaweb.org/.

Anderson, J.R. (1984). Ethology and ecology of sleep in monkeys and apes. *Advances in the Study of Behaviour* 14, 165–229.

Aubret, F. (2004). Aquatic locomotion and behaviour in two disjunct populations of Western Australian tiger snakes, *Notechis ater occidentalis*. *Australian Journal of Zoology* 52, 357–368.

Archer, M., Hand, S., and Godthelp, H. (1994). Patterns in the history of Australia's mammals and inferences about palaeohabitats. In *History of Australian Vegetation: Creataceous to Recent*. Hill, R. [ed.] pp. 80–103. Cambridge University Press, Cambridge, UK.

Ayers, J.M., and Clutton-Brock, T.H. (1992). River boundaries and species range size in Amazonian primates. *The American Naturalist* 140, 531–537.

Baird, W. (1999) "The Co-management of Mekong River Inland Aquatic Resources in Southern Lao PDR." Pakse, Lao PDR. http://www.co-management.org/download/ianbaird.pdf.

Banguera-Beebee, T.J.C., and Griffiths, R.A. (2005). The amphibian decline crisis: a watershed for conservation biology? *Biological Conservation* **125**, 271–285.

Banguera-Hinestroza, E., Cárdenas, H., Ruiz-García, M., Marmontel, M., Gaitán, E., Vázquez, R., and García-Vallejo, F. (2002). Molecular identification of evolutionarily significant units in the Amazon river dolphin *Inia* sp. (Cetacea: Iniidae). *The Journal of Heredity* **93**, 312–322.

Barclay, R.M.R., Fenton, M.B., Tuttle, M.D., and Ryan, M.J. (1981). Echolocation calls produced by *Trachops cirrhosus* (Chiroptera: Phyllostomidae) while hunting for frogs. *Canadian Journal of Zoology* **59**, 750–753.

Bennett, D. (1995). "A Little Book on Monitor Lizards." Viper Press, Glossop, UK.

Bergallo, H.G., and Magnusson, W.E. (1999). Effects of climate and food availability on four rodent species in southeastern Brazil. *Journal of Mammalogy* **80**, 472–486.

Berger, S.R. (2000). Global distribution of chytridiomycosis in amphibians. http://www.jcu.edu.au/school/phtm/PHTM/frogs/chyglob.htm.

Berger, S.R., Speare, R., Daszak, P., Green, D.E., Cunningham, A.A., Goggin, C.L., Slocombe, R., Ragan, M.A., Hyatt, A.D., McDonald, K.R., Hines, H.B., Lips, K.R., Marrantelli, G., and Parkes, H. (1998). Chytridiomycosis causes amphibian declines in the rainforest of Australia and central America. *Proceedings of the National Academy of Sciences* **95**, 9031–9036.

Bertoluci, J., and Rodrigues, T.M. (2002). Seasonal patterns of breeding activity of Atlantic rainforest anurans at Boracéia, Southeastern Brazil. *Amphibia-Reptilia* **23**, 161–167.

Best, R.C. (1981). Foods and feeding habits of wild and captive Sirenia. *Mammal Review* **11**, 3–29.

Best, R.C. (1984). The aquatic mammals and reptiles of the Amazon. *In* "The Amazon: Limnology and Landscape Ecology of a Mighty Tropical River and its Basin" (H. Sioli, Ed.), pp. 371–412. Dr W. Junk Publishers, Dordrecht, The Netherlands.

Best, R.C., and da Silva, V.M.F. (1989). Amazon river dolphin, Boto, *Inia geoffrensis* (de Blainville, 1817). *In* "Handbook of Marine Mammals" (S.H. Ridgway and R.J. Harrison, Eds), pp. 1–23. Academic Press, London, UK.

Blanc, C.P., and Fretey, T. (2004). Repartition écologique des Amphibiens dans La Réserve de Faune de La Lope et La Station Biologique de La Makandé (Gabon). *Bulletin de la Société Zoologique de France* **129**, 297–315.

Blouch, R.A. (1984). "Current Status of the Sumatran Rhino and Other Large Mammals in Southern Sumatra." Unpublished Project 3303 Report No. 4 to IUCN/WWF Indonesia Programme, Bogor, 39 pp.

Bodie, J.R. (2001). Stream and riparian management for freshwater turtles. *Journal of Environmental Management* **62**, 443–455.

Bodie, J.R., Semlitsch, R.D., and Renken, R.B. (2000). Diversity and structure of turtle assemblages: associations with wetland characters across a floodplain landscape. *Ecography* **23**, 444–456.

Bodmer, R. (1990). Responses of ungulates to seasonal inundations in the Amazon floodplain. *Journal of Tropical Ecology* **6**, 191–201.

Bodmer, R. (1991). Strategies of seed dispersal and seed predation in Amazonian ungulates. *Biotropica* **23**, 255–261.

Boon, P.J. (1998). River restoration in five dimensions. *Aquatic Conservation: Marine and Freshwater Ecosystems* **1**, 257–264.

Boonratana, R. (2000). Ranging behaviour of proboscis monkeys (*Nasalis larvatus*) in the lower Kinabatabgan (northern Borneo). *International Journal of Primatology* **21**, 497–518.

Boruah, S., and Biswas, S.P. (2002). Application of ecohydrology in the Brahmaputra river. *Ecohydrology and Hydrobiology* **2**, 79–87.

Branch, B. (1998). "Field Guide to Snakes and Other Reptiles of Southern Africa." Ralph Curtis Books, Sanibel Island, FL, USA.

Bravo, S.P., and Sallenave, A. (2003). Foraging behaviour and activity patterns of *Alouatta caraya* in the northeastern Argentinean flooded forest. *International Journal of Primatology* **24**, 825–846.

Brazaitis, P., Rebelo, G.H., Yamashita, C., Odierna, E.A., and Watanabe, M.E. (1996). Threats to Brazilian crocodilian populations. *Oryx* **30**, 275–284.

Brites, V.L.C., and Rantin, F.T. (2004). The influence of agricultural and urban contamination on leech infestation of freshwater turtles, *Phrynops geoffroanus*, taken from two areas of Uberanbinha River. *Environmental Monitoring and Assessment* **96**, 273–291.

Brooks, D.M., Bodmer, R.E, and Matola S. (Eds) (1997). "Tapirs – Status Survey and Conservation Action Plan." IUCN/SSC Tapir Specialist Group, IUCN, Gland, Switzerland. http://www.tapirback.com/tapirgal/iucn-ssc/tsg/action97/cover.htm.

Burbidge, A.A., Kirsch, J.A.W., and Main, A.R. (1974). Relationships within the Chelidae (Testudines: Pleurodira) of Australia and New Guinea. *Copeia* **1974**, 392–409.

Caceres, N. (2003). Use of the space by the opossum *Didelphis aurita* Wied-Neuwied (Mammalia, Marsupialia) in a mixed forest fragment of southern Brazil. *Revista Brasileira de Zoologia* **20**, 315–322.

Caldecott, J.O., Blouch, R.A., and Macdonald, A.A. (1993). The Bearded pig, *Sus barbatus*. *In* "Pigs, Peccaries and Hippos: Status Survey and Conservation Action Plan" (W.L.R. Oliver, Ed.), pp. 136–145. IUCN, Gland, Switzerland.

Campbell I.C., Poole C., Giesen W., and Valbo-Jorgensen J. (2006). Species diversity and ecology of Tonle Sap Great Lake, Cambodia. *Aquatic Sciences-Research Across Boundaries*, 68, 355–373.

Carr, T., and Bonde, R.K. (2000). Tucuxi (*Sotalia fluviatilis*) occurs in Nicaragua, 800 km north of its previously known range. *Marine Mammal Science* 16, 447–452.

Carter, S.K., and Rosas, F.C.W. (1997). Biology and conservation of the Giant Otter, *Pteronura brasiliensis*. *Mammal Review* 27, 1–26.

Cerdeño, E. (1998). Diversity and evolutionary trends of the family Rhinocerontidae (Perissodactyla). *Palaeogeography, Palaeoclimatology, Palaeoecology* 141, 13–34.

Cheung, S.M., and Dudgeon, D. (2006). Quantifying the Asian turtle crisis: market surveys in southern China, 2000–2003. *Aquatic Conservation: Marine and Freshwater Ecosystems* 16, 751–770.

Chippindale, P.T., Bonett, R.M., Baldwin, A.S., and Wiens, J.J. (2004). Phylogenetic evidence for a major reversal of life-history evolution in plethodontid salamanders. *Evolution* 58, 2809–2822.

Churchfield, S. (1998). Habitat use by water shrews, the smallest of amphibious mammals. *In* "Behaviour and Ecology of Riparian Mammals" (N. Dunstone and M.L. Gorman, Eds), pp. 49–68. Symposia of the Zoological Society of London 71, Cambridge University Press, Cambridge, UK.

Cogger, H.G. (1996). "Reptiles and Amphibians of Australia." Reed Books, Chatswood, New South Wales, Australia.

Colares, I.G., and Colares, E.P. (2002). Food plants eaten by Amazonian manatees (*Trichechus inunguis*, Mammalia: Sirenia). *Brazilian Archives of Biology and Technology* 45, 67–72.

Collins, A.C., and Dubach, J.M. (2000). Biogeographic and ecological forces responsible for speciation in Ateles. *International Journal of Primatology* 21, 421–444.

Collins, J.P., and Storfer, A. (2003). Global amphibian declines: sorting the hypotheses. *Diversity and Distributions* 9, 89–98.

Colyn, M.M. (1991). L'importance zoogèographique du Bassin du fleuve Zaire pour la spéciation: le cas des primates scimiens. *Annalen Zoologische Wetenschappen. Koninklijk Museum voor Midden-Afrika Tervuren, Belgie* 264.

Colwell, R.K. (2000). A barrier runs through it…or maybe just a river. *Proceedings of the National Academy of Sciences* 97, 13470–13472.

Cordeiro, N.J., and Howe, H.F. (2001). Low Recruitment of trees dispersed by animals in African forest fragments. *Conservation Biology* 15, 1733–1739.

Craighead, F.C. (1968). The role of the alligator in shaping plant communities and maintaining wildlife in the southern Everglades. *Florida Naturalist* 41, 2–7.

da Silva, C.J., Wantzen, K.M., da Cunha, C.N., and de Arruda Machado, F. (2001). Biodiversity in the Pantanal wetland, Brazil. *In* "Biodiversity of Wetlands: Assessment, Function and Conservation, Vol. 2" (B. Gopal, W.J. Junk, and J.A. Davis, Eds), pp. 187–215. Backhuys Publishers, Leiden, The Netherlands.

da Silveira, R., and Thorbjarnarson, J.B. (1999). Conservation implications of commercial hunting of black and spectacled caiman in the Mamirauá sustainable development reserve, Brazil. *Biological Conservation* 88, 103–109.

de Buffrénil, V., and Hémery, G. (2002). Variation in longevity, growth, and morphology in exploited Nile monitors (*Varanus niloticus*) from Sahelian Africa. *Journal of Herpetology* 36, 419–426.

de Oliveira Santos, M..C., Rosso, S., Aguiar dos Santos, R., Bulizani Lucato, S.H., and Bassoi, M. (2002). Insights on small cetacean feeding habits in southeastern Brazil. *Aquatic Mammals* 28, 38–45.

Delgado, R., and van Schaik, C.P. (2000). The behavioral ecology and conservation of the Orangutan (*Pongo pygmaeus*): a tale of two islands. *Evolutionary Anthropology* 9, 201–218.

d'Huart, J.-P. (1993). The Forest hog (*Hylochoerus meinertzhageni*). *In* "Pigs, Peccaries and Hippos: Status Survey and Conservation Action Plan" (W.L.R. Oliver, Ed.), pp. 84–93. IUCN, Gland, Switzerland. http://www.iucn.org/themes/ssc/sgs/pphsg/APchap4-3.htm (accessed on 17 April, 2007).

Dirzo, R., and Miranda, A. (1991). Altered patterns of herbivory and diversity in the forest understorey: a case study of the possible consequences of contemporary defaunation. *In* "Plant–Animal Interactions: Evolutionary Ecology in Tropical and Temperate Regions" (P.W. Price, G.W. Lewinsohn, G. Fernandes, and W.W. Benson, Eds), pp. 273–287. Wiley, New York, USA.

Dodd, C.K., and Smith, L.L. (2003) Habitat destruction and alteration: historical trends and future prospects for amphibians. *In* "Amphibian Conservation" (R.D. Semlitsch, Ed.), pp. 94–112. Smithsonian Institution, Washington, DC, USA.

Domning, D.P. (1982). Evolution of manatees: a speculative history. *Journal of Paleontology* 56, 599–619.

Doody, J.S., Young, J.E., and Georges, A. (2002). Sex differences in activity and movements in the pig-nosed turtle *Carettochelys insculpta*, in the wet–dry tropics of Australia. *Copeia* 2002, 98–108.

Doody, J.S., Georges, A., and Young, J.E. (2003). Twice every second year: reproduction in the pig-nosed turtle, *Carettochelys insculpta*, in the wet–dry tropics of Australia. *Journal of Zoology (London)* 259, 179–188.

Doran, D.M., and Mc Neilage, A. (1998). Gorilla ecology and behaviour. *Evolutionary Anthropology* 6, 120–131.

Dudgeon, D. (2000a). Large-scale hydrological changes in tropical Asia: prospects for riverine biodiversity. *BioScience* 50, 793–806.

Dudgeon, D. (2000b). Riverine wetlands and biodiversity conservation in tropical Asia. *In* "Biodiversity of Wetlands: Assessment, Function and Conservation, Vol. 1" (B. Gopal, W.J. Junk, and J.A. Davis, Eds), pp. 35–60. Backhuys Publishers, Leiden, The Netherlands.

Dudgeon, D., Choowaew, S., and Ho, S.-C. (2000). River conservation in south-east Asia. *In* "Global Perspectives on River Conservation: Science, Policy and Practice" (P.J. Boon, B.R. Davies, and G.E. Petts, Eds), pp. 281–310. John Wiley & Sons Ltd., Chichester, UK.

Dunstone, N. (1998). Adaptations to the semi-aquatic habit and habitat.*In* "Behaviour and Ecology of Riparian Mammals" (N. Dunstone and M.L. Gorman, Eds), pp. 1–16. Symposia of the Zoological Society of London 71, Cambridge University Press, Cambridge, UK.

Earl of Cranbrook (Ed.) (1988) "Key Environments: Malaysia." Pergamon Press, Oxford, UK.

Edwards, H.H., and Schnell, G.D. (2001). Status and ecology of *Sotalia fluviatilis* in the Cayos Misktos Reserve, Nicaragua. *Marine Mammal Science* 17, 445–472.

Ehmann, H.F.W. (1992). "Encyclopaedia of Australian Animals: Reptiles." Angus & Robertson, Sydney, Australia.

Eltringham, S.K. (1993). The Common Hippopotamus, *Hippopotamus amphibius*. *In* "Pigs, Peccaries and Hippos: Status Survey and Conservation Action Plan" (W.L.R. Oliver, Ed.), pp. 32–145. IUCN, Gland, Switzerland. http://www.iucn.org/themes/ssc/sgs/pphsg/APchap3-2.htm.

Emmons, L.H. (1997). "Neotropical Rain Forest Mammals, a Field Guide (2nd Edition)." The University of Chicago Press, Chicago, IL, USA.

Eterovick, P.C., Oliveira de Queiroz Carnaval, A.C., Borges-Nojosa, D.M., Leite Silvano, D., Segalla, M.V., and Sazima, I. (2005). Amphibian declines in Brazil: an overview. *Biotropica* 37, 166–180.

Fay, J.M., Agnagna, M., Moore, J., and Oko, R. (1989). Gorillas (*Gorilla gorilla gorilla*) in the Likouala Swamp forests of north central Congo: preliminary data on populations and ecology. *International Journal of Primatology* 10, 477–486.

Fernandes, M.E.B. (1991). Tool use and predation of oysters by the tufted capuchin in brackish water mangrove swamps. *Primates* 32, 529–531.

Ferreira, Jr P.D., and Castro, P.T.-A. (2003). Geological control of *Podocnemis expansa* and *Podocnemis unifilis* nesting areas in Rio Javaés, Bananal Island, Brazil. *Acta Amazonica* 33, 445–468.

Fish, F.E., and Baudinette, R.V. (1999). Energetics of locomotion by the Australian water rat (*Hydromys chrysogaster*): a comparison of swimming and running in a semi-aquatic mammal. *Journal of Experimental Biology* 202, 353–363.

Findley, J.S., and Wilson, D.E. (1982). Ecological significance of chiropteran morphology. *In* "Ecology of Bats" (T.H. Kunz, Ed.), pp. 243–260. Plenum Press, New York, USA

Fittkau, E.J. (1970). Role of caimans in the nutrient regime of mouth-lakes of Amazon affluents (an hypothesis). *Biotropica* 2, 138–142.

Fittkau, E.J. (1973). Crocodiles and the nutrient metabolism of Amazonian waters. *Amazoniana* 4, 103–133.

Fonseca, G.A.B., Herrmann, G., Leite, Y.L.R., Mittermeier, R.A., Rylands, A.B., and Patton, J.L. (1996) Lista anotada dos Mamiferos do Brasil. *Occasional Papers in Conservation Biology* 4, 1–38.

Fordham, D., Hall, R., and Georges, A. (2004). Aboriginal harvest of long-necked turtles in Arnhem Land, Australia. *Turtle and Tortoise Newsletter* 7, 20–21.

French, B. (1997). False vampires and other carnivores: a glimpse at this select group of bats reveals efficient predators with a surprisingly gentle side. *Bats* 15, 11–14.

Funk, W.C., Fletcher-Lazo, G., Nogales-Sornosa, F., and Almeida-Renoso, D. (2004) First description of a clutch and nest site for the genus *Caecilia* (Gymnophiona: Caeciliidae). *Herpetological Review* 35, 128–130.

Gascon, C., Malcolm, J.R., Patton, J.L., da Silva, M.N.F., Bogart, J.P., Lougheed, S.C., Peres, C.A., Neckel, S., and Boag, P.T. (2000). Riverine barriers and the geographic distribution of Amazonian species. *Proceedings of the National Academy of Sciences* 97, 13672–13677.

Geist, V. (1999). "Deer of the World: Their Evolution, Behaviour and Ecology." Swan Hill Press, Airlife Publishing Ltd., Shrewsbury, UK.

Gibbons, J.W., Scott, D.E., Ryan, T.J, Buhlmann, K.A, Tuberville, T.D., Metts, D.S., Greene, J.L, Mills, T., Leiden, Y., Poppy, S., and Winne, C.T. (2000). The global decline of reptiles, déja vu amphibians. *BioScience* 50, 653–666.

Gingerich, P.D. (2003). Land to sea transition in early whales: evolution of Eocene Archaeoceti (Cetacea) in relation to skeletal proportions and locomotion in living semiaquatic mammals. *Paleobiology* 29, 429–454.

Goldspink, C.R., Holland, R.K., Sweet, G., and Stjernstedt, R. (1998). A note on the distribution and abundance of puku, *Kobus vardonii* Livingstone, in Kasanka National Park, Zambia. *African Journal of Ecology* 36, 23–33.

Gordos, M.A., Franklin, C.E., and Limpus, C.J. (2004). Effect of water depth and water velocity upon the surfacing frequency of the bimodally respiring freshwater turtle, *Rheodytes leukops*. *Journal of Experimental Biology* 207, 3099–3107.

Grabert, H. (1984). Migration and speciation of the South American Iniidae (Cetacea, Mammalia). *Zeitschrift Saugetierkunden* 49, 334–341.

Greene, H.W. (1997). "Snakes: the Evolution of Mystery in Nature." The University of California Press, Berkeley, CA, USA.

Grigg, G.C., Johansen, K., Harlow, P., Beard, L.A, and Taplin, L.E. (1986). Facultative aestivation in a tropical freshwater turtle *Chelodina rugosa*. *Comparative Biochemistry and Physiology A* **83**, 321–323.

Groves, C. (1981). "Ancestors for the Pigs: Taxonomy and Phylogeny of the Genus *Sus*." Technical Bulletin No. 3, Australian National University Press, Canberra, Australia, 96 pp.

Grubb, P. (1993). The Afrotropical suids: *Potamochoerus*, *Hylochoerus* and *Phacochoerus*. *In* "Pigs, Peccaries and Hippos: Status Survey and Conservation Action Plan" (W.L.R. Oliver, Ed.). IUCN, Gland, Switzerland. http://www.iucn.org/themes/ssc/sgs/pphsg/APchap4-1.htm.

Grubb, P. (2001). Endemism in African rainforest mammals. *In* "African Rain Forest Ecology and Conservation" (W. Weber, L.J.T. White, A. Vedder, and L. Naughton-Treves, Eds), pp. 88–100. Yale University Press, New Haven, CT, USA.

Grubb, P., Butynski, T.M., Oates, J.F., Bearder, S.K., Disotell, T.R., Groves, C.P., and Struhsaker, T. (2003). Assessment of the diversity of African primates. *International Journal of Primatology* **24**, 1301–1357.

Guichón, M.L., Doncaster, C.P., and Cassini, M.H. (2003). Population structure of coypu (*Myocastor coypus*) in their region of origin and comparison with introduced populations. *Journal of Zoology (London)* **261**, 265–272.

Gupta, A. (2002). The beleaguered chelonians of northeastern India. *Turtle and Tortoise Newsletter* **6**, 16–17.

Gutleb, A.C., Schenck, C., and Staib, E. (1997). Giant otter (*Pteronura brasiliensis*) at risk? Total mercury and methyl–mercury levels in fish and otter scats, Peru. *Ambio* **26**, 511–514.

Gyi, K.K. (1970). A revision of colubrid snakes of the subfamily Homalopsinae. *University of Kansas Publications, Museum of Natural History*, Lawrence **20**, 47–223.

Haffer, J., and Prance, G.T. (2001) Climatic forcing of evolution in Amazonia during the Cenozoic on the refuge theory of biotic differentiation. *Amazoniana* **16**, 579–607.

Hamilton, A.C. (2001). Hotspots in African forests as Quaternary refugia.*In* "African Rain Forest Ecology and Conservation" (W. Weber, L.J.T. White, A. Vedder, and L. Naughton-Treves, Eds), pp. 57–67. Yale University Press, New Haven, CT, USA.

Harcourt, C.S., and Sayer J.A. (Eds) (1996). "The Conservation Atlas of Tropical Forests: the Americas." Simon & Schuster, New York, USA.

Harper, D.M., and Mavuti, K.M. (1996). Freshwater wetlands and marshes. *In* "East African Ecosystems and their Conservation" (T.R. McClanahan and T.P. Young, Eds), pp. 217–239. Oxford University Press, Oxford, UK.

Hero, J.-M., and Morrison, C. (2004). Frog declines in Australia: global implications. *Herpetological Journal* **14**, 175–186.

Hero, J.-M., Magnusson, W.E., Rocha, C.F.D., and Catterall, C.P. (2001) Antipredator defenses influence the distribution of amphibian prey species in the Central Amazon rainforest. *Biotropica* **33**, 131–141.

Herron, J.C. (1991). Growth rates of Black caiman *Melanosuchus niger* and spectacled caiman *Caiman crocodilus*, and the recruitment of breeders in hunted caiman populations. *Biological Conservation* **55**, 103–113.

Heymann, E.W., Encarnación, F.C., and Canaquin, J.E.Y (2002). Primates of the Río Curaray, northern Peruvian Amazon. *International Journal of Primatology* **23**, 191–201.

Hussain, S.A., and Choudhury, B.C. (1998). Feeding ecology of the smooth-clawed otter *Lutra perspicillata* in the National Chambal Sanctuary, India.*In* "Behaviour and Ecology of Riparian Mammals" (N. Dunstone and M.L. Gorman, Eds), pp. 229–249. Cambridge University Press, Cambridge, UK.

Inger, R.F., and Iskandar, D.T. (2005) A collection of amphibians from West Sumatra, with descriptions of a new species of *Megophrys* (Amphibia: Anura). *The Raffles Bulletin of Zoology* **53**, 133–142.

Istituto Ecologia Applicata (1999). "A Databank for the Conservation and Management of the African Mammals." Istituto Ecologia Applicata, Rome, Italy, 1156 pp.

IUCN (1998). "Deer: Status Survey and Conservation Action Plan." Compiled by C. Wemmer and the IUCN/SSC Deer Specialist Group, International Union for the Conservation of Nature, Gland, Switzerland, 106 pp.

IUCN (2000). "The 2000 IUCN Red List of Threatened Species." IUCN, Gland, Switzerland.

IUCN (2006). "IUCN Red List of Threatened Species." IUCN, Gland, Switzerland. http://www.iucnredlist.org (accessed on 1 February, 2007).

IUCN, CI, and SN (2004). "Global Amphibian Assessment." International Union for the Conservation of Nature (IUCN), Conservation International (CI), and NatureServe (NS), Washington, DC, USA. http://www.globalamphibians.org (accessed on 15 October, 2004).

Jackson, D.C. (2000). Living without oxygen: lessons from the freshwater turtle. *Comparative Biochemistry and Physiology A* **125**, 299–315.

Jacobsen, N.H.G., and Kleynhans, C.J. (1993). The importance of weirs as refugia for hippopotami and crocodiles in the Limpopo River, South Africa. *Water SA* **19**, 301–306.

Jeffery, R.C.V. (1993). Wise use of floodplain wetlands in the Kafue flats of Zambia. *In* "Towards the Wise Use of Wetlands" (T.J. Davis, Ed.). Wise Use Project, Ramsar Convention Bureau, Gland, Switzerland. http://www.ramsar.org/lib/lib_wise.htm#cs17.

Junk, W.J. (2001). The flood-pulse concept of large rivers: learning from the tropics. *Verhandlungen der Internationale Vereinigung für Theoretische und Angewandte Limnologie* **27**, 3950–3953.

Karns, D.R., Voris, H.K., and Goodwin, J.C. (2002). Ecology of oriental-Australian rear-fanged water snakes (Colubridae: Homalopsinae) in the Pasir Ris Park Mangrove Forest, Singapore. *The Raffles Bulletin of Zoology* 50, 487–498.

Kennett, R.M. (1996). Growth models for two species of freshwater turtle, *Chelodina rugosa* and *Elseya dentata*, from the wet–dry tropics of northern Australia. *Herpetologica* 52, 383–395.

Kennett, R.M., Christian, K., and Bedford, G. (1998). Underwater nesting by the Australian freshwater turtle *Chelodina rugosa*: effect of prolonged immersion and eggshell thickness on incubation period, egg survivorship, and hatchling size. *Canadian Journal of Zoology* 76, 1019–1023.

Kennett, R.M., Georges, A., Thomas, K., and Georges, T.C. (1992). Distribution of the long-necked freshwater turtle *Chelodina novaeguineae* and new information on its ecology. *Memoirs of the Queensland Museum* 32, 179–182.

Kern, JA (1964). Observations on the habits of the proboscis monkey, *Nasalis larvatus* (Wurmb), made in the Brunai Bay area, Borneo. *Zoologica* 49, 183–192.

King, L.C. (1978). The geomorphology of central and southern Africa. *In* "Biogeography and Ecology of Southern Africa" (M.J.A. Werger and A.C. van Bruggen, Eds), pp. 1–18. Dr W. Junk Publishers, The Hague, The Netherlands.

Kingdon, J. (1971–77). "East African Mammals: An Atlas of Evolution in Africa. Volumes 1–3." Academic Press, San Diego, CA, USA.

Kingdon, J. (1990). "Island Africa: The Evolution of Africa's Rare Animals and Plants." Harper Collins Publishers, London, UK, 287 pp.

Klingel, H. (1991). The social organization and behaviour of *Hippopotamus amphibius*. *In* "African Wildlife: Research and Management" (F.I.B. Kayanja and E.L. Edroma, Eds), pp. 72–75. International Council of Scientific Unions, Paris, France.

Kofron, C.P. (1992). Status and habitats of the three African crocodiles in Liberia. *Journal of Tropical Ecology* 8, 265–273.

Kolbe, J.J., and Janzen, F.J. (2002) Spatial and temporal dynamics of turtle nest predation: edge effects. *Oikos* 99, 538–544.

Kottelat, M., and Whitten, A. (1996). "Freshwater Biodiversity in Asia." World Bank Paper No. 343, The World Bank, Washington, DC, USA, 59 pp.

Köhler, J., Vieites, D.R., Bonett, R.M., García, F.H., Glaw, F., Steinke, D., and Vences, M. (2005). New amphibians and global conservation: a boost in species discoveries in a highly endangered vertebrate group. *BioScience* 55, 693–696.

Kruuk, H. (1995). "Wild Otters: Predation and Populations." Oxford University Press, Oxford, UK.

Kruuk, H., Carss, D.N., Conroy, J.W.H., and Durbin, L. (1993). Otter (*Lutra lutra* L.) numbers and fish productivity in rivers in north-east Scotland. *In* "Mammals as Predators" (N. Dunstone and M.L. Gorman, Eds), pp. 171–191. Symposium of the Zoological Society of London 65, Clarendon Press, Oxford, UK.

Kuchling, G. (1988). Population structure, reproductive potential and increasing exploitation of the freshwater turtle *Erymnochelys madagascariensis*. *Biological Conservation* 43, 107–113.

Leatherwood, S., Reeves, R.R., Würsig, B., and Shearn, D. (2000). Habitat Preferences of river dolphin in the Peruvian Amazon. *In* "Biology and Conservation of Freshwater Cetaceans in Asia" (R.R. Reeves, B.D. Smith, and T. Kasuya, Eds), pp. 131–144. IUCN/SSC Occasional Paper No. 23, Gland, Switzerland and Cambridge, UK.

Legler, J.M., and Georges, A. (1993). Family chelidae. *In* "Fauna of Australia, Vol. IIA, Amphibia and Reptilia" (C.J. Clasby, G.J.B. Ross, and P.L. Beesely, Eds), pp. 142–152. Australian Government Publishing Service, Canberra, Australia.

Lehtinen, R.M., Ramanamanjato, J.B., and Raveloarison, J.G. (2003). Edge effects and extinction proneness in a herpetofauna from Madagascar. *Biodiversity and Conservation* 12, 1357–1370.

Lenz, S. (1995). Zur Biologie und Ökologie des Nilwarans, *Varanus niloticus* (Linnaeus 1766) in Gambia, Westafrika. *Mertensiella* 5, 1–256.

Leslie, A.J., and Spotila, J.R. (2000). Osmoregulation of the Nile crocodile, *Crocodylus niloticus*, in Lake St. Lucia, Kwazulu/Natal, South Africa. *Comparative Biochemistry and Physiology A* 126, 351–365.

Leslie, A.J., and Spotila, J.R. (2001). Alien plant threatens Nile crocodile (*Crocodylus niloticus*) breeding in Lake St. Lucia, South Africa. *Biological Conservation* 98, 347–355.

Lock, J.M. (1972). The effects of hippopotamus grazing on grasslands. *Journal of Ecology* 60, 445–467.

Lougheed, S.C., Gacon, C., Jones, D.A., Bogart, J.P., and Boag, P.T. (1999). Ridges and rivers: a test of competing hypotheses of Amazonian diversification using dart-poison frog (*Epipedobates femoralis*). *Proceedings of the Royal Society of London (B)* 266, 1829–1835.

Luiselli, L. (2001). The ghost of a recent invasion in the reduced feeding rates of spitting cobras during the dry season in a rainforest region of tropical Africa? *Acta Oecologica* 22, 311–314.

Luiselli, L. (2003). Do snakes exhibit shifts in feeding ecology associated with the presence or absence of potential competitors? a case study from tropical Africa. *Canadian* 81, 228–236.

Luiselli, L., and Akani, G.C. (2002). An investigation into composition, complexity and functioning of snake communities in the mangroves of southern Nigeria. *African Journal of Ecology* 40, 220–227.

Luiselli, L., and Akani, G.C. (2003). An indirect assessment of the effects of oil pollution on the diversity and functioning of turtle communities in the Niger Delta, Nigeria. *Animal Biodiversity and Conservation* 26, 57–65.

Luiselli, L., Akani, G.C., and Capizzi, D. (1998). Food resource partitioning of a community of snakes in a swamp-rainforest of south-eastern Nigeria. *Journal of Zoology (London)* 246, 125–133.

Luiselli, L., Akani, G.C., and Capizzi, D. (1999). Is there any interspecific competition between dwarf crocodiles (*Osteolaemus tetraspis*) and Nile monitors (*Varanus niloticus ornatus*) in the swamps of central Africa? a study from south-eastern Nigeria. *Journal of Zoology (London)* 247, 127–131.

Luiselli, L., Angelici, F.M., and Akani, G.C. (2001). Food habits of *Python sebae* in suburban and natural habitats. *African Journal of Ecology* 39, 116–118.

Luiselli, L., Akani, G.C., Politano, E., Odegbune, E., and Bello, O. (2004) Dietary shifts of sympatric freshwater turtles in pristine and oil-polluted habitats of the Niger Delta, southern Nigeria. *Herpetological Journal* 14, 57–64.

Luiselli, L., Akani, G.C., Rugiero, L., and Politano, E. (2005). Relationships between body size, population abundance and niche characteristics in the communities of snakes from three habitats in southern Nigeria. *Journal of Zoology (London)* 265, 207–213.

MacDonald, D.W. (2001). "The New Encyclopedia of Mammals." Oxford University Press, Oxford, UK.

Madsen, T., and Shine, R. (1996). Seasonal migration of predator and prey – a study of pythons and rats in tropical Australia. *Ecology* 77, 149–156.

Magnusson, W.E., da Silva, V.E., and Lima A.P. (1987). Diets of Amazonian crocodilians. *Journal of Herpetology* 21, 85–95.

Magnusson, W.E., and Lima, A.P. (1991). The ecology of a cryptic predator, *Paleosuchus trigonatus*, in a tropical rainforest. *Journal of Herpetology* 25, 41–48.

Maley, J. (2001). The impact of arid phases on the African rainforests through geological history. *In* "African Rain Forest Ecology and Conservation" (W. Weber, L.J.T. White, A. Vedder, and L. Naughton-Treves, Eds), pp. 68–87. Yale University Press, New Haven, CT, USA.

Manlius, N. (2000). Historical ecology and biogeography of the hippopotamus in Egypt. *Belgian Journal of Zoology* 130, 59–66.

March, I.J. (1993) The white-lipped peccary (*Tayassu pecari*). *In* "Pigs, Peccaries and Hippos: Status Survey and Conservation Action Plan" (W.L.R. Oliver, Ed.). IUCN, Gland, Switzerland. http://www.iucn.org/themes/ssc/sgs/pphsg/APchap2-3.htm.

Marcus, M.J. (1985). Feeding associations between capybaras and jacanas: a case of interspecific grooming and possibly mutualism.*Ibis* 127, 240–243.

Marsh, C.W. (1985). A resurvey of the Tana River primates and their forest habitat. *Primate Conservation* 7, 72–81.

Mason, C.F., and Rowe-Rowe, D.T. (1992). Organochlorine pesticide residues and PCBs in otter scats from Natal. *South African Journal of Wildlife Resources* 22, 29–31.

McCarthy, T.S., Ellery, W.N., and Bloem, A. (1998). Some observations on the geomorphological impact of hippopotamus (*Hippopotamus amphibius* L.) in the Okavango Delta, Botswana. *African Journal of Ecology* 36, 44–56.

McGuire, T.L. (2002). "Distribution and Abundance of River Dolphins in the Peruvian Amazon." Ph.D. Dissertation, Wildlife and Fisheries Sciences, Texas, A&M University, College Station, TX, USA.

Meijaard, E. (1997). The importance of swamp forest for the conservation of the orangutan (*Pongo pygmaeus*) in Kalimantan. *In* "Tropical Peatlands" (J.O. Rieley and S.E. Page, Eds), pp. 243–254. Samara Publishing Ltd., Cardigan, UK.

Miyamoto, M.M., Kraus, F., and Ryder, O.A. (1990). Phylogeny and evolution of antlered deer determined from mitochondrial DNA sequences. *Proceedings of the National Academy of Sciences* 87, 6127–6131.

Moll, D., and Moll, E.O. (2004). "The Ecology, Exploitation, and Conservation of River Turtles." Oxford University Press, Oxford, UK.

Monteiro-Filho, E.L.A. (2000). Group organization in the dolphin *Sotalia guianensis* in an estuary of southeastern Brazil. *Ciência i Cultura* 52, 97–101.

Monteiro-Neto, C., Itavo, R.V., and Moraes, L.E. (2003). Concentrations of heavy metals in *Sotalia fluviatilis* (Cetacea: Delphinidae) off the coast of Ceara, northeast Brazil. *Environmental Pollution* 123, 319–324.

Morrison, C., and Hero J.-M. (2003). Geographic variation in the life-history characteristics of amphibians: a review. *Journal of Animal Ecology* 72, 270–279.

Murphy, J.C., and Voris, H.K. (2005). A new Thai *Enhydris* (Serpentes: Colubridae: Homalopsinae). *The Raffles Bulletin of Zoology* 53, 143–147.

Murphy, J.C., Voris, H.K., and Karns, D. R. (2005). Rainbows in the mud. *Reptiles Magazine*.

Naiman, R.J., and Rogers, K.H. (1997). Large animals and system-level characteristics in river corridors. *BioScience* 47, 521–529.

Nefdt, R.J.C. (1996). Reproductive seasonality in Kafue lechwe antelope. *Journal of Zoology (London)* 239, 155–166.

Neiff, J.J. (2001). Diversity in some tropical wetland systems of South America. *In* "Biodiversity of Wetlands: Assessment, Function and Conservation, Vol. 2" (B. Gopal, W.J. Junk and J.A. Davis, Eds), pp. 157–186. Backhuys Publishers, Leiden, The Netherlands.

Newton, I. (2003). "The Speciation and Biogeography of Birds." Academic Press, San Diego, CA, USA.

Nishiwaki, M. (1984). Current status of the African manatee. *Acta Zoologica Fennica* 172, 135–136.

Nishihara, T. (1995). Feeding ecology of western lowland gorillas in the Nouabale-Ndoki National Park, Congo. *Primates* 36, 151–168.

Nowak, R.M. (1999). "Walker's Mammals of the World: Volumes 1 and 2 (6th Edition)." Johns Hopkins University Press, Baltimore, MD, USA.

Oates, J.F. (1996). "African Primates: Status Survey and Conservation Action Plan." IUCN/SSC Primate Specialist Group, IUCN, Gland, Switzerland.

O'Connor, T.G., and Campbell, B.M. (1986). Hippopotamus habitat relationships on the Lundi River, Gonarezhou National Park, Zimbabwe. *African Journal of Ecology* 24, 7–26.

Ohler, A., Swan, S.R., and Daltry, J.C. (2002). A recent survey of the amphibian fauna of the Cardamom Mountains, Southwest Cambodia with the description of three new species. *The Raffles Bulletin of Zoology* 50, 465–481.

Oliver, W.L.R., and Deb Roy, S. (1993). The Pigmy Hog (*Sus salvanius*). *In* "Pigs, Peccaries and Hippos: Status Survey and Conservation Action Plan" (W.L.R. Oliver, Ed.). IUCN, Gland, Switzerland. http://www.iucn.org/themes/ssc/sgs/pphsg/APchap5-3.htm.

O'Shea, T.J. (1994). Manatees. *Scientific American* 271, 66–72.

Pardini, R., and Trajano, E. (1999). Use of shelters by the Neotropical otter (*Lontra longicaudis*) in an Atlantic forest stream, southeastern Brazil. *Journal of Mammalogy* 80, 600–610.

Parris, K.M., and McCarthy, M.A. (1999). What influences the structure of frog assemblages at forest streams? *Australian Journal of Ecology* 24, 495–502.

Patton, J.L., da Silva, M.N.F., and Malcolm, J.R. (1994). Gene genealogy and differentiation among spiny rats (Rodentia: Echimyidae) of the Amazon basin: a test of the riverine barrier hypothesis. *Evolution* 48, 1314–1323.

Peck, S.B. (1975). Amphipod dispersal in the fur of aquatic mammals. *Canadian Field-Naturalist* 89, 181–182.

Peres, C.A., and Carkeek, A.M. (1993). How caimans protect fish stocks in western Brazilian Amazonia – a case for maintaining the ban on caiman hunting. *Oryx* 27, 225–230.

Peres, C.A., Patton, J.L., and da Silva, M.N.F. (1996). Riverine barriers and gene flow in Amazonian saddle-back tamarins. *Folia Primatologica* 67, 113–124.

Perrin, M.R., and Carugati, C. (2000). Habitat use by the Cape clawless otter and the spotted-necked otter in the KwaZulu-Natal, Drakensberg, South Africa. *South African Journal of Wildlife Research* 30, 103–113.

Perrin, M.R., Carranza, I.D., and Linn, I.J. (2000). Use of space by the spotted-necked otter in the KwaZulu-Natal, Drakensberg, South Africa. *South African Journal of Wildlife Research* 30, 15–21.

Peterhans, J.C.K., and Patterson, B.D. (1995). The Ethiopian water mouse *Nilopegamys* Osgood, with comments on semi-aquatic adaptations in African Muridae. *Zoological Journal of the Linnean Society* 113, 329–349.

Pineda, E., and Halffter, G. (2004). Species diversity and habitat fragmentation: frogs in a tropical montane landscape Mexico. *Biological Conservation* 117, 499–508.

Platt, S.G., Kalyar, W., and Ko, W.K. (2000). Exploitation and conservation status of tortoises and freshwater turtles in Myanmar. *In* "Proceedings of a Workshop on Conservation and Trade of Freshwater Turtles and Tortoises in Asia" (P.P. van Dijk, B.L. Stuart, and G.J. Rhodin, Eds), pp. 95–100. Chelonian Research Monographs No. 2, Chelonian Research Foundation, Lunenburg, MA, USA.

Pluto, T.G., and Bellis, E.D. (1988). Seasonal and annual movements of riverine map turtles, *Graptemys geographica.Journal of Herpetology* 22, 152–158.

Poulsen, J.R., and Clark, C.J. (2004).Densities, distributions and seasonal movements of gorillas and chimpanzees in swamp forest in northern Congo. *International Journal of Primatology* 25, 285–306.

Pringle, C.M., Scatena, F.N., Paaby-Hansen, P., and Nuñez-Ferreira, M. (2000). River conservation in Latina America and the Caribbean. *In* "Global Perspectives on River Conservation: Science, Policy and Practice" (P.J. Boon, B.R. Davies and G.E. Petts, Eds), pp. 41–77. John Wiley & Sons Ltd., Chichester, UK.

Rashid, S.M.A. (2004). "Population Ecology and Management of Water Monitors, *Varanus salvator* (Laurenti 1768), at Sungei Buloh Wetland Reserve, Singapore." Unpublished Ph.D. Thesis, National Institute of Education, Nanyang Technological University, Singapore.

Rebelo, G.H., and Lugli, L. (2001). Distribution and abundance of four caiman species (Crocodylia: Alligatoridae) in Jau National Park, Amazonas, Brazil. *Revista de Biologia Tropical* 49, 1095–1109.

Rebelo, G.H., and Magnusson, W.E. (1983). An analysis of effects of hunting on caiman crocodiles and *Melanosuchus niger* based on the sizes of confiscated skins. *Biological Conservation* 26, 95–104.

Reizer, C. (1984). Heurs et malheurs du régime alimentaire dans les aménagements hydro-agricoles. *Annales de Gembloux* 90, 217–241.

Riley, J., and Huchzermeyer, F.W. (1999). African dwarf crocodiles in the Likouala swamp forests of the Congo basin: habitat, density and nesting. *Copeia* 1999, 313–320.

Rodrigues, M., Olmos, F., and Galetti, M. (1993). Seed dispersal by tapir in southeastern Brazil. *Mammalia* 57, 460–461.

Rosas, F.C.W. and Monteiro-Filho, E.L.A. (2002). Reproduction of the estuarine dolphin (*Sotalia guianensis*) on the coast of Paraná, southern Brazil. *Journal of Mammalogy* 83, 507–515.

Ross, J.P. (Ed.) (1996). "Crocodiles: Proceedings of the 13th Working Meeting of the Crocodile Specialist Group." IUCN, Gland, Switzerland.

Ross, J.P. (Ed.) (1998). "Crocodiles: Status Survey and Conservation Action Plan (2nd Edition)." IUCN/SSC Crocodile Specialist Group, IUCN, Gland, Switzerland. http://www.flmnh.ufl.edu/natsci/herpetology/act-plan/plan1998a.htm.

Rowe-Rowe, D.T. and Somers, M.J. (1998). Diet, foraging behaviour and coexistence of African otters and the water mongoose.*In* "Behaviour and Ecology of Riparian Mammals" (N. Dunstone and M.L. Gorman, Eds), pp. 214–227. Symposia of the Zoological Society of London 71, Cambridge University Press, Cambridge, UK.

Samedi, A.R., and Iskandar, D.T. (2000). Freshwater turtle and tortoise conservation and utilization in Indonesia. *In* "Proceedings of a Workshop on Conservation and Trade of Freshwater Turtles and Tortoises in Asia" (P.P. van Dijk, B.L. Stuart, and G.J. Rhodin, Eds), pp. 106–111. Chelonian Research Monographs No. 2, Chelonian Research Foundation, Lunenburg, MA, USA.

Schenck, C., and Staib, E. (1998). Status, habitat use and conservation of the giant otter in Peru. *In* "Behaviour and Ecology of Riparian Mammals" (N. Dunstone and M.L. Gorman, Eds), pp. 359–370. Symposia of the Zoological Society of London 71, Cambridge University Press, Cambridge, UK.

Seebacher, F., Elsworth, P.G., and Franklin, C.E. (2003). Ontogenic changes of swimming kinematics in a semi-aquatic reptile (*Crocodylus porosus*). *Australian Journal of Zoology* 51, 15–24.

Semlitsch, R.D., and Bodie, J.R. (2003). Biological criteria for buffer zones around wetlands and riparian habitats for amphibians and reptiles. *Conservation Biology* 17, 1219–1228.

Shea, G.M. (1999). The distribution and identification of dangerously venomous Australian terrestrial snakes. *Australian Veterinary Journal* 77, 791–798.

Shepherd, C.R. (2000). Export of live freshwater turtles and tortoises from North Sumatra and Riau, Indonesia: a case study. *In* "Proceedings of a Workshop on Conservation and Trade of Freshwater Turtles and Tortoises in Asia" (P.P. van Dijk, B.L. Stuart, and G.J. Rhodin, Eds), pp. 112–119. Chelonian Research Monographs No. 2, Chelonian Research Foundation, Lunenburg, MA, USA.

Shine, R. (1988). Constraints on reproductive investment: a comparison between aquatic and terrestrial snakes. *Evolution* 42, 17–27.

Shine, R. (1991a). Strangers in a strange land: ecology of the Australian colubrid snakes. *Copeia* 1991, 120–131.

Shine, R. (1991b). "Australian Snakes, a Natural History." Cornell University Press, Ithaca, New York, USA.

Shine, R., and Madsen, T. (1997). Prey abundance and predator reproduction: rats and pythons on a tropical Australian floodplain. *Ecology* 78, 1078–1086.

Shine, R., and Shetty, S. (2001). Moving in two worlds: aquatic and terrestrial locomotion in sea snakes (*Laticauda colubrina*, Laticaudidae). *Journal of Evolutionary Biology* 14, 338–346.

Shine, R., Cogger, H.G., Reed, R.R., Shetty, S. and Bonnet, X. (2003). Aquatic and terrestrial locomotor speeds of amphibious sea-snakes (Serpentes, Laticaudidae). *Journal of Zoology* 259: 261–268.

Slowinski, J.B., and Keogh, J.S. (2000). Phylogenetic relationships of elapid snakes based on Cytochrome *b* mtDNA sequences. *Molecular Phylogenetics and Evolution* 15, 157–174.

Somers, M.J., and Purves, M.G. (1996). Trophic overlap between three syntopic semi-aquatic carnivores: Cape clawless otter, spotted-necked otter and water mongoose. *African Journal of Ecology* 34, 158–166.

Sousa Lima, R.S., Paglia, A.P., and Fonseca, G.A.B. (2002). Signature information and individual recognition in the isolation calls of Amazonian manatees, *Trichechus inunguis. Animal Behaviour* 63, 301–310.

Souza, F.L., and Abe, A.S. (2000). Feeding ecology, density and biomass of the freshwater turtle, *Phrynops geoffroanus*, inhabiting a polluted urban river in south-eastern Brazil. *Journal of Zoology (London)* 252, 437–446.

Souza, F.L., Cunha, A.F., Oliveira, M.A., Pereira, A.G., and dos Reis, S.F. (2002). Estimating dispersal and gene flow in the Neotropical freshwater turtle *Hydromedusa maxmiliani* (Chelidae) by combining ecological and genetic methods. *Genetic and Molecular Biology* 25, 151–155.

Spawls, S., and Branch, B. (1995). "The Dangerous Snakes of Africa." Blandford Press, London, UK.

Spencer, R.J. (2002). Experimentally testing nest site selection: fitness trade-offs and predation risk in turtles. *Ecology* 83, 2136–2144.

Starret, A., and Fisler, G.F. (1970). Aquatic adaptations of the water mouse, *Rheomys underwoodi. Los Angeles Country Museum Contributions in Science* 182, 1–14.

Storey, K.B. (1996). Metabolic adaptations supporting anoxia tolerance in reptiles: recent advances. *Comparative Biochemistry and Physiology B* 113, 23–35.

Sunquist, M., Karanth, K.U., and Sunquist, F. (1999). Ecology, behaviour and resilience of the tiger and its conservation needs. *In* "Riding the Tiger: Tiger Conservation in Human Dominated Landscapes" (J. Seidensticker, S. Christie, and P. Jackson, Eds), pp. 5–18. Cambridge University Press, Cambridge, UK.

Tana, T.S., Hour, P.L., Thach, C., Sopha, L., Sophat, C., Piseth, H., and Kimchay, H. (2000). Overview of turtle trade in Cambodia. *In* "Proceedings of a Workshop on Conservation and Trade of Freshwater Turtles and Tortoises in Asia" (P.P. van Dijk, B.L. Stuart, and G.J. Rhodin, Eds), pp. 55–58. Chelonian Research Monographs No. 2, Chelonian Research Foundation, Lunenburg, MA, USA.

Thirakhupt, K., and van Dijk, P.P. (1994). Species diversity and conservation of turtles in western Thailand. *Natural History Bulletin of the Siam Society* 42, 207–259.

Thorbjarnarson, J.B., Perez, N., and Escalona, T. (1993). Nesting of *Podocnemis unifilis* in the Capanaparo River, Venezuela. *Journal of Herpetology* 27, 344–347.

Timberlake, J. (1998). "Biodiversity of the Zambesi Basin Wetland: Review and Preliminary Assessment of Available Information. Phase I. Final Report (February 1998)." Consultancy Report for the World Conservation Unit (IUCN). http://www.biodiversityfoundation.org/documents/BFA%20No. 3_Wetlands%20Biodiversity_1.pdf.

Tucker, J.K., Limpus, C.J., Priest, T.E., Cay, J., Glen, C., and Guarino, E. (2001). Home ranges of Fitzroy turtles (*Rheodytes leukops*) overlap riffle zones: potential concerns related to river regulation. *Biological Conservation* 102, 171–181.

Vallan, D. (2000). Influence of forest fragmentation on amphibian diversity in the nature reserve of Ambohitantely, highland Madagascar. *Biological Conservation* 96, 31–43.

van Dijk, P.P. (2000). The turtles of Asia. *In* "Proceedings of a Workshop on Conservation and Trade of Freshwater Turtles and Tortoises in Asia" (P.P. van Dijk, B.L. Stuart, and G.J. Rhodin, Eds), pp. 15–23. Chelonian Research Monographs No. 2, Chelonian Research Foundation, Lunenburg, MA, USA.

van Dijk, P.P., Stuart, B.L., and Rhodin, G.J. (Eds) (2000) "Proceedings of a Workshop on Conservation and Trade of Freshwater Turtles and Tortoises in Asia." Chelonian Research Monographs No. 2, Chelonian Research Foundation, Lunenburg, MA, USA.

Vannote, R.G., Minshall, W., Cummins, K.W., Sedell, J.R., and Cushing, E. (1980). The river continuum concept. *Canadian Journal of Fisheries and Aquatic Sciences* 37, 130–137.

Vidal, O. (1993). Aquatic mammal conservation in Latin America: problems and perspectives. *Conservation Biology* 7, 788–795.

Viljoen, P.C., and Biggs, H.C. (1998). Population trends of hippopotami in the rivers of the Kruger National Park, South Africa. *In* "Behaviour and Ecology of Riparian Mammals" (N. Dunstone and M.L. Gorman, Eds), pp. 251–279. Symposia of the Zoological Society of London 71, Cambridge University Press, Cambridge, UK.

Wahungu, G.M. (2001). Common use of sleeping sites by two primate species in Tana River, Kenya. *African Journal of Ecology* 39, 18–23.

Waitkuwait, W.E. (1989). Present knowledge on the West African slender-snouted crocodile, *Crocodylus cataphractus* Cuvier 1824 and the West African dwarf crocodile, *Osteolaemus tetraspis*, Cope 1861. *In* "Crocodiles: Their Ecology, Management and Conservation", pp. 259–275. Special Publication of the IUCN/SSC Crocodile Specialist Group, IUCN, Gland, Switzerland.

Wallace, A.R. (1852). On the monkeys of the Amazon. *Proceedings of the Zoological Society of London* 20, 107–110.

Wallace, A.R. (1876). "The Geographic Distribution of Animals, Vol. 1." Harper & Brothers Publishers, New York, USA.

Wang, E., Ferreira, V.L., and Himmelstein, J. (2004). "Amphibians and Reptiles of the Southern Pantanal." Earthwatch expedition proposal, Earthwatch Institute, Boston, MA, USA. http://www.earthwatch.org/site/pp.asp?c=8nJELMNkGiF&b=1986507.

Ward, J.V., Bretschko, G., Brunke, M., Danielopol, D., Gibert, J., Gonser, T., and Hildrew, A.G. (1998). The boundaries of river systems: the metazoan perspective. *Freshwater Biology* 40, 531–569.

Watts, C.H.S., and Baverstock, P.R. (1994). Evolution in New Guinean Muridae (Rodentia) assessed by microcomplement fixation of albumin. *Australian Journal of Zoology* 42, 295–306.

Webb, J.K., Brown, G.P., and Shine, R. (2001). Body size, locomotor speed and antipredator behaviour in a tropical snake (*Tropidonophis mairii*, Colubridae): the influence of incubation environments and genetic factors. *Functional Ecology* 15, 561–568.

Welcomme, R.L. (1985). "River Fisheries." FAO Technical Paper No. 262, FAO, Rome.

Weldon, C., du Preez, L.H., Hyatt, A.D., Muller, R., and Speare, R. (2004). Origin of the amphibian chytrid fungus. *Emerging Infectious Diseases* 10, 2100–2105.

Williams, S.E., and Hero, J.M. (2001). Multiple determinants of Australian tropical frog biodiversity. *Biological Conservation* 98, 1–10.

Williamson, D.T. (1990). Habitat selection by Red lechwe (*Kobus leche lleche* Gray, 1950). *African Journal of Ecology* 28, 89–101

Williamson, D.T. (1994). Social behaviour and organisation of red lechwe in Linyati swamp. *African Journal of Ecology* 32, 130–141.

Wright, S.J., Zeballos, H., Dominguez, I., Gallardo, M.M., Moreno, M.C., and Ibanez, R. (2000). Poachers alter mammal abundance, seed dispersal and seed predation in a Neotropical forest. *Conservation Biology* 14, 227–239.

Zhang, X., Wang, D., Liu, R., Wei, Z., Hua, Y., Wang, Y., Chen, Z., and Wang, L. (2003). The Yangtze River dolphin or baiji (*Lipotes vexillifer*): population status and conservation in the Yangtze River, China. *Aquatic Conservation: Marine and Freshwater Ecosystems* 13, 51–64.

# 7

# *Riparian Wetlands of Tropical Streams*

Karl M. Wantzen, Catherine M. Yule, Klement Tockner, and Wolfgang J. Junk

Riparian wetlands are temporarily or permanently inundated and/or water-logged zones along the margins of streams and rivers. They link permanent aquatic habitats with upland terrestrial habitats, and surface-water with groundwater. This chapter focuses on riparian wetlands associated with low-order tropical streams, which have been lesser studied than equivalent ecotones associated with large rivers. We demonstrate that (a) these wetlands provide valuable habitats for diverse and highly specialized flora and fauna; (b) these serve as important longitudinal and transversal corridors for exchange of material and dispersal of biota; and (c) these perform important ecosystem functions locally as well as at the catchment scale. For example, headwater wetlands are key sites for mutual subsidies between terrestrial and aquatic systems, and are pivotal areas for the transformation of nutrients and organic matter. All riparian wetlands are subject to significant modification by humans, which compromises their functional integrity. However, riparian ecotones along low-order streams often occupy limited areas beyond the banks, and awareness of their ecological importance is limited in comparison to the extensive wetlands and floodplains associated with large lowland rivers. Creating awareness of the need for their sustainable management will be a challenging task.

## I. INTRODUCTION

Riparian wetlands have been defined as '*lowland terrestrial ecotones which derive their high water tables and alluvial soils from drainage and erosion of adjacent uplands on the one side or from periodic flooding from the other*' (McCormick, 1979). In contrast to an 'idealized river corridor', where vast riparian wetlands develop primarily along downstream sections, riparian wetlands associated with tropical headwater streams occur in a variety of forms, but mainly as floodplains of variable width. Like more extensive, larger ecotones associated with higher order rivers, riparian wetlands provide habitats for specific, diverse, and often endangered flora and fauna, and are thus fundamental to maintaining high biodiversity in and along streams (Naiman *et al.*, 1998, 2005; see also Chapter 6 of this volume). Riparian wetlands are also responsible for multiple ecological functions: for instance, these serve as important hydrological buffers and key retention areas for sediments, agricultural pesticides, and fertilizers (Brinson, 1993; Tockner and Stanford, 2002; Naiman *et al.*, 1998, 2005). These are sites of high primary and secondary productivity, and also act as migration corridors and/or microclimatic retreats for many taxa (Naiman *et al.*, 2005). Several studies have emphasized the significance of riparian wetlands for stream organic-matter budgets (Wantzen and Junk, 2000), solute balance (McClain *et al.*, 1994; McClain and Elsenbeer, 2001), and the structure of benthic invertebrate assemblages (Smock, 1994; Arscott *et al.*, 2005).

We are relatively well informed about the structure and function of large floodplains, and recent reviews by Junk and Wantzen (2004) and Tockner and Stanford (2002) indicate the present state of our knowledge and offer a detailed synopsis, which we do not wish to repeat. Comprehensive accounts of the ecology of temperate riparian zones are given by Naiman *et al.* (2005) and Wantzen and Junk (in press). By contrast, our understanding of riparian wetlands along headwater streams in the tropics is still in its infancy. Accordingly, it is the ecology of these systems that we focus in this chapter. While headwater wetlands may not extend far beyond the stream channel, their importance becomes very evident from a calculation of their cumulative area within a catchment (Table I; see also Brinson, 1993; Tockner and Stanford, 2002). As Table I shows, the extent of floodplain associated with low-order streams can match or exceed that of large rivers. Unfortunately, these wetlands may not be considered or included in management or planning strategies. For example, 1 : 250 000 scale maps employed in a landscape planning study in Mato Grosso (Brazil) omitted half of the first-order-streams in the catchment of interest, and thus no consideration was given to their associated wetlands (Wantzen *et al.*, 2006).

There are several reasons for the paucity of information about riparian wetlands along low-order streams. Firstly, the investigation of these wetlands is laborious because they are integral components of a larger landscape, and are thus influenced by processes acting at a variety of scales in aquatic and terrestrial environments. Secondly, the processes acting within riparian wetlands may occur seasonally or intermittently with limited spatial extent. Their functional performance is, therefore, often wider appreciated, despite frequently incorporating important biogeochemical and metazoan-driven turnover processes, i.e. so-called 'hot spots' and 'hot moments' (*sensu* McClain *et al.*, 2003; Wantzen and Junk, 2006). Thirdly, riparian wetlands of all types have been widely affected by sustained human impacts, such as conversion for agriculture, channelization, and flood-control structure, and these changes have already modified the structure and functional integrity of these ecotones. Fourthly, many wetlands and riparian zones in the tropics are unpleasant study sites that harbor poisonous snakes, stinging

*TABLE I* Stream Order, Estimated Number of Streams, Average and Total Length of Rivers and Streams, Average Riparian Width and Total Floodplain Surface Area in the USA (Brinson, 1993; after Leopold *et al.*, 1964)

| Stream order | Number | Average length (km) | Total length (km) | Estimated floodplain width (m) | Floodplain surface area (km²) |
|---|---|---|---|---|---|
| 1 | 1,570,000 | 1.6 | 2,526,130 | 3 | 7,578 |
| 2 | 350,000 | 3.7 | 1,295,245 | 6 | 7,771 |
| 3 | 80,000 | 8.5 | 682,216 | 12 | 8,187 |
| 4 | 18,000 | 19.3 | 347,544 | 24 | 8,341 |
| 5 | 4,200 | 45.1 | 189,218 | 48 | 9,082 |
| 6 | 950 | 103.0 | 97,827 | 96 | 9,391 |
| 7 | 200 | 236.5 | 47,305 | 192 | 9,082 |
| 8 | 41 | 543.8 | 22,298 | 384 | 8,562 |
| 9 | 8 | 1,250.2 | 10,002 | 768 | 7,681 |
| 10 | 1 | 2,896.2 | 2,896 | 1536 | 4,449 |

insects such as mosquitoes, and dense, thorny vegetation; when combined with their sometimes restricted spatial extent, these features help to account for the lack of attention headwater wetlands have received from researchers.

The term 'tropical', as used throughout this book, does not refer to a single set of conditions but can be subdivided into different regional landscapes and climatic types. This caveat is important especially for the present chapter since the inundation frequency and duration of riparian wetlands depends greatly on climate, which will, for example, differ between humid equatorial regions and seasonal monsoonal latitudes with distinct wet and dry seasons. In this chapter, we provide examples of headwater wetlands from different climatic areas including seasonal savannahs (Cerrado, Brazil), inland rainforests (Amazonia, Brazil), coastal lowland forests (peatswamp forest, Malaysia), and (sub)tropical Africa and Australia. Due to a wide range of wetland types, we first summarize the general features shared by riparian wetlands and then introduce some representative wetlands from different tropical regions. We conclude with recommendations for conservation and management of these habitats, and outline some priorities for future research.

## II. TYPES OF RIPARIAN WETLANDS

The term 'wetland' is rather broad and includes a range of habitat types (see Tiner, 1999). It includes temporary wetlands along the margins of 'flashy', spate-prone streams as well as flood-pulsing wetlands arranged as 'pearls on a string' along river or steam corridors (Fig. 1; see also Ward *et al.*, 2002; Wantzen, 2003; Junk and Wantzen, 2004). Permanent, moist wetlands are dominant in areas with high precipitation, low drainage, and shallow groundwater tables. In anastomosing river sections, these often consist of 'terrestrialized' meanders and anabranches. The extent of the connection of these wetlands with the stream channel during pulsed flow

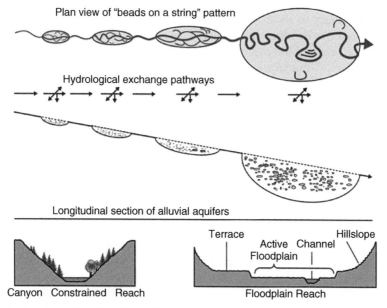

**FIGURE 1** Idealized configuration of a river corridor as an alternating sequence of constrained reaches and floodplain sections. Predominant hydrological exchange pathways are indicated for longitudinal (horizontal arrows), lateral (oblique arrows), and vertical (vertical arrows) dimensions (after Ward *et al.*, 2002; with permission from Blackwell Publications).

events has a profound influence on their ecology, because it determines the distribution and dispersal of aquatic biota and the exchange of energy and material. In this chapter, we describe headstream wetlands by attempting to harmonize the existing terminology derived mainly from larger river systems with the characteristics of riparian habitats along low-order streams. The three main criteria used for categorizing these wetlands are water permanency, exchange pathways of water and material, and substrate type.

## A. Hygropetric Zones

A hygropetric zone is a thin layer of water covering a rock surface. It depends on the permanent recharge of the water film, which is usually generated from near-surface ground-water. Due to extremely shallow water depth, hygropetric zones provide ideal conditions for primary production, although UV light exposure and risk of drying provide harsh environmental conditions that limit the occurrence of macrobiota. Hygropetric zones are among the least studied habitats associated with tropical and temperate streams, although they can be expected to harbor many as yet undescribed species, particularly aquatic Lepidoptera, midges, and other Diptera (Fischer *et al.*, 1998).

## B. Rock Pools

Rock pools develop after floods recede along streams that flow over impermeable bedrock. The duration of water-logging depends on climatic conditions and size of the pool, but it can be extended by regular 'recharging' either by stream spates or by rain water. Rock pools can be considered as 'ecological batch experiments'. When first filled, the number of predators is low, and the presence of flood-borne organic matter and developing algal biomass makes rock pools favorable habitats for an array of invertebrates, such as chironomids (Diptera) and microcrustaceans (e.g. Jocque *et al.*, 2006). Competition for resources in pools is initially low, but experiments in Amazonia have shown that there is fast development from individual-rich, species-poor assemblages to more diverse assemblages (Nolte, 1988, 1989), and the number of predators increases over time (Azevedo-Ramos and Magnusson, 1999). Specialized rock-pool inhabitants have short life cycles and adaptations that allow them to survive droughts between filling phases. The chironomid genus *Apedilum*, for example, has an egg-to-egg life cycle duration of less than 1 week (Nolte, 1995). In the rock pools of a South American Cerrado stream, the mayfly *Cloeodes hydation* (Baetidae) can survive up to 9 h of desiccation and can even cope with drying/rehydration by reiterative moulting (Nolte *et al.*, 1996). Rock pools along streams can be important breeding habitats for medically important dipteran vectors of human parasites such as malaria and elephantiasis.

## C. Para- and Orthofluvial Ponds

In riparian floodplains, cut-and-fill alluviation, coupled with groundwater–surfacewater interactions, creates a complex array of shallow lentic habitats (ponds, backwaters, and abandoned channels). Ponds are either located in the active region (=parafluvial ponds) or in the adjacent riparian forest (=orthofluvial ponds). Often, floodplain ponds originate from 'abandoned' anabranching or anastomosing channels, formed by processes similar to those observed along large rivers (Wantzen *et al.*, 2005b). During non-flood conditions, the exchange of water between ponds and the mainstream depends upon the presence of floodplain channels and the porosity of alluvium. In coarse-grained floodplains, for example, ponds are primarily fed by alluvial and hill-slope groundwaters. If the riparian zone is dominated by swamp forest, rotting tree roots or plant debris and decomposing organic matter lead to failures within the soil that can act as subterranean macropores delivering oxygen-rich stream water into floodplain ponds and water bodies (K. M. Wantzen, unpublished observations).

Aquatic organisms colonize para- and orthofluvial ponds from the permanent channel via porous interstitial flow paths through the stream hyporheic zone. The hyporheic zone may be extensive, and stream invertebrates have been found hundreds of metres from the channel of temperate rivers with coarse gravel floodplains (Stanford and Ward, 1988). In such situations, para- and orthofluvial ponds contribute disproportionately to aquatic biodiversity (Arscott *et al.*, 2005). In tropical latitudes, similarly, high contribution of lateral habitats to total stream biodiversity can be anticipated. In sites where sediments are fine-grained, floodplain ponds tend to be more permanent or persistent, and are connected to the mainstream by secondary channels (i.e. direct connectivity) or via surface flow during flood events. Both the density and diversity of benthic invertebrates in Neotropical riparian forest ponds of Brazil are positively related to the degree of hydrological connectivity with mainstream but negatively correlated with leaf litter input (Wantzen *et al.*, 2005a). The importance of riparian ponds can be inferred from the fact that Smock (1994) documented a net export of invertebrates from the floodplain to the main channel of streams in a subtropical forested wetland.

## D. Anabranches and Floodplain Channels

In dynamic floodplain streams, and especially in braided sections, the flow becomes divided among several channels, comprising a main channel (transporting the largest portion of water) and numerable smaller anabranches. These channels, together with the floodplain channels that link the main channel with floodplain ponds (see Section II-C), offer habitat conditions that are generally similar to the main channel, although in seasonal tropical climates they may cease flowing during the drier months. In large river-floodplain systems, such as the Paraguay–Paraná system (see Wantzen *et al.*, 2005b), flow in the floodplain channels and anabranches is sufficient to sustain rheophilic species in the riparian zone (e.g. Marchese *et al.*, 2002). These habitats may also serve as refuges and potential sources for recolonization of the stream channel after major disturbances in temperate streams (Robinson *et al.*, 2004). It is not known if the same phenomenon occurs in tropical streams, and analyses of the distribution of biota between the main channel and anabranches will be needed to resolve this.

## E. Moist Zones on the Riparian Floodplain

Wetlands of this type are present along most unregulated streams. These comprise the zones fringing the stream channel that are wetted during spates or seasonal high-water periods, with expansion during flooding depending on the steepness of the stream channel and lateral slopes. The period for which surface-water is retained after a flooding event will vary according to the hydraulic conductivity of the substrate in the riparian zone: coarse sediments dry out within hours or a few days, while soils that are fine grained and well supplied with organic matter or stranded debris may maintain moisture long enough to allow aquatic or semi-aquatic insects such as tipulid and limoniid crane flies (Diptera) to complete their life cycles. Some terrestrial invertebrates inhabiting the floodplains of large tropical rivers have adaptations and strategies such as physical gills, anaerobiosis, and water-proof retreats, to survive wetted (and sometimes oxygen-free) periods, (Adis and Junk, 2002). Thus far, their counterparts living along tropical headwater streams have hardly been studied. The extent to which periodically-inundated riparian zones along these streams serve as refugia for aquatic organisms during spates is not known, but passive drift or active migration into these low-flow marginal sites can take place. Moist riparian wetlands may also serve as refugia for upland terrestrial animals during droughts or seasonally dry periods.

Small tropical rain forest streams receive considerable inputs of leaf litter (see Chapter 3); in Amazonia, this amounts to 6–10 t ha yr$^{-1}$ (Klinge, 1977; Adis *et al.*, 1979; Luizão, 1982). Litter accumulates in banks on the inner side of lowland stream meanders. Litter banks extend from the riparian zone to the stream channel connecting the stream bed and the floodplain. Minor water-level fluctuations maintain humidity within litter accumulations, and steep oxygen gradients may build up. These banks provide microhabitats for stenotopic invertebrates and small fish species; dense and stable banks of litter may serve as refuges especially during floods (Walker, 1985). Feeding experiments with $^{32}$P-labeled leaves indicate that the animals inhabiting litter banks are sedentary and territorial (Walker *et al.*, 1991; Walker and Henderson, 1998). Torrential flow conditions during intense spates disrupt litter banks, forcing the specialized litter-associated fauna to drift and colonize new accumulations downstream and exposing them to predation by open-water species during the 'reset mechanism' (Wantzen and Junk, 2006).

## F. Unforested Streamside Swamps

These headwater wetlands are permanently moist due to low drainage, high precipitation, and/or shallow groundwater tables. The water-logged soils generally do not support woody vegetation. In seasonal savannah climates, such as those typical of the Brazilian Cerrado, these riparian wetlands are known as 'veredas' and extend 30–150 m from the stream edge. Veredas are characterized by grassy or graminoid vegetation and the presence of palms of the genus *Mauritia*. These support a rich array of plants and animals, but are extremely vulnerable to erosion (Wantzen *et al.*, 2006, and citations therein). Similar wetlands can be found throughout the seasonal tropics on periodically moist, nutrient-poor soils; for example, in northern Madagascar and in the African *Andropogon* (Poaceae) savannah. 'Dambos', which is a Bantu word for seasonally wet, grass-covered, flat areas, forms characteristic water-logged wetlands that cover large areas of drainage basins in several parts of Africa (Balek and Perry, 1973; Acres *et al.*, 1985). Almost 30% of the Central Zambezi Miombo Woodland, one of the largest ecoregions in Africa, is occupied by dambos and other wetlands. Both perennial and non-perennial headwater swamps influence water resource availability in Africa because they act as reservoirs absorbing water during rainy season and releasing it during dry season. In addition, dambos are useful for agriculture and for grazing during dry season.

In the Upper Nile valley (e.g. along the major tributaries flowing into Lake Victoria), permanent swamps occupy areas of thousands of square kilometers. The vegetation consists of dense stands of grasses, reeds and sedges, such as *Vossia cuspidate* (Hippo grass), *Phragmites*, and *Cyperus*, and exotic water hyacinth (*Eichhornia crassipes*: Pontederiaceae). These are extremely productive habitats; the fresh biomass of Water hyacinth alone can exceed 70 kg m$^{-2}$ (Denny, 1984).

## G. Forested Streamside Swamps

The number of tree species adapted to long-term soil water logging along the streams is higher in the tropics than temperate regions. The vegetation associated with such conditions in savannahs is well described, because the riparian or gallery forest contrasts clearly with the surrounding grassland, and it is characteristically speciose (Felfili, 1995; Budke *et al.*, 2004; see below). The demarcation of riverine forests from terrestrial forests is less clear, and floristic analyses of strictly riparian vegetation in the tropical forest regions are rare. In the Amazon Basin, researchers have differentiated the vegetation of large river floodplains from that of upland forests, but have not distinguished the composition of riparian forests along low-order streams from that of terrestrial forests.

Riparian trees are adapted to permanently saturated soils provide organic matter to the streams in the form of leaf litter, bark, etc.; these also act as a retention zone for drifting sediments and organic matter during spates. The chemical and physical recalcitrance of leaf litter produced can retard decomposition, reduce oxygen composition, and influence other aspects of soil and water conditions in these wetlands (see also Chapter 3). The tree canopies provide good shade so that these swamps experience less extreme microclimates than herbaceous or grassy wetlands; however, it can reduce the growth of algae and aquatic macrophytes (Fittkau, 1967). In dry or seasonal tropics, the wetland forest vegetation is highly dependent on groundwater, and the outer landward fringe tends to be more abrupt than in wetter areas. These narrow bands of forest facilitate animal migration along rivers, and their productivity exerts a major influence on adjacent drier and more terrestrial upland areas, making them a valuable natural resource (Hughes, 1988).

## H. Peatswamp Forests

Peatswamp forests resemble forested streamside swamps, but have a completely different genesis, hydrology, and lateral extent. Unlike the wetlands described earlier, where the hydraulic forces of streams shape the riparian wetlands, peatswamps are primarily a product of plant succession on water-logged soils. These develop in areas of swamp forest where leaf litter and organic debris have accumulated in layers of peat that may reach 20 m in depth. For this reason, peatswamps are generally raised above nearby rivers, and the degree of interaction between these and rivers and wetlands may not be great, although they are hydrologically connected to them via numerous small streams draining the peat and flowing into larger rivers. Water is stored within peat accumulations in the wet season and released during the dry season, thus mitigating floods and droughts in the downstream parts of drainage. In low-lying coastal regions, peatswamps may develop behind mangrove swamps as the latter advance seawards on newly deposited alluvium. New plant communities develop on the land side, replacing the mangrove trees.

Southeast Asia presently covers the largest area of tropical peat lands in the world (Rieley and Page, 1997; Page *et al.*, 2004), and peatswamp forests are the most extensive inland wetlands in Peninsular Malaysia and Borneo (particularly Sarawak and Kalimantan). Peatswamp forests also occur widely in the low-lying regions near large rivers in Sumatra, New Guinea, and Thailand. Peat in these swamp forests is largely composed of leaves and woody debris from trees (Anderson, 1964; Rieley and Page, 1997), and is thus quite different from peat occurring in bogs at higher latitudes that is built up mainly of *Sphagnum* moss, herbaceous plants, and grasses. Furthermore, most tropical peatswamp forests are relatively new, and have typically evolved within the past 4000 years following the last glacial maximum. Despite this, and their nutrient-poor soils, peatswamps can contain very many species of plants (Anderson, 1964; Rieley *et al.*, 1997; see also below), including those that are endemic to these habitats, such as the commercially important tree *Gonystylus bancanus* (Thymelaceae) (Rieley *et al.*, 1997; Dudgeon, 2000a, and references therein).

The difference between forested streamside swamps and peatswamps may not always be clear, but seems to be reflected in the classification of forested swamplands used by Rieley *et al.* (1997). They recognized topogenic and ombrogenous peatswamps. The former are riparian swamps that contain shallow accumulations of organic matter (much less than 50 cm) that are flooded by rising water levels during wet season. These are equivalent to the forested streamside swamps described earlier. The latter are true peatswamps with thick (>50 cm up to 20 m) accumulations of organic matter, which exist at some distance away from rivers and tend to be dome shaped or gently sloping. The domes of ombrogenous peatswamps are situated above the highest level of wet season flooding, but the water table is close to or above the surface for much of the year, and some of these peatswamps may be flooded for several months annually. Water drains outward from the center of the peatswamp dome in a series of streams that empty into rivers fringed with topogenic swamp forest.

Regardless of the extent of seasonal inundation, conditions in the peatswamps and streams draining them may be quite extreme and, at the center of peatswamps in Asia, the forest may be dominated by few species – typically species in the dipterocarp genus *Shorea* (Rieley and Page, 1997). Dissolved oxygen and nutrient levels are low, the waters are acidic (typically pH 2.9–3.7) with limited autotrophic production, and high levels of tannins and other secondary compounds leached from plant litter typify these 'blackwaters'. The inhabitants are mostly stenotopic species that are tolerant to – or dependent upon – these conditions, and – as discussed below – include an array of fishes and invertebrates that are usually not found in other types of wetlands (Dudgeon, 2000a, 2000b; Ng *et al.*, 1994; Ng, 2004).

## III. HYDROLOGY

A natural inundation regime is regarded as the key driver of instream and riparian ecosystem processes and plays an important role in sustaining biodiversity (e.g. Junk *et al.*, 1989; Poff *et al.*, 1997; Junk and Wantzen, 2004). The magnitude, frequency, duration, and timing of the inundation regime influence biotic communities and ecosystem processes, either directly or indirectly, through their effects on other primary regulators, and modification of the inundation regime thus has cascading effects on the ecological integrity of river-floodplain systems (Bunn *et al.*, 2005). Natural, unregulated streams exhibit two kinds of hydrologic behaviors with respect to increase in discharge: those that flood instream habitats within bankfull limits – i.e. 'flow-pulses' (*sensu* Tockner *et al.*, 2000) – and those that spill over the bankfull limits and inundate the surrounding land – i.e. 'flood pulses' (*sensu* Junk *et al.*, 1989; see also Junk and Wantzen, 2004). The 'flashiness' of the hydrograph depends on the size and topography of the catchment, on the precipitation regime, and on the hydrological buffering capacity of the riparian zone. The size of the area that may become flooded and the duration of flooding depend on the landscape gradient and obstacles that reduce the runoff in the channel. Steep valleys generally have intense, but brief, flood events, while plains and forested valleys have smaller, long-lasting floods. Often, spates can occur within moments of heavy rainstorm,which decrease over a few hours as rains abate. In areas that are either arid or have strong seasonal rainfall, stream discharge remains high relative to baseflow conditions (Fig. 2) as water percolates through the catchment into the channel (Poff *et al.*, 1997; Bunn *et al.*, 2005; Yoshimura *et al.*, 2005).

It has been generally assumed that the correlation between the riparian zone and the stream channel during floods is insufficient time to permit consumers to exploit flood-borne resources, and that the unpredictability of individual spates would pose a risk of density-independent mortality, limiting the potential of biota to become adapted to them (Junk *et al.*, 1989). However, in tropical monsoonal regions or those with seasonal climate, the number of spates and the discharge volume in excess of baseflow increase predictably during rainy season (Wantzen and Junk, 2000; see Fig. 2). The gradual increase in flow leads to a progressive increase in the availability of riparian zone to aquatic fauna for several months in the year, resulting in greater habitat diversity. Larvae and juveniles of aquatic animals can escape from predation into permanently flooded habitats of the riparian zone, and many stream organisms benefit from flood-borne resources (see below).

Extended flooding in streams, lasting for several days or weeks, occurs during backflooding when the water level of the receiving river is so high that the discharge of tributary streams cannot enter and accumulate within these smaller channels. This phenomenon is common in streams draining the floodplains of large rivers (e.g. Rueda-Delgado *et al.*, 2006) or streams that become blocked by sediments carried by rivers (Wantzen, 2003). Due to the impounding effect of sediment or water, the variation of current within streams is much lower during backfloods than during spates or flood events (Rueda-Delgado *et al.*, 2006).

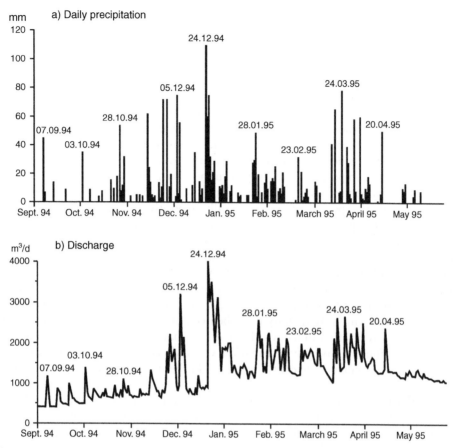

FIGURE 2 Precipitation and discharge of Tenente Amaral stream, a tributary of the São Lourenço River, Brazil (data from Wantzen, 2003).

In the seasonal tropics, intense rainfall can cause high surface runoff, which can lead to erosion, especially in small, low-order streams (e.g. Douglas, 1968). The mean annual maximum and minimum discharge ratio is a good indicator of the extent of flooding in riparian floodplains and of the strength of erosive forces. In northern Neotropics, this ratio is around 20 for rivers that drain savannah and dry forests, and below 10 for rivers that drain moist forests (Lewis *et al.*, 1995; see also Chapter 1). Severe spates can remove or deliver particulate organic matter and reorganize the position of boulders and large woody debris, which in turn influence the flow velocity and patterns of erosion and deposition that affect channel architecture and may create new wetland patches (Edwards *et al.*, 1999; Gurnell *et al.*, 2005). Fine sediment deposits of the dry season that clog interstices between gravel and boulders become washed out during the first storms of the rainy season, which increases habitat diversity (Dudgeon, 1982). When floods recede, fine particles and significant quantities of organic matter (up to $1 \text{kg DW m}^2$: K. M. Wantzen, personal observations) become deposited among tree trunks and roots in the riparian zones and swamp forests, which may enhance soil formation. Alternate periods of wetting and drying in the riparian zones have important consequences on the dynamics of decomposition of such organic matter (Mathooko *et al.*, 2000; see also Chapter 3).

## IV. BIODIVERSITY

Although riparian wetlands are important for the biodiversity of aquatic and terrestrial species, our knowledge of the diversity of plant and animal species, as well as their specific adaptations to survive or to benefit from the wetting and drying cycle, is still scant (Wantzen and Junk, 2000). In most tropical areas, the species-effort-curves have yet to reach saturation, and the number of reported species tends to reflect sampling intensity. There is an urgent need to improve our taxonomic knowledge of riparian species and, given the rate of human modification of riparia (see below), there is increasing risk that many species may become extinct before even being described. The benthic stream fauna exhibit traits adapted to the hydraulic patterns that control substrate composition and oxygen supply (Statzner and Higler, 1986; Wagner and Schmidt, 2004). This applies also to the inhabitants of adjacent wetlands. Although substrate preferences vary among species, there is generally a positive relationship between substrate diversity, especially the presence of very coarse substrates, and species richness. The fringing wetlands not only represent an extension of stream habitats but can offer additional habitats with distinctive hydraulic and substrate conditions. The ecotonal nature of fringing riparian wetlands would lead to the supposition that they support much higher biodiversity than in permanent aquatic habitats or purely terrestrial habitats (Wantzen and Junk, 2000), including migratory species from both (Wantzen and Junk, 2006). Habitat dynamics in riparian ecotones are intermediate between the frequently disturbed conditions in the main stream channel and those in the more terrestrial portions, which are driven mainly by plant succession and periodic inundation/desiccation. Biodiversity in riparian wetlands is thus likely to be subject to the processes that tend to enhance diversity embodied in the intermediate-disturbance hypothesis (Ward *et al.*, 2002).

When analyzing the real-world conditions, one finds that the expectation that biodiversity will peak in the riparian wetlands of tropical streams is not fulfilled by all taxonomic groups studied thus far. What factors may limit biodiversity in these wetlands? Some of the factors include, little or no water, high concentrations of dissolved organic matter leached from plant litter, and periodic scarcity or absence of oxygen. Studies on isolated wetland ponds in a swamp forest in Mato Grosso (Brazil) suggest that colonization by benthic invertebrates was inhibited by high levels of plant secondary compounds in water (Wantzen *et al.*, 2005a), and similarly low colonization of seasonal pools in an Australian stream has been reported (Bunn, 1988). Acidic waters and lack of dissolved calcium limit the occurrence of decapod crustaceans and most mollusks in many tropical stream wetlands, e.g. Amazonia (Junk and Robertson, 1997) and Southeast Asian peatswamps (Ng *et al.*, 1994; Dudgeon 2000a).

Despite apparently unfavorable conditions, some specialists do thrive. The water in streams that drain peatswamps is clear, but dark in colour due to dissolved phenolics, and is highly acidic (pH 2.5–4.5) and low in oxygen and nutrients. These habitats nonetheless support diverse, highly specialized and well-adapted endemic flora and fauna of global significance, as reported from peatswamp forests of Malaysia, Indonesia, and Thailand (Ng *et al.*, 1994; Page *et al.*, 1997; Ng, 2004). Fish biodiversity is especially high: 20% of the estimated 250 species of freshwater fishes in Peninsular Malaysia have been recorded from a single peatswamp, while the total number of fish species from such swamps in Malaysia, Borneo, and Sumatra is in the order of 200–300 (Ng, 2004). Over 200 plant species are known from the peatswamp forests of Peninsular Malaysia, while Anderson (1964) recorded 927 plant species from the peat swamp forests of Brunei and Sarawak (see also Rieley *et al.*, 1997). Southeast Asian peat swamp forests are also home to an array of mammals, including primates and species of global conservation significance (Page *et al.*, 1997; Dudgeon 2000a; see also Chapter 6). Because of the vast extent of plant and animal diversity that they support, Dudgeon (2000a) has proposed that swamp forests and peatswamps be treated as 'keystone habitats' of conservation priority in Asia.

Plant biodiversity of riparian wetlands in seasonal Neotropics is very high. Felfili (1995) recorded 93 tree species in 81 genera and 44 families from 64 ha of undisturbed gallery forests in Brazilian Cerrado, while Budke *et al.* (2004) have identified 57 species in 47 genera and 26 families from 1 ha of riverine forest in subtropical southern Brazil. Samples from riparian vereda wetland in Minas Gerais (Brazil) hosted 136–361 species, in which as many as 168 were confined to this habitat type (Araújo *et al.*, 2002). The invertebrate fauna of these ecotonal systems is still poorly known, although an exceptional diversity of copepod microcrustacean has been documented (Reid, 1984), but the composition and habitats of the stream-side and semi-aquatic mammals and most of the other vertebrates are not well known (see Chapter 6). Likewise, information on larger riparian animals in the Afrotropics is fragmentary (see Chapter 6 and references therein), but there have been some studies of dragonfly (Odonata) assemblages (Samways and Steytler, 1996) and the floristics of riparian forests (Natta *et al.*, 2002) in Africa. More information on the composition and ecology of biodiversity in tropical stream wetlands is needed urgently, so that the population status of threatened species – especially those riparian obligates – can be assessed adequately and their habitat requirements better understood.

## V. AQUATIC-TERRESTRIAL LINKAGES

Food webs in riparian wetlands are generally made up of short chains with a high degree of omnivory. Their structure is influenced by hydrology and usually depends on the maintenance of hydrological connectivity. Although these generalizations have been developed from research in the Australian wet–dry tropics (Douglas *et al.*, 2005), these may apply generally for tropical streams and associated wetlands (see also Chapter 2).

As ecotones between the stream channel and upland, more terrestrial sites, riparian wetlands exhibit very steep ecological gradients, which are likely to enhance a variety of biological processes (Naiman, 1998). These include transfer of energy, nutrients, and material from the riparian zone to the stream (e.g. litter fall and resultant litter processing); transfer of energy, nutrients and material from the stream to surrounding terrestrial habitats (e.g. emergence of insects, predation upon adult aquatic insects); and production, storage, and processing of autochthonous biomass within the riparian wetland (Fig. 3). The intensity and timing of these

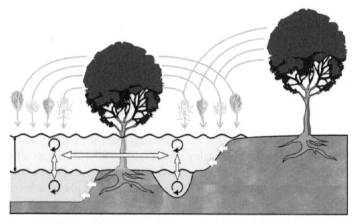

*FIGURE 3* Wetland pools as 'bioreactors' for organic matter processing and temporary storage. During low water levels, wetland pools receive direct allochthonous inputs from trees (e.g. leaves, seeds, flowers, fallen insects), and they exchange this material with the stream during floods. Between floods, the relatively long residence time (compared with the stream channel) and ease of access by both aquatic and terrestrial consumers (during drying phases) enhance organic matter processing.

processes may vary considerably, in part due to the variable nature and characteristics of riparian wetlands and also in response to stream hydrology. The frequency and extent of flooding are key factors in organic matter processing and exchange between main channel and floodplain, and in turn the relative contribution of organic matter derived from the catchment, the channel, and the floodplain, changes along the river course (Junk and Wantzen, 2004). Organic matter processing does not occur in an equilibrated time-space pattern, but rather at 'hot spots' and/or during 'hot moments' (*sensu* Wantzen and Junk, 2006). This temporal and spatial unpredictability challenges our ability to understand and quantify organic matter budgets in riparian wetlands.

Most riparian wetlands are highly productive, providing material and energy for adjacent ecosystems. The efficiency of the transfer rate of energy across the aquatic-terrestrial boundary depends on the ratio of shoreline length to stream area and the permeability of the boundary along the margins (Naiman *et al.*, 1998). Flooding creates semi-aquatic conditions colonized by aquatic and terrestrial biota that benefit from the 'reciprocal subsidies' on the ecotone between the two bordering systems (Nakano and Murakami, 2001). For example, many terrestrial organisms depend on water-borne resources, such as flood wracks (Bastow *et al.*, 2002) or emerging aquatic insects (Paetzold and Tockner, 2005), and both the stream channel and the riparian zone can be an important source of flying insects for terrestrial insectivores (Douglas *et al.*, 2005). Conversely, aquatic consumers may move into newly inundated areas of riparian zones to feed, and fruits, seeds, and terrestrial insects can be important components of aquatic food webs (Nakano and Murakami, 2001). For example, 40% of the diet of freshwater crocodiles (*Crocodylusa johnstoni*: Crocodylidae) in Australia is of terrestrial origin, and terrestrial insects make seasonally variable contributions to the diet of tropical stream fishes (e.g. Chan and Dudgeon, 2006; see also Chapter 4).

Seasonal floods promote exchange between streams and their riparian zones, and facilitate reciprocal land-water subsidies, the most conspicuous of these being the transport of terrestrial (allochthonous) plant litter (see also Chapter 3). The residence time of such organic matter in riparian wetlands is intermediate to the stream channel and the adjacent terrestrial environment. As a result, pools associated with the riparian zone or floodplains of tropical streams are ideal 'bioreactor' habitats for the processing of terrestrial organic matter (see Chapter 3). For example, larvae of the trichopteran shredder *Phylloicus* (Calamoceratidae), which builds a case of dead leaves, occur in large numbers in the pools near Neotropical streams, where they shred litter that has been conditioned by microbes. Such larvae are uncommon in the mainstream leaf litter. Leaf litter turnover is high, and larvae and their food are likely to be transported downstream (Wantzen and Wagner, 2006).

With decreasing frequency of flooding, permanently wetted riparian wetland habitats tend to accumulate organic matter. This is most evident in tropical Asian peatswamp forests, where organic matter accumulates at the base of trees and forms hummocks. During rainy season, surface-water accumulates in depressions between the hummocks to create seasonally available open-water habitats for fishes and invertebrates. Other peatswamp forests may be more extensively inundated to some depth for months together, and these sites often have deep accumulations of peat (for details, see Rieley and Page, 1997). Decomposition of leaf litter is extremely slow in peatswamps because nutrient scarcity results in the production of leaves and other litter that are tough and toxic, and contain high levels of plant defensive compounds to deter terrestrial herbivores. Unpublished studies by C. Yule *et al.* suggest that accumulation of litter as peat takes place because it is not processed by microbes or aquatic invertebrates; however, *in situ* experiments indicate that aquatic bacteria and fungi are abundant in peatswamp streams and readily break down the leaves of less well-defended plants. These findings contradict the assumption (e.g. Whitmore, 1984) that peat builds up because of extreme conditions of acidity and low oxygen (see Chapter 3 for more details). The same authors also suggest that dissolved organic carbon from newly shed leaves is taken up by bacteria to provide the basis for a food web of invertebrate primary consumers and their predators.

## VI. HUMAN IMPACTS, CONSERVATION, AND SUSTAINABLE MANAGEMENT

Removal of riparian vegetation may alter the morphology of stream and river channels and the ecosystem services (Sweeney et al., 2004). Unfortunately, conservation of riparian wetlands is hampered by the fact that they are often ephemeral, and thus their importance may not be evident during baseflow conditions. Even permanent riparian wetlands generally occupy only narrow strips adjacent to the channel of low-order streams, and thus they are inconspicuous despite the fact, as discussed earlier (see Table I), that their combined area within a catchment may be substantial (e.g. Wantzen et al., 2006). Such wetlands are perceived by humans as favorable sites for colonization, agriculture, or pasture. The factors that make these systems biologically productive and diverse also make them attractive to humans and subject to anthropogenic alteration of stream environments and land-use changes in the surrounding drainage basin. The latter include cattle trampling and more general impacts arising from land-use changes and deforestation, which release large amounts of sediments and can lower the groundwater table in wetlands (Wantzen, 2006; K. M. Wantzen et al., unpublished observations). Proximity to water will facilitate the construction of aquaculture enterprises in streamside wetlands, with the associated risk of releasing exotic fishes and crustaceans into the environment (Orsi and Agostinho, 1999; other examples are given in Chapter 3).

Channel rectification and channel deepening of streams have destroyed most riparian wetlands along headwater streams in agricultural and urban landscapes, and they are often subject to water extraction or impoundments for irrigation (see Chapter 10 for details of other threats to tropical streams). The increasing use of small-scale dams for hydropower is a particular threat for low-order streams; here, a solution could be to 'sacrifice' certain streams and to maintain the natural hydrological dynamics and longitudinal connectivity of some portion of each drainage basin or stream of particular ecological importance (Greathouse et al., 2006). Virtually all human activities in the riparian wetlands reduce the area, amplitude, and frequency of flood events, which are precisely the conditions needed to sustain wetlands (Richter et al., 1996). Changes in hydrological regime are not only deleterious for the native biota of streams and their wetlands, but they favor the distribution of exotic plants (Bunn and Arthington, 2002; see also Chapter 10).

Of all the riparian wetlands along tropical streams, the threats to tropical peatswamp forest are of particular concern, and this is one of most threatened wetland habitat types of the world (Dudgeon, 2000a; Ng et al., 1994). For instance, by the year, 2000, 77% of the peat swamp forests of Peninsular Malaysia had been cleared of the remainder, only half is considered to be in pristine condition (Mohamed Idris, 2001), and much of the rest – as well as huge swathes in Kalimantan – are scheduled for logging or conversion to agriculture (especially oil palm). These wetlands have all but disappeared from Java (Whitten et al., 1997). The conversion of peatswamp forests for agriculture is generally difficult and expensive, and often takes place after the forest has been burned. The resulting soils are acidic, low in nutrients and bulk density, hydrophobic, and subject to subsidence (Notohadiprawiro, 1996), and many areas that were drained and cleared have become unproductive wasteland. Draining and selective logging of peatswamps increases the frequency of fires. In 1994, 1997, and 2002, ENSO-related drought events combined with peat forest clearing and draining resulted in widespread, devastating fires across Borneo, Sumatra, and Irian Jaya, which released enormous amounts of carbon into the atmosphere, contributing to global warming (Page et al., 2002).

Conservation and restoration schemes for riparian wetlands along all streams and rivers depend on the maintenance or re-establishment of hydrological diversity and landscape dynamism within drainage basins (Ward *et al.*, 1999; Zalewski *et al.*, 2001; Wantzen and Junk, 2004; Douglas *et al.*, 2005). This is especially difficult in tropical latitudes inhabited by large human populations with reasonable expectations of socioeconomic development and livelihood improvements, in a context where environmental legislation is often lacking or weakly enforced (Dudgeon *et al.*, 2000), and financial support for conservation initiatives is poor. On priority is the need to communicate the economic values of functioning riparian wetlands (e.g. Sweeney *et al.*, 2004; Naiman *et al.*, 2005) so as to help raise awareness among policy makers. These values include buffering hydrological variability and maintenance of continuous spring flows or stream discharge during low-flow periods; amelioration of regional climatic fluctuations, and carbon sequestration/storage; storage and sequestration of agrochemicals and heavy metals; soil protection and control of erosion; maintenance of biodiversity through habitat provision; and provision of goods of economic and subsistence importance, such as fish, timber, and forest products (e.g. fruits, resins, and medicinal plants).

The indirect economic value of the functions provided by tropical riparian wetlands is difficult to assess, although it is an issue that needs to be addressed with some urgency. One global benefit of increasing importance is storage and sequestration of carbon, especially in peatswamps (Page *et al.*, 2004). Parish (2002) estimated that a 10-m-deep layer of tropical peat stores $5800\,t\,C\,ha^{-1}$, as compared with $300–500\,t\,ha^{-1}$ for other tropical forests. For this reason the destruction of Asian peatswamps is a matter of global concern as conversion for agriculture may result in rapid decomposition of peat and liberation of stored carbon. While the amounts of carbon that might be released are not known with certainty, it is evident that soils that are high in organic matter have the highest rates of carbon loss (Bellamy *et al.*, 2005). Similarly, high carbon storage and potential losses can be anticipated in other riparian wetlands, such as the Neotropical Cerrado of Mato Grosso (K.M. Wantzen *et al.*, unpublished observations).

While global carbon emissions are gaining increasing global attention, small-scale or local conservation schemes that yield immediate economic returns for stakeholders are needed to protect riparian headwater wetlands. This can be best achieved by the combination of law enforcement and involvement of local communities with the resources they depend upon. For instance, Wantzen *et al.* (2006) have suggested that planting buffer zones of native trees between the agricultural areas and riparian wetlands in the Brazilian Cerrado might reduce the degradation of streamside habitat. The commercial use of non-wood products from the trees would help offset the costs of planting, and provide an additional income to the farmers. This system could also be used to restore or enhance the existing erosion gullies caused by streamside agriculture. Care must be taken: such schemes do not offer 'one size fits all', which can be applied across a region (certainly not across the tropics as a whole), and the outcome of projects intended to enhance riparian ecotones is bound to depend on the current state of alteration or degradation of the system. Particular attention needs to be paid to the use of appropriate approaches and technologies for a given region, including the use of native species for re-vegetation and habitat restoration.

## VII. CONCLUSIONS AND PROSPECTS

Riparian wetlands associated with tropical headwater streams fulfill an important and essential ecosystem functions, yet they are severely endangered by a range of non-sustainable and, perhaps, irreversible human activities. Conservationists and landscape managers in the tropics face the challenge that detailed studies are needed to better understand the functions and values of riparian wetlands while, at the same time, the rate of degradation and habitat alterations continues to increase and requires immediate action (Moulton and Wantzen, 2006; see also Chapter 10). While this review does not allow us to take a prescriptive stance upon the additional research that will be needed to facilitate decision-making, some lacunae in our knowledge have became evident during its preparation. For example, it became apparent that we are less informed about floodplains in Africa and Asia, and future research should include these areas. However, even within a given region, our knowledge of different wetland types was fragmentary, and we lack the complete picture of these ecotones for any part of the tropics. Some other conspicuous and potentially-important information gaps are set out below, and indicate some research priorities for stream ecologists and colleagues from cognate disciplines.

- Where and when do geochemical and biotic 'hot spots' and 'hot moments' occur in riparian wetlands? What are the locations and periods of significant nutrient inputs and organic matter turnover?
- How and when are riparian wetlands colonized? Which are the organisms that depend upon them as a migration corridor?
- What are the interactions between terrestrial and aquatic organisms within riparian wetlands? To what extent there is 'reciprocity' in the transfer of energy and material between land and water, and how is it mediated by trophic interactions in ecotones?
- Which life stages are dependent on particular habitats within land–water ecotones along headwater streams? How does connectivity across the ecotone or along stream networks influence biodiversity in tropical riparian wetlands?
- How do aquatic and terrestrial invertebrates as well as fishes adapt to, exploit, or tolerate with periodic flooding or desiccation of the riparian zone? To what extent do these strategies make them vulnerable to anthropogenic modification of seasonal flow regimes?

## REFERENCES

Acres, B.D., Blair Rains, A., King, R.B., Lawton, R.M., Michell, A.J.B., and Rackham, L.J. (1985). African dambos: their distribution, characteristics and use. *Zeitschrift für Geomorphologie N.F., Supplement* **52**, 63–86.

Adis, J., and Junk, W.J. (2002). Terrestrial invertebrates inhabiting lowland river floodplains of Central Amazonia and Central Europe: a review. *Freshwater Biology* **47**, 711–731.

Adis, J., Furch, K., and Irmler, U. (1979). Litter production of a Central Amazonian inundation forest. *Tropical Ecology* **20**, 236–245.

Anderson, J.A.R. (1964). The structure and development of the peatswamps of Sarawak and Brunei. *Journal of Tropical Geography* **18**, 7–16.

Araújo, G.M., Barbosa, A.A.A., Arantes, A.A., and Amaral, A.F. (2002). Composição florística de veredas no Município de Uberlândia, MG. *Revista Brasileira de Botânica* **25**, 475–493.

Arscott, D.B., Tockner, K., and Ward, J.V. (2005). Lateral organization of aquatic invertebrates along the corridor of a braided floodplain river. *Journal of the North American Benthological Society* **24**, 934–954.

Azevedo-Ramos, C., and Magnusson, W.E. (1999). Predation as the key factor structuring tadpole assemblages in a savanna area in Central Amazonia. *Copeia* **1**, 33.

Balek, J., and Perry, J.E. (1973). Hydrology of seasonally inundated African headwater swamps. *Journal of Hydrology* **19**, 227–249.

Bastow, J.L., Sabo, J.L.M., Finlay, J.C., and Power, M.E. (2002). A basal aquatic-terrestrial trophic link in rivers: algal subsidies via shore-dwelling grasshoppers. *Oecologia* **131**, 261–268.

Bellamy, P.H., Loveland, P.J., Bradley, R.I., Lark, R.M., and Kirk, G.J.D. (2005). Carbon losses from all soils across England and Wales, 1978–2003. *Nature* **437**, 245–248.

Brinson, M.M. (1993). Changes in the functioning of wetlands along environmental gradients. *Wetlands* **13**, 65–74.

Budke, J.C., Giehl, E.L.H., Athayde, E.A., Eisinger, S.M., and Záchia, R.A. (2004). Florística e fitossociologia do componente arbóreo de uma floresta ribeirinha, arroio Passo das Tropas, Santa Maria, RS, Brasil. *Acta Botânica Brasileira* **18**, 581–589.

Bunn, S.E. (1988). Processing of leaf litter in two northern Jarrah Forest streams, Western Australia: II. The role of macroinvertebrates and the influence of soluble polyphenols and inorganic sediment. *Hydrobiologia* **162**, 211–223.

Bunn, S.E., and Arthington, A.H. (2002). Basic principles and ecological consequences of altered flow regimes for aquatic biodiversity. *Environmental Management* 30, 492–507.

Bunn, S.E., Thoms, M.C., Hamilton, S.K., and Capon, S.J. (2005). Flow variability in dryland rivers: boom, bust and the bits in between. *River Research and Applications* 22, 179–186.

Chan, E.K.W., and Dudgeon, D. (2006). Riparian vegetation affects the food supply of stream fish in Hong Kong. *In* "Sustainable Management of Protected Areas for Future Generations" (C.Y. Jim and R.T. Corlett, Eds), pp. 219–231. World Conservation Union (IUCN) and World Commission on Protected Areas (WPCA), Gland, Switzerland.

Denny, P. (1984). Permanent swamp vegetation of the Upper Nile. *Hydrobiologia* 110, 79–90.

Douglas, I. (1968). Erosion in the Sungai Gombak catchment, Selangor, Malaysia. *Journal of Tropical Geography* 26, 1–16.

Douglas, M.M., Bunn, S.E., and Davies, P.M. (2005). River and wetland food webs in Australia's wet-dry tropics: general principles and implications for management. *Marine and Freshwater Research* 56, 329–342.

Dudgeon, D. (1982). Spatial and temporal changes in the sediment characteristics of Tai Po Kau Forest Stream, New Territories, Hong Kong, with some preliminary observations upon within-reach variations in current velocity. *Archiv für Hydrobiologie, Supplement* 64, 36–64.

Dudgeon, D. (2000a). Riverine wetlands and biodiversity conservation in tropical Asia. *In* "Biodiversity of Wetlands: Assessment, Function and Conservation, Vol. 1" (B. Gopal, W.J. Junk and J.A. Davis, Eds), pp. 35–60. Backhuys Publishers, Leiden, The Netherlands.

Dudgeon, D. (2000b). The ecology of tropical Asian rivers and streams in relation to biodiversity conservation. *Annual Review in Ecology and Systematics* 31, 239–263.

Dudgeon, D., Choowaew, S., and Ho, S.-C. (2000). River conservation in South-east Asia. *In* "Global Perspectives on River Conservation: Science, Policy and Practice" (P.J. Boon, B.R. Davies and G.E. Petts, Eds.), pp. 281–310. John Wiley and Sons, Chichester, UK.

Edwards, P.J., Kollmann, J., Gurnell, A.M., Petts, G.E., Tockner, K., and Ward, J.V. (1999). A conceptual model of vegetation dynamics on gravel bars of a large Alpine river. *Wetlands Ecology and Management* 7, 141–153.

Felfili, J.M. (1995). Diversity, structure and dynamics of a gallery forest in central Brazil. *Vegetatio* 117, 1–15.

Fischer, J., Fischer, F., Schnabel, S., Wagner, R., and Bohle, H.W. (1998). The biology of springs and springbrooks. *In* "Studies in Crenobiology" (L. Botosaneanu, Ed.), pp. 181–199. Backhuys Publishers, Leiden, The Netherlands.

Fittkau, E.J. (1967). On the ecology of Amazonian rain-forest streams. *In* "Atas do Simpósio sôbre a Biota Amazônica, Vol. 3 (Limnologia)" (H. Lent, Ed.), pp. 97–108. Rio de Janeiro: Conselho Nacional des Pesquisas, Rio de Janeiro, Brazil.

Greathouse, E.A., Pringle, C., and Holmquist, J.E. (2006). Conservation and management of migratory fauna: dams in tropical streams of Puerto Rico. *Aquatic Conservation* 16, 695–712.

Gurnell, A.M., Tockner, K., Edwards, P.J., and Petts, G. (2005). Effects of deposited wood on biocomplexity of river corridors. *Frontiers in Ecology and Environment* 3, 377–382.

Harner, M.J., and Stanford, J.A. (2003). Differences in cottonwood growth between a losing and a gaining reach of an alluvial floodplain. *Ecology* 84, 1453–1458.

Hughes, F.M.R. (1988). The ecology of African floodplain forests in semi-arid and arid zones: a review. *Journal of Biogeography* 15, 127–140.

Jocque, M., Martens, K., Riddoch, B., and Brendonck, L. (2006). Faunistics of ephemeral rock pools in southeastern Botswana. *Archiv für Hydrobiologie* 165, 415–431.

Junk, W.J., and Robertson, B.A. (1997). Aquatic invertebrates. *In* "The Central Amazon Floodplain – Ecology of a Pulsing System" (W.J. Junk, Ed.), pp. 279–294. Springer Verlag GMbH, Berlin.

Junk, W.J., and Wantzen, K.M. (2004). The flood pulse concept. New aspects, approaches, and applications – an update. *In* "Proceedings of the 2nd Large River Symposium (LARS), Phnom Penh, Cambodia" (R. Welcomme and T. Petr, Eds), pp. 117–149. Food and Agriculture Organization and Mekong River Commission, RAP Publication, 2004/16, Bangkok, Thailand.

Junk, W.J., Bayley, P.B., and Sparks, R.E. (1989). The Flood Pulse Concept in river-floodplain-systems. *Canadian Special Publication of Fisheries and Aquatic Sciences* 106, 110–127.

Klinge, H. (1977). Fine litter production and nutrient return to the soil in three natural forest stands of Eastern Amazonia. *Geology and Ecology of the Tropics* 1, 159–167.

Lewis, W.M., Jr, Hamilton, S.K., and Saunders, J.F., III (1995). Rivers of Northern South America. *In* "River and Stream Ecosystems" (C.E. Cushing, K.W. Cummins and G.W. Minshall, Eds), pp. 219–256. Elsevier, Amsterdam, The Netherlands.

Luizão, F. (1982). "Produção e Decomposição de Liteira em Floresta de Terra Firme da Amazônia Central. Aspectos Químicos e Biológicos da Llixiviação e Remorção de Nutrientes da Liteira". Unpuplished MSc. Thesis, FUA/INPA, Manaus, Brazil.

Marchese, M.R., Escurra de Drago, I., and Drago, E.C. (2002). Benthic macroinvertebrates and physical habitat relationships in the Paraná River-floodplain system. *In* "The Ecohydrology of South American Rivers and Wetlands" (M.E. McClain, Ed.), pp. 111–132. International Association of Hydrological Sciences, Wallingford, UK.

Mathooko, J.M., M'Erimba, C.M., and Leichtfried, M. (2000). Decomposition of leaf litter of *Dombeya goetzenii* in the Njoro River, Kenya. *Hydrobiologia* **418**, 147–152.

McClain, M.E., and Elsenbeer, H. (2001). Terrestrial inputs to Amazon streams and internal biogeochemical processing. *In* "The Biogeochemistry of the Amazon Basin" (M.E. McClain, R.L. Victoria and J.E. Richey, Eds.), pp 185–208. Oxford University Press, New York.

McClain, M.E., Richey, J.E., and Pimentel, T.P. (1994). Groundwater nitrogen dynamics at the terrestrial-lotic interface of a small catchment in the Central Amazon Basin. *Biogeochemistry* **27**, 113–127.

McClain, M.E., Boyer, E.W., Dent, C.L., Gergel, S.E., Grimm, N.B., Groffman, P., Hart, S.C., Harvey, J., Johnston, C., Mayorga, E., Mcdowell, W.H., and Pinay, G. (2003). Biogeochemical hot spots and hot moments at the interface of terrestrial and aquatic ecosystems. *Ecosystems* **6**, 301–312.

McCormick, J.F. (1979). A summary of the national riparian symposium. *In* "Strategies for Protection and Management of Floodplain Wetlands and other Riparian Ecosystems" (R.R. Johnson and J.F. McCormick, Eds), pp. 362–363. United States Department of Agriculture, Forest Service (General Technical Report WO 0197-6109: 12), Washington, DC, USA.

Mohamed Idris, S.M. (2001). "Malaysian Environment Alert, 2001". Sahabat Alam Malaysi, Pulau Pinang, Malaysia.

Moulton, T., and Wantzen, K.M. (2006). Conservation of tropical streams – special questions or conventional paradigms? *Aquatic Conservation* **16**, 659–663.

Naiman, R.J. (1988). Animal influences on ecosystem dynamics. *Bioscience* **38**, 750–752.

Naiman, R.H., Décamps, H., Pastor, J., and Johnston, C.A. (1988). The potential importance of boundaries to fluvial systems. *Journal of the North American Benthological Society* **7**, 289–306.

Naiman, R.J., Décamps, H., and McClain, M.E. (2005). "Riparia: Ecology, Conservation, and Management of Streamside Communities". Elsevier, New York.

Nakano, S., and Murakami, M. (2001). Reciprocal subsidies. Dynamic interdependence between terrestrial and aquatic food webs. *Proceedings of the National Academy of Sciences* **98**, 166–170.

Natta, A.K., Sinsin, B., and Van der Maesen, L. (2002). Riparian forests, a unique but endangered ecosystem in Benin. *Botanische Jahrbuecher fuer Systematik, Pflanzengeschichte und Pflanzengeographie* **55**, 55–69.

Ng, P.K.L. (2004). A tragedy with many players. *Nature* **430**, 396–398.

Ng, P.K.L., Tay, J.B., and Lim, K.K.P. (1994). Diversity and conservation of blackwater fishes in Peninsular Malaysia, particularly the North Selangor peat swamp forest. *Hydrobiologia* **285**, 203–218.

Nolte, U. (1988). Small water colonization in pulse stable varzea and constant terra firme biotopes on the Neotropics. *Archiv für Hydrobiologie* **113**, 541–550.

Nolte, U. (1989). Observations on Neotropical rainpools with emphasis on Chironomidae (Diptera). *Studies on Neotropical Fauna and Environment* **24**, 105–120.

Nolte, U. (1995). From egg to imago in less than seven days, *Apedilum eachistum* (Chironomidae). *In* "Chironomids: from Genes to Ecosystems" (P.S. Cranston, Ed.), pp. 177–184. CSIRO Publications, Melbourne, Australia.

Nolte, U., Tietböhl, R.S., and McCafferty, W.P. (1996). A mayfly from tropical Brazil capable of tolerating short-term dehydration. *Journal of the North American Benthological Society* **15**, 87–94.

Notohadiprawiro, T. (1996). Constraints to achieving the agricultural potential of tropical peatlands – an Indonesian perspective. *In* "Tropical Lowland Peatlands of Southeast Asia. Proceedings of a Workshop on Integrated Planning and Management of Tropical Lowland Peatlands held at Cisarua, Indonesia, 3–8 July, 1992" (E. Maltby, C.P. Immirzi and R.J. Safford, Eds), pp. 139–154. IUCN, Gland, Switzerland.

Orsi, M.L., and Agostinho, A.A. (1999). Introdução de espécies de peixes por escapes acidentais de tanques de cultivo em rios da bacia do Rio Paraná, Brasil. *Revista Brasileira de Zoologia* **16**, 557–560.

Paetzold, A., and Tockner, K. (2005). Effects of riparian arthropod predation on the biomass and abundance of aquatic insect emergence. *Journal of the North American Benthological Society* **24**, 395–402.

Page, S.E., Rieley, J.O., Doody, K., Hodgson, S., Husson, S., Jenkins, P., Murrough-Bernard, H., Otway, S., and Wilshaw, S. (1997). Biodiversity of tropical peatswamp forest: a case study of animal diversity in the Sungai Sebangau catchment of Central Kalimantan, Indonesia. *In* "Tropical Peatlands" (J.O. Rieley and S.E. Page, Eds), pp. 231–242. Samara Publishing Ltd., Cardigan, UK.

Page, S.E., Siegert, F., Rieley, J.O., Boehm, H.-D., Jaya, A., and Limin, S. (2002). The amount of carbon released from peat and forest fires in Indonesia during, 1997. *Nature* **420**, 6165.

Page, S.E., Wust, R.A.J., Weiss, D., Rieley, J.O., Shotyk, W., and Limin, S.H. (2004). A record of Late Pleistocene and Holocene carbon accumulation and climate change from an equatorial peat bog (Kalimantan, Indonesia): implications for past, present and future carbon dynamics. *Journal of Quaternary Sciences* **19**, 625–635.

Parish, F. (2002). Overview on peat, biodiversity, climate change and fire. *In* "Prevention and Control of Fire in Peatlands. Proceedings of the Workshop on Prevention and Control of Fire in Peatlands, 19–21 March, 2002, Kuala Lumpur" (F. Parish, E.C. Padmanabhan, L. Lee and H.C. Thang, Eds), pp. 320–331. Global Environment Centre and Forestry Department, Kuala Lumpur, Malaysia.

Poff, N.L., Allan, J.D., Bain, M.B., Karr, J.R., Prestegaard, K.L., Richter, B.D., Sparks, R.E., and Stromberg, J.C. (1997). The natural flow regime – a paradigm for river conservation and restoration. *Bioscience* **47**, 769–784.

Reid, J.W. (1984). Semiterrestrial meiofauna inhabiting a wet campo in Central Brasil, with special reference to the Copepoda (Crustacea). *Hydrobiologia* **118**, 95–111.

Richter, B.D., Baumgartner, J.V., Powell, J., and Braun, D.P. (1996). A method for assessing hydrologic alteration within ecosystems. *Conservation Biology* **10**, 1163–1174.

Rieley, J.O., and Page, S.E. (Eds) (1997). "Tropical Peatlands". Samara Publishing Ltd., Cardigan, UK.

Rieley, J.O., Page, S.E., Limin, S.H., and Winarti, S. (1997). The peatland resource of Indonesia and the Kalimantan Peat Swamp Forest Research Project. *In* "Tropical Peatlands" (J.O. Rieley and S.E. Page, Eds), pp. 37–44. Samara Publishing Ltd., Cardigan, UK.

Robinson, C.T., Tockner, K., and Burgherr, P. (2004). Drift benthos relationships in the seasonal colonization dynamics of alpine streams. *Archiv für Hydrobiologie* **160**, 447–470.

Rueda-Delgado, G., Wantzen, K.M., and Beltrán, M. (2006). Leaf litter decomposition in an Amazonian floodplain stream: impacts of seasonal hydrological changes. *Journal of the North American Benthological Society* **25**, 231–247.

Samways, M.J., and Steytler, N.S. (1996). Dragonfly (Odonata) distribution patterns in urban and forest landscapes, and recommendations for riparian management. *Biological Conservation* **78**, 279–288.

Smock, L.A. (1994). Movements of invertebrates between stream channels and forested floodplains. *Journal of the North American Benthological Society* **13**, 524–531.

Stanford, J.A., and Ward, J.V. (1988). The hyporheic habitat of river ecosystems. *Nature* **335**, 64–66.

Statzner, B., and Higler, B. (1986). Stream hydraulics as a major determinant of benthic invertebrate zonation patterns. *Freshwater Biology* **16**, 127–139.

Sweeney, B.W., Bott, T.L., Jackson, J.K., Kaplan, L.A., Newbold, J.D., Standley, L.J., Hession, W.C., and Horwitz, R.J. (2004). Riparian deforestation, stream narrowing, and loss of stream ecosystem services. *Proceedings of the National Academy of Sciences* **101**, 14132–14137.

Tiner, R.W. (1999). "Wetland Indicators. A Guide to Wetland Identification, Delineation, Classification, and Mapping". Lewis Publishers, Boca Raton, USA.

Tockner, K., and Stanford, J.A. (2002). Riverine flood plains. present state and future trends. *Environmental Conservation* **29**, 308–330.

Tockner, K., Malard, F., and Ward, J.V. (2000). An extension of the Flood Pulse Concept. *Hydrological Processes* **14**, 2861–2883.

Wagner, R., and Schmidt, H.-H. (2004). Yearly discharge patterns determine species abundance and community diversity: analysis of a 25 year record from the Breitenbach. *Archiv für Hydrobiologie* **16**, 511–540.

Walker, I. (1985). On the structure and ecology of the micro-fauna in the Central Amazonian forest stream Igarapé da Cachoeira. *Hydrobiologia* **122**, 137–152.

Walker, I., and Henderson, P.A. (1998). Ecophysiological aspects of Amazonian blackwater litterbank fish communities. *In* "Physiology and Biochemistry of Fishes of the Amazon" (A.L. Val, V.M.F. Almeida-Val and D.J. Randall, Eds), pp. 7–22. INPA, Manaus, Brazil.

Walker, I., Henderson, P.A., and Sterry, P. (1991). On the patterns of biomass transfer in the benthic fauna of an amazonian black-water river, as evidenced by $^{32}$P label experiment. *Hydrobiologia* **215**, 153–162.

Wantzen, K.M. (2003). Cerrado streams – characteristics of a threatened freshwater ecosystem type on the tertiary shields of South America. *Amazoniana* **17**, 485–502.

Wantzen, K.M. (2006). Physical pollution: effects of gully erosion in a tropical clear-water stream. *Aquatic Conservation* **16** (7), 733–749.

Wantzen, K.M., and Junk, W.J. (2000). The importance of stream-wetland-systems for biodiversity: a tropical perspective. *In* "Biodiversity of Wetlands: Assessment, Function and Conservation, Vol. 1" (B. Gopal, W.J. Junk and J.A. Davis, Eds), pp. 311–334. Backhuys Publishers, Leiden, The Netherlands.

Wantzen, K.M., and Junk, W.J. (2006). Aquatic-terrestrial linkages from streams to rivers: biotic hot spots and hot moments. *Archiv für Hydrobiologie Supplement* **158**, 595–611.

Wantzen, K.M., and Junk, W.J. (2008). Riparian wetlands. *In* "Encyclopaedia of Ecology" (B. Ronan, Ed.), in press. Elsevier, New York.

Wantzen, K.M., and Wagner, R. (2006). Detritus processing by shredders: a tropical-temperate comparison. *Journal of the North American Benthological Society* **25**, 214–230.

Wantzen, K.M., Da Rosa, F.R., Neves, C.O., and Nunes Da Cunha, C. (2005a). Leaf litter addition experiments in riparian ponds with different connectivity to a Cerrado Stream in Mato Grosso, Brazil. *Amazoniana* **18**, 387–396.

Wantzen, K.M., Drago, E., and da Silva, C.J. (2005b). Aquatic habitats of the Upper Paraguay River-Floodplain-System and parts of the Pantanal (Brazil). *Ecohydrology and Hydrobiology* **21**, 1–15.

Wantzen, K.M., Sá, M.F.P., Siqueira, A., and Nunes da Cunha, C. (2006). Conservation scheme for forest-stream-ecosystems of the Brazilian Cerrado and similar biomes in the seasonal tropics. *Aquatic Conservation* **16**, 713–732.

Ward, J.V., Tockner, K., and Schiemer, F. (1999). Biodiversity of floodplain river ecosystems: ecotones and connectivity. *Regulated Rivers: Research and Management* **15**, 125–139.

Ward, J.V., Tockner, K., Arscott, D.B., and Claret, C. (2002). Riverine landscape diversity. *Freshwater Biology* **47**, 517–539.

Whitmore, T.C. (1984). "Tropical Rainforests of the Far East. 2nd Edition". Clarendon Press, Oxford, UK.

Whitten, T., Soeiaatmajda, R.E., and Afiff, S.A. (1997) "The Ecology of Java and Bali". Oxford University Press, Oxford.

Yoshimura, C., Omura, T., Furumai, H., and Tockner, K. (2005). Present state of rivers and streams in Japan. *River Research and Applications* **21**, 93–112.

Zalewski, M., Bis, B., Frankiewicz, P., Lapinska, M., and Puchalski, W. (2001). Riparian ecotone as a key factor for stream restoration. *Ecohydrology and Hydrobiology* **1**, 245–251.

# *8*

# *Tropical High-Altitude Streams*

Dean Jacobsen

Streams in the tropics are not necessarily warm, nor do they invariably contain a high diversity of species. Mountain ranges, highland plateaus and volcanoes reach high elevations in some parts of the tropics and these areas are drained by streams as cold as their counterparts at higher latitudes. Consequently, the composition and diversity of the biota of tropical high-altitude streams differs substantially from those in the tropical lowlands and, in many ways, tropical high-altitude rhithrals, krenals, kryals and lake-outlets are similar to north-temperate alpine streams. One prominent difference between the conditions in high-altitude or alpine streams at different latitudes is the less pronounced seasonality in the tropics. Otherwise, tropical alpine streams are distinguished by their extreme altitude. Because the timber and permanent snowline lie much higher in the tropics than at latitudes further north and south tropical alpine streams extend to altitudes above 5000 m asl. A particularly interesting consequence of such extreme altitude is the low availability of dissolved oxygen for stream inhabitants due to the low atmospheric partial pressure of oxygen.

This chapter synthesizes the very sparse information on tropical high-altitude streams, including physico-chemical conditions, primary producers and macrophytes, fishes and, especially, macroinvertebrates. A particular aim was to describe similarities and differences in the environment, community composition and functioning of these streams compared to streams in tropical lowlands as well as those at higher latitudes.

Human population density generally decreases at higher altitudes, and alpine or glacier-fed streams may be a vital source of water for rural communities as well as far-distant floodplain and urban communities. However, because of the relatively benign climate, human activities can extend to higher altitudes than is possible in temperate latitudes, and high-altitude tropical streams may be affected by a number of human impacts such as organic pollution. Anthropogenic activities, including the threat of global warming, pose a particular threat to tropical high-altitude streams because of the susceptibility of their biota to oxygen depletion, and the changes in discharge regimes due to increased rates of glacial ablation and altered rain patterns.

## I. INTRODUCTION

Most of what we know about alpine streams has been learned from mountains at temperate latitudes, which have recently attracted considerable research (see special issues of *Freshwater Biology* **46**(12), 2001, and *Hydrobiologia* **562**(1), 2006). In contrast, high-altitude streams in the tropics are probably one of the least-studied ecosystems on earth (Ward, 1994). Mountain ranges, highland plateaus, and volcanoes in the tropics reaching very high altitudes are drained by streams as cold as their counterparts at higher latitudes. However, because the forest margins and permanent snow lines lie much higher in the tropics than at temperate and arctic latitudes, tropical alpine streams *sensu stricto* are distinguished from their counterparts by their extreme altitude.

The terms mountain, alpine, highland or upland stream generally refer to streams with high slopes, fast and turbulent flow and coarse substratum, but say little about the actual altitude. It is impossible to define a universal lower altitude limit for tropical high-altitude streams, because the changes from low to high-altitude ecosystems are gradual. However, for reasons given later in this chapter, streams above 3000 m asl can generally be denoted 'high altitude', even though they might be situated on a low-relief plateau not considered as mountainous (Meybeck *et al.* 2001). Tropical high-altitude streams vary in channel form and appearance, ranging from high-gradient torrents with cascades, to soft-bottomed lake outlet streams, or low gradient, braided channels on high plateaus (Fig. 1), but they have features in common that distinguish them from tropical streams at lower altitudes as well as from temperate streams at high and low altitudes.

Human population density generally decreases with increasing altitude, and tropical streams originating at high altitudes often have better water quality than streams further down. In addition, because rainfall is usually less seasonal at high altitudes, these streams also provide a predictable and reliable water resource and may be of immense importance to rural communities and cities at lower elevations. Nonetheless, alpine streams in the tropics are not necessarily near-pristine, as we often expect from temperate regions (Ward, 1994; but see Füreder *et al.*, 2002). Human settlements, agriculture, and other anthropogenic activities in the tropics can extend to relatively high elevations because of the benign climate. Thus, tropical high-altitude streams can be affected by organic pollution, agrochemicals, sedimentation, and mining wastes. In addition, these streams are particularly threatened by dams and water abstraction (see also Chapter 10 of this volume), and may be particularly sensitive to the consequences of global warming (McGregor *et al.*, 1995; Castella *et al.*, 2001).

(a)                          (b)                          (c)

*FIGURE 1*   (a) A step-cascade stream running through high Andean forest at 3200 m asl elevation in southern Ecuador. (b) A deep, narrow moorland stream channel with submerged macrophytes draining páramo at 4000 m asl elevation in Central Ecuador. (c) A braided stream on the Bolivian Altiplano (4000 m asl elevation) draining the Cordillera Real. Huayna Potosí (6091 m asl) is seen in the background (Photos by D. Jacobsen) (see colour plate section).

## A. Tropical High-Altitude Regions

On a global scale, 1.52% (= 746 700 km$^2$) of the land area of the tropics is situated above 3000 m elevation (Table I). However, 5.24% (= 704 600 km$^2$) of tropical South America is above 3000 m, and the extensive Central and Northern Andes thus constitute 94% of all tropical lands above this elevation (Table I). Most studies of tropical high-altitude streams have been conducted in this region, and data from them comprise the bulk of the information presented in this chapter. High-altitude streams also occur in Central America, mostly on the extensive Mexican tableland, and to a much lesser extent in Guatemala and Costa Rica (Table I). Africa has quite extensive high-altitude areas mainly on the Ethiopian tableland, which comprises 80% of all land above 3000 m elevation in Africa (Harrison and Hynes, 1988), plus the Ruwenzori and Virunga ranges in central Africa and a few isolated volcanoes in East Africa. The most extensive high-altitude areas in South-east Asia and Oceania are in New Guinea, in addition to isolated volcanoes and mountains on many islands in the Indonesian archipelago (especially Borneo). Unfortunately, virtually nothing is known about ecology of streams draining these highlands. The world's highest streams are located in the Himalayas (Ormerod *et al.*, 1994; Brewin *et al.*, 1995), but even the southernmost part of this region has a pronounced seasonal climate and lies north of the tropical belt. The highest tropical streams of glacial origin, 4000–5000 m asl, are located primarily in the Central Andes of Peru and Bolivia, but they are also numerous in the equatorial, Northern Andes of Ecuador, Colombia and Venezuela. In Africa, such equatorial, glacial streams are confined to the Ruwenzori Mountains and in highlands around Mt Kenya and Mt Kilimanjaro volcanoes in East Africa. Only the tropics have permanently flowing streams at such extreme altitudes. Streams at similar altitudes outside the tropics freeze, or at least get much colder, during winter.

## B. Tropical High-Altitude Environments

In wet tropical mountains, air temperature decreases at an average rate of about 0.5–0.6°C per 100 m elevation gain (= the vertical lapse rate), with slight variations according to local conditions, resulting in a mean air temperature of about 10°C at 3000 m asl in humid equatorial regions (Sarmiento, 1986). This mean temperature is quite similar to temperate lowland regions. The frost boundary, where night temperatures below freezing may occur year round, is also situated around 3000 m asl (Sarmiento, 1986). Frost is an important ecological boundary for tropical, warm-adapted life (Dudgeon and Corlett, 2005) and is thought to limit terrestrial adult stages of some stream insects (Van Someren, 1952).

At the equator, elevations of 3000 m asl usually coincide with the upper margins of montane rainforest or 'cloud forest', a forest of lower stature but similar complexity to lower-altitude forests (Sarmiento, 1986). As a result, tropical streams above the 3000-m tree line are usually surrounded by vegetation comprising grass and scrub extending from the forest fringes, where it is usually quite dense and complex, up to the permanent snow line 4800–5000 m asl, where the vegetation is low and discontinuous. Mean air temperature at these higher elevations is about 0°C and freezing occurs every night making this the approximate upper altitudinal limit of equatorial streams. Accordingly, tropical streams above 3000 m elevation are truly high-altitude, and occupy an environment that corresponds approximately to the alpine 'life zone'. Moving away from the equator or toward dryer climates (such as the Altiplano of southern Bolivia at latitude 17–18°S), the frost and forest limits shift to lower elevations although the limit of permanent snow may be as high as 6000 m asl (Troll, 1968). It should be noted that during the last Ice Age, some 20 000 years ago, air temperatures in high tropical mountains were about 6°C lower than today, with permanent snow cover probably extending down to about 3500 m asl (Colinvaux, 1987; Schubert, 1988). Consequently, tropical (as well as temperate) high-altitude environments are of more recent origin than the surrounding lowlands.

TABLE 1 Total Tropical Land Areas (between the Tropics of Capricorn and Cancer) and Proportions at Indicated Altitudinal Classes on Different Continents. (Data Compiled by Flemming Skov, National Environmental Research Institute, Denmark.)

| Altitudinal class (m asl) | Central America | | South America | | Africa | | Asia | | Oceania | | World | |
|---|---|---|---|---|---|---|---|---|---|---|---|---|
| | km² | % | km² | % | km² | % | km² | % | km² | % | km² | % |
| 0–1000 | 1.0914 | 68.18 | 11.8605 | 88.30 | 17.9631 | 78.49 | 6.6673 | 89.74 | 3.7422 | 95.54 | 41.3245 | 83.88 |
| 1000–2000 | 0.3246 | 20.27 | 0.5739 | 4.27 | 4.6472 | 20.31 | 0.7007 | 9.43 | 0.0995 | 2.54 | 6.3459 | 12.88 |
| 2000–3000 | 0.1782 | 11.13 | 0.2925 | 2.18 | 0.2562 | 1.12 | 0.0613 | 0.82 | 0.0636 | 1.62 | 0.8518 | 1.73 |
| 3000–4000 | 0.0058 | 0.36 | 0.4009 | 2.98 | 0.0187 | 0.08 | 0.0005 | 0.01 | 0.0114 | 0.29 | 0.4418 | 0.90 |
| 4000–5000 | 0.0009 | 0.06 | 0.2925 | 2.18 | 0.0003 | <0.01 | Summits | | Summits | | 0.2937 | 0.60 |
| 5000–6000 | Summits | | 0.0112 | 0.08 | Summits | | | | | | 0.0112 | 0.02 |
| 6000–7000 | | | Summits | | | | | | | | | |
| Total area (million km²) | 1.6009 | | 13.4319 | | 22.8862 | | 7.4298 | | 3.9167 | | 49.2655 | |

FIGURE 2   (a) A stream surrounded by *Espletia* shrubs on the El Angel páramo at 3800 m asl in northern Ecuador. (b) A stream enclosed by *Polylepis* woodlands at 3600 m asl in Central Ecuador covered by dense *Polylepis* woodlands. (c) Low flow in a wide stream channel on the Bolivian Altiplano (3800 m asl). (d) A stream at 3000 m asl in a Dry Inter-Andean valley close to La Paz, Bolivia (Photos by D. Jacobsen) (see colour plate section).

The relatively humid and species-rich vegetation above the tree line in the Northern Andes is called 'páramo'. Some páramos are characterized by spectacular alpine Asteraceae of the genus *Espletia* that can reach heights of 3–4 m (Fig. 2). Patches of dense *Polylepis* (Rosaceae) woodlands can extend up to about 4000 m asl in sheltered areas, and form part of the páramo environment, so that even streams higher than 3000 m asl are occasionally to be enclosed by forest (Fig. 2). Alpine vegetation of similar physiognomy is found at high altitudes in the mountains of East Africa, but with other giant Asteraceae (*Senecio*) and Campanulaceae (*Lobelia*) replacing *Espletia*. High-elevation regions of the drier Central Andes lack páramo and support the less diverse 'puna' grassland (Monasterio and Vuilleumier, 1986).

The relationship between precipitation and altitude varies among tropical regions and is determined by latitude, local topography, wind patterns, and continentality (Sarmiento, 1986; Beniston, 2006). As a general rule, however, precipitation at very high altitudes is less than that at lower elevations. Precipitation may fall as drizzle, hail or snow, but not usually as the torrential downpours typical of tropical lowlands. Seasonal variation in precipitation becomes more pronounced with increasing distance from the equator. In contrast to precipitation, incident solar irradiance may be very intense at high altitudes, although this is obviously dependent on local cloudiness. In the Andes, for example, irradiance is up to 50% higher than at sea level for an equivalent atmospheric moisture regime (Lewis *et al.*, 1995).

High temperatures and precipitation ensure that tropical lowland soils are normally highly leached and weathered so that streams draining them are poor in dissolved ions. In contrast, the drier and cooler climate at high altitudes often results in dark brown soils that have a higher organic content and are richer in bases and more fertile than most lowland soils (Buringh, 1979). Common soil types in the high-elevation tropics include humic variants of red soils (humic nitisols and ferralsols) and various weathering stages of andosols in regions with historic or recent volcanic activity (Reading *et al.*, 1995). Tropical highlands often show a particularly high diversity of soil types in volcanic areas.

## II. STREAM PHYSICAL AND CHEMICAL CHARACTERISTICS

### A. Temperature and Oxygen

Low water temperature is the most characteristic feature of high-altitude streams. Streams in the Ruwenzori Mountains, Uganda, closely follow a vertical lapse rate of 0.5°C per 100 m altitude gain, giving a mean temperature of 5°C at 4000 m asl (Busulwa and Bailey, 2004). Slopes of the mean water temperature versus altitude regressions in both Ecuadorian (T = 20.5–0.00328 m) (Fig. 3) and Bolivian streams (T = 22.1–0.00397 m) (Wasson *et al.*, 1989) were slightly lower than in the study from Uganda, yielding mean temperatures of 10.7°C (Ecuador) and 10.2°C (Bolivia) at 3000 m asl and 7.4°C (Ecuador) and 6.2°C (Bolivia) at 4000 m asl . The variability around the regression line seen in Fig. 3 indicates that temperature is not determined solely by altitude or air temperature, but some of it may be due to differences in site characteristics such as insolation, groundwater input, and heat loss to evaporation (Ward, 1985; Walling and Webb, 1992; Caissie, 2006). Streams at the lower end of the altitude gradient (2500–3000 m asl) tend to be slightly cooler than the air, while streams at the highest altitudes tend to be slightly warmer than the air (Fig. 3). The difference between mean water and air

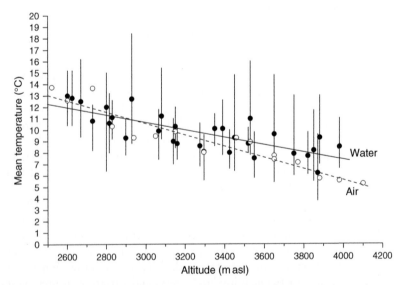

*FIGURE 3* Mean values (filled circles) and maximum–minimum ranges of water temperature (measured during two-week period) in streams along an altitudinal gradient from 2500 to 4000 m asl in Ecuador, as summarized by the continuous regression line (T°C = 20.48–0.00328 m asl). Mean air temperatures (also over two-week period) are shown as open circles, and summarized by the dashed regression line (T°C = 25.39–0.00494 m asl) (D. Jacobsen, unpublished observations).

temperature is probably due to denser vegetation cover (thus lower insolation) and higher heat loss due to evaporation at lower altitudes.

In equatorial streams of the Ecuadorian páramo, the amplitude of diel temperature change is rather small, but nonetheless exceeds the limited annual variation in monthly mean temperature (Table II). In contrast, open streams on the Bolivian Altiplano (Spanish for 'high plain') and in dry inter-Andean valleys show considerable amplitude in both diel and annual temperature variations (Table II). During the dry season, when skies over the Altiplano are clear, extreme insolation can increase daytime water temperatures to 32°C (although 21–25°C is more common) whereas heat loss at night can result in ice formation; diel temperature fluctuations in such streams may reach 20°C (Wasson *et al.*, 1989). For example, during 2 days in the dry season, mean maximum temperatures in eight Altiplano steams reached 21.3°C with mean diel fluctuations of 13.3°C (Table III). In the more humid, cloudy, and densely vegetated Yungas on the eastern slopes of the Andes toward the Bolivian Amazon, diel and seasonal temperature fluctuations are small, resembling those of similar streams in Ecuador (Table II).

The effects of direct daytime insolation on water temperatures is clearly demonstrated by comparing two sites in the same stream over a two-week period: an open páramo site and one 3-km downstream (and just 200 m asl lower) shaded by *Polylepis* forest (Fig. 4). Rising temperatures during the day lowered oxygen concentrations at both sites, especially at the lower site, reaching a minimum during early afternoon. However, at the open site, biological processes (respiration and especially photosynthesis) produce a slight and opposite variation in oxygen saturation so that there were distinct diel fluctuations in percent oxygen saturation that were not apparent at the shaded site (Fig. 4). Likewise, unshaded streams on the Bolivian Altiplano that contained high biomass of filamentous algae also showed considerable diel fluctuations in oxygen saturation (Table III).

The greater solubility of oxygen in water at lower temperatures has given rise to the general assumption that cool, high-altitude streams are always rich in oxygen. This is not the case. Oxygen concentration is almost constant with increasing altitude because the decreasing atmospheric pressure of oxygen with altitude offsets the effect of decreasing temperature (and thus increased solubility). Further, this result has the interesting implication that tropical streams that are close to air-saturation (i.e. most unpolluted streams without excessive plant biomass) have oxygen concentrations of 7–8 mg $O_2$ $L^{-1}$ irrespective of altitude (Fig. 5). This is lower than that normally prevailing in air-saturated streams in temperate lowlands (9–12 mg $O_2$ $L^{-1}$). However, the decrease in atmospheric partial pressure of oxygen with increasing altitude leads to a decline in percent saturation of dissolved oxygen (Fig. 5) and, as will be discussed later, percent saturation is more important to the stream biota than oxygen concentration.

## B. Channels and Flow

Highland drainage basins are generally of a more limited extent than in the lowlands, and precipitation is generally less, and thus tropical high-altitude streams have a relatively narrow range of sizes in terms of area or discharge. Río Desaguadero (3700–3800 m asl), which has a discharge of about 100 m³ s⁻¹ and drains Lake Titicaca and a large part of the extensive Bolivian-Peruvian Altiplano, is an exception to this generalization. Channel morphology and related features of tropical high-altitude streams are not substantially different from alpine streams at higher latitudes (Ward, 1994) in which channel morphology is determined by discharge regime, sediment load, catchment soils, topography and gradient, and riparian vegetation and land use (Church, 1992). These factors vary independently among and along streams in complex tropical mountainscapes creating a variety of different channel types (e.g. step-pool cascades, constrained, braided, and meandering: see Figs 1 and 2) similar to those described from the European Alps (Füreder *et al.*, 2002). Likewise, habitats such as riffles, rapids, glides, scour

TABLE II  Annual Temperature Records from Andean Streams. Ecuadorian Data from D. Jacobsen (Unpublished Observations); Bolivian data from Wasson et al. (1989)

| | Ecuador (1999) | | | | Bolivia (1988) | | | | |
|---|---|---|---|---|---|---|---|---|---|
| Region | Eastern cordillera, open páramo | | Upper Amazonia, cloud forest | | Altiplano | Intra-Andean dry valley | Cordillera Real, inner slopes | Eastern outer slopes 'Yungas' | |
| Altitude (m asl) | 3850 | 3880 | 2050 | 2210 | 3820 | 2600 | 4300 | 4150 | 2900 |
| Mean annual | 8.2 | 6.2 | 14.4 | 12.8 | 11.1 | 15.8 | 9.4 | 7.7 | 10.9 |
| Min., absolute | 5.5 | 3.8 | 12.9 | 11.6 | 0.0 | 4.0 | 0.0 | 0.0 | 8.0 |
| Max., absolute | 11.7 | 9.6 | 15.3 | 14.1 | 22.0 | 30.0 | 21.0 | 15.5 | 17.0 |
| Range, overall | 6.2 | 5.8 | 2.4 | 2.5 | 22.0 | 26.0 | 21.0 | 15.5 | 9.0 |
| Max. monthly mean | 9.0 | 6.8 | 14.9 | 13.2 | 14.0 | 20.0 | (10.3) | (9.2) | 11.5 |
| Min. monthly mean | 7.4 | 5.6 | 13.9 | 12.2 | 7.0 | 12.0 | (8.3) | (6.0) | 10.5 |
| Δ Monthly mean | 1.6 | 1.2 | 1.0 | 1.0 | 7.0 | 8.0 | (2.0) | (3.2) | 1.0 |

Values in parentheses are based on only 7 months of data.

TABLE III    Oxygen Saturation (Relative to Pressure at Sea Level) and Water Temperatures Measured at 15-min intervals over Two Days During November 2005 in eight Streams on the Bolivian Altiplano (Jacobsen and Marín, 2007)

|  | Oxygen saturation (%) | | | Water temperature (°C) | | |
|---|---|---|---|---|---|---|
|  | Min | Max | Range | Min | Max | Range |
| Minimum | 33 | 66 | 9 | 6.2 | 17.2 | 9.8 |
| Maximum | 57 | 121 | 84 | 9.6 | 23.6 | 17.2 |
| Mean | 46 | 91 | 44 | 8.0 | 21.3 | 13.3 |

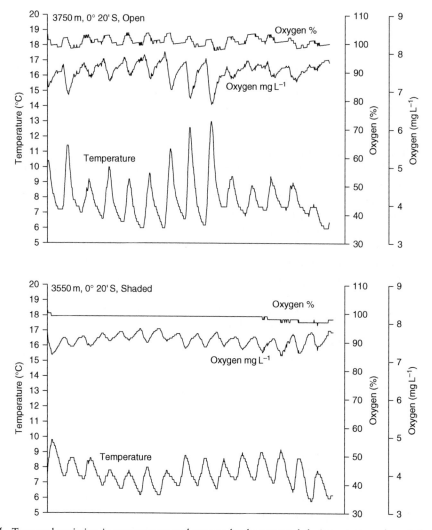

FIGURE 4    Temporal variation in temperature and oxygen levels measured during a two-week period at two sites along the same stream in Ecuador: an open páramo site (upper graph), and a site shaded by *Polylepis* woodlands (lower graph). Note: Oxygen saturation is given in relation to atmospheric equilibrium at the altitude of each site (D. Jacobsen, unpublished observations).

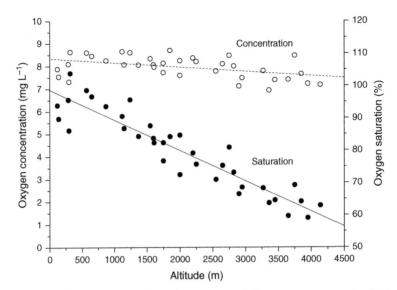

FIGURE 5  Oxygen concentrations (open circles) and saturations relative to pressure at sea level (filled circles) in 36 small streams along an altitudinal gradient from sea level to over 4000 m asl in Ecuador. Values are point measurements made between 1000 and 1200 h, with best-fit linear regression lines. Streams containing high algal or macrophyte biomass were not included [Data are from Jacobsen (2000)].

pools, cascades, and plunge pools are common. Nonetheless, while tropical high-altitude streams are often hydraulically rough, erosional habitats with steep slopes, straight channels, turbulent flow, and coarse substratum, they do not always conform to the typical characteristics of alpine streams. Deep, organically-rich soils on low-gradient highland plateaus, such as the Andean páramos, give rise to streams with narrow, deeply incised channels and low turbulence flowing over beds made up of fine substratum (Fig. 1b).

Substrate stability and hydrological regime of tropical high-altitude streams have received little study. Since precipitation is lower, spates might be less frequent than at lower elevations, but steep catchment topography and reduced vegetation cover may make upland streams more spate-prone. Streams on the Bolivian Altiplano experience substantial discharge differences between the dry and wet seasons that are responsible for the disproportionately wide stream channels in relation to discharge during low-flow periods (Fig. 2c). Streams in the Bolivian Andes exhibit a considerable range of seasonal and daily variability in flow parameters, but the extent of daily variability is not closely related to seasonal variability (Wasson *et al.*, 1998) and those draining basins with metamorphic geology seem to be more stable, regardless of channel form, than streams draining basins with sedimentary geology (Wasson *et al.*, 1998). Although Ecuadorian páramo streams exhibit large seasonal and short-term variations in discharge (Fig. 6), and generally have more stable and confined channels than streams on the Altiplano. This may reflect lower seasonal variation in precipitation in the equatorial highlands where the páramo streams originate, and these streams also have higher stability indices than nearby Amazonian piedmont streams (Jacobsen, 2003).

Stream discharge regimes vary systematically according to the main water source, and can generally be classified into one of three types (for details, see Ward, 1994; Füreder et al., 2002; Hieber *et al.*, 2002; Brown *et al.*, 2003; Maiolini *et al.*, 2006). Krenal streams or springs are groundwater fed with stable discharge; kryal streams are fed by glacial melt water and have predictable discharge cycles. The flow of rhithral streams is derived from snowmelt and rain, and more temporally variable or unpredictable because of the influence of local precipitation. The features of Ecuadorian examples of each of these stream types are given in Table IV. Kryals and

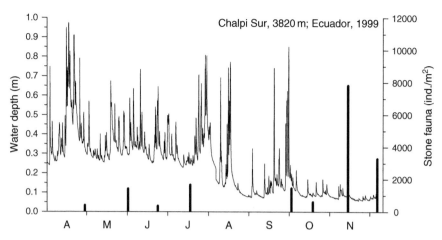

FIGURE 6   Eight months of continuous measurements of water depth in a small (1 m wide) Ecuadorian páramo stream (3820 m asl). Bars denote macroinvertebrate densities per m² cobble surface (D. Jacobsen, unpublished observations).

rhithrals in Europe have diel variations in discharge, as well as marked seasonal peaks during the late-summer glacial ablation (kryals: e.g. Milner and Petts, 1994; Robinson *et al.*, 2001; Rott *et al.*, 2006) or due to spring and summer snowmelt (rhithrals: e.g. Schütz *et al.*, 2001; Smith *et al.* 2001; Brown *et al.*, 2003). Tropical highland kryals have pronounced diel discharge fluctuations (Fig. 7) because high daytime temperatures cause glacial ablation while night-time freezing leads to low discharge, but these tropical glacial streams lack the dramatic seasonal discharge variations typical of temperate-arctic kryals and are more stable with respect to flow, sediment transport, and channel shape. In addition, snow rarely accumulates at high altitudes in the tropics, and seasonal variation in the discharge of rhithrals is relatively unpredictable with short-term fluctuations driven by rainfall, although this may be more or less seasonal in incidence (Fig. 6). In general, equatorial highland kryals and rhithrals differ from their temperate counterparts with respect to their lower seasonal predictability; Table V summarizes this and other features of tropical high-altitude streams.

## C. Suspended Solids

The amount of suspended solids in streams depend on a number of factors such as soil type, topography, precipitation, vegetation, and human activities in the catchment (Walling and Webb, 1992). Even within very short distances, rhithral, krenal, and kryal streams may show very different mean levels and ranges of suspended organic and inorganic particles (Table IV). Despite this, alpine streams are generally thought to have clear water (Ward, 1994). This is, in fact, the case for highland streams in Ecuador although differences between the highlands and lowlands in this regard are slight. Data from 12 páramo streams (3450–4100 m asl) revealed mean suspended-solid concentrations of 1.3 mg L$^{-1}$ (range: 0.7–2.1) (Encalada, 1997) compared to 2.5 mg L$^{-1}$ (0.9–9.1) in 12 lowland (130–600 m asl) streams (Schultz, 1997). This pattern was confirmed by comparison of mean monthly–bimonthly suspended-solid loads in three páramo streams (3.0–3.8 mg L$^{-1}$) with those in three Amazonian piedmont streams (6.6–11.3 mg L$^{-1}$) in Ecuador (D. Jacobsen, unpublished observations). Lencioni and Rossaro (2005) found that kryal streams close to glaciers in the Italian Alps had suspended solid loads of 49.1 mg L$^{-1}$ (SD = 71.7), which is within the range reported from a turbid Ecuadorian kryal about 5 km from the glacial snout (Table IV).

Suspended-sediment loads in small tropical streams are highly variable because they are influenced by local precipitation and mass wasting of sediments more than by steady

*TABLE IV*  Physical, Chemical, and Biological Features of one Example of Each of three Stream Types Situated within 6 km of Each Other in the Highlands of the Ecuadorian Eastern Cordillera. (D. Jacobsen, unpublished observations)

| | *Rhithral* (3980 m asl) | *Krenal* (4080 m asl) | *Kryal* (4240 m asl) |
|---|---|---|---|
| **Physical habitat** | | | |
| Distance from source (m) | 5500 | 150 | 5000 |
| Width (cm) | 90 | 130 | 120 |
| Depth (cm) | 23 | 15 | 5 |
| Current (cm s$^{-1}$) | 38 | 31 | 27 |
| Water level fluct. (cm) | 16.1 | 0.7 | 12.9 |
| Mean temp. (°C) | 8.5 | 9.8 | 6.1 |
| Temp range | 4.4 (6.6–11.0) | 0.4 (9.6–10.0) | 16.0 (0.4–16.4) |
| Substratum | Moderately embedded pebble and cobble | Loose gravel and pebble | Firmly embedded pebble and cobble |
| **Water quality** | | | |
| Conductivity ($\mu$S cm$^{-1}$) | 172–190 | 246–249 | 26–31 |
| pH | 6.78–6.84 | 6.38–6.42 | 7.28–7.37 |
| Oxygen conc. (mg L$^{-1}$) | 5.97–6.98 | 5.91–6.38 | 5.97–7.71 |
| Oxygen sat. (%) | 52–60 (equil. = 61.0 %) | 53–57 (equil. = 60.5 %) | 54–69 (equil. = 59.5 %) |
| Suspended POM (mg L$^{-1}$) | 3.0–11.2 | 0.3–0.4 | 1.5–1.9 |
| Suspended PIM (mg L$^{-1}$) | 8.2–76.2 | 0.3–0.8 | 49.2–81.2 |
| Water colour | Slightly brownish (humic) | Crystal clear | Highly milky (turbid) |
| **Primary producers** | | | |
| Periphyton (mg ch *a* m$^{-2}$) | 35 (*Nostoc*) | 19 | 16 |
| Filamentous green algae | None | Little (*Microspora, Ulotrix*) | Abundant (*Microspora, Ulotrix*) |
| Moss | None | Common | Common |
| Rooted macrophytes | Common (*Callitriche, Lilaeopsis*) | Abundant (*Callitriche, Myriophyllum*) | None |
| **Macroinvertebrates** | | | |
| Groups (no.) | 20 | 17 | 18 |
| Density (ind. m$^{-2}$) | 5408 | 5500 | 2288 |
| Richness (Fisher's $\alpha$) | 3.13 | 2.73 | 3.54 |
| Dominant (sub)families | Orthocladiinae | Hyallelidae | Orthocladiinae |
| | Baetidae | Baetidae | Elmidae |
| | Simuliidae | Orthocladiinae | Baetidae |
| | Elmidae | Limnephilidae | Simuliidae |
| | Planariidae | Planariidae | Podonominae |

POM, particulate organic matter; PIM, particulate inorganic matter.

'background' erosion (Sioli, 1975). Indeed, suspended solid levels of 125 mg L$^{-1}$ have been measured during a spate in a small Ecuadorian páramo stream (D. Jacobsen, unpublished observations). That stream lacked any human activities in the upstream portion of the catchment, although this factor can often give rise to elevated suspended-solid loads in tropical highland streams, especially during spates. Severe erosion and sediment transport during rain showers have been reported from Ethiopian streams (2800–3000 m asl) draining catchments that had been overgrazed and farmed (Harrison and Hynes, 1988). Likewise, streams in the heavily cultivated and densely populated Ecuadorian Central Valley (2600–3100 m asl) had wet-season suspended-solid loads that were five times higher than in the dry season (49.7 versus 8.7 mg L$^{-1}$),

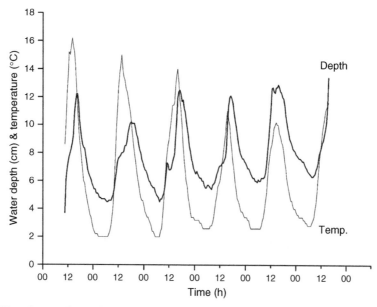

FIGURE 7 Half-hourly recordings of water depth and water temperature over a six-day period in a small glacial stream at 4240 m asl on the flanks of Mt Antisana, Ecuador (D. Jacobsen, unpublished observations).

TABLE V Features of the Physicochemical Environment in Equatorial High-altitude Streams Compared with Equatorial Lowland, Temperate Alpine, and Temperate Lowland Streams. Only Attributes that are Expected to Differ Systematically among the Four Stream Groups are Included

| Attribute | Equatorial high-altitude | Equatorial lowland | Temperate alpine | Temperate lowland |
|---|---|---|---|---|
| Temperature, mean | Medium | High | Low | Medium |
| Temp, seasonal variation | Low | Low | High | High |
| Temp, diel variation | High | Low | High | Medium |
| Oxygen concentration | Medium, stable | Medium, stable | High, variable | High, variable |
| Oxygen availability | Low | High | Medium | High |
| Discharge, seasonal var. | Low | Low | High | Medium |
| Discharge, diel variation | Medium | High | Medium | Medium |
| Channel heterogeneity | High | Low | High | Medium |

although there was a great deal of fluctuation around these mean values (D. Jacobsen, unpublished observations). Streams draining catchments with sedimentary geology on the Bolivian Altiplano and in inter-Andean valleys have naturally high suspended-solid loads: mean values from $300\,mg\,L^{-1}$ (dry season) to $18\,000\,mg\,L^{-1}$ (wet season) have been measured in an inter-Andean valley stream (2500 m asl); by contrast streams on metamorphic rocks have wet-season loads below $26\,mg\,L^{-1}$ (Wasson *et al.*, 1998). An extensive study by Maldonado and Goitia (2003) in Cochabamba Province, Bolivia, also showed large differences in suspended-solid loads among high-altitude streams (3150–4200 m asl) during the wet season ($6–326\,mg\,L^{-1}$), but values from nearby piedmont streams were within a similar range ($25–186\,mg\,L^{-1}$). Together, these data seem to suggest that there is no systematic difference in concentrations of suspended solids between tropical high-altitude, tropical lowland and temperate alpine streams.

## D. Water Chemistry

Most alpine streams at temperate latitudes are relatively nutrient poor compared to streams at lower altitudes (Ward, 1994), but there is a shortage of comparable data from the tropics. An exception is information from streams draining the Ruwenzori mountains in Uganda (Busulwa and Bailey, 2004), where conductivity increases downstream from $20-60\,\mu S\ cm^{-1}$ (4000 m asl) to $60-220\,\mu S\ cm^{-1}$ (1000 m asl). In streams of the Venezuelan Sierra Nevada, conductivity increased from $35-95\,\mu S\ cm^{-1}$ (3180–3735 m asl) to $20-880\,\mu S\ cm^{-1}$ at lower elevations (830–1650 m asl): pH increased from 6.5–7.7 to 6.8–8.5, and alkalinity rose from 7–32 to $9-519\,mg\ L^{-1}CaCO_3$ over the same altitudinal range (Segnini and Chacón, 2005). Although a similar downstream increase has been described for the Purari River, Papua New Guinea (Petr, 1983), such trends are far from always the case in the tropics. For instance, Maldonado and Goitia (2003) did not find a consistent difference in the mean ionic composition of stream water along an altitudinal gradient of 225 to over 4000 m asl in Cochabamba Province, Bolivia. Similarly, a survey of 45 streams in the Ecuadorian Andes by Monaghan *et al.* (2000) indicated that geographic region explained more of the variation in water chemistry among streams than altitude or human modification of catchments.

Broad-scale regional and altitudinal differences in water chemistry have been found in Ecuadorian streams (Fig. 8), with nutrients decreasing from the highlands toward the typically more leached and nutrient-poor soils of the tropical lowlands. In addition, geologically complex highlands with great heterogeneity of soil types and recent volcanic activity can exhibit high variability in stream water chemistry over short distances and, with a few notable exceptions in the Amazon lowlands, mean values and variability of pH and conductivity increase from lowland to highland streams (Fig. 8). Central Valley streams have considerably higher conductivity and pH values than the other regions, as well greater mean concentrations of nitrate ($240\,\mu g\ L^{-1}$) and phosphate ($200\,\mu g\ L^{-1}$), relative to páramo streams (100 and $230\,\mu g\ L^{-1}$, respectively), Amazonian piedmont streams (65 and $39\,\mu g\ L^{-1}$) or coastal piedmont streams (191 and $22\,\mu g\ L^{-1}$) (Schultz, 1997; Bojsen and Jacobsen, 2003; Ríos, 2004). These differences are likely due to higher human population densities and more intensive uses of fertilizers in the Central Valley. Overall, the limited data suggest that nutrient levels in tropical high-altitude streams are not necessarily different from lowland streams, but they may be elevated as a result of human activities or the sometimes naturally-fertile highland soils.

## III. PRIMARY PRODUCTION AND DETRITUS

## A. Benthic Algae

Autochthonous production by benthic periphytic algae is probably the basis of food webs in alpine streams at temperate latitudes (Ward, 1994; Füreder *et al.*, 2003; Rott *et al*, 2006), and this most likely is also true for unshaded high-altitude streams in the tropics (Harrison, 1995) although measures of primary production are lacking. Estimates of benthic algal biomass (expressed as mg chlorophyll *a* $m^{-2}$) do exist, however, and reveal relatively high values. Encalada (1997) reported that mean biomass on cobbles in 12 Ecuadorian páramo streams (3400–4100 m asl) was $33\,mg\ m^{-2}$ (range $= 2-175$), the higher values being attributable to accumulations of filamentous algae. Wasson and Marín (1988) report benthic algal biomasses of $10-40\,mg\ m^{-2}$ from Bolivian Andean streams, where there was no consistent seasonal difference. However, Bolivian Altiplano streams in general support high standing stocks of filamentous algae during low-flow periods (D. Jacobsen, personal observations), and filamentous algae dominating an Ecuadorian kryal stream attained a mean biomass of $16\,mg\ m^{-2}$ (Fig. 9; Table IV). In nutrient-enriched streams 2600–3100 m asl in the Ecuadorian Central Valley, benthic algal

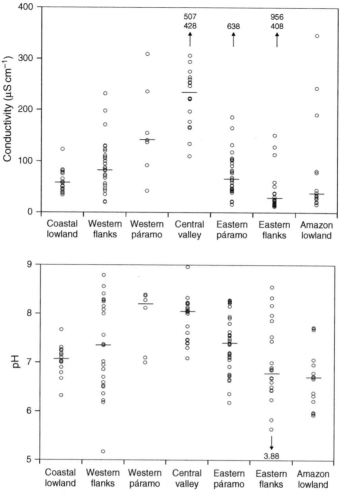

*FIGURE 8*   Point measurements of conductivity (at 25°C) and pH in streams from several geographical regions along the transect from west (left) to east (right) in northern Ecuador. Horizontal lines denote the mean value for each region. Altitudes in each region (m asl) are as follows – Coastal lowlands: 100–300 m; Western flanks: 600–2300 m; Western páramo: 3300–4200 m; Central Valley: 2600–3100 m; Eastern páramo: 3400–4100 m; Eastern flanks: 1300–2200 m; and Amazon lowland: 300–600 m (D. Jacobsen, unpublished observations).

standing stocks of 212 mg m$^{-2}$ have been recorded during stable-flow conditions in the dry season (Jacobsen and Encalada, 1998), much of this algae consisting of filamentous forms. These are flushed out during rain, when biomass falls to 20–50 mg m$^{-2}$. These values are higher than have been reported in shaded lowland streams in Bolivia and Ecuador (4–23 mg m$^{-2}$), but are quite comparable to those in open lowland streams in Ecuador (17–72 mg m$^{-2}$; Bojsen and Jacobsen, 2003); they are also higher than those generally found in glacial streams of the European Alps (up to 3.6–3.9 mg m$^{-2}$: Lencioni and Rossaro, 2005), although higher values (80–130 mg $a$ m$^{-2}$) can be attained during low flow conditions (Robinson and Uehlinger, 2002; Rott *et al.*, 2006).

When high biomass of periphyton is recorded in high-altitude tropical streams, it is usually associated with accumulations of filamentous algae or other macroscopic forms. *Cladophora* (Chlorophyta) is the most common genus and represents much of the algal biomass in Ecuadorian high-altitude streams, but *Vaucheria* (Chrysophyta), and *Microspora* (Chlorophyta) are also common and occasionally abundant (Fig. 3). *Ulotrix*, *Spirogyra*, and *Oedegonium*

FIGURE 9   Thick mats of *Microspora* (Chlorophyta) covering cobbles in a milky, kryal stream 4240 m asl in Central Ecuador. Additional data from this stream are given in Table IV (Photo by D. Jacobsen) (see colour plate section).

(Chlorophyta) plus *Batrachospermum* (Rhodophyta) and mats of *Oscillatoria* (Cyanophyta) are also common, but constitute less biomass. Other algae, such as the hollow, pea-like nodules of *Nostoc* (Cyanophyta) and the sea lettuce-like *Prasiola* (Chlorophyta), also occur in many high-altitude streams in Ecuador, while in Bolivian Altiplano streams *Ulotrix, Spirogyra, Oedegonium,* and *Cladophora* seem to dominate (D. Jacobsen, personal observations). *Hydrurus* (Chrysophyta), a characteristic macroalga of Holarctic high-mountain streams (Ward, 1994; Rott *et al.*, 2006), has not been reported from tropical high-altitude streams. *Lemanea* (Rhodophyta) is also typical of temperate alpine streams (Rott *et al.* 2006) and is abundant in the streams at high altitude in Ethiopia (Harrison and Hynes, 1988) and on Mt Elgon in East Africa (Williams and Hynes, 1971), but has not been reported from the high Andes.

There is, as yet, little that can be concluded with respect to the factors limiting or regulating algal biomass and production in tropical high-altitude streams, although temperatures should retard algal growth and the limited information on nutrient concentrations suggests possible nitrogen limitation. High ambient light levels above the tree-line could promote primary production, but photo-inhibition as well as damage due to high UV radiation are potential consequences of high irradiances (Vinebrooke and Leavitt, 1996; Rader and Belish, 1997). At present, little is known about potential adaptations of high-altitude autotrophs to excess UV radiation (Sommaruga and García-Pichel, 1999; Rott *et al.*, 2006).

## B. Macrophytes

Mosses and liverworts are able to withstand high current velocities, and are characteristic of mountain streams with coarse, stable substrata. Mosses seem to be particularly common at sites receiving little direct sunlight, such as streams shaded by forest or deep, narrow gorges. Rooted macrophytes are not normally associated with high-altitude streams since they generally have steep gradients and coarse substratum, although the Podostemaceae genus *Dicraea* reaches 2000 m asl on Mt Elgon in East Africa (Williams and Hynes, 1971). Unlike most highlands, the extensive high Andean plateaus have many relatively low-gradient streams with limited turbulence and fine, stable substratum. These streams and, in particular, lake outlets and springs support abundant and diverse aquatic macrophytes (Fig. 10). The proportion of the bed of Ecuadorian páramo streams that is covered by rooted macrophytes, mosses, and filamentous green algae is generally greater for lake outlets than other streams (mean: 54% versus 16%) (Encalada, 1997), and mean species richness of rooted macrophytes is higher (14 versus 11 species) than in streams without lakes (Jacobsen and Terneus, 2001), reflecting the more stable

*FIGURE 10* (a) A springbrook in Ecuador (4100 m asl) dominated by *Myriophyllum* and *Callitriche*. Mt Antisana (5758 m asl) in the background. (b) A lake outlet stream in Central Ecuador (3600 m asl) dominated by *Lilaeopsis*, *Myriophyllum*, and *Potamogeton*. (c) An unstable, braided stream on the Bolivian Altiplano (3900 m asl) with dense patches of *Hydrocotyle* and *Elodea* along the banks (Photos by D. Jacobsen) (see colour plate section).

discharge regime of lake outlets. The páramo lakes may also provide a source of recruitment for macrophytes in downstream reaches. Charophyceae macroalgae were also important in some lake outlet streams (Jacobsen and Terneus, 2001), and a further feature of the páramo stream flora was the presence of the fern allies *Isoetes* (five spp.) at a number of sites (Table VI); elsewhere, they are usually confined to oligotrophic lakes (Fig. 11; see also Jacobsen and Terneus, 2001).

Total plant cover on the bed of a sample of 12 Ecuadorian páramo streams ranged from 28 to 85%, with a mean species richness of 12.5 (range = 9–19), and a total of 64 plant species were collected (Table VII; see also Jacobsen and Terneus, 2001). *Callitriche* (five species) was the most common genus, but nine genera were very common and found at half or more of the surveyed stream sites (Table VI). Most of the 64 species had semiaquatic or amphibious life forms, and totally-submerged plants (i.e. obligate aquatic plants) represented only four species. Total species richness of these streams is lower than comparable streams in Denmark, due to the greater richness of submerged taxa (12 versus 4) in the Danish streams (Jacobsen and Terneus, 2001). While submerged plants did not contribute greatly to overall macrophyte diversity in Ecuadorian páramo streams some, such as *Myriophyllum*, contributed most to total plant cover (Table VI). Many of the same plant genera occur on the Bolivian Altiplano where even quite unstable streams usually have significant patches of the amphibious *Hydrocotyle* and occasionally *Callitriche* in shallow water along the banks, many have *Elodea* (Hydrocharitaceae; this was less common in Ecuador) and some have *Potamogeton* and *Myriophyllum* growing submerged in deeper parts (Fig. 10).

The flora associated with Ecuadorian streams is made up of primarily temperate genera (63%), followed by cosmopolitan (27%), Neotropical (7%), and circumtropical (1%) plants, with no genera endemic to the páramo (Cleef and Chaverri, 1992; Cook, 1996). All three genera of submerged plants (*Myriophyllum, Potamogeton,* and *Elodea*) are cosmopolitan with a mainly north-temperate distribution (Cook, 1996). Even when taking account of genera such as *Isoetes*, *Callitriche, Elatine* (Elatinaceae), and *Crassula* (Crassulaceae) that were common in Ecuadorian páramo streams, the aquatic or semi-aquatic flora remains dominated by cosmopolitan cool-adapted species, with a primarily temperate distribution.

*TABLE VI*  Occurrence, Number of Species and % Cover of Aquatic Macrophyte Genera that were Present in at least 3 of 12 Surveyed Ecuadorian Páramo Streams (3400–4100 m asl). Data from Jacobsen and Terneus (2001)

|  | *No. of localities* | *No. of species* | *% Cover* | *Life form* |
|---|---|---|---|---|
| *Callitriche* (Callitrichaceae) | 11 | 5 | 17 (0.1–39) | Amphibious |
| *Cotula* (Asteraceae) | 10 | 2 | 4 (0.1–9) | Semiaquatic |
| *Juncus* (Juncaceae) | 9 | 3 | 11 (0.1–59) | Amphibious |
| *Gunnera* (Gunneraceae) | 8 | 1 | 19 (2–41) | Semiaquatic |
| *Myriophyllum* (Haloragidaceae) | 8 | 1 | 52 (32–87) | Submerged |
| *Lilaeopsis* (Apiaceae) | 7 | 1 | 16 (0.1–36) | Amphibious |
| *Isoetes* (Isoetaceae) | 6 | 5 | 3 (0.1–10) | Amphibious |
| *Isolepis* (Cyperaceae) | 6 | 2 | 6 (0.1–16) | Amphibious |
| *Equisetum* (Equisetaceae) | 6 | 1 | 8 (1–14) | Semiaquatic |
| *Carex* (Cyperaceae) | 5 | 4 | 11 (0.1–17) | Semiaquatic |
| *Ranunculus* (Ranunculaceae) | 5 | 2 | 9 (1–19) | Amphibious |
| *Crassula* (Crassulaceae) | 4 | 1 | 5 (0.1–9) | Amphibious |
| *Elatine* (Elatinaceae) | 4 | 1 | 0.4 (0.1–1) | Amphibious |
| *Lachemilla* (Rosaceae) | 4 | 1 | 5 (0.1–12) | Semiaquatic |
| *Mimulus* (Scrophulariaceae) | 4 | 1 | 8 (2–17) | Amphibious |
| *Hydrocotyle* (Araliaceae) | 3 | 4 | 8 (1–19) | Semiaquatic |
| *Graminea* genus 'a' (Poaceae) | 3 | 3 | 14 (0.1–38) | Semiaquatic |
| *Cardamine* (Brassicaceae) | 3 | 2 | 1 (0.1–2) | Amphibious |
| *Potamogeton* (Potamogetonaceae) | 3 | 2 | 33 (1–74) | Submerged |
| *Plantago* (Plantaginaceae) | 3 | 2 | 0.1 (–0.1) | Semiaquatic |
| *Rorippa* (Brassicaceae) | 3 | 1 | 11 (0.1–29) | Amphibious |
| *Azorella* (Apiaceae) | 3 | 1 | 0.1 (–0.1) | Semiaquatic |

Ranges are given in parentheses.

(a)    (b)    (c)

*FIGURE 11*   (a) *Isoetes* dominating a small rhithral stream at 4000 m asl in Ecuador. (b) A springbrook at 4100 m asl in Ecuador with patches of *Callitriche* and *Myriophyllum*. (c) Dense growth of *Potamogeton* in a turbid, sluggish Altiplano stream in Bolivia at 3900 m asl (Photos by D. Jacobsen) (see colour plate section).

## C. Detritus

Input of allochthonous particulate organic matter (POM) in the form of plant litter or detritus decreases with increasing altitude at temperate latitudes, especially above tree line where woody debris is absent (Ward, 1994; Uehlinger and Zah, 2003); inputs to glacial streams are extremely low (Zah and Uehlinger, 2001). The same general pattern seems likely to apply along

*TABLE VII*   Contribution to Total Plant Cover, Mean Species Richness and Total Species Richness for Three Life Forms of Aquatic Macrophytes Surveyed in 12 Ecuadorian Páramo Streams (3400–4100 m asl). Data from Jacobsen and Terneus (2001)

|  | *Contribution to plant cover (%)* | *Species richness per stream* | *Total species richness* |
|---|---|---|---|
| Submerged | 38 (1–72) | 1.1 (0–3) | 4 |
| Amphibious | 32 (13–49) | 5.3 (4–8) | 22 |
| Semiaquatic | 31 (2–52) | 6.1 (2–12) | 38 |
| Overall | 100 | 12.5 (9–19) | 64 |

Ranges are given in parentheses.

altitudinal gradients in the tropics, but inputs and stocks of benthic and suspended POM have seldom been quantified in high-altitude streams. A comprehensive account of organic matter processing in tropical (lowland) streams is given in Chapter 3 of this volume.

Streams draining highland areas that support dense growths of herbs and shrubs receive quantities of allochthonous litter, but there are few instances where this has been quantified. Wasson and Marín (1988) measured standing stocks of benthic POM in five morphologically different high-altitude streams in the Bolivian Andes, reporting values from a low of $<2\,\mathrm{g}\ \mathrm{m}^{-2}$ to high of $40\,\mathrm{g}\ \mathrm{m}^{-2}$ with three of the streams containing $5–10\,\mathrm{g}\ \mathrm{m}^{-2}$ of POM (whether the units are DW or AFDW is not stated). Very coarse ($>5\,\mathrm{mm}$), coarse ($>1–5\,\mathrm{mm}$) and fine ($>0.25–1\,\mathrm{mm}$) POM contributed about equally to total POM stocks. Comparative data from 12 Ecuadorian lowland streams showed POM stocks ranged from $3–25\,\mathrm{g}$ AFDW $\mathrm{m}^{-2}$ (mean $=$ $10\,\mathrm{g}$ AFDW $\mathrm{m}^{-2}$), which is well within the same order of magnitude (Bojsen and Jacobsen, 2003). These preliminary values indicate that allochthonous detritus is present in tropical highland streams in amounts that are broadly comparable with equivalent lowland streams, and could be a significant energy source for stream consumers. This may be supplemented by detritus derived from autochthonous sources following death and decomposition of aquatic macrophytes.

Concentrations of suspended organic matter in streams in the Venezuelan Andes of Sierra Nevada ranged from $0.87–4.50\,\mathrm{mg}\ \mathrm{L}^{-1}$ at high altitudes (3180–3735 m asl) to $0.80–1.61\,\mathrm{mg}\ \mathrm{L}^{-1}$ at lower elevations (830–1650 m asl) (Segnini and Chacón, 2005). Similar levels were found in open Ecuadorian páramo streams ($0.7–2.0\,\mathrm{mg}\ \mathrm{L}^{-2}$) (Encalada, 1997), but levels of $11.2\,\mathrm{mg}\ \mathrm{L}^{-2}$ have been recorded (Table IV), and there are presently insufficient data to draw any robust generalizations. POM levels did not differ between lake outlet streams and streams without lakes (Encalada, 1997) indicating that, in contrast to most lowland lakes but in common with European alpine lakes, high Andean lakes are not significant sources of suspended organic matter for their effluent streams (Hieber *et al.*, 2002; Maiolini *et al.*, 2006).

## IV. THE FAUNA

### A. The Fish Fauna

The difference in the fish fauna between lowland and highland streams in the tropics is striking, and there is a marked decline in species richness with altitude (Chapter 3 of this volume gives details). Consequently, the fish fauna of tropical high-altitude streams is very species poor, as is generally the case for alpine streams at higher latitudes. In South America, the high Andes are inhabited by two genera of benthic, insectivorous catfishes: *Astroblepus* (Astroblepidae), which occurs from Venezuela to northern Bolivia, and *Trichomycterus* (Trichomycteridae),

FIGURE 12   Left: *Astroblepus* sp. (length 9 cm) from a large, swiftly flowing boulder-filled stream at 2300 m, Ecuador. Right: *Trichomycterus* sp. (length 4 cm) from a stream at 3900 m asl on the Bolivian Altiplano (Photos by D. Jacobsen) (see colour plate section).

which occurs from Colombia to Bolivia (Fig. 12). In Bolivia, both families are found on the eastern slopes of the Andes (the Yungas) where *Astroblepus* occur up to about 3000 m asl and *Trichomycterus* considerably higher, while on the Altiplano only *Trichomycterus* occur (R. Marín, Instituto de Ecologia, La Paz, personal communication). Densities of *Trichomycterus* in streams at 2700–3500 m asl in the Bolivian Torotoro national park are high (0.5–5 individuals per m²; Miranda and Pouilly, 1999), and they are common in Bolivian Altiplano streams where they may inhabit quite extreme habitats such as tiny rivulets and streams loaded with suspended sediments (D. Jacobsen, unpublished observations). In Ecuador, *Trichomycterus* is not known from altitudes exceeding about 1600 m asl (R. Barriga, Escuela Politechnica Nacional de Quito, Ecuador, personal communication).

*Astroblepus* is the most common and species-rich group in the Ecuadorian highlands (where it is known as Preñadilla), and the only native fish that is known to live in streams up to 3200 m asl (R. Barriga, personal communication). At least six species of *Astroblepus* are recognized in Ecuador, but there may be as many as 20 (R. Barriga, personal communication). All *Astroblepus* use underground refugia, digging into the substratum if the stream dries up. Since pre-Colombian times, *Astroblepus* has been an important food source for the rural people of the highlands (R. Barriga, personal communication). However, degradation of streams, fragmentation of suitable habitats, and pollution have drastically reduced numbers and occurrence of *Astroblepus* in Ecuador, and these fish are now found as small isolated populations (Vélez-Espino, 2005). Although *Astroblepus* may reach a length of over 25 cm, individuals larger than 5–6 cm are now rare.

The benthic armored catfish genus *Chaetostoma* (Loricariidae), which is a scraping herbivore, also occurs in the Andes, reaching about 1500 m asl in Ecuador and 2000 m asl in The Yungas of Bolivia. In addition, Cyprinodontidae (*Orestias* spp.), which are most commonly found in lentic habitats occasionally reach more than 4000 m asl in Bolivian Andean streams (R. Barriga and R. Marín, personal communication).

In Africa, loach catfishes (*Amphilius* spp.: Amphiliidae) are found in Ethiopian highland streams and on the mountains of East Africa (Ward, 1994). Their upper altitudinal limits are not well known, but *Amphilius lampei* occur at 2500 m asl in Ethiopia (Harrison and Hynes, 1988). Small benthic cyprinids (*Garra* spp.) inhabit pools in Ethiopian streams up to 3000 m asl (Harrison and Hynes, 1988), and superficially-similar Kneriidae (or shellears: *Kneria* spp.) are known from streams at 1800 m asl in the southeastern part of Zaïre (Banister, 1986). Small *Barbus* (Cyprinidae) also occur at unspecified 'high altitudes' in Zaïre (Banister, 1986), and were found in Ethiopian streams at 2000 m asl by Harrison and Hynes (1988).

The most common fishes in most tropical high-altitude streams are probably introduced Brown trout (*Salmo trutta*), Brook trout (*Salvelinus fontinalis*) and, especially, Rainbow trout (*Oncorhynchus mykiss*) (Segnini and Bastardo, 1995; Mathooko, 1996). In addition, the Pejerrey, *Basilichthys bonariensis* (Atherinidae) was introduced to the Lake Titicaca-Poopó system in Bolivia in 1955, and is now found in most streams on the Bolivian Altiplano where it is widely caught by locals (Rubén Marín, personal communication). Like trout, larger Pejerrey are insectivorous and piscivorous. In many streams, one or more introduced species are the only fishes present, and Ward (1994) has speculated whether many of the highest tropical streams were historically devoid of fishes or had been displaced by trout. On Mt Elgon (East Africa), for example, trout occupy the high-altitude streams, while the native catfish *Amphilius jacksonii* reaches only 1400 m asl (Williams and Hynes, 1971). In the Venezuelan Andes, Flecker (1992) reported no native drift-feeding fish above 1500 m, but trout were common above this altitude.

## B. Macroinvertebrate Richness at High Altitudes

The decrease of taxonomic richness of stream fishes with increasing altitude is a well-known phenomenon, but the pattern for invertebrate species richness is less clear perhaps because of a lack of data from streams that have been sampled throughout their course (Jacobsen, 2004). Cressa (2000) reported richness and diversity in 28 streams from sea level to 2700 m asl in Venezuela, but found no relationship with altitude, while Monaghan *et al.* (2000) reported little relationship between richness and altitude in 45 Ecuadorian streams between 780 and 3940 m asl, possibly due to the confounding effects of land use and ecoregions. Jacobsen (2004), however, found that local family richness of macroinvertebrates in Ecuadorian streams decreased linearly with altitude. Zonal richness showed a more complex pattern, remaining constant from sea level up to around 1800 m asl, but decreasing at higher altitudes (Fig. 13; Jacobsen, 2004). Family richness at the scale of the stream site (local) and at that of discrete altitudinal zones were both twice as high at sea level than at 4000 m elevation. Further, at both extremes of the altitude gradient, local richness (at individual sites) was approximately 50% that of zonal (or regional) richness. Few families were lost or gained from zonal richness between sea level and 1800 m but, between 1800 and 3800 m asl, families present in lowland streams were lost with few new families gained. Most families present at high altitudes were thus 'euryzonal' with a very broad altitudinal distribution (Jacobsen, 2004). A similar pattern has been observed in temperate streams (Ward, 1994). Individual genera and species have narrower altitudinal ranges than families, as shown for Simuliidae (Diptera) and Elmidae (Coleoptera) on Mt Elgon, East Africa (Williams and Hynes, 1971). All of the few families restricted to higher altitudes

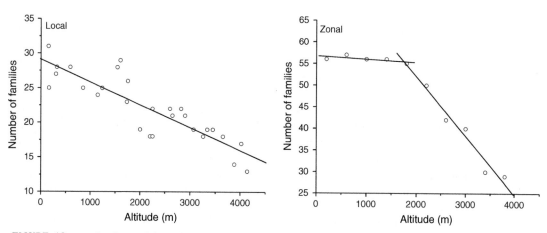

FIGURE 13   Local richness (left) and zonal richness (right) of stream macroinvertebrate groups (mostly families) in relation to altitude in Ecuadorian streams [Redrawn from Jacobsen (2004) courtesy of Blackwell Publishing].

in Ecuadorian streams (Hyallelidae, Gripopterygidae, Limnephilidae, Anomalopsychidae and Blephariceridae), except one (Blephariceridae) were apparently represented by a single genus only (Jacobsen, 2004); other information on their biology is given in Table VIII.

Bolivian and Ethiopian highland streams seem somewhat impoverished with respect to taxon richness relative to those in Ecuador (Table IX). If these differences are real (and not just sampling artifacts), they may reflect differences in the orientation, location, and isolation of highland areas, as well as their individual biogeographical histories. The Ethiopian highlands, as well as other high altitude areas in eastern equatorial Africa, are relatively small and isolated mountain 'islands' (*sensu* Hedberg, 1961; Ward, 1994) surrounded by warmer lowlands, and their insular nature strongly influences the distribution of cold-adapted taxa (Ward, 1994) and may promote endemism. Indeed, the Ethiopian highland stream fauna seems to have a high percentage of endemic species, and the Red Sea rift has apparently limited more recent immigration of Palaearctic elements (Harrison and Hynes, 1988). Faunal exchange can be higher along extensive mountain ranges, especially those with a north–south orientation (Illies, 1969; Ward, 1994) such as the Andes and the Central American highlands. The Andean cold-adapted stream fauna is derived from early Palaeantarctic elements and later Nearctic immigrants (Illies, 1969), and this mixture has enriched the Ecuadorian Andean fauna with life forms or 'types' equivalent to the family level. In contrast, the Neotropical high-altitude fauna may not be particularly diverse at the species level. The north-south orientation of the Andes permitted the stenothermal fauna to retreat to appropriate regions during Quaternary climatic variations – for example to the equatorial regions during the Pleistocene glaciation – thereby avoiding the isolation that promoted speciation in west-east oriented mountain ranges in areas such as Europe (Illies, 1969).

The seemingly low richness of some Bolivian high-altitude streams may be due to contemporary ecological conditions. For example, streams on the Bolivian Altiplano around Lake Titicaca are notably poor in macroinvertebrate groups compared to streams at the same altitude on the Ecuadorian páramo (Table X), especially EPT taxa (mainly Trichoptera). This impoverishment might be due to extreme diel variations in temperature and oxygen saturation (Table III). Low taxon richness in streams west of Oruro has been attributed to erosion and the possible influence of heavy metals (Hamel and Van Damme, 1999), but this does not seem to be a general feature of Bolivian streams as relatively high richness has been reported from two streams at around 4300 m asl, with 22–23 insect families and non-insect orders/classes (Wasson and Marín, 1988; Franken and Marín, 1992).

Comparing richness of tropical and temperate highland streams is difficult due to scarcity of data, as well varying sampling methods and taxonomic resolution, but it is possible to adjust some published data from Alpine streams in Europe to a level of identification that allows comparison with some of the research on tropical highland stream invertebrates described above. On this basis, Hieber *et al.* (2005) collected an average of 26 taxa during bimonthly sampling of two rhithral streams (2300 m asl) over 2 years, but Rüegg and Robinson (2004) collected just 7–14 taxa per site after sampling three streams (2600 m asl) during an entire summer. Füreder *et al.* (2005) sampled several sites in two streams (2250–2450 m asl) for a year, recording 19 taxa in one stream and 22 in the other, while Maiolini *et al.* (2006) collected 14–26 taxa in groups containing 6–13 streams between 1600 and 2700 m asl. While the sampling intensity of these studies varies somewhat, it is generally greater than has been applied to tropical highland streams. Taking this into account, the latter appear to support richer macroinvertebrate communities (Tables IX and X), and yielded more taxa than the European streams.

Data on macroinvertebrate densities in tropical highland streams are scarcer than information on taxon richness. Densities in Ecuadorian páramo streams (3300–4200 m asl) were around 6000 individuals per m$^2$ (Table VIII). This is somewhat higher than that recorded from

*TABLE VIII*   The Macroinvertebrate Taxa Characteristic of Ecuadorian High-altitude Páramo Streams (Not Including Lake Outlets). The Data are Based on Surber Samples (Mesh Size 500 μm) from 15 Streams (3300–4200 m asl). Occurrence = % of locations present; FGG = presumed functional feeding group (D. Jacobsen, unpublished observations)

| | Family/subfamily | Occurrence | Density (ind. m$^{-2}$) | Mean proportion (% of density) | FFG |
|---|---|---|---|---|---|
| **Non-insects** | | | | | |
| Turbellaria | Planariidae | 100 | 406 | 6.8 | Predator |
| Nematoda | | 40 | 6 | 0.1 | Predator |
| Oligochaeta | | 100 | 331 | 5.5 | Collector |
| Hirudinea | | 33 | 3 | 0.0 | Predator |
| Gastropoda | Lymnaeidae | 7 | 3 | 0.0 | Scraper |
| | Physidae | 7 | 0 | 0.0 | Scraper |
| | Planorbidae | 7 | 0 | 0.0 | Scraper |
| Bivalvia | Pisiidae | 33 | 45 | 0.8 | Filter |
| Amphipoda | Hyallelidae | 100 | 277 | 4.6 | Col/shredder |
| Hydracarina | | 93 | 22 | 0.4 | Predator |
| **Insects** | | | | | |
| Plecoptera | Perlidae* | 47 | 35 | 0.6 | Predator |
| | Gripopterygidae | 13 | 10 | 0.2 | Scraper/Col |
| Ephemeroptera | Baetidae | 100 | 690 | 11.6 | Col/scraper |
| | Leptophlebiidae | 20 | 21 | 0.4 | Collector |
| Coleoptera | Elmidae | 100 | 461 | 7.7 | Scraper |
| | Scirtidae | 73 | 15 | 0.3 | Scraper |
| | Staphylinidae | 13 | 1 | 0.0 | Predator |
| Trichoptera | Anomalosychidae | 20 | 6 | 0.1 | Scraper/Col |
| | Calamoceratidae† | 7 | 2 | 0.0 | Shredder |
| | Glossosomatidae | 60 | 123 | 2.1 | Scraper |
| | Helicopsychidae | 20 | 11 | 0.2 | Scraper |
| | Hydrobiosidae | 73 | 21 | 0.4 | Predator |
| | Hydropsychidae | 20 | 67 | 1.1 | Filter |
| | Hydroptilidae | 73 | 517 | 8.7 | Scraper/piercer |
| | Leptoceridae | 47 | 6 | 0.1 | Shredder? |
| | Limnephilidae | 87 | 129 | 2.2 | Shredder |
| Diptera | Blephariceridae | 53 | 16 | 0.3 | Scraper |
| | Ceratopogonidae | 80 | 174 | 2.9 | Predator |
| | Chironominae | 40 | 30 | 0.5 | Collector |
| | Empididae | 87 | 83 | 1.4 | Predator |
| | Ephydridae | 13 | 1 | 0.0 | Collector |
| | Diamesinae | 40 | 12 | 0.2 | Collector |
| | Dolichoridae | 20 | 1 | 0.0 | Predator |
| | Limoniidae | 80 | 43 | 0.7 | Predator |
| | Muscidae | 20 | 2 | 0.0 | Predator |
| | Orthocladiinae | 100 | 2118 | 35.5 | Collector |
| | Podonominae | 87 | 82 | 1.4 | Collector |
| | Psychodidae | 40 | 28 | 0.5 | Collector |
| | Simuliidae | 100 | 147 | 2.5 | Filter |
| | Tanypodinae | 73 | 13 | 0.2 | Predator |
| | Tipulidae | 20 | 1 | 0.0 | Shredder/collector |
| **Total** | | | 5958 | 100 | |

* Not above approximately 3500 m asl.

† Not above approximately 3300 m asl.

*TABLE IX*   Taxon Richness (Insect Families and Non-insect Orders/Classes) of Macroinvertebrates in different tropical highland regions of comparable altitude

| Country | Study region | Altitude range | Location | Taxa | Author |
|---------|--------------|----------------|----------|------|--------|
| Ecuador | Eastern páramos | 3600–4000 | 8 | 34 | Jacobsen (unpublished observations) |
| Ecuador | Central Valley | 2600–3100 | 8 | 34 | Jacobsen and Encalada (1998) |
| Ecuador | Stream Carihuayco | 2600–3780 | 9 | 29 | Jacobsen (unpublished observations) |
| Bolivia | Altiplano, La Paz – Titicaca | 3820–4000 | 8 | 26 | Jacobsen and Marín (2007) |
| Bolivia | Cordillera Oriental, Cochabamba | 3730–4200 | 8 | 19 | Maldonado and Goitia (2003) |
| Bolivia | Prepuna and Puna, Cochabamba | 2500–3580 | 9 | 15 | Maldonado and Goitia (2003) |
| Ethiopía | Around southern Rift Valley | 2500–3500 | 10 | 22 | Harrison and Hynes (1988) |

*TABLE X*   Macroinvertebrate Assemblage Parameters (Based on Surber Samples with Mesh Size 500 μm) from eight streams on the Ecuadorian Páramo (3600–4000 m asl) and on the Bolivian Altiplano (3800–4000 m asl) (Jacobsen and Marín, 2007)

|  | *Ecuador* | *Bolivia* |
|---|---|---|
| Total no. of families | 34 | 26 |
| Mean no. of families per locality | 18.9 | 11.3 |
| Richness (mean Fisher's α) | 3.05 | 1.46 |
| Total no. of trichopteran families | 8 | 2 |
| % EPT individuals | 25.7 | 0.7 |

EPT: Ephemeroptera, Plecoptera and Trichoptera.

Ecuadorian lowland streams (Jacobsen *et al.*, 1997; Jacobsen and Bojsen, 2002), but slightly less than in streams above the tree line in the European Alps (Füreder *et al.*, 2002).

## C. Composition of the Macroinvertebrate Fauna

The general resemblance of stream macroinvertebrate faunas worldwide has been noted by many stream ecologists (e.g. Illies, 1964; Hynes, 1970), and is particularly evident when it comes to high-altitude streams within the tropics and more generally. For example, the cold-adapted chironomid (Diptera) genus *Diamesa* is the most characteristic component of the fauna in glacial streams world wide in streams where temperatures do not exceed 2°C (Ward, 1994; Milner *et al.*, 2001; Füreder *et al.*, 2002). The genus is primarily Holarctic, but also occurs at very high altitudes in tropical Africa and the Neotropics (Ward, 1994). The resemblance in appearance, functional role, and composition of macroinvertebrate communities between tropical streams at high altitudes and temperate streams at high latitudes is perhaps even more striking (Jacobsen *et al.*, 1997). In a comparative study of stream faunas at three altitudes in Ecuador and lowland streams in Denmark, Jacobsen *et al.* (1997) showed that the composition of the Ecuadorian highland fauna (at family level) was more similar to the Danish lowland fauna than to the Ecuadorian lowland fauna. This finding supports the view of Illies (1964) that a rhithron biocoenosis adapted to cold conditions is found at both temperate and tropical latitudes, but occurs at progressively higher altitudes in streams closer to the equator. For example, some widely-distributed Trichoptera genera such as *Cheumatopsyche* (Hydropsychidae) in the high mountains in Africa and *Helicopsyche* (Helicopsychidae) in the Andes are confined to much lower elevations at higher latitudes (Ward, 1994).

One difference between high-altitude streams in tropical and temperate latitudes is the tendency for non-insects to constitute a considerable proportion of the macroinvertebrate fauna at some tropical sites (Ward, 1994). In a small lake outlet stream at 3300 m asl in southern Ecuador, Hydracarina (25%), Oligochaeta (20%), Copepoda (19%), Turbellaria (3%), Nematoda (2%), Amphipoda (2%), and Ostracoda (2%) made up 73% of the fauna (Turcotte and Harper, 1982b). At 4115 m asl, 25% of macroinvertebrates in the Naro Moru River, Mt Kenya, were Hydracarina and Turbellaria (Van Someren, 1952). Samples from Ecuadorian streams contained on average 18% non-insects, mainly Turbellaria, Oligochaeta, and Amphipoda (Table VIII). In contrast, highland streams in Ethiopia had neither Amphipoda nor Oligochaeta, and Turbellaria and Hydracarina were not especially abundant (Harrison and Hynes, 1988). Likewise, non-insects contributed only a small percentage of the macroinvertebrate faunas of high-altitude streams in Cochabamba Province, Bolivia (Maldonado and Goitia, 2003). These non-insect groups are much less common in tropical lowland streams (see Chapter 4 of this volume), where Amphipoda (Hyallelidae) do not occur at all. In contrast, Decapoda, which are widespread and common in tropical lowland streams, are absent at higher altitudes; the single exception being a report of crabs of the genus *Potamon* (Potamidae) close to the 4321 m summit of Mt Elgon in East Africa (Hynes *et al.*, 1961).

Despite the occasional importance of non-insects, the macroinvertebrate fauna in tropical high-altitude streams is usually taxonomically and numerically dominated by insects. The insect fauna of Ecuadorian páramo streams comprises five orders: Plecoptera, Ephemeroptera, Coleoptera, Trichoptera, and Diptera (Table VIII; Fig. 14). Dipteran family richness remains similar to that in streams at lower altitudes, and this group generally has the greatest density and family richness of the five orders in highland streams. For instance, in an open páramo stream (3800 m asl) with abundant periphyton, Orthocladiinae (74%) and Podonominae (8%) chironomids, plus *Bezzia* (Ceratopogonidae: 6%) and Empididae (4%) made up over 90% of benthic macroinvertebrates (D. Jacobsen, unpublished observations). The richness and numerical importance of the four remaining orders decreases with increasing altitude (Fig. 14), although some Trichoptera families (Glossosomatidae, Hydroptilidae, and Limnephilidae) are widespread and sometimes abundant (Table VIII). Only two Ephemeroptera families (Baetidae and Leptophlebiidae) occur at high altitudes, but they can be numerically important, especially the cosmopolitan genus *Baetis*. The Coleoptera is represented Elmidae and Scirtidae which are both common and often abundant. Elmidae beetles dominated the macroinvertebrate fauna of a stream at 4320 m asl in the Bolivian Cordillera Real (Wasson and Marín, 1988) and appear to dominate Bolivian Altiplano streams in general (D. Jacobsen and R. Marín, unpublished observations). By contrast, Coleoptera are not usually present in temperate alpine streams, and even Elmidae are rarely abundant (Ward, 1994). Also unlike temperate alpine streams, the Plecoptera contribute little to both density and richness in the tropics, comprising only two families (Perlidae and Gripopterygidae) in high Neotropical streams, and only one (Gripopterygidae) above 3400 m asl. Odonata and Heteroptera are diverse and abundant in lowland streams, but seem to be absent from streams above 3000 m asl in Ecuador. However, Aeshnidae and Coenagrionidae (Odonata) have been collected in streams on the Peruvian-Bolivian Altiplano at 3800–4000 m asl (Roback *et al.*, 1980) while the Notonectidae and Corixidae (Heteroptera) have been recorded at 4320 m asl in a Bolivian Cordillera Real stream (Wasson and Marín, 1988) and are generally common in Altiplano streams (D. Jacobsen, unpublished observations). Megaloptera and Lepidoptera have apparently not been recorded in streams above 2500 m.

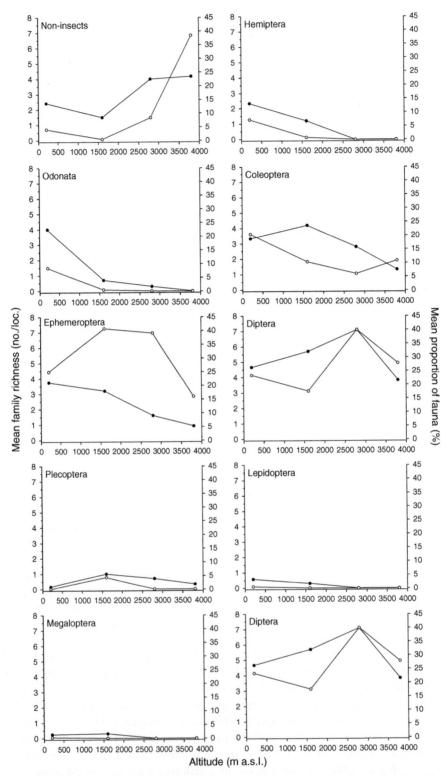

FIGURE 14  Mean family richness per locality (filled circles) and mean relative abundance (% of total density) (open circles) for macroinvertebrate groups (mostly insect orders) in 10 streams of each at four altitudes in Ecuador [Data are from Jacobsen (2004) and D. Jacobsen (unpublished observations)].

The macroinvertebrate fauna of high-altitude Neotropical streams is mostly characterized by a limited number of families. Among a total of 40 taxa found in 15 Ecuadorian streams (Table VIII), 20 occurred in at least half of all sites, and just four were found at a single site. Seven taxa were present in all streams sampled (Planariidae, Oligochaeta, Hyallelidae, Baetidae, Elmidae, Orthocladiinae and Simuliidae), and 10 more were found in at least two-thirds of all streams (Hydracarina, Scirtidae, Hydrobiosidae, Hydroptilidae, Limnephilidae, Ceratopogonidae, Empididae, Limoniidae, Podonominae, and Tanypodinae). Ten of these 17 taxa are both very widespread and locally dominant (Table VIII), representing core components of the Ecuadorian páramo stream fauna. The core taxa are reduced slightly in the relatively depauperate Bolivian Altiplano streams (Table X) where Elmidae, Chironomidae, Baetidae, Oligochaeta, and Hirudinea occur everywhere (except at the most contaminated sites), while Hyallelidae, Corixidae, Dytiscidae, Hydroptilidae and Ostracoda occur at most sites (Jacobsen and Marín, 2007). In terms of abundance, Altiplano streams are dominated by Elmidae (44%), Oligochaeta (32%), and Chironomidae (16%).

## D. Spatial Variability of Macroinvertebrate Assemblages

Macroinvertebrate assemblages in tropical high-altitude streams are spatially variable at small scales in response to contemporary ecological conditions. In Ecuadorian streams, local richness ($\alpha$-diversity) decreased slightly more with increasing altitude than regional ($\gamma$-diversity), leading to a higher family turnover ($\beta$-diversity) among streams at high altitudes (Jacobsen, 2003, 2004). This effect was reflected in a higher mean Bray-Curtis similarity among lowland streams than among high-altitude streams (67 versus 52%; Jacobsen, 2003). The higher heterogeneity of assemblages at high-altitudes could reflect greater environmental heterogeneity among streams but might also be caused by reduced dispersal of adult insects in the cold windy conditions (Deshmukh, 1986) and topographic barriers presented by mountains.

As is the case for tropical streams in general, flow and disturbance regimes seem to be prime factors governing the spatial variation of macroinvertebrate assemblages in high-altitude streams. Wasson *et al.* (1998) compared macroinvertebrate richness in five Bolivian Andes streams across a range of hydrological, temperature, and substrate stabilities and suspended sediment loads. Local environmental conditions affected richness and disrupted the general trend toward decreasing richness with increasing altitude. The richest streams were those at highest altitudes, which might have been due to attainment of maximum richness at intermediate disturbance levels (Fig. 15) as predicted by the Intermediate Disturbance Hypothesis (Townsend *et al.*, 1997). Suspended solids were also important, as the most impoverished site had high loadings (Wasson *et al.*, 1998). Soil and bed erosion lead to high suspended-sediment loads in the Ethiopian highland streams, and affected sites have considerably fewer macroinvertebrate taxa than non-affected streams at the same altitude, with Turbellaria, Hydropsychidae, and Elmidae, being especially vulnerable (Harrison and Hynes, 1988). Fossati *et al.* (2001) report that suspended solid loads above 100 mg L$^{-1}$ have a negative effect on macroinvertebrate assemblages in Bolivian low Andean streams also.

Several site characteristics that are related to and regulated by disturbance regime affect the variability of macroinvertebrate assemblages. For example, Encalada (1997) found that the percentage of the stream bed covered by plants (filamentous algae, mosses, and rooted macrophytes) explained a large proportion of the variability in richness and assemblage composition in Ecuadorian páramo streams (3400–4100 m asl), with stable hydrological conditions in lake outlet streams leading to higher plant cover and differing macroinvertebrate assemblages. Compared to other páramo streams, lake outlet assemblages had higher densities (10 207 versus 4068 individuals per standard kick sample), fewer families (11 versus 13) and were dominated by non-insects (58% versus 20%). As well as affecting and reflecting habitat stability, plant

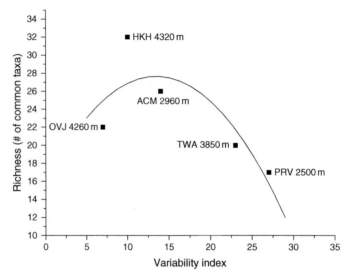

FIGURE 15  Macroinvertebrate richness (in terms of number of common taxa) in five Bolivian Andes streams in relation to a stream variability index based on a range of physical habitat parameters. Stream altitudes and a second-order polynomial regression line are shown also [Data are from Wasson *et al.* (1998)].

cover may also affect macroinvertebrate assemblage structure if respiration by the high plant biomass depletes dissolved oxygen concentrations at night.

The longitudinal succession of fauna along glacier-fed kryal streams has been well documented in Europe but not in the tropics (Ward, 1994). Preliminary qualitative data from Ecuador (Fig. 16; Tables IV) reveal that the Podonominae chironomids, which are common in all high-altitude streams in the Neotropics, dominated assemblages within 1 km of the glacier snout, where Diamesinae larvae were much less abundant. Taxon richness increased

FIGURE 16  Changes along the upper course of a small, milky kryal draining a glacier on Volcán Antisana (5758 m asl), Ecuador. (a) Close to the glacier snout at 4850 m asl; $T_{max} = 0.5°C$, and there are no benthic macrofauna. (b) Descending the moraine at 4770 m asl and 0.5 km from the glacier snout; $T_{max} = 6°C$, with benthic macrofauna dominated by Podonominae; only two other Diptera taxa (Diamesinae and Muscidae) present. (c) At 4590 m asl and 1.2 km from the glacier snout; $T_{max} \approx 10°C$, and two taxa (Simuliidae and Hydrobiosidae) have been added to the macrofauna. (d) At 4430 m asl and 2.2 km from the glacier; $T_{max} = 13.4°C$, and eight taxa are present due to the addition of Baetidae, Elmidae and Turbellaria (Photos by D. Jacobsen) (see colour plate section).

downstream to 18 taxa (families/subfamilies) 5 km away, where maximum stream temperature reached 16°C. These values are similar to glacial streams in the European Alps (16–20 taxa), at the same temperature (~15°C) and distance from the glacier snout (3.6–4.6 km: Castella *et al.*, 2001). These data also support the hypothesis of Jacobsen *et al.* (1997) that, irrespective of altitude and latitude, taxon richness of stream macroinvertebrates is related to maximum stream temperature. Apart from Podonominae replacing Diamesinae chironomids, the change in assemblage structure along the Ecuadorian kryal agreed well with the model of Milner *et al.* (2001) which assumes that increases in channel stability and maximum temperature with greater distance from the glacier are the main parameters governing longitudinal changes in macroinvertebrate assemblages. However, tropical kryals are likely to be more physically stable than seasonal kryals at temperate latitudes (see above) and this suggests that temperature is relatively more important in structuring faunal assemblages along tropical glacier-fed streams.

## E. Functional Organization of the Macroinvertebrate Fauna

Compared to small north-temperate streams, the most striking feature of tropical streams appears to be the general paucity of shredders in terms of species and abundance (Dudgeon, 1989, 2000), although this may not be true for some tropical Australian streams (Cheshire *et al.*, 2005). Irons *et al.* (1994) suggested that the scarcity of shredders is due to high temperatures in tropical streams that increase microbial activity and decomposition rates of leaf litter, so that coarse detritus is not readily available to shredders (for more discussion, see Chapters 3 and 4 of this volume). According to this microbial activity hypothesis, more shredders should be expected in cooler forest streams at higher altitudes. In a study of streams from 2000 to 2700 m asl in Kenya, Dobson *et al.* (2002) found partial support for this hypothesis, as both shredders and coarse detritus were most abundant at the highest (and coolest) site. However, shredders were scarcer than might be expected based on standing stock of detritus, and the amounts of detritus were comparable to those found in European streams. They concluded that scarcity of shredders does not seem to be caused by limited availability of coarse detritus (Dobson *et al.*, 2002), and proposed several alternative explanations. Harrison (1995) reported that relative proportions of macroinvertebrate functional feeding groups (FFG) were the same in lowland streams and mountain torrents in tropical Africa, and that the proportions of shredders were always low, even in forested highland streams. Shredders were also the least important FFG in cool Ecuadorian páramo streams in terms of both richness and abundance (Table VIII), as well as in streams in dense vegetation at lower altitudes in the Ecuadorian Central Valley (Jacobsen and Encalada, 1998).

Most macroinvertebrate taxa in tropical high altitude streams belong to either the scraper (grazer) or collector FFG (Table VIII). Filter-feeders are occasionally very abundant but, even at lake outlets, they do not generally dominate (Encalada, 1997). This relatively low percentage of filters is in contrast to the temperate lowland lake outlets, but agrees with the findings from European alpine lake outlets (Hieber *et al.*, 2002; Maiolini *et al.*, 2006). Overall, predators constitute the most diverse FFG, accounting for almost 30% of all taxa in Ecuadorian páramo streams (Table VIII). This surprisingly high diversity of predators has also been observed in Neotropical lowland streams (Fox, 1977; Jacobsen *et al.*, 1997) as well as in tropical Asian (Dudgeon, 1989) and Australian streams (Cheshire *et al.*, 2005).

## F. Life Histories, Temporal Variability, and Drift of Macroinvertebrates

Tropical high-altitude streams have hydrological dynamics and temperature regimes that differ from those prevailing in lowland streams. Despite this, the very few studies available indicate that the timing of life-cycle events and overall faunal dynamics are similar in high-altitude

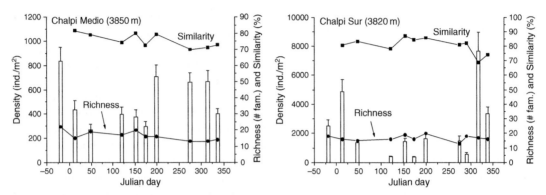

FIGURE 17   Seasonal variation in mean density (+SE: columns), family richness, and Bray–Curtis similarity with first sample of macroinvertebrates on cobbles in two small (1–2 m wide) páramo streams in Ecuador during 1999 (D. Jacobsen, unpublished observations).

and lowland streams, and that synchronized life histories (e.g. timing of adult emergence) are the exceptions (for more discussion on seasonality, see Chapter 4). In a study of Ephemeroptera (three species), Plecoptera (two species), and Trichoptera (six species) in a small lake-outlet stream at 3300 m asl in Ecuador, only *Anacroneuria* sp. (Perlidae: Plecoptera) had a well-defined annual life history (Turcotte and Harper, 1982b). Likewise, of five dominate macroinvertebrates species investigated in a glacier-fed stream at 3600–4400 m asl close to La Paz, Bolivia, only *Claudioperla tigrina* (Gripopterygidae: Plecoptera) exhibited any evidence of a synchronized life history, while the smallest larvae of the other four species were present throughout the year (Molina, 2004). Apart from these studies, published accounts of life-history parameters such as generation times, turnover, and secondary production of macroinvertebrates in tropical high-altitude streams are lacking.

The apparent absence of synchronized life-history events does not produce temporal stability in stream macroinvertebrate assemblages. For instance, total macroinvertebrate densities varied considerably over a year of sampling (Fig. 17) in two small Ecuadorian streams (3800 m asl); family richness and faunal similarity among samples, however, showed little temporal variation and certainly no evidence of a seasonal pattern. Similar observations have been made in a small lake-outlet stream (3300 m asl) in southern Ecuador (Turcotte and Harper, 1982b) and in a stream (2035 m asl) draining Mt Kenya (Mathooko and Mavuti, 1992). Most authors agree that hydrological events (spates and droughts) are the main cause of temporal variations in macroinvertebrate assemblages in tropical high-altitude streams (Turcotte and Harper, 1982b; Mathooko and Mavuti, 1992; Flecker and Feifarek, 1994; Jacobsen and Encalada, 1998). The primary role of discharge is illustrated by a small first-order Ecuadorian páramo stream (3820 m asl) where densities of the stone-surface macroinvertebrates (mainly Orthocladiinae: 37–98%; mean = 78%) are shown with 8 months of continuous measurements of water depth (Fig. 6); clearly, the highest densities occurred after prolonged periods of low flow.

Other studies of tropical highland streams have revealed manifest differences between dry and wet season macroinvertebrate assemblages. The number of taxa, density and biomass of macroinvertebrates in five Bolivian Andean streams were all markedly lower in the wet season (Wasson and Marín, 1988), and the percentage of seasonal variation in these three parameters was positively related to an index of seasonal variability in environmental factors (Fig. 18). Similarly, mean densities of macroinvertebrates in Ecuadorian Central Valley streams (2500–3100 m asl) were approximately 10-fold higher and mean family richness was also greater during the dry season (25.1) than the wet season (17.6: Jacobsen and Encalada, 1998). Assemblage structure during the wet season seemed highly stochastic, with biotic factors

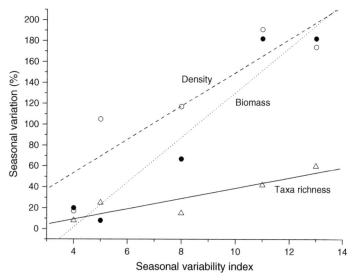

*FIGURE 18* Percent seasonal variation in macroinvertebrate taxa richness (triangles and solid regression line), density (open circles and dashed regression line), and biomass (filled circles and dotted regression line) in five Bolivian Andes streams in relation to a seasonal variability index based on various physical habitat parameters. Seasonal variation in assemblage parameters were calculated as the difference between the wet and dry season samples divided by the mean value. Macroinvertebrate data are from Wasson and Marín (1988), while data on the variability index are from Wasson *et al.* (1998).

probably having relatively little influence. During stable, low-flow periods in dry season, fast colonizers, such as Baetidae, Simuliidae, and Chironomidae, became dominant members of the assemblage.

Drift of macroinvertebrates (i.e. the downstream transport of organisms in the water column) in most tropical lowland streams is mainly nocturnal, with higher drift rates at night probably being related to the presence of day-active predatory fishes. Studies from Neotropical streams have provided support for this hypothesis by showing a positive relationship between night : day drift ratios and species richness and/or densities of day-active drift feeding fish (Flecker, 1992; Pringle and Ramírez, 1998). Physical disturbance by nocturnal, nonpredatory benthic scrapers, such as atyid shrimps and the armored catfishes (Loricariidae), may also trigger an escape reaction by macroinvertebrates thereby increasing nocturnal drift rates (Jacobsen and Bojsen, 2002). In contrast, drift studies in streams at 3300 m asl in Ecuador (Turcotte and Harper, 1982a), 1700–2200 m asl in Venezuela (Flecker, 1992), 1700–2700 m asl in Costa Rica (Pringle and Ramírez, 1998), 1900 m asl in Colombia (Quiñones *et al.*, 1998), and 2100–3800 m asl in Ecuador (Jacobsen and Bojsen, 2002) have all recorded aperiodic drift of aquatic macroinvertebrates, with either no difference between day and night drift densities or higher drift densities in daytime. Most of these highland streams either lacked fish entirely or supported only benthivorous catfish (*Astroblepus*), although some contained introduced Brook, Brown, or Rainbow trout. In fact, aperiodic drift seems to be the general pattern in tropical and temperate mountain streams, regardless of whether they contain trout or other predatory fish (Hieber *et al.*, 2003). Drift densities of tropical highland streams under normal flow conditions are in the range 0.22–0.36 individuals per $m^3$ in Kenya (Mathooko and Mavuti, 1992) and 0.13–0.70 individuals per $m^3$ in Ecuador (Jacobsen and Bojsen, 2002). These values are broadly comparable to rates in tropical lowland streams. However, in Costa Rican lowland streams, where drift was dominated by larval shrimps (50–70% of individuals), mean drift densities may reach 7–13 individuals per $m^3$, greatly exceeding values in highland streams where shrimps were absent (Pringle and Ramírez, 1998; Ramírez and Pringle, 1998).

## G. Faunal Adaptations to High-Altitude Streams

The atmosphere is the main source of dissolved oxygen in most aquatic systems, and low atmospheric oxygen pressure at high altitudes thus has the potential to affect life in water as well as on land. Early studies suggested a causal relationship between the assumed greater oxygen concentrations in high-altitude streams and the distribution of stream macroinvertebrates along altitudinal gradients (Illies, 1964; Hynes, 1970). It was further proposed that macroinvertebrates had no need to evolve special respiratory adaptations in the cool, turbulent, and oxygen-saturated waters of mountain streams (Illies, 1964; Ward and Stanford, 1982). However, Jacobsen *et al.* (2003) have argued that, at very high altitudes, oxygen may be a limiting factor for macroinvertebrates. Oxygen uptake in aquatic macroinvertebrates occurs as passive diffusion over the body surface, and if present, through thin-walled gills. This process is driven by differences in oxygen partial pressure between the inside of the animal and the surrounding water, and not by an oxygen concentration difference *per se* (Spicer and Gaston, 1999; Jacobsen, 2000). As described above, oxygen saturation (∼ oxygen partial pressure) in tropical streams decreases with increasing altitude (Fig. 5). In addition to the partial pressure of oxygen, the potential supply of oxygen to macroinvertebrates is influenced by two temperature-dependent physical factors. These are the diffusion coefficient of oxygen in water, which increases with increasing temperature, and the kinematic viscosity of water, which decreases with increasing temperature, thereby reducing the thickness of the laminar sub-layer surrounding the body surface and enhancing the diffusive influx of oxygen. Based on these considerations, Jacobsen *et al.* (2003) have related the potential oxygen supply for macroinvertebrates to temperature (and thus altitude), suggesting that, at 4000 m asl, the potential oxygen supply will be only one fifth of that at sea level.

In poikilothermic animals, such as macroinvertebrates, respiratory demand for oxygen declines with decreasing temperature. Aquatic macroinvertebrates will confront increasing oxygen deficiency at higher altitudes if their metabolic demand decreases less with altitude (because of lower temperature) than does the oxygen supply. Rostgaard and Jacobsen (2005) measured *in situ* the respiration of Ephemeroptera, Plecoptera, and Trichoptera larvae in streams at altitudes from 400 to 3800 m asl in Ecuador. The mean weight-specific respiration rate of larvae at 3800 m asl (∼9.5°C during daytime) was approximately half that at 400 m asl (∼23.5°C), and thus the community-level respiration rate decreased considerably less over the altitudinal range studied than the potential oxygen supply. This apparent 'gap' between oxygen supply and demand suggests that macroinvertebrates in high-altitude streams may be living close to, or under conditions of oxygen deficiency. Morphological adaptations to optimize oxygen uptake, such as increased surface-volume ratio achieved by developing small body size or large gills, might be expected to be present among invertebrates living in high-altitude streams. Despite this, Jacobsen (2000) found no systematic difference in relative gill sizes of Trichoptera larvae in streams along a 4000-m altitudinal gradient in Ecuador. In addition, there was also no difference in the maximum body size of 17 families of macroinvertebrates in streams below 1000 m asl and in streams above 3500 m asl, but it was notable that the lowland streams contained a greater proportion of large-bodied families, such as Corydalidae, Gomphidae, Libellulidae, Euthyplociidae, Ptilodactylidae, and Calamoceratidae (Jacobsen *et al.*, 2003).

Behavioral adaptations to oxygen-poor, high-altitude stream environments could include increased ventilatory capacity and a preference for microhabitats with faster current, but these have yet to be investigated. Furthermore, slow growth and low activity could be expected as a consequence of existence under such conditions. Exclusion of poorly-adapted species could lead to macroinvertebrate assemblages with low species richness containing few species sensitive to low oxygen saturation. Assemblages may also have lower resistance and higher sensitivity to further lowering of oxygen pressure, for example by organic pollution, than assemblages in lowland streams (Jacobsen *et al.* 2003).

## V. ORGANIC POLLUTION AND OTHER ENVIRONMENTAL IMPACTS

High-altitude streams in the tropics are affected by a number of human activities, which are less common in alpine environments at higher latitudes. These include contamination with nutrients, pesticides, and sediments from fields; contamination with nutrients and organic matter from livestock grazing; receipt of wastewaters from human settlements and cities; acidification and contamination with heavy metals from mining; sedimentation arising from erosion of gravel pits and construction works; reductions in stream discharge (or complete disappearance of flow) due to water abstraction; and fragmentation of longitudinal connectivity caused by dams. The effects of these impacts in high-altitude tropical streams are not expected to differ from those described at lower altitudes (see also Chapter 10 of this volume), and all have consequences for physical habitat, water quality and hence community structure and ecological function. Of particular interest here is pollution of streams by untreated organic waste matter from domestic, agricultural, and industrial sources. The effects of organic pollution and characteristic changes in the assemblage composition in tropical streams are discussed in more detail in Chapter 4, and are quite similar to those that have been well described in temperate streams (e.g. Hynes, 1960), but there is evidence that these impacts are accentuated by low oxygen availability in tropical high-altitude streams. In a study of the effect of organic pollution in streams in the densely-populated Ecuadorian Inter-Andean Valley (2600–3100 m asl), Jacobsen (1998) found a shift from a fauna dominated by Ephemeroptera, Plecoptera, and Trichoptera to assemblages comprising tolerant groups such as tubificid worms, *Psychoda* (Psychodidae: Diptera) and *Chironomus* (Chironominae) at about 56% oxygen saturation relative to sea level. This oxygen saturation would be critical for many European stream insects (see references in Jacobsen, 1998), and would normally lead to pronounced changes in the fauna of tropical and temperate lowland streams (D. Jacobsen, personal observations). It must be stressed that the partial pressure of atmospheric oxygen at 3000 m asl is 70% of that at sea level, and thus water in equilibrium with the atmosphere at 3000 m asl has a saturation of only 70% relative to sea level. Because it takes less organic pollution to reduce oxygen saturation from 70 to 56% in tropical high-altitude streams than to reduce it from 100 to 56% in a lowland stream, the former are much closer to potentially-critical levels of oxygen. It should also be noted that water temperatures in equatorial high-altitude streams are similar to those of temperate lowland streams during summer, and the metabolic oxygen demands of macroinvertebrates are unlikely to differ greatly. The relatively high water temperatures in combination with low oxygen pressure may make macroinvertebrate assemblages in high-altitude tropical streams particularly sensitive to organic pollution, and considerably more so than in equivalent temperate lowland streams.

## VI. FUTURE RESEARCH DIRECTIONS

As will be apparent from the summary of our knowledge of the ecology of tropical high-altitude streams contained in this chapter, we have still much to learn about these habitats. Spatial and temporal variations in basic chemical and physical characteristics of high-altitude streams, and the changes in nutrient concentrations, dissolved organic carbon in relation to hydrologic regimes should be elucidated. More information on the composition of the biota is needed also. Data on composition and diversity of algae, macrophytes, invertebrates, fish, and other vertebrates are particularly scarce for African and – most especially – Asian high-altitude streams. More focused studies could lead to a better understanding of annual cycles in growth rate and biomass of stream macrophytes, and their consequences for patch dynamics. Submerged macrophyte assemblages in tropical high-altitude streams are usually dominated by cosmopolitan taxa that could show growth and temporal changes comparable to those observed in streams at higher latitudes. In the tropics, these dynamics are more likely to be caused by stochastic flow fluctuations than by predictable seasonal changes in temperature, light intensity, and water level that characterize temperate streams. Long-term studies of patch dynamics, plant cover, shoot growth, and mortality in relation to environmental factors could provide an understanding of these dynamics.

A number of factors peculiar to high altitudes could limit primary production and thus the biomass of benthic algae and other primary producers. Elevated UV radiation at high altitudes, especially those in the tropics, may affect metabolic and photosynthetic enzymes and thus constrain primary production. In contrast, the atmospheric partial pressure of carbon dioxide at high altitudes is low, also potentially limiting primary production. Aquatic plants at very high altitudes could have evolved adaptations optimizing the uptake of carbon produced by microbial processes in the sediment or geologically-derived carbonates and bicarbonates. The potential effects of these factors and potential adaptations are presently unknown.

Similarly, little is known about life histories and virtually nothing about growth rates, population dynamics, and production of macroinvertebrates in tropical highland streams. Lower temperatures at high altitudes, and possible oxygen limitation, would lead to the expectation that growth is slower and secondary production is less than in tropical lowland streams. Nevertheless, there may be scope for evolutionary temperature compensation that optimizes enzyme reactions because temperatures at high altitudes in the tropics do not show the extent of seasonal variation prevailing in temperate streams. The food sources (e.g. the relative importance of autochthonous producers and allochthonous detritus) and feeding behavior of some of the commonest macroinvertebrate taxa in tropical high-altitude streams have yet to be determined. Moreover, due to the relatively low species richness in high-altitude streams, studies of food-web structure and energy flow could be more feasible than in streams at lower elevations, and might provide insights into community organization that cannot be gained in more complex systems.

The absence of a clear diel drift periodicity of macroinvertebrates in tropical high-altitude streams, even in the presence of native and introduced insectivorous fish, deserves attention. Does drift periodicity occur under any circumstances in high-altitude streams and, if so, what triggers such periodicity? In addition, it seems probable that introduced fish, such as trout, have impacted native fish populations and macroinvertebrate assemblages in high-altitude streams, but these effects await documentation. The general decrease in macroinvertebrate taxon richness with increasing altitude warrant further investigation; in particular, the question of whether this trend is driven by low dissolved oxygen availability at very high altitudes. The results could have important implications for secondary production in these streams and the resistance of macroinvertebrate assemblages to organic pollution.

Finally, high-altitude ecosystems are expected to be particularly affected by ongoing global climate change, and tropical high-altitude streams could serve as highly sensitive indicators of these impacts. Altitudinal distribution patterns of organisms may be affected, and the upper limit of certain cold-adapted taxa may have to move upward several hundred meters. Increased stream temperatures and changed run-off regimes due to altered rainfall and increased ablation of glaciers will also affect key processes and habitat characteristics, such as channel stability, sediment transport, gas exchange, life-history parameters, assemblage dynamics, organic matter processing, and primary and secondary production. The implications for biodiversity in high-altitude streams will be hard to predict with accuracy, but they may well be profound.

# REFERENCES

Banister, K.E. (1986). Fish of the Zaire system. *In* "The Ecology of River systems" (B.R. Davies and K.F. Walker, Eds), pp. 215–224. Dr. W. Junk Publishers, Dordrecht, The Netherlands.

Beniston, M. (2006). Mountain weather and climate: a general overview and a focus on climatic change in the Alps. *Hydrobiologia* 562, 3–16.

Bojsen, B.H., and Jacobsen, D. (2003). Effects of deforestation on macroinvertebrate diversity and assemblage structure in Ecuadorian Amazon streams. *Archiv für Hydrobiologie* 158, 317–342.

Brewin, P.A., Newman, T.M.L., and Ormerod, S.J. (1995). Patterns of macroinvertebrate distribution in relation to altitude, habitat structure and land use in streams of the Nepalese Himalaya. *Archiv für Hydrobiologie* 135, 79–100.

Brown, L.E., Hannah, D.M., and Milner, A.M. (2003). Alpine stream habitat classification: an alternative approach incorporating the role of dynamic water source contributions. *Arctic, Antarctic and Alpine Research* 35, 313–322.

Buringh, P. (1979). "Introduction to the Study of Soils in Tropical and Subtropical Regions." Centre for Agricultural Publishing and Documentation, Wageningen, The Netherlands.

Busulwa, H.S., and Bailey, R.G. (2004). Aspects of the physicochemical environment of the Rwenzori rivers, Uganda. *African Journal of Ecology* 42, 87–92.

Caissie, D. (2006). The thermal regime of rivers: a review. *Freshwater Biology* 51, 1389–1406.

Castella, E., Adalsteinsson, H., Brittain, J.E., Gislason, G.M., Lehmann, A., Lencioni, V., Lods-Crozet, B., Maiolini, B., Milner, A.M., Olafsson, J.O., Saltveit, S.J., and Snook, D.L. (2001). Macrobenthic invertebrate richness and composition along a latitudinal gradient of European glacier-fed streams. *Freshwater Biology* 46, 1811–1831.

Cheshire, K., Boyero, L., and Pearson, R.G. (2005). Foods webs in tropical Australian streams: shredders are not scarce. *Freshwater Biology* 50, 748–769.

Church, M. (1992). Channel morphology and typology. *In* "The Rivers Handbook, Vol. 1" (P. Calow and G.E. Petts, Eds), pp. 126–143. Blackwell Publishing, Oxford, UK.

Cleef, A.M., and Chaverri, A. (1992). Phytogeography of the páramo flora of Cordillera de Talamanca, Costa Rica. *In* "Páramo – an Andean Ecosystem under Human Influence" (H. Balslev and J.L. Luteyn, Eds), pp. 45–60. Academic Press, London, UK.

Colinvaux, P. (1987). Amazon diversity in light of the paleoecological record. *Quaternary Science Reviews* 6, 93–114.

Cook, C.D.K. (1996). "Aquatic Plant Book." SPB Academic Publishing, Amsterdam, The Netherlands.

Cressa, C. (2000). Macroinvertebrate community structure of twenty-eight Venezuelan streams. *Verhandlungen der Internationale Vereinigung für Theoretische und Angewandte Limnologie* 27, 2511–2518.

Deshmukh, I. (1986). "Ecology and Tropical Biology." Blackwell Publishing, Oxford, UK.

Dobson, M., Magana, A., Mathooko, J.M., and Ndegwa, F.K. (2002). Detritivores in Kenyan highland streams: more evidence for the paucity of shredders in the tropics? *Freshwater Biology* 47, 909–919.

Dudgeon, D. (1989). The influence of riparian vegetation on the functional organization of four Hong Kong stream communities. *Hydrobiologia* 179, 183–194.

Dudgeon, D. (2000). The ecology of tropical Asian rivers and streams in relation to biodiversity conservation. *Annual Review of Ecology and Systematics* 31, 239–263.

Dudgeon, D., and Corlett, R.T. (2005). "The Ecology and Biodiversity of Hong Kong." Joint Publishing Company, Hong Kong, China.

Encalada, A. (1997). Diversidad y abundancia de macroinvertebrados en relación a factores físico-químicos y de fuentes de alimento en dos tipos de ríos de páramos del Ecuador. MSc. Thesis, Pontificia Universidad Católica del Ecuador, Quito, Ecuador.

Flecker, A.S. (1992). Fish predation and the evolution of invertebrate drift periodicity: evidence from Neotropical streams. *Ecology* 73, 438–448.

Flecker, A.S., and Feifarek, B. (1994). Disturbance and the temporal variability of invertebrate assemblages in two Andean streams. *Freshwater Biology* 31, 131–142.

Fossati, O., Wasson, J.G., Héry, C., Salinas, G., and Marín, R. (2001). Impact of sediment releases on water chemistry and macroinvertebrate communities in clear water Andean streams (Bolivia). *Archiv für Hydrobiologie* 151, 33–50.

Fox, L.R. (1977). Species richness in streams – an alternative mechanism. *The American Naturalist* 111, 1017–1021.

Franken, M., and Marín, R. (1992). Influencia de una fábrica de estuco sobre un ecosistema acuático en la ciudad de La Paz. *Ecología en Bolivia* 19, 73–96.

Füreder, L., Vacha, C., Amprosi, K., Bühler, S., Hansen, C.M.E., and Moritz, C. (2002). Reference conditions of alpine streams: physical habitat and ecology. *Water, Air and Soil Pollution: Focus* 2, 275–294.

Füreder, L., Welter, C., and Jackson, J.K. (2003). Dietary and stable isotope ($\delta^{13}C$, $\delta^{15}N$) analyses in alpine streams. *International Review of Hydrobiology* 88, 314–331.

Füreder, L., Wallinger, M., and Burger, R. (2005). Longitudinal and seasonal pattern of insect emergence in alpine streams. *Aquatic Ecology* 39, 67–78.

Hamel, C., and Van Damme, P.A. (1999). Acidificación de ríos por contaminación con metales pesados en la zona altoandina boliviana: indicadores bentónicos. *Revista Boliviana de Ecología y Conservación Ambiental* 6, 191–201.

Harrison, A.D. (1995). Northeast African rivers and streams. *In* "River and Stream Ecosystems" (C.E. Cushing, K.W. Cummins, and G.W. Minshall, Eds), pp. 507–517. Elsevier, Amsterdam, The Netherlands.

Harrison, A.D., and Hynes, H.B.N. (1988). Benthic fauna of Ethiopian mountain streams and rivers. *Archiv für Hydrobiologie Supplement* 81, 1–36.

Hedberg, O. (1961). The phytogeographical position of the Afroalpine flora. *Recent Advances in Botany* 1, 914–919.

Hieber, M., Robinson, C.T., Uehlinger, U., and Ward, J.V. (2002). Are alpine lake outlets less harsh than other alpine streams? *Archiv für Hydrobiologie* 154, 199–223.

Hieber, M., Robinson, C.T., and Uehlinger, U. (2003). Seasonal and diel patterns of invertebrate drift in different alpine stream types. *Freshwater Biology* 48, 1078–1092.

Hieber, M., Robinson, C.T., Uehlinger, U., and Ward, J.V. (2005). A comparison of benthic macroinvertebrate assemblages among different types of alpine streams. *Freshwater Biology* 50, 2087–2100.

Hynes, H.B.N. (1960). "The Biology of Polluted Waters." Liverpool University Press, Liverpool, UK.

Hynes, H.B.N. (1970). "The Ecology of Running Waters." Liverpool University Press, Liverpool, UK.

Hynes, H.B.N., Williams, T.R., and Kershaw, W.E. (1961). Freshwater crabs and *Simulium neavei* in East Africa. *Annals Tropical Medicine and Parasitology* 55, 197–201.

Illies, J. (1964). The invertebrate fauna of the Huallaga, a Peruvian tributary of the Amazon river, from the sources down to Tingo Maria. *Verhandlungen der Internationale Vereinigung für Theoretische und Angewandte Limnologie* 15, 1077–1083.

Illies, J. (1969). Biogeography and ecology of Neotropical freshwater insects, especially those from running waters. *In* "Biogeography and Ecology in South America" (E.J. Fittkau, J. Illies, H. Klinge, G.H. Schwabe, and H. Sioli, Eds), pp. 685–708. Dr. W. Junk Publishers, The Hague, The Netherlands.

Irons, J.G. III, Oswood, M.W., Stout, R.J., and Pringle, C.M. (1994). Latitudinal patterns in leaf litter breakdown: is temperature really important? *Freshwater Biology* 32, 401–411.

Jacobsen, D. (1998). Influence of organic pollution on the macroinvertebrate fauna of Ecuadorian highland streams. *Archiv für Hydrobiologie* 143, 179–195.

Jacobsen, D. (2000). Gill size of trichopteran larvae and oxygen supply in streams along a 4000-m gradient of altitude. *Journal of the North American Benthological Society* 19, 329–343.

Jacobsen, D. (2003). Altitudinal changes in diversity of macroinvertebrates from small streams in the Ecuadorian Andes. *Archiv für Hydrobiologie* 158, 145–167.

Jacobsen, D. (2004). Contrasting patterns in local and zonal family richness of stream invertebrates along an Andean altitudinal gradient. *Freshwater Biology* 49, 1293–1305.

Jacobsen, D., and Bojsen, B. (2002). Macroinvertebrate drift in Amazon streams in relation to riparian forest cover and fish fauna. *Archiv für Hydrobiologie* 155, 177–197.

Jacobsen, D., and Encalada, A. (1998). The macroinvertebrate fauna of Ecuadorian highland streams and the influence of wet and dry seasons. *Archiv für Hydrobiologie* 142, 53–70.

Jacobsen, D., and Marín, R. (2007). Bolivian Altiplano streams with the low richness of macroinverterabates and large diel fluctuations in temperature and dissolved oxygen. *Aquatic Ecology*, DOI 10. 007/s10452-007-9127-x.

Jacobsen, D., and Terneus, E. (2001). Aquatic macrophytes in cool aseasonal and seasonal streams: a comparison between Ecuadorian highland and Danish lowland streams. *Aquatic Botany* 71, 281–295.

Jacobsen, D., Schultz, R., and Encalada, A. (1997). Structure and diversity of stream macroinvertebrates assemblages: the effect of temperature with altitude and latitude. *Freshwater Biology* 38, 247–261.

Jacobsen, D., Rostgaard, S., and Vásconez, J.J. (2003). Are macroinvertebrates in high altitude streams affected by oxygen deficiency? *Freshwater Biology* 48, 2025–2032.

Lencioni, V., and Rossaro, B. (2005). Microdistribution of chironomids (Diptera: Chironomidae) in alpine streams: an autoecological perspective. *Hydrobiologia* 533, 61–76.

Lewis, W.M. Jr, Hamilton, S.K., and Saunders, J.F. III. (1995). Rivers of Northern South America. *In* "River and stream ecosystems" (C.E. Cushing, K.W. Cummins, and G.W. Minshall, Eds), pp. 219–256. Elsevier, Amsterdam, The Netherlands.

Maiolini, B., Lencioni, V., Boggero, A., Thaler, B., Lotter, A.F., and Rossaro, B. (2006). Zoobenthic communities of inlets and outlets of high altitude Alpine streams. *Hydrobiologia* 562, 217–229.

Maldonado, M., and Goitia, E. (2003). Las hidroregiones del departamento de Cochabamba. *Revista Boliviana de Ecología* 13, 117–141.

Mathooko, J.M. (1996). Rainbow trout (*Oncorhynchus mykiss* Walbaum) as a potential natural "drift sampler" in a tropical lotic ecosystem. *Limnologica* 26, 245–254.

Mathooko, J.M., and Mavuti, K.M. (1992). Composition and seasonality of benthic invertebrates, and drift in the Naro Moru River, Kenya. *Hydrobiologia* 232, 47–56.

McGregor, G., Petos, P.E., Gurnell, A.M., and Milner, A.N. (1995). Sensitivity of Alpine stream ecosystems to climatic change and human impacts. *Aquatic Conservation* 5, 233–247.

Meybeck, M., Green, P., and Vörösmarty, C. (2001). A new typology for mountains and other relief classes. *Mountain Research and Development* **21**, 34–45.

Milner, A.M., and Petts, G.E. (1994). Glacial rivers: physical habitat and ecology. *Freshwater Biology* **32**, 295–307.

Milner, A.M., Brittain, J.E., Castella, E., and Petts, G.E. (2001). Trends of macroinvertebrate community structure in glacier-fed rivers in relation to environmental conditions: a synthesis. *Freshwater Biology* **46**, 1833–1847.

Miranda, G., and Pouilly, M. (1999). Ecología comparative de poblaciones superficiales y cavernícolas de *Trichomycterus* spp. (Siluriformes) en el parque Nacional de Toro Toro. *Revista Boliviana de Ecología y Conservación Ambiental* **6**, 163–171.

Molina, C.I. (2004). Estudio de los rascos biologicos y ecologicos en poblaciones de los ordenes Ephemeroptera, Plecoptera y Trichoptera (clase Insecta), en un río al pie del glaciar Mururata. MSc. Thesis, Universidad Mayor de San Andrés, Facultad de ciencias puras y naturales, La Paz, Bolivia.

Monaghan, K.A., Peck, M.R., Brewin, P.A., Masiero, M., Zarate, E., Turcotte, P., and Ormerod, S.J. (2000). Macroinvertebrate distribution in Ecuadorian hill streams: the effect of altitude and land use. *Archiv für Hydrobiologie* **149**, 421–440.

Monasterio, M., and Vuilleumier, F. (1986). Introduction: high tropical mountain biota of the world. *In* "High Altitude Tropical Biogeography" (F. Vuilleumier and M. Monasterio, Eds), pp. 3–7. Oxford University Press, Oxford, UK.

Ormerod, S.J., Rundle, S.D., Wilkinson, S.M., Daly, G.P., Dale, K.M., and Juttner, I. (1994). Altitudinal trends in the diatoms, bryophytes, macroinvertebrates and fish of a Nepalese river system. *Freshwater Biology* **32**, 309–322.

Petr, T. (Ed.) (1983). "The Purari – Tropical Environment of a High Rainfall River Basin." Dr. W. Junk Publishers, The Hague, The Netherlands.

Pringle, C.M., and Ramírez, A. (1998). Use of both benthic and drift sampling techniques to assess tropical stream invertebrate communities along an altitudinal gradient, Costa Rica. *Freshwater Biology* **39**, 359–373.

Quiñones, M.L., Ramírez, J.J., and Díaz, A. (1998). Estructura numérica de la comunidad de macroinvertebrados aquáticos derivadores en la zona ritral del río Medellín. *Actualidades Biológicas* **20**, 75–86.

Rader, R.B., and Belish, T.A. (1997). Effects of ambient and enhanced UV-B radiation on periphyton in a mountain stream. *Journal of Freshwater Ecology* **12**, 615–628.

Ramírez, A., and Pringle, C.M. (1998). Invertebrate drift and benthic community dynamics in a lowland Neotropical stream, Costa Rica. *Hydrobiologia* **386**, 19–26.

Reading, A.J., Thompson, R.D., and Millington, A.C. (1995). "Humid Tropical Environments." Blackwell Publishing, Oxford, UK.

Ríos, B. (2004). Las comunidades de macroinvertebrados bentónicos de dos cuencas altoandinos de Ecuador. MSc. Thesis, Departamento de Ecología, Universidad de Barcelona, Spain.

Roback, S.S., Berner, L., Flint, O.S. Jr, Nieser, N., and Spangler, P.J. (1980). Results of the Catherwood Bolivian–Peruvian Altiplano expedition Part 1: aquatic insects except Diptera. *Proceedings of the Academy of Natural Sciences of Philadelphia* **132**, 176–217.

Robinson, C.T., and Uehlinger, U. (2002). Glacial streams in Switzerland: a dominant feature of alpine landscapes. *EAWAG News* **54**, 6–8.

Robinson, C.T., Uehlinger, U., and Hieber, M. (2001). Spatio-temporal variation in macroinvertebrate assemblages of glacial streams in the Swiss Alps. *Freshwater Biology* **46**, 1663–1672.

Rostgaard, S., and Jacobsen, D. (2005). Respiration rate of stream insects measured *in situ* along a large gradient of altitude. *Hydrobiologia* **549**, 79–98.

Rott, E., Cantonati, M., Füreder, L., and Pfister, P. (2006). Benthic algae in high altitude streams in the Alps – a neglected component of the aquatic biota. *Hydrobiologia* **562**, 195–216.

Rüegg, J., and Robinson, C.T. (2004). Comparison of macroinvertebrate assemblages of permanent and temporary streams in an Alpine flood plain, Switzerland. *Archiv für Hydrobiologie* **161**, 489–510.

Sarmiento, G. (1986). Ecological features of climate in high tropical mountains. *In* "High Altitude Tropical Biogeography" (F. Vuilleumier and M. Monasterio, Eds), pp. 13–45. Oxford University Press, Oxford, UK.

Schubert, C. (1988). Climatic changes during the last glacial maximum in northern South America and the Caribbean: a review. *Interciencia* **13**, 128–137.

Schultz, R. (1997). Biologisk struktur i ecuadorianske lavlandsvandløb med forskellig grad af riparisk skygning. MSc. Thesis, Freshwater Biological Laboratory, University of Copenhagen, Copenhagen, Denmark.

Schütz, C., Wallinger, M., Burger, R., and Füreder, L. (2001). Effects of snow cover on the benthic fauna in a glacier-fed stream. *Freshwater Biology* **46**, 1691–1704.

Segnini, S., and Bastardo, H. (1995). Cambios ontogenéticos en la dieta de la Trucha Arcoiris (*Oncorhynchus mykiss*) en un rio andino neotropical. *Biotropica* **27**, 495–508.

Segnini, S. and Chacón, M.M. (2005). Caracterización fisicoquímica del hábitat interno y ribereño de ríos Andinos en la Cordillera de Mérida, Venezuela. *Ecotropicos* **18**, 38–61.

Sioli, H. (1975). Tropical river: The Amazon. *In* "River Ecology" (B.A. Whitton, Ed.), pp. 461–488. Blackwell Publishing, Oxford, UK.

Smith, B.P.G., Hannah, D.M., Gurnell, A.M., and Petts, G.E. (2001). A hydrogeomorphological context for ecological research on alpine glacial rivers. *Freshwater Biology* **46**, 1579–1596.

Sommaruga, R., and García-Pichel, F. (1999). UV-absorbing mycosporine-like compounds in planktonic and benthic organisms from a high-mountain lake. *Archiv für Hydrobiologie* **144**, 255–269.

Spicer, J.I., and Gaston, K.J. (1999). Amphipod gigantism dictated by oxygen availability? *Ecology Letters* **2**, 397–403.

Townsend, C.R., Scarsbrook, M.R., and Dolédec, S. (1997). The intermediate disturbance hypothesis, refugia and biodiversity in streams. *Limnology and Oceanography* **42**, 938–949.

Troll, C. (1968). The cordilleras of the tropical Americas: aspects of climatic, phytogeographical and agrarian ecology. *In* "Geo-ecology of the Mountainous Regions of the Tropical Americas" (C. Troll, Ed.), pp. 15–56. Ferdinand Dümmlers Verlag, Bonn, Germany.

Turcotte, P., and Harper, P.P. (1982a). Drift patterns in a high Andean stream. *Hydrobiologia* **89**, 141–151.

Turcotte, P., and Harper, P.P. (1982b). The macro-invertebrate fauna of a small Andean stream. *Freshwater Biology* **12**, 411–419.

Uehlinger, U., and Zah, R. (2003). Organic matter dynamics. *In* "Ecology of a Glacial Floodplain" (J.V. Ward and U. Uehlinger, Eds), pp. 199–215. Kluwer Academic Publishers, Dordrecht, The Netherlands.

Van Someren, V.D. (1952). "The Biology of Trout in Kenya Colony." Government Printer, Nairobi, Kenya.

Vélez-Espino, L.A. (2005). Population viability and perturbation analyses in remnant populations of the Andean catfish *Astroblepus ubidiai*. *Ecology of Freshwater Fish* **14**, 125–138.

Vinebrooke, R.D., and Leavitt, P.R. (1996). Effects of ultraviolet radiation on periphyton in an alpine lake. *Limnology and Oceanography* **41**, 1035–1040.

Walling, D.E., and Webb, B.W. (1992). Water quality I: physical characteristics. *In* "The Rivers Handbook Vol. 1" (P. Calow and G.E. Petts, Eds), pp. 48–72. Blackwell Publishing, Oxford, UK.

Ward, J.V. (1985). Thermal characteristics of running waters. *Hydrobiologia* **125**, 31–46.

Ward, J.V. (1994). Ecology of alpine streams. *Freshwater Biology* **32**, 277–294.

Ward, J.V., and Stanford, J.A. (1982). Thermal responses in the evolutionary ecology of aquatic insects. *Annual Review of Entomology* **27**, 97–117.

Wasson, J.G., and Marín, R. (1988). Tipología y potencialidades biológicas de los ríos de altura en la región de La Paz (Bolivia): metodología y primeros resultados. *Memoria Sociedad de Ciencias Naturales La Salle* **48**, 97–122.

Wasson, J.G., Guyot, J.L., Dejoux, C., and Roche, M.A. (1989). "Régimen térmico de los ríos de Bolivia." Orstrom IHH-UMSA PHCAB IIQ-UMSA SENAIM Hidrobiología-UMSA, La Paz, Bolivia.

Wasson, J.G., Marín, R. Guyot, J.L., and Maridet, L. (1998). Hydro-morphological variability and benthic community structure in five high altitude Andean streams (Bolivia). *Verhandlungen der Internationale Vereinigung für Theoretische und Angewandte Limnologie* **26**, 1169–1173.

Williams, T.R., and Hynes, H.B.N. (1971). A survey of the fauna of streams on Mount Elgon, East Africa, with special reference to the Simuliidae (Diptera). *Freshwater Biology* **1**, 227–248.

Zah, R., and Uehlinger, U. (2001). Particulate organic matter inputs to a glacial stream ecosystem in the Swiss Alps. *Freshwater Biology* **46**, 1597–1608.

# 9

# Are Tropical Streams Ecologically Different from Temperate Streams?

Andrew J. Boulton, Luz Boyero, Alan P. Covich, Michael Dobson, Sam Lake, and Richard Pearson

If tropical streams differ ecologically from temperate ones, we must be cautious in our extrapolation of ecosystem models developed in temperate-zone streams. Similarly, approaches and techniques used routinely in management of temperate streams may not be applicable in the tropics. Despite considerable variability in geological history, flow regime and geomorphology, streams in the tropics typically receive higher insolation and more intense rainfall, with warmer water and often relatively predictable floods. For many groups of aquatic taxa, tropical streams also harbour higher biodiversity than their temperate equivalents. Nonetheless, there is little published evidence for consistent differences in food-web structure, productivity, organic-matter processing and nutrient dynamics, or responses to disturbance which would indicate that the term 'tropical' has special significance when applied to stream ecology. Instead, ecological processes in tropical streams appear to be driven by the same variables that are important in temperate ones. For example, biotic responses to drought and flooding are similar to those in temperate streams while in-stream productivity is limited by the same factors: nutrients, shading, disturbance, and trophic structure. Shredders are reputed to be rare in many tropical streams but this also is the case in many southern temperate streams, implying that models of leaf breakdown developed in the north-temperate zone may not have the universal applicability often assumed. Biome comparisons among temperate and tropical streams are confounded by the immense inherent variability of streams within both these zones, and the wide range of climatic and hydrological conditions – even in the tropics. Valid extrapolation of models and management strategies may be less a matter of tropical versus temperate streams but, instead, of ensuring comparability at appropriate scales and fuller understanding of ecological mechanisms, plus recognition of the magnitude and complexity of spatial and temporal variation in stream ecosystems at all latitudes.

## I. INTRODUCTION

Are tropical streams ecologically different from temperate streams? Against the backdrop of 'ecology's oldest pattern' (Hawkins, 2001) that diversity in the tropics far exceeds that of temperate zones, posing this question may seem superfluous. However, the observed latitudinal differences in biodiversity are not consistent across broad taxonomic groups and have multiple explanations (e.g. Rohde, 1992; Rosenzweig, 1997). When present, these patterns of biodiversity may generate significant differences in food-web structure and productivity that are at odds with predictions from models developed in temperate streams (Covich, 1988). Wide variation in climate, geomorphology, landscape evolution, and geological history across tropical zones potentially masks patterns that might be considered 'unique' to tropical streams (Wantzen *et al.*, 2006). It is also likely that the local variation among streams found within tropical and temperate zones is so great that latitudinal trends in regional ecological drivers such as temperature, photoperiod, and seasonality seldom are detectable. Here, we address the following questions: is the perception that tropical streams are very different ecologically from temperate ones based on facts or assumptions? Does such a perception result from inappropriate comparisons confounded by differences in geology, flow regime, disturbance history, sampling procedures and taxonomic resolution, biogeography, anthropogenic influence, and uneven data sets?

We follow the latitudinal definition of 'tropical' used elsewhere in this book (see Chapter 1) to encompass streams lying between 23°N and 23°S but, where appropriate, extend this to include areas where monsoonal seasonal rains fall. While it is futile to attempt to draw firm boundaries across a climatic continuum, for ease of discussion and from widespread precedence in the scientific literature we use the terms 'tropical' and 'temperate' as adjectives for regional comparisons of their stream ecology. Despite the impossibility of drawing generalizations about an idealized 'tropical' or 'temperate' stream, there is a need to compare ecological processes and diversity in equivalent-sized streams from different parts of tropical and temperate regions. This comparison is necessary because the vast amount of research undertaken on north temperate streams has given rise to ecological models (e.g. the River Continuum Concept: Vannote *et al.*, 1980; the Riverine Productivity Model: Thorp and Delong, 1994) that are now being used to guide research questions and management approaches in rivers worldwide. The flood pulse concept (Junk *et al.*, 1989; Junk and Wantzen, 2004), developed in tropical rivers, is a notable exception to this generalization. However, if tropical streams are ecologically different from temperate ones, uncritical extrapolation of these models may be seriously misleading (Covich, 1988; Dudgeon, 1999a; Dudgeon *et al.*, 2000; Pringle *et al.*, 2000a; Bass, 2003; Wantzen *et al.*, 2006).

Systematic differences between the ecology of tropical and temperate streams could also invalidate commonly-used management strategies that have been developed and tested in temperate streams. For example, faecal coliforms are used routinely as indicators of recent faecal contamination in waterways in the United States, Europe and Australia. However, in some tropical streams relatively undisturbed by human activity, Carillo *et al.* (1985) found high counts of faecal coliforms including *Escherichia coli*, apparently promoted by the natural tropical patterns of high water temperatures and humidity. Another example is the failure in many tropical rivers of the common biological index for stream health derived from counts of the sensitive insect orders of Ephemeroptera, Plecoptera and Trichoptera (EPT) because representatives of the Plecoptera are naturally absent or rare (Dudgeon, 1999a; Vinson and Hawkins, 2003).

Many papers describing ecological phenomena from tropical streams compare their results with those from temperate counterparts (Greathouse and Pringle, 2006) but sometimes ideas developed in the tropics are applied to temperate stream ecosystems. For example, Benke *et al.* (2000) applied the flood pulse concept developed for the Amazon basin (Junk *et al.*, 1989) to a smaller north-temperate drainage system in Georgia. In another example, Dobson *et al.* (2002) make the case that apparently low populations of leaf-shredding invertebrates in many tropical streams may demonstrate that the popular model of leaf breakdown with its emphasis on mediation by shredders could actually be an unusual feature, more representative of streams in the north temperate zone and not a global mechanism of detrital processing. Shredders have also been shown to be uncommon in many streams in south-temperate latitudes (e.g. Winterbourn *et al.*, 1981; Lake *et al.*, 1986) and this global pattern deserves further attention so that it can be elucidated more clearly. Examples of the limited generality of concepts will likely increase as more ecologists explore the similarities and differences among streams across latitudes. Given that tropical catchments cover the largest area of the Earth's climatic zones, we need broader, more representative studies from these potentially different regions (Wantzen *et al.*, 2006).

This chapter explores the evidence for and against the proposition that there are fundamental differences in the ecology of temperate and tropical streams. First, we examine the evidence for qualitative differences in climate, hydrology and geomorphology between tropical and temperate streams that might generate a habitat templet for ecological differences. We then compare aspects of their biodiversity and biogeography, food-web organization and trophic structure, primary productivity and organic matter dynamics, and the role of disturbance. We seek to determine whether trends in these variables are merely quantitative variations across latitudes, and whether the term 'tropical' can be ecologically informative when discussing streams and rivers. We also speculate on potentially productive research directions in instances where the search for comparisons has revealed substantial knowledge gaps or generated broad-scale hypotheses that could be tested by inter-biome comparisons.

## II. TROPICAL SETTINGS: INTERACTIONS BETWEEN CLIMATE AND LANDSCAPE EVOLUTION

### A. Geological History and Geomorphology

Tropical land masses span a broad range of geological history ranging from the relatively young islands of the South Pacific created by tectonic upheaval and volcanism (Craig, 2003) to the ancient Precambrian drainage basins of many tropical continental regions, such as the Niger in Africa (Alagbe, 2002), where effects of sea-level change and plate movement have been minimal. Furthermore, most tropical streams in low-elevation drainage basins are older than those found in the temperate zones of North America and Europe where glaciation created 'lake districts' drained by a network of rivers of similar ages. Their location near the equator, and well away from the polar caps, has meant that some of the Earth's largest rivers (e.g. Amazon, Orinoco, and Mekong) are also among the oldest, with major biogeographical and biodiversity implications. For example, the Amazon and Orinoco have been estimated to support (respectively) 3000 and 2000 fish species (Lowe-McConnell, 1987), although the actual totals may be higher (see also Chapter 5 of this volume). At the other end of the tropical spectrum in age and stream size, the geologically recent and insular streams of the Caribbean and South Pacific have fewer than 10 fish species, and this is typical of tropical oceanic islands (Pringle *et al.*, 2000a; Donaldson and Myers, 2002).

Differences in geological histories have significant implications for the geomorphology of their rivers. Physical controls on stream-channel dynamics include formation of steep waterfalls in areas of tectonically uplifted terrain (Craig, 2003), along with potentially rapid erosion, formation of landslides and extensive deposition of sediments (Ahmand *et al.*, 1993), especially prevalent in young drainage basins. In these types of stream channel, sediments are typically coarse, riffles and glides are common, and pools are relatively shallow. Hydrological responses to heavy rain in these systems are flashy, with scouring spates commonly disturbing the stream biota and mobilizing organic matter and sediments (Covich *et al.*, 1991, 2006; Johnson *et al.*, 1998). The same disturbances occur in equivalent temperate streams on steep mountains (Benda and Dunne, 1997; Woodward and Hildrew, 2002) with similar physical impacts on stream ecosystem properties (e.g. Kiffney *et al.*, 2004).

Older rivers have more weathered catchments and well-developed alluvial lowland plains where floodplain ecosystems are long-established and diverse (Chestnut and McDowell, 2000). The drainage boundaries of many large tropical rivers are physically defined by ridges and valleys of different geologic ages, yielding different combinations of weathered elements and macronutrients (Markewitz *et al.*, 2001). Not surprisingly, most of the processes of abiotic weathering and leaching match those observed in temperate rivers (e.g. McDowell, 1998), with the main difference that such processes have been occurring for longer in many ancient tropical drainages and, therefore, the sediments would be more weathered (White *et al.*, 1998). However, an important distinction must be made between the weathering processes in the wet–dry tropics versus those in the humid tropics where the differences are more driven by hydrological patterns than historical variables. Osterkamp (2002) claims that features typical of the humid tropics (e.g. frequent and intense storms; intense biochemical weathering; hill slope creep and landslides favoured by a deep, chemically weathered regolith) promote more rapid weathering and subsequent water transport of the products of weathering than in equivalent-aged temperate catchments. Conversely, in the wet–dry tropics, alternation of inundation and desiccation may expedite release of weathered nutrients that are then taken up by geochemical and microbial processes, resulting in the dilute chemical composition of many tropical rivers (Stallard, 2002).

Selective leaching of silica (White *et al.*, 1998), the main constituent of most rocks, degrades mineral structure and leads to silica-poor clays in tropical river basins. Successive cycles of wetting and baking of these clays cause physical stress, swelling and cracking of soils, and increases the mobilization of nutrients (Twinch, 1987). Temperature also plays a major role, through its action on physical disintegration of rocks and thermal enhancement of chemical and biological weathering processes. The ionizing power of water increases steeply with temperature, being four times higher at 25°C than at 10°C so that chemically aggressive solutions can penetrate deep into rock (Faniran and Areola, 1978). Intense storms, typical of the tropics (see below), contribute to the accelerated erosion and transport of weathered sediments. Overall, patterns of weathering and geomorphological processes are likely to be complex derivatives of geological history, rock type, climate, and hydrology – any influence of latitude is probably largely via the effects of the generally higher temperatures. As shown in Chapter 1 of this volume, maximum water temperatures of 25–35°C occur at low to moderate elevations in the tropics, and this would be expected to influence weathering rates as well as oxygen saturation and rates of metabolism (see below).

## B. Climate and Hydrology

Within both tropical and temperate zones, there is great variability in climate from arid to humid, and in topography from plains to mountain ranges. However, at the regional scale, far more precipitation falls on the land in the tropics than in the temperate zones (Dai and Trenberth, 2002), and runoff in terms of precipitation minus evaporation is highest (ca. 2000 mm on average) in the tropics. In the southern temperate zone, runoff is also quite high (ca. 1500 mm) due to orographic rain on the west coasts of New Zealand and South America. In the northern temperate zone, it is consistently lower than in the tropics (<1200 mm, Baumgartner and Reichel, 1975). Regionally, tropical climates are dominated by the high inputs of solar radiation which cause rapid evaporation from terrestrial and aquatic surfaces. This, coupled with the typically high rainfall intensity, characterizes the Intertropical Convergence Zone between the trade wind systems on either side of the equator. When large amounts of water vapour that have evaporated from warm tropical oceans condense into heavy rain, latent heat energy is released, increasing turbulence and generating the violent storms emblematic of much of the tropics. The Intertropical Convergence Zone does not form a continuous belt but, instead, fills and reforms to produce a chain of major disturbances, causing variable rainfall and alternating patterns of flooding and droughts throughout the equatorial tropics (Pereira, 1989; see also Chapter 1 of this volume).

Most large tropical rivers experience predictable seasonal floods of great magnitude (Balek, 1983, Dudgeon, 1999a, 2000; Arrington and Winemiller, 2006). Monsoon systems operate in the tropics over Asia, northern and western Africa, northern Australia, and in the Americas as annual cycles of wet and dry seasons (Webster *et al.*, 1998). The rainy season occurs when warm, moist winds move inland from the warm oceans whilst the dry season eventuates when cold, dry winds blow from temperate lands locked in winter. However, unpredictable and damaging floods may arise from hurricanes, typhoons or cyclones (Balek, 1983; Covich *et al.*, 2006) which, although mainly tropical, occasionally affect subtropical and temperate zones. Monsoon intensities, especially in Asia, are linked ('teleconnected') with the El Niño-Southern Oscillation phenomenon, and it is largely the strength of ENSO events that governs the strength of the monsoons (Jury and Melice, 2000). Strong and persistent ENSO events may give rise to devastating droughts in Africa, Southeast Asia, and Australia while causing floods in parts of China, Europe and the Americas (Davis, 2001).

Of the 12 greatest river drainages on Earth, ranked in terms of discharge ($>415\,km^3\,yr^{-1}$), six are tropical, with the highest being the Amazon with a mean annual discharge of $6642\,km^3\,yr^{-1}$, followed by the Congo at $1308\,km^3\,yr^{-1}$ (Dai and Trenberth, 2002; Table I). After proportional correction for drainage area, all tropical river drainages except the Congo have higher runoff than temperate rivers. Another outlier, the Brahmaputra [lat. 25.2°N], probably should be grouped with the tropical river systems on the basis of runoff per unit area of catchment; Table I). At least as far as large rivers are concerned, annual flow variability tends to be only marginally lower in the tropics than in temperate zones (McMahon *et al.*, 1992). This lack of marked difference is important because claims of a 'more predictable' flow environment in the tropics (e.g. Junk and Welcomme, 1990) must be treated with caution in the context of considering a wide range of tropical climates and inter-annual variability at the scale of whole drainage networks. Even non-seasonal streams in humid tropical areas can be subject to occasional drought, with concomitant impacts on the aquatic biota (Covich *et al.*, 2003, 2006). The primary issue here is that of stream size; even where rainfall is generally highly seasonal, flows will follow predictable patterns in large rivers but remain flashy and relatively unpredictable in the headwaters (see Section V). While much of the present chapter focuses on comparing tropical and temperate stream ecosystems because that is where the majority of research has been done, there are obvious ecological differences associated with river size (e.g. Stanley and Boulton, 2000). Some of these will be exaggerated in wide tropical rivers

*TABLE I* Top 12 River Basins on Earth, Ranked Based on Discharge at River Mouth (from Dai and Trenberth, 2002)

| River basin | Discharge $(km^3\,yr^{-1})$ | Drainage area $(1000\,km^2)$ | Discharge/drainage area | Latitude | Location |
|---|---|---|---|---|---|
| **Amazon** | **6642** | **5854** | **1.135** | **2** | **Obidos, Brazil** |
| **Congo** | **1308** | **3699** | **0.354** | **4.3** | **Kinshasa, Congo** |
| **Orinoco** | **1129** | **1039** | **1.087** | **8.1** | **Pte Angostu, Venezuela** |
| Yangtze (Chang Jiang) | 944 | 1794 | 0.527 | 30.8 | Datong, China |
| Brahmaputra | 628 | 583 | 1.078 | 25.2 | Bahadurabad, Bangladesh |
| Mississippi | 610 | 3203 | 0.190 | 32.3 | Vicksburg, Mississippi, USA |
| Yenisey | 599 | 2582 | 0.232 | 67.4 | Igarka, Russia |
| Paran | 568 | 2661 | 0.213 | 32.7 | Timbues, Argentina |
| Lena | 531 | 2418 | 0.220 | 70.7 | Kusur, Russia |
| **Mekong** | **525** | **774** | **0.678** | **15.1** | **Pakse, Laos** |
| **Tocantins** | **511** | **769** | **0.665** | **3.8** | **Tucurui, Brazil** |
| **Tapajos** | **415** | **502** | **0.827** | **5.2** | **Jatoba, Brazil** |

The ratio of volume to drainage area indicates runoff. Tropical rivers are in bold.

where higher water temperatures and insolation can enhance plant productivity and microbial respiration (as described below).

River drainages also may be characterized by their annual flow regimes. Using a global database of monthly flows in 969 streams, Haines *et al.* (1988) identified 15 different river flow regimes. In the Asian-Australian monsoonal regions, they found streams with flow peaks in mid- or late-summer (Groups 6 and 7 respectively) whereas rivers with peak flows in autumn (Groups 9, 10 and 11) occur in Africa, central and southern America, and Southeast Asia. Conversely, streams in the cold temperate zones with a snow season tend to have flow peaks in late spring and early summer (Groups 2, 3 and 4) and streams of the warm temperate zones have flow peaks in autumn, winter, and early spring (Groups 11, 12, 13, and 14). While descriptors from temperate zones such as 'autumn' or 'spring' cannot have exact equivalents in tropical latitudes (and so should be interpreted with caution), this monthly analysis reveals distinct differences in the annual flow regimes between tropical- and temperate-zone rivers at a broad scale, related to latitudinal climatic patterns and the timing of peak discharge.

Puckridge *et al.* (1998) explored patterns of flow variability in more detail, assessing 23 measures of hydrological variability for 52 large river drainages worldwide, with approximately equal numbers of gauging stations in arid, temperate, tropical, and continental climatic regions. They identified groups of river systems along a spectrum from 'tropical' to 'dryland'. However, some river drainages from continental climates occurred at both extremes of the range, illustrating the limitations of even this quite complex analysis. For example, Central African river drainages from wet tropical climates (Oubangu [Ubangui], Sanaga, and Niger) shared the lowest variability scores with those from continental and/or polar climatic influences (Neva, Fraser, and Idigirka). Puckridge *et al.* (1998) emphasize that each river system appeared to have its own 'signature' of flow regime that was masked by efforts to combine the collective measures of hydrological variability in their analysis. Nonetheless, a significant association between the vectors of flow variability and climate remained and, on most measures of hydrological variability, tropical river drainages tended to be significantly lower than their equivalent-sized temperate counterparts (Puckridge *et al.*, 1998).

Do different patterns in hydrology result in ecological differences between tropical and temperate river systems? Is the relatively predictable flow regime, even in the seasonal wet-dry tropics, likely to have repercussions on biodiversity and ecological processes? Tropical and temperate regions are both drained by large-river drainages that, in the unregulated and unconstrained state, have large seasonal floods (Dudgeon, 2000; Arrington and Winemiller, 2006). This 'flood pulse' may be more marked and predictable in some tropical rivers due to the extended wet season, leading Junk *et al.* (1989) to propose the flood pulse concept which asserts that regular pulses of river discharge largely govern the dynamics of river-floodplain systems. The flood pulse concept has been criticized for focusing on over-bank flows and for understating the significance of spatial and temporal flow variability in rivers within other climatic regions (e.g. Walker *et al.*, 1995b; but see also Junk and Wantzen, 2004). This is an interesting reversal of the usual case when models established in the north-temperate zone are applied to tropical streams. Floods play central roles in ecosystem replenishment and disturbance, regardless of climatic zone (Dodds, 2002; Woodward and Hildrew, 2002), and the extent to which these differ between tropical and temperate streams is discussed below (see Section V).

## III. BIODIVERSITY AND ENDEMISM

### A. Different Scales of Biodiversity

Latitudinal gradients of diversity have been discussed for more than a century (Hawkins, 2001). Most groups of organisms are more diverse in the tropical oceans and on land; only a few groups peak in diversity in temperate zones (e.g. cambarid and astacid crayfishes, Plecoptera, simuliid blackflies, salmonid fish) or show no evident latitudinal gradient (e.g. stream macrophytes; Crow, 1993; Jacobsen and Terneus, 2001). For many groups, the latitudinal gradient has been persistent and is clearly evident in fossil assemblages (Buzas *et al.*, 2002). While it is widely accepted that diversity is highest in the tropics, little is known about the reasons underlying this pattern or the variability within and among groups and at different scales of local and regional diversity. Understanding the mechanisms responsible for this difference might shed light on ecological differences between tropical and temperate streams.

Diversity gradients in fresh water have received much less attention than marine or terrestrial environments (Boyero, 2002). Comparisons of diversity have been hampered by the imbalance in research effort between these climatic zones. For example, lists of species are scarce for the tropics – especially for macroinvertebrates – and the identification of tropical species has been difficult for non-specialists. Freshwater assemblages comprise a diverse mix of taxa that have invaded fresh waters independently, following different biogeographical patterns, which may well explain why overall latitudinal diversity gradients for these organisms are obscure (Heino *et al.*, 2002).

Diversity can be viewed at two scales – local (alpha diversity) and regional (gamma diversity) – that are usually strongly related across different taxa and continents (Caley and Schluter, 1997). Typically, a larger regional pool of species is reflected in greater species richness at the local site scale. Where site richness does not increase with regional richness, some form of ecological limitation is suggested (Craig, 2003). A direct relationship between local and regional diversity is apparent in stream fishes (Angermeier and Winston, 1998) and Odonata (dragonflies and damselflies: Caley and Schluter, 1997). Frequent and/or intense disturbance, which is a prevailing feature in most streams (Flecker and Feifarek, 1994; Dodds, 2002), might limit the number of species coexisting locally, irrespective of regional diversity. On the other hand, the high capability for dispersal of aquatic insects (by flying or drifting) suggests that their species richness should be primarily under regional control (Heino *et al.*, 2002). It is therefore convenient to examine local and regional diversity separately.

## B. Latitudinal Gradients in Local Diversity in Streams

Published comparisons of diversity among streams have typically been based on small numbers of samples, and have failed to show consistent latitudinal patterns. For example, Patrick (1964) suggested that tropical American streams were no more species-rich than their temperate counterparts, but Stout and Vandermeer (1975) arrived at the opposite conclusion based on an analysis of taxonomic units rather than biological species. Flowers (1991) found high variability in species richness between streams in Panama and concluded that, when this variation is more thoroughly documented, it may swamp any 'latitudinal gradient' in aquatic insect diversity. Comparisons among detailed studies suggest that diversity in the tropics is greater (e.g. Bishop, 1973; Pearson *et al.*, 1986; Smith and Pearson, 1987; Table II) or lower (Arthington, 1990) but are hindered by lack of standardization of methods used.

A rare example of the application of standard methods across a large latitudinal gradient is Lake *et al.* (1994) who sampled Australian temperate and tropical streams with the specific aim of comparing species richness. They used identical methods to sample stones at each site, and focused on streams that were very similar in geomorphology. Their conclusion was that the tropical streams did have greater species richness of aquatic insects. Although the contrast was statistically significant, the difference was neither large nor consistent among taxa. Interestingly, the sample units (stones) supported no more species in the tropical than in the temperate streams. However, in the tropics, as each new stone was sampled, species continued to be added beyond the limit in temperate streams such that an asymptote in the species versus area curve

*TABLE II*  Conclusions Drawn by Ecologists Comparing Biota and Ecosystem Processes between Temperate and Tropical Streams

| Response variables | Comparative trend | Reference |
|---|---|---|
| *Diversity, richness and community composition* | | |
| Fungal communities on submerged wood | Higher fungal diversity on submerged wood in tropical streams; distinct fungal communities occur on wood in tropical, subtropical and temperate streams | Hyde and Goh (1999); Ho *et al.* (2001) |
| Blackfly (Simuliidae) species richness | Greater species richness in temperate streams | Hamada *et al.* (2002); McCreadie *et al.* (2005) |
| Species richness of Chironomidae (Diptera) | About 1.8 times more chironomid species, on average, in tropical streams | Coffman and de la Rosa (1998) |
| Chironomidae species richness in acid streams | Higher species richness of chironomid midges in tropical acid streams than equivalent temperate streams | Cranston *et al.* (1997) |
| Decapod diversity and ubiquity | Decapod crustaceans such as prawns and crabs are more diverse and widespread in tropical streams | Covich (1988); Dudgeon (1999a) |
| Representation of Plecoptera (stoneflies) | Lower abundance and diversity of Plecoptera in tropical streams | Vinson and Hawkins (2003) |
| Invertebrate species richness and water temperature | Similar relationship between insect species richness and maximum water temperature but two to fourfold higher species richness in lowland streams | Jacobsen *et al.* (1997) |
| Invertebrate species diversity on rocks | More rock-dwelling invertebrates in tropical Australian streams than in Australian temperate streams | Pearson *et al.* (1986); Lake *et al.* (1994) |
| *Habitat selection and behaviour* | | |
| Food exploitation behaviour of shredders | No difference in feeding behaviour or preference for conditioned leaves between shredders from tropical and temperate streams | Graça *et al.* (2001) |
| Importance of woody debris to fish communities | As in temperate streams, addition of woody debris to pools in a tropical stream promoted fish species richness and abundance | Wright and Flecker (2004) |
| Size-based predator avoidance behaviour | As in temperate streams, multiple prey species in a tropical stream showed predator avoidance behaviour to large fish | Layman and Winemiller (2004) |
| Eel growth and behaviour | Elvers (*Anguilla marmorata*: Anguillidae) in a tropical stream grew faster and migrated sooner than in temperate streams | Robinet *et al.* (2003) |

(continued)

TABLE II  (*continued*)

| Response variables | Comparative trend | Reference |
|---|---|---|
| *Macroinvertebrate life histories and production* | | |
| Larval development times for insects | All 35 tropical species examined had faster total development times than (typically univoltine) life histories of temperate stream counterparts | Jackson and Sweeney (1995) |
| Insect emergence from streams | No differences in community composition at order level of insects emerging from temperate and tropical streams | Freitag (2004) |
| Seasonal regulation of life cycles of Chironomidae | Life-cycles of tropical chironomids are much less seasonally regulated than temperate species | Coffman and de la Rosa (1998) |
| Gross primary production | Relatively high in tropical rivers given low nutrient concentrations, likely to be extremely rapid rates of nutrient cycling | Cotner *et al.* (2006) |
| Secondary production of benthic insects | Abundance, biomass and secondary production in a tropical stream were low compared with temperate streams but in some tropical streams population turnover is rapid and larval growth may be very fast | Ramírez and Pringle (1998); Salas and Dudgeon (2002, 2003) |
| *Stream metabolism and organic matter processing* | | |
| Stream metabolism and organic matter dynamics | Net ecosystem production correlated with photosynthetically active radiation and both tropical and temperate streams acted as net sinks of organic matter | Mulholland *et al.* (2001) |
| Responses of macroinvertebrate functional feeding groups to changes in food availability | In tropical streams in New Guinea and Hong Kong, responses of population density and relative abundance of some functional feeding groups to changes in riparian conditions and algal and detrital food availability were weaker than described for temperate streams | Dudgeon (1988, 1994) |
| Use of allochthonous foods by Ephemeroptera nymphs | Based on carbon and nitrogen stable-isotope ratios, mayflies in tropical streams assimilate allochthonous food sources more than in temperate streams | Salas and Dudgeon (2003) |
| Pathways of leaf litter breakdown in streams | Leaf litter breakdown is facilitated by shredders in some temperate streams but as shredders are often rare in tropical streams, a greater role may be played by microbial activity, enhanced by high water temperatures | Irons *et al.* (1994); Dudgeon and Wu (1999); Bass (2003) |

(*continued*)

TABLE II    (*continued*)

| Response variables | Comparative trend | Reference |
|---|---|---|
| Proportion of herbivorous fish | Herbivorous fish species represent a greater proportion of the total fish community in tropical than temperate streams | Wootton and Oemke (1992) |
| Primary energy source for metazoan consumers | Stable-isotope data show that autochthonous primary production entering food webs via algal-grazer or decomposer pathways is the primary annual energy source for consumers in some tropical and temperate streams larger than third order | Thorp and Delong (2002) |
| Trophic guild representation | Tropical and temperate streams contain the same trophic guilds but some tropical streams have relatively lower densities of macroinvertebrate shredders | Bass (2003) |

For further discussion and examples, see text.

(i.e. cumulative species by stones) was achieved at a greater number of stones, and therefore species, in the tropics (Lake *et al.*, 1994).

In another study that used standardized methods, Jacobsen *et al.* (1997) found that lowland tropical streams in Ecuador had more taxa than temperate lowland streams in Denmark whereas highland Ecuadorian and Danish upland streams shared very similar richness (see also Chapter 8 of this volume). The number of insect orders and families increased with maximum stream temperature, and therefore decreased with altitude and latitude. Although this study did not involve a latitudinal gradient of the type investigated by Lake *et al.* (1994), a compilation of published data (Jacobsen *et al.*, 1997) confirmed the relationship with temperature, which may result from a direct temperature effect on speciation, combined with geological history and/or climatic changes (Jacobsen *et al.*, 1997).

It would be expensive to replicate comparisons of taxon richness at many sites across a broad latitudinal gradient. Consequently, sampling at multiple sites has usually been confined to comparisons within major biogeographic regions. Only when the numbers of samples are high enough to encompass within-region variations do such comparisons become convincing. Vinson and Hawkins (2003) have produced the most comprehensive global study of stream invertebrate diversity to date, based on a compilation of data from 495 published estimates of local generic richness of several orders of aquatic insects. They used strict criteria to ensure reliability of diversity estimates, adopting only studies with substantial sampling effort. They restricted their analyses to numbers of genera and to the EPT orders (Ephemeroptera, Plecoptera and Trichoptera), because most studies identified these three taxa to genus and generic identifications are more frequently used than species-level identifications. The results revealed substantial site-to-site variation in richness within realms and biomes. However, the general pattern of richness, and especially the maxima from each region and biome, is quite clear. The summary of findings of Vinson and Hawkins (2003) are set out below:

- Generic richness of Ephemeroptera has peaks at 30°S, 10°N, and 40°N
- Generic richness of Plecoptera peaks at 40°N and 40°S
- Generic richness of Trichoptera is less variable but highest near the equator and 40°N and 40°S

- Richness of EPT taxa declines with increasing elevation, except for Plecoptera
- Maximum richness of EPT taxa increases with estimates of net primary productivity and declines with disturbance
- EPT orders have large variation in richness from site to site within latitudinal realms, but highest richness is recorded in broad-leaf forest biomes
- There is no consistency in the diversity of EPT orders across realms: for example, diversity of Ephemeroptera is highest in the Afrotropical realm; diversity of Plecoptera is highest in the Nearctic; and diversity of Trichoptera is highest in Australia.

These trends reveal numerous biogeographic patterns that potentially confound comparisons of latitudinal gradients (Craig, 2003), and mean that taxonomic comparisons of latitudinal patterns among continents must be made cautiously. However, in some regions, detailed knowledge of insect taxa within continents enables valid comparisons. For example, in Australia, apparently low diversity in Ephemeroptera and Plecoptera contrasts with higher diversity of Trichoptera and Diptera (Lake *et al.*, 1986). In streams of the Queensland Wet Tropics bioregion, Trichoptera are very diverse (Benson and Pearson, 1988; Walker *et al.*, 1995a), as are Chironomidae (Diptera) at the local scale (Pearson *et al.*, 1986) although their species richness is not exceptionally high at the regional scale and does not follow any latitudinal gradient (Cranston, 2000; McKie *et al.*, 2005). In a study of 394 Nearctic and 138 Neotropical sites, McCreadie *et al.* (2005) reported a greater richness of blackflies (Simuliidae) in temperate streams. As regional richness rose, local diversity reached an asymptote where further increases in regional richness did not match increases in local diversity (McCreadie *et al.*, 2005).

Where local site diversity and regional diversity are uncorrelated, then resolution of the issue of whether species richness is higher in the tropics becomes confounded by the interaction between evolutionary/biogeographic processes and local ecological and/or stochastic processes (Craig, 2003). Local diversity is a subset of regional diversity. Regional and historical processes can be important determinants of local patterns of diversity because they determine the characteristics of the species pools from which local communities can be assembled, and they set the upper limit on local species richness (Caley and Schluter, 1997). However, local diversity also depends on a hierarchical framework of multiple spatio-temporal scales (Boyero, 2003), each with an array of physical and biotic factors that can limit diversity (Angermeier and Winston, 1998). Tonn *et al.* (1990) proposed a conceptual framework of environmental 'filters' that represent the processes acting at different spatio-temporal scales that sequentially reduce the regional pool of species to a subset that occur locally. In streams, these filters might include flow regime and associated habitat characteristics (Poff and Ward, 1989) that determine the habitat templet (Townsend and Hildrew, 1994), as well as historical and biogeographic factors such as temporal and spatial isolation (Covich, 1988; Craig, 2003).

## C. Latitudinal Gradients in Regional Diversity in Streams

With increasing taxonomic work in tropical regions, it is becoming feasible to use cumulative records of species number to compare between regions. For example, Boyero (2002) was able to compare species lists for different major regions from North to South America, and found much higher diversity in Central America despite the fact that taxonomic effort is likely to be less in Central America than North America. Such data on species per unit area could be biased by large differences in the area of the regions, but plotting richness data against area of the region and examining the deviations from the regression line (i.e. residuals) is informative (Rosenzweig, 1997). Plots of combined American and Australian data for Ephemeroptera indicate that the main positive outlier is Central America, confirming that it has higher richness (Fig. 1). Further support for this approach can be drawn from the fact that the end-points (Tasmania and North America) have well-described faunas whereas two regions with

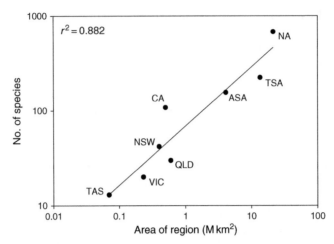

FIGURE 1 Species–area relationship for Ephemeroptera from different regions in the Americas and Australia (R. G. Pearson and L. Boyero, unpublished observations). Region abbreviations: ASA, Austral South America; CA, Central America; NA, North America; NSW, Coastal New South Wales (Australia); QLD, Coastal Queensland (Australia); TAS, Tasmania (Australia); TSA, Tropical South America; VIC, Victoria (Australia).

negative residuals (Queensland and tropical South America) have incompletely described faunas (Christidis, 2003). It is probable that the pattern would vary somewhat among different taxa; for example, Trichoptera and Diptera are particularly diverse in Australia (Pearson *et al.*, 1986; Lake *et al.*, 1994). Using the same approach, Odonata yield a similar trend to Ephemeroptera with tropical regions showing positive deviations from the regression line whereas the negative deviations are represented by temperate data sets (Fig. 2). Relative to its size, Central America apparently has the greatest species richness of these insect orders among the regions studied, and the data from each order represent an independent test of the hypothesis that tropical regions have higher richness of freshwater biodiversity.

Latitudinal trends in regional species diversity may also be asymmetrical. Boyero (2002) compiled total family, generic, and species richness data for Odonata and Ephemeroptera in four

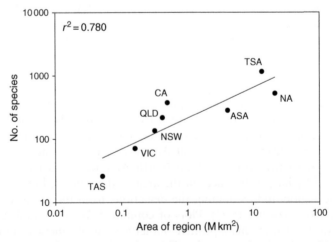

FIGURE 2 Species–area relationship for Odonata from different regions in the Americas and Australia (R. G. Pearson and L. Boyero, unpublished observations). For abbreviations, see Fig. 1.

large zones in the New World: temperate North America, tropical Central America, tropical South America, and temperate South America. The data were then standardized by total area. At all taxonomic levels, the richest zone was Central America with tropical and temperate South America being richer than temperate North America, implying asymmetrical diversity gradients. Clearly, more information is needed from a wider range of geographic regions and a greater variety of taxa, especially non-insect forms since it is clear that some freshwater groups (e.g. decapod crustaceans) are predominately tropical whereas others (isopod and amphipod crustaceans) are mainly temperate.

The detection of regional diversity gradients in riverine fish is easier than for macroinvertebrates due to the relative lack of taxonomic problems, the availability of many good data sets, and the ability to standardize comparisons. Fish species richness, either within continents or globally, is significantly correlated with total surface area of drainage basins and mean annual discharge (Welcomme, 1985; Oberdorff *et al.*, 1995) and also with net primary productivity, using estimates of terrestrial primary productivity as a surrogate for aquatic primary productivity (Oberdorff *et al.*, 1995). The slope of the line relating fish species richness to drainage-basin area is steeper for tropical rivers than temperate rivers (Amarasinghe and Welcomme, 2002), perhaps reflecting the greater per-unit-area richness of stream fishes in the tropics. Stream frog assemblages in New Guinea appear to have a limited number of species (usually about five), but different streams have different complements of species (S. J. Richards, James Cook University, personal communication). Therefore, frog diversity is not particularly high at the local site level, but moving from site to site greatly increases the cumulative number of species. This parallels the higher stone-to-stone accumulation of invertebrate species in the tropics compared to temperate streams (Lake *et al.*, 1994). Overall, these studies indicate regionally high biodiversity but not necessarily local maxima in tropical streams. If functional redundancy is high (i.e. many different species can fill the same niche), it is possible that these differences in biodiversity do not translate into ecological differences between tropical and temperate streams.

## D. Endemism in Tropical and Temperate Streams

Endemism is an inherent feature of patchy environments at a range of scales, regardless of latitude, and is common in streams in mountainous areas where altitudinal climatic gradients may restrict dispersal. Endemism might be hypothesized to be greater in the tropics because tropical mountains span a greater variety of biomes than temperate mountains of equivalent altitude. Even relatively low mountain systems can affect biogeography and endemism. For example, mountains in the Australian tropics are not high (maximum 1700 m asl) but do display high endemism in some groups such as the parastacid crayfish and marked gradients in diversity of Trichoptera (Pearson, 2005). Endemism may be high among many aquatic invertebrate groups in montane regions of Africa, particularly because they are geographically isolated from each other (e.g. Dobson, 2004), but our current poor taxonomic understanding of the region makes definitive statements premature. In contrast, the stream fauna of the Andes exhibits a relatively lower degree of endemism, probably because the mountains lie in a single range with a direct connection to the temperate zone of southern South America (for more information, see Chapter 8).

Geological habitat fragmentation also promotes endemism across oceanic islands. For example, endemism is high on Madagascar and the Caribbean islands. Madagascar is one of the 12 'megadiverse' countries that together harbour 70% of the Earth's plant and animal species (McNeeley *et al.*, 1990), and approximately 90% of the island's freshwater species are endemic (Benstead *et al.*, 2000). Indonesia is also megadiverse, with over 900 species of amphibians and more dragonflies (over 660 species) than any other country (Dudgeon, 2000). The causes of endemicity are probably the same across temperate and tropical latitudes, but the higher regional species diversity typical of many freshwater groups in the tropics probably enhances opportunities for adaptive speciation, leading to higher levels of endemicity. This hypothesis can only be tested adequately as taxonomic knowledge of tropical island stream biotas increases.

## IV. FOOD-WEB ORGANIZATION

### A. Trophic Structure and Taxonomic Representation

Many invertebrate taxa that dominate temperate streams are either rare or absent from streams in the tropics (Table II), and it is likely that such taxonomic 'gaps' may influence the trophic structure of stream food webs. For example, stoneflies (Plecoptera) are very diverse in the Nearctic and Palaearctic regions, including 17 European genera in northwest Africa north of the Sahara, and they also show a remarkable diversity in the Cape region of South Africa. However, across the rest of Africa, this order is represented by only a single genus, *Neoperla* (Perlidae). This low diversity is matched elsewhere in the tropics (Dudgeon, 1999a; Freitag, 2004) and raises the issue as to whether other groups expand to fill the trophic niches typically occupied by stoneflies, or whether the food webs of some of these tropical streams differ from their temperate counterparts – for instance, in terms of their dependence on inputs of riparian leaf litter (see also Chapter 3 of this volume).

Stoneflies are important across all feeding guilds in the Holarctic region where the mayfly family Baetidae, despite its high species richness, is dominated by grazers. However, in tropical Africa, the Baetidae has diversified into a wide variety of feeding guilds not occupied by them in the temperate zone, apparently exploiting the niches unfilled by Plecoptera. For example, there are several tropical genera of predatory baetids (e.g. Lugo-Ortiz and McCafferty, 1998) and at least one shredding genus (Dobson *et al.*, 2002). Evidently some taxa can adapt to vacant niches with the consequence that trophic structure of food webs is fundamentally the same in tropical and temperate streams (Table II), although 'trophic replacement' may involve different groups. The increased diversity of tropical Odonata mentioned earlier could also offer the possibility that members of this predatory order could occupy the niche vacated by the scarcity of large, carnivorous Plecoptera.

The same type of trophic group replacement in tropical streams has been shown for tadpoles. Amphibians are abundant and diverse in the humid tropics, and have been shown to directly influence the availability of algal resources and density of primary consumers (Ranvestel *et al.*, 2004). Studies of tropical stream communities have documented the role of tadpoles as 'ecosystem engineers' and shown how different species serve as detritivores (Flecker *et al.*, 1999) or as grazers that modify benthic habitats used by other species. Wootton and Oemke (1992) experimentally manipulated fish grazing pressure to determine the importance of herbivory by fishes in a Costa Rican stream, reporting that fish reduced macrophyte abundance and periphyton biomass. They concluded that because herbivorous fish species generally represent a larger proportion of the total fish community in tropical streams, they will have a more significant influence on plant and animal abundance than is the case in equivalent temperate habitats. The generality of these results has yet to be demonstrated, but it is notable that tropical streams contain fish functional groups, including specialized benthic herbivores such as Loricariidae (Neotropics) and Balitoridae (Asia) and certain Cyprinidae (e.g. *Garra* spp. in Asian and Africa) that are absent from temperate streams (for details, see Chapter 5 of this volume). It is not yet clear whether these different trophic pathways through fish and tadpoles in tropical streams alter rates of nutrient recycling or other functional processes, and this topic warrants more research (see McIntyre *et al.*, 2007). It also raises questions about functional redundancy and resilience to disturbance; we were unable to find studies comparing these between tropical and temperate streams and rivers, despite the widespread concern over declining populations of amphibians in many regions (Stallard, 2001).

The detrital pathway is the primary source of energy in most heterotrophic streams (Dodds, 2002), and has received substantial attention in temperate systems. In tropical streams, it would be expected that relatively high plant productivity would increase the supply of plant detritus, with consequent effects on detritivorous species that provide essential ecosystem services (e.g. decomposition of organic matter, recycling of nutrients). In this regard, it is significant that when Rosemond *et al.* (1998) used underwater electric fields to exclude macroconsumers feeding on leaf packs in a stream in Costa Rica, an increase in invertebrate densities (mainly collector-gatherers) on leaf packs was not accompanied by any change in decay rates. This result is opposite to what has been observed in temperate streams where insect predators have been shown to control shredder densities and hence leaf pack breakdown rates (e.g. Oberndorfer *et al.*, 1984; Malmqvist, 1993). These observations support the hypothesis that leaf decay in tropical streams may be driven primarily by factors such as microbial activity rather than insect shredding (Rosemond *et al.*, 1998, Table II), and this seems borne out by the apparent paucity of insect shredders in many tropical streams (but see Section V). The matter is discussed in more detail in Chapter 3 of this volume.

Some investigators take the view that the process of leaf breakdown in tropical and temperate streams is similar (Benstead, 1996), despite the fact that macroconsumers eating plant material are much more common in tropical streams (Flecker, 1992; Wootton and Oemke, 1992); this might be interpreted as another example of the trophic replacement described earlier. Omnivorous fish (e.g. Characidae, Cichlidae and Poeciliidae) and shrimps (e.g. Atyidae, Xiphocarididae, and Palaemonidae) play major roles in leaf breakdown in some tropical streams where insect shredders are scarce or absent, as in some Puerto Rican (March *et al.*, 2001, Wright and Covich, 2005) and Costa Rican streams (Pringle and Hamazaki, 1998), while freshwater crabs have a similar function in Kenyan streams (Dobson *et al.*, 2002, Dobson, 2004). Insect shredders are reported to be numerically important in some other tropical streams, such as in Panama (L. Boyero, personal observations), Costa Rica (Benstead, 1996) and the Australian wet tropics (Cheshire *et al.*, 2005), where these insects (especially the Leptoceridae and Calamoceratidae) make a potentially important contribution to litter breakdown (Bastian *et al.*, 2007).

## B. Food-Web Compartmentalization in Tropical and Temperate Streams

As we saw earlier for the hydrological comparisons, generalizations across tropical and temperate streams must take into account the spatial scales and sizes of lotic systems, and this need for appropriate scaling is equally true for trophic structure and food webs. In many tropical streams, steep waterfalls limit fish access to headwaters in coastal drainages (Covich *et al.*, 2006). These forested streams are dominated both numerically and in biomass by diverse species of omnivorous decapods (Crowl and Covich, 1994, Covich *et al.*, 2003) that carry out a range of trophic roles and whose activities also influence rates of organic-matter breakdown and primary productivity in ways not evident in temperate streams. Low-elevation tropical rivers have a high number of fish species, which tend to displace shrimps and have partially radiated to fill their niches. Along these tropical streams, the landscape imposes a type of food-web compartmentalization, with predation from fish having large impacts on invertebrates at low elevations whereas invertebrates dominate energy flow at middle and upper elevations, above barriers to fish dispersal. Although such distinct compartmentalization has yet to be described for temperate streams, and it may not be a special feature of tropical streams (Layman and Winemiller, 2004; see also Table II), the importance of decapods and greater functional diversity of fish at lower latitudes could give rise to some thus far unperceived differences in stream function.

## V. PRODUCTIVITY, ORGANIC MATTER DYNAMICS, AND OTHER ECOSYSTEM PROCESSES

### A. Productivity

Productivity in permanently-flowing tropical streams might be anticipated to be higher than that in equivalent temperate streams because continuously high temperatures allow year-round growth of both aquatic and surrounding terrestrial vegetation, while the persistent presence of water in many permanent streams reduces the impact of rainfall seasonality. Consistent high levels of insolation lead to continual primary production. In contrast, productivity in permanent temperate streams occurs mainly in spring and summer, and so total annual production is predicted to be less in equivalent-sized streams. In a comparison of stream metabolism among biomes in North America using directly comparable methods (Mulholland *et al.*, 2001), gross primary production was correlated most strongly with photosynthetically-active radiation, but was also limited by phosphorus. Highest primary production was measured in a temperate desert stream and productivity was least in a tropical stream (Quebrada Bisley, Puerto Rico). Although this stream had a closed canopy, shaded temperate streams included in the comparison study achieved higher gross primary productivities (Mulholland *et al.*, 2001). Other studies in tropical and temperate streams confirm that primary productivity is typically limited by light and nutrients, regardless of latitude (Pringle and Triska, 1991; Pearson and Connolly, 2000; Montoya *et al.*, 2006). Primary production in some well-lit tropical streams may be limited by intense herbivory that strongly suppresses autotrophic biomass (Ortiz-Zayas *et al.*, 2005). In such situations, even in full sunlight, community metabolism depends substantially upon allochthonous detritus from upstream forested stream reaches.

Secondary production in tropical streams is infrequently researched, making comparisons with temperate zones difficult. Macroinvertebrate abundance is often high in tropical streams, especially where fish predators are excluded by waterfalls (Covich, 1988; Covich *et al.*, 2006). When these high densities are combined with the more rapid turnover generally encountered in tropical freshwaters (e.g. Jackson and Sweeney, 1995, Ramírez and Pringle, 1998), secondary production would typically be expected to be higher than in temperate systems. At first glance, this appears to be not the case. In one of the few such studies, Ramírez and Pringle (1998) demonstrated that secondary insect production in Sábalo Stream, Costa Rica, estimated at 363.7 mg AFDM $m^{-2}$ $yr^{-1}$, was low compared with that in similar-sized temperate streams. They speculated that this was due to the low abundance of insects (a maximum of 2188 individuals per $m^2$) which was not typical of tropical streams, but a consequence of the high predation intensities at their study site where three species of omnivorous *Macrobrachium* (Palaemonidae) shrimps were abundant (Pringle and Hamazaki, 1998). Besides reducing insects by predation and competition (Ramírez and Hernandez-Cruz, 2004), these shrimps would also have made a significant contribution to macroinvertebrate secondary production. This has been shown to be the case elsewhere in the tropics (e.g. Bright, 1982; Mantel and Dudgeon, 2004; Yam and Dudgeon, 2006), and shrimp production may be a very significant proportion of total benthic production, especially in comparison with temperate streams where shrimps are absent. There have been few other studies of secondary benthic productivity in tropical streams, but there is some evidence of inter-year taxon-specific variation in production related to rainfall (Dudgeon, 1999b). A detailed account of macroinvertebrate production in tropical streams is included in Chapter 4 of this volume.

Many tropical stream insects are multivoltine and have rapid growth, short generation times, and high production: biomass ratios (e.g. Jackson and Sweeney, 1995; Salas and Dudgeon, 2002, 2003). This suggests that estimates of secondary production based on assumptions about univoltine life histories or uncritical application of standard methods for temperate streams will significantly underestimate production of tropical stream insects. However, high turnover rates do not always result in high production. Benke (1998) recorded that extremely high turnover rates of chironomids in a subtropical river were matched by high net production. This could reflect short larval life spans and may be interpreted as a strategy for species living in streams frequently disturbed by floods, where mortality is high due to abiotic factors (Huryn and Wallace, 2000). This type of adaptation has been reported in tropical lowland rivers by Ramírez and Pringle (2006), and parallels the situation observed in frequently-disturbed streams in other climatic zones (e.g. desert streams in Arizona: Grimm and Fisher, 1989).

## B. Organic Matter Dynamics: Seasonal Inputs and Instream Litter Retention

In forested tropical streams, as in the temperate zone, allochthonous inputs of detritus are typically the major source of energy. Leaf-litter dynamics in all streams are determined by seasonality of leaf fall and by retention within river channels. Leaves entering the stream are conditioned by microbial colonists, and broken down by physical abrasion and the feeding activities of invertebrates, especially 'shredders' that are capable of comminuting coarse organic matter (>1 mm in size, Cummins and Klug, 1979). This model of leaf breakdown is widely supported by studies in the north-temperate zone but does it hold true in humid and wet-dry tropical streams?

In deciduous forests of the north-temperate zone, leaf litter input is strongly seasonal, falling during a short period of time in the autumn, often associated with increased flows as rainfall increases (Dodds, 2002). In evergreen forests of the southern temperate zone, leaf fall is less seasonal although for many eucalypt species there is a distinct peak of litter drop in summer, coinciding with low or zero flows (Bunn, 1986). When flow resumes or increases, litter that has accumulated on the streambed is carried in a pulse and is now available to aquatic biota (Boulton, 1991). In the humid tropics, however, inputs via leaf fall may continue throughout the year (Benson and Pearson, 1993; Larned, 2000) and, coupled with the relatively stable flows, allochthonous organic matter supply is not as pulsed as in many temperate streams (see also Chapter 3 of this volume).

Conversely, in the wet-dry tropics, leaf fall often occurs at the beginning of the dry season when river flows are declining (Douglas, 1999), resembling the southern temperate streams described above. Again, this may leave large volumes of leaf litter sitting in the river channel for several months, until the rains return. Terrestrial breakdown processes including microbial conditioning and invertebrate attack are likely to be accelerated by the high temperatures typical of the tropics but fire may also be a significant factor, consuming the leaf litter but also enhancing litter fall through 'scorching'. For example, fires in the savannahs of the wet-dry tropics of northern Australia killed vines and woody species, reduced the canopies of surviving trees, and caused a shedding of leaves several days after the fire had passed (Douglas *et al.*, 2003). Although the effect was to increase the amount of leaf litter in the riparian zone and on the dry streambed, standing crops of leaf litter were still lower than in unburnt streams. However, in the burnt streams, the lack of vines and other retentive structures meant that litter was readily carried into the stream so that, after flow resumed, standing crops of organic matter in burnt streams exceeded those in unburnt streams. By the end of the flow phase, standing crops of benthic organic matter were similar between burnt and unburnt streams because the destruction of debris dams during the fire had reduced in-stream retention of litter, enhancing export from the burnt streams (Douglas *et al.*, 2003).

In-stream retention dynamics of leaf litter are expected to show no differences between tropical and temperate latitudes because of the overriding influences of flow patterns on export and retention (Brookshire and Dwire, 2003; see also Chapter 3 of this volume). Mathooko *et al.* (2001) demonstrated the importance of rocky outcrops, the stream edge, roots and particularly debris dams in capturing entrained leaves in the Njoro River, Kenya. These features have been highlighted as key retention structures in temperate streams (Speaker *et al.*, 1984, Webster *et al.*, 1987), and the same holds true for lateral transport and deposition on tropical and temperate floodplains (Junk *et al.*, 1989, Jones and Smock, 1991; see also Chapter 7 of this volume). Similarly, Afonso and Henry (2002) identified the same relationship between retention and discharge in a Brazilian tropical stream as has been found in equivalent temperate streams, noting the loss of organic matter that occurs during high flows when retentive structures are submerged or carried downstream. Unfortunately, there is a dearth of studies of transport distance and deposition velocity of coarse particulate organic matter in tropical streams (see Table 4 in Brookshire and Dwire, 2003), which limits our ability to compare litter dynamics in tropical and temperate streams. The limited published information is summarized in Chapter 3 of this volume.

## C. Organic Matter Dynamics: Biologically Mediated Breakdown

The prevailing model of leaf breakdown in streams emphasizes the central role played by microbes in 'conditioning' the litter and improving its palatability for detritivorous invertebrates, including shredders such as some tipulid larvae (Diptera), particular Plecoptera and Ephemeroptera nymphs, and certain case-building Trichoptera larvae (Dodds, 2002). Irons *et al.* (1994) proposed that microbial activity was the main determinant of processing rate at low latitudes whereas macroinvertebrates were more important at higher latitudes. The greater efficiency of microbial activity as a consequence of higher water temperatures therefore leads to higher breakdown rates in the tropics than in temperate regions (Irons *et al.*, 1994; see also Chapter 3 of this volume).

An apparent fundamental difference between tropical and temperate streams is the paucity of shredding invertebrate detritivores in many parts of the tropics. This is a widespread phenomenon, having been identified from Hong Kong (Dudgeon and Wu, 1999) and elsewhere in tropical Asia (Dudgeon, 1999a), New Guinea (Dudgeon, 1994; Yule, 1996), Central America (Rosemond *et al.*, 1998), Brazil (Walker, 1987), Colombia (Mathuriau and Chauvet, 2002) and East Africa (Tumwesigye *et al.*, 2000; Dobson *et al.*, 2002), although it is not true of tropical Australian rainforest streams (Pearson and Tobin, 1989; Pearson *et al.*, 1989; Nolen and Pearson, 1993; Cheshire *et al.*, 2005). This apparent absence of a feeding guild from much of the tropics – despite its importance in temperate streams – requires an explanation. The shredders that dominate the north temperate zone are concentrated into relatively few higher taxa, most notably certain Plecoptera and Trichoptera, plus amphipod crustaceans. These are taxa that probably evolved in cool running waters so one hypothesis to explain their absence from the tropics is a lack of tolerance for consistently high water temperatures (Irons *et al.*, 1994).

Shredders may be scarce or absent from tropical streams for several reasons that are not mutually exclusive. First, the biomass of detritus may be too low or its pattern of input too unpredictable to support viable shredder populations. Although very few estimates of leaf litter biomass and dynamics have been carried out in tropical streams, all suggest that detritus is at least as abundant as in equivalent temperate streams (review in Dobson *et al.*, 2002; see also Chapter 3 of this volume). Secondly, increased microbial activity may also be a reason for reduced shredder numbers in the tropics (Irons *et al.*, 1994). If microorganisms and macroinvertebrates are in direct competition for a limiting resource, then consistently high temperatures may give the microbial component the competitive advantage. Although there are few data to support or refute this idea, Mathuriau and Chauvet (2002) demonstrated the high biomass of hyphomycete fungi in leaf litter decomposing in Colombia and, crucially, the rapid rate of accrual of this biomass. However, this hypothesis that microbes and macroinvertebrates are competing for resources would only apply if leaf litter inputs were low or highly pulsed; high microbial activity combined with a constant supply of fresh leaf litter would surely be advantageous to shredding invertebrates, as there would be a constant supply of conditioned detritus. Thus far, the evidence does not support the notion that competition from microbes limits the abundance of shredders in tropical streams (see also Chapter 3).

A third explanation is that the inputs are lower in quality than those in temperate zones because there are high concentrations of toxic compounds in the leaves of many tropical species (Wantzen *et al.*, 2002; see Chapters 3 and 7). The quality of leaf litter as a food resource for shredders can be estimated by its breakdown rate (Petersen and Cummins, 1974). Breakdown rates from the tropics are generally high (e.g. Dudgeon and Wu, 1999; Larned, 2000; Dobson *et al.*, 2004) and include the highest rates recorded in the literature (Irons *et al.*, 1994, Mathuriau and Chauvet, 2002). The relative palatability of leaf types in temperate and tropical zones has been compared using reciprocal feeding experiments (Graça *et al.*, 2001). In that study, *Alnus glutinosa* (Betulaceae) leaves from Germany and *Hura crepitans* (Euphorbiaceae) leaves from Venezuela were offered to European and South American shredders. All animals preferred conditioned over unconditioned leaves, and although the South American shredders preferred the temperate *Alnus* leaves, *Gammarus* amphipods from Germany did not distinguish between the leaf types. Furthermore, all shredder species grew significantly irrespective of the leaf species upon which they were fed. However, the German leaves were derived from a forest dominated by only two tree species, whereas the Venezuelan forest contained at least seven other riparian species (Graça *et al.*, 2001). Total shredder activity will be determined by the overall quality of leaf litter entering the channel, so the palatability of all common species needs to be assessed. Dobson *et al.* (2004) found that all the riparian leaf types decomposed rapidly in the Njoro River, Kenya, and hence were probably highly palatable. In an Australian stream, however, palatability has been shown to vary substantially among species (Pearson and Tobin, 1989; Nolen and Pearson, 1993; Bastian *et al.*, 2007), as would be expected in a system where there may be scores of species contributing litter to the stream since each may have various defences against terrestrial herbivores (Benson and Pearson, 1993). Palatability can also vary depending on whether the leaves come from the canopy or from the understorey, as the latter may be less well defended by secondary compounds (Downum *et al.*, 2001). In any case, whether leaf secondary compounds alter the relative importance or role of shredding macroinvertebrates in leaf decomposition in tropical streams is still unresolved (Wright and Covich, 2005).

A fourth explanation is that shredding taxa really are present, but have been overlooked. This hypothesis deserves more research because there is some evidence that some tropical taxa found on leaf packs are not normally associated with shredding elsewhere. For example, the baetid mayflies of the genus *Acanthiops* in the East African highlands are shredders (Dobson *et al.*, 2002), although baetids in north-temperate streams are grazers. More widespread are freshwater crabs, generally overlooked or significantly under-recorded by standard benthic sampling methods, are often extremely common, at least in sub-Saharan Africa, where they can account for 80% or more of the entire benthic invertebrate biomass (Dobson, 2004). Analyses of gut contents of several species of freshwater crabs have demonstrated that their diet is dominated by shredded detritus (Dobson, 2004). These figures are derived from a very small number of studies but if this pattern is consistent across the continent, then shredders, far from being rare, would be significantly more important in tropical Africa than in most temperate zones. Shrimps and fishes in tropical streams may also play key roles in processing leaf litter, and atyid and palaemonid shrimps which are abundant in almost all tropical streams may be especially important in this regard. For example, in Puerto Rican streams, leaf litter decomposes rapidly in the presence of shrimps (Crowl *et al.*, 2001; March *et al.*, 2001; see also Chapter 4 of this volume). In Australia, parastacid crayfish are important consumers of leaf litter, including green leaves that are washed into rainforest streams during severe storms (J. Coughlan and R. G. Pearson, unpublished data).

Finally, it should be noted that the paucity of even 'typical' shredders in the tropics is not universal. For example, Graça *et al.* (2001) reported that two shredding caddisflies are abundant in streams in northern Venezuela, while Benstead (1996) recorded high densities of insect shredders in Costa Rican streams, albeit of only three taxa. In Queensland rainforest streams in the Australian tropics, shredders also seem to be quite common (Cheshire *et al.*, 2005), and much litter processing is carried out by macroinvertebrates (Pearson *et al.*, 1989; Nolen and Pearson, 1993; Bastian *et al.*, 2007). These observations give some support to the hypothesis that shredders (especially crabs and shrimps) have been overlooked, partly because of expectations of the taxa that might be involved have been based solely on the ubiquitous model of leaf-litter processing developed in northern temperate systems.

## D. Nutrient Cycling and Limitation

Most of our knowledge about nutrient cycling and limitation is derived from work in temperate streams. Effort has focused on phosphorus because this has been shown to limit primary production in many rivers (e.g. Fairchild *et al.*, 1985; Pan and Lowe, 1994), including tropical streams (Pringle *et al.*, 1986; Larned and Santos, 2000). Although concentrations of nutrients in tropical streams are typically considered to be very low due to intensive weathering (as described earlier), geothermal activity in some streams such as those at La Selva, Costa Rica, increases phosphorus concentrations of in-welling groundwater, enhancing rates of microbial respiration (Ramírez *et al.*, 2003) but not necessarily affecting insect assemblage composition (Ramírez *et al.*, 2006). Where phosphorus enrichment of tropical streams has taken place, respiration rates can exceed those in the temperate streams (Webster *et al.*, 1995). Such differences in nutrient concentrations and their effects on microbial respiration arise from local factors such as groundwater inputs and catchment geology, and do not seem to be attributable to latitude.

Like phosphorus, nitrogen has been shown to limit or co-limit primary productivity in temperate (Hill and Knight, 1988), subtropical (Mosisch *et al.*, 2001), and tropical (Ramos-Escobedo and Vazquez, 2001) streams. Downing *et al.* (1999) suggest that nitrogen limitation is more common in the tropics, but nitrogen dynamics appear to be similar across temperate and tropical streams. The few studies undertaken with $^{15}$N tracers in temperate streams (Mulholland *et al.*, 2000; Tank *et al.*, 2000) yield uptake rates of ammonium and nitrate comparable to those reported from Quebrada Bisley stream in Puerto Rico (Merriam *et al.*, 2002). Data are too scarce to generalize but differences in $^{15}$N balance appear to be more related to stream retentiveness (lack of woody debris in the tropical stream) and size than to latitude. A similar conclusion was drawn by Webster *et al.* (2003): a comparison of ammonium uptake among streams at different latitudes in North America produced no statistically-significant relationships.

## E. Responses to Natural Disturbances and Introduced Species

Many tropical streams experience cyclones and hurricanes (e.g. Covich *et al.*, 1991) as well as drought (Covich *et al.*, 2003, 2006), and these natural disturbances seem just as prevalent in temperate systems where their effects are broadly similar. For example, floods resulting from torrential rains during cyclones scour benthic plants and animals from headwater streams (Maltchik and Pedro, 2001) and redistribute sediments, whereas droughts cause differential mortality, localized crowding due to habitat contraction (Crowl and Covich, 1994), and a prolonged decrease in reproductive output of some taxa (Covich *et al.*, 2003). Rates of faunal and biogeochemical recovery also appear to be similar and rather swift (McDowell *et al.*, 1996; Shivoga, 2001), reflecting the adaptations of the local biota to the prevailing disturbance regime. Inhabitants of riffles in upland streams in the tropics experience a similar range of disturbances to those in temperate zones, although the intensity of tropical storms may reduce the predictability of events compared with, say, streams fed by snowmelt (Rosser and Pearson, 1995). Streams in the aseasonal tropics may experience a significant range of disturbances on a daily basis (Yule and Pearson, 1996).

There have been very few studies of disturbance in large tropical rivers although there are grave concerns about potential effects of dams and other hydrological alteration (Dudgeon,

2000; Pringle *et al.*, 2000b; Wantzen *et al.*, 2006). Ribbink (1994) assessed the role of natural disturbance in large tropical river basins, and showed that African river systems generally support low fish diversity, with the exception of the Congo. While this is partially due to the high habitat diversity of the Congo system, it also arises from its equatorial location. Dry seasons in one section of the catchment are countered by wet seasons in the other, resulting in a relatively constant flow in the main catchment area. In contrast, other major river systems beyond the equator exhibit greater seasonality and major fluctuations in discharge, so fishes must be opportunists adapted to frequent large-scale disturbance (Ribbink, 1994).

While the negative effects of exotic or introduced species on biota and ecosystem processes in temperate streams are well known (e.g. Allan and Flecker, 1993), responses to species introduced into tropical streams have been less documented although they appear to be similar. Processes such as litter decomposition that rely upon living organisms are susceptible to changes wrought by the introduction of alien species. For example, in eastern and southern Africa, Rainbow trout (*Oncorhynchus mykiss*: Salmonidae) was widely introduced into highland streams during the first half of the 20th century, often into streams that were previously fishless (see also Chapter 8 for discussion of trout in South America). Anecdotal evidence suggests that the arrival of trout led to a large reduction in density of crabs as a result of predation (see Dobson, 2004). As crabs appear to be key members of the detritus-processing guild (see above), depletion in their numbers would inevitably reduce processing efficiencies of organic matter. Conversely, the North American crayfish *Procambarus clarkii* (Cambaridae), introduced for aquaculture in various parts of Africa, has escaped into the wild and is gradually extending its range (Dobson, 2004). *P. clarkii* acts as a shredder in Hawaiian mesocosm studies and is more effective than native shrimps (Larned *et al.*, 2003). Other concerns over the introduction of *P. clarkii* include its effect in out-competing endemic crustaceans in locations where it has become abundant, such as the shrimp *Macrobrachium grandimanus* (Palaemonidae) in Hawaii (Larned *et al.*, 2003).

Theories that diverse communities are more resistant to invasions by introduced species are being challenged (Levine, 2000), and it seems that biodiverse tropical streams are no less susceptible than their temperate counterparts. Thus far, the impacts of invasive species on communities have been most severe on low-diversity systems including islands and highly disturbed habitats (Kolar and Lodge, 2001), although there are exceptions to this (e.g. Lake Victoria). Tropical streams face a number of threats from human disturbance (as described elsewhere in this volume, especially in Chapter 10) and their repercussions are likely to be at least as serious as those reported for temperate systems.

## VI. CONCLUSIONS

Comparisons of temperate and tropical streams are confounded by immense variability inherent in these systems within these latitudes, and the wide range of climatic and hydrological conditions. Few robust generalizations can be made, and those that can are rather self-evident. Streams in the tropics typically receive higher insolation and more intense rainfall, have warmer water and often large predictable floods, and for many taxa, harbour higher biodiversity than their temperate equivalents. However, our review of the literature fails to reveal consistent differences in food-web structure, productivity, organic-matter processing or nutrient dynamics, and response to disturbance (Table II), and there is no persuasive evidence that the adjective 'tropical' has particular significance when applied to stream ecology. Instead, ecological processes in tropical streams tend to be driven by the same variables that are important in temperate ones. For example, biotic responses to drought and flooding are similar to those in temperate streams while productivity is limited by the same variables of nutrients, shading, and trophic structure. Consequently, valid extrapolation of models and management strategies may not be so much an issue of latitude (tropical versus temperate) but of ensuring suitable comparability at an appropriate (stream) scale. This review demonstrates clearly that whereas ecological mechanisms may be similar, the organisms involved in ecological processes in tropical and temperate streams can and do differ. It will be necessary to progress beyond the taxonomic biases arising from the results of investigations undertaken in the temperate (particularly north-temperate) latitudes when trying to explain phenomena observed in tropical streams.

## ACKNOWLEDGMENTS

We thank David Dudgeon, Alonso Ramírez, and an anonymous reviewer for useful comments that improved the final text. We are also grateful to our respective institutions for support and funding during the preparation of this manuscript.

## REFERENCES

Afonso, A.A.O., and Henry, R. (2002). Retention of particulate organic matter in a tropical headwater stream. *Hydrobiologia* 482, 161–166.

Ahmand, R., Scatena, F.N., and Gupta, A. (1993). Morphology and sedimentation in Caribbean montane streams: examples from Jamaica and Puerto Rico. *Sed. Geol.* 85, 57–169.

Alagbe, S.A. (2002). Groundwater resources of River Kan, Gimi Basin, north-central, Nigeria. *Env. Geol.* 42, 404–413.

Allan, J.D., and Flecker, A.S. (1993). Biodiversity conservation in running waters. *BioScience* 43, 32–43.

Amarasinghe, U.S., and Welcomme, R.L. (2002). An analysis of fish species richness in natural lakes. *Env. Biol. Fish.* 65, 327–339.

Angermeier, P.L., and Winston, M.R. (1998). Local *vs.* regional influences on local diversity in stream fish communities of Virginia. *Ecology* 79, 911–927.

Arrington, D.A., and Winemiller, K.O. (2006). Habitat affinity, the seasonal flood pulse, and community assembly in the littoral zone of a Neotropical floodplain river. *J. N. Am. Benthol. Soc.* 25, 126–141.

Arthington, A.H. (1990). Latitudinal gradients in insect species richness of Australian lotic systems: a selective review. *Trop. Freshwat. Biol.* 2, 179–196.

Balek, J. (1983). "Hydrology and Water Resources in Tropical Regions." Elsevier, Amsterdam, The Netherlands.

Bass, D. (2003). A comparison of freshwater macroinvertebrate communities on small Caribbean islands. *BioScience* 53, 1094–1100.

Bastian, M., Boyero, L., Jackes, B.R., and Pearson, R.G. (2007). Leaf litter diversity and shredder preferences in an Australian tropical rain-forest stream. *J. Trop. Ecol.* 23, 219–229.

Baumgartner, A., and Reichel, E. (1975). "The World Water Balance. Mean Annual Global, Continental and Maritime Precipitation, Evaporation and Run-off." Elsevier, Amsterdam, The Netherlands.

Benda, L., and Dunne, T. (1997). Stochastic forcing of sediment supply to channel networks from landsliding and debris flow. *Water. Resour. Res.* 32, 2849–2863.

Benke, A.C. (1998). Production dynamics of riverine chironomids: extremely high biomass turnover rates of primary consumers. *Ecology* 79, 899–910.

Benke, A.C., Chaubey, I., Ward, G.M., and Dunn, E.L. (2000). Flood pulse dynamics of an unregulated river floodplain in the southeastern US coastal plain. *Ecology* 81, 2730–2741.

Benson, L.J., and Pearson, R.G. (1988). Diversity and seasonality of adult Trichoptera captured in a light-trap at Yuccabine Creek, a tropical Australian rainforest stream. *Aust. J. Ecol.* 13, 337–344.

Benson, L.J., and Pearson, R.G. (1993). Litter inputs to a tropical Australian upland rainforest stream. *Aust. J. Ecol.* 18, 377–383.

Benstead, J.P. (1996). Macroinvertebrates and the processing of leaf litter in a tropical stream. *Biotropica* 28, 367–375.

Benstead, J.P., Stiassny, M.L.J., Loiselle, P.V., Riseng, K.J., and Raminosoa, N. (2000). River conservation in Madagascar. *In* "Global Perspectives on River Conservation: Policy and Practice" (P.J. Boon, B.R. Davies and G.E. Petts, Eds), pp. 205–231, John Wiley & Sons, Chichester, UK.

Bishop, J.E. (1973). "Limnology of a Small Malayan River Sungai Gombak." Dr. W. Junk Publishers, The Hague, The Netherlands.

Boulton, A.J. (1991). Eucalypt leaf decomposition in an intermittent stream in south-eastern Australia. *Hydrobiologia* 211, 123–136.

Boyero, L. (2002). Insect biodiversity in freshwater ecosystems: is there any latitudinal gradient? *Mar. Freshwat. Res.* 53, 753–755.

Boyero, L. (2003). Multiscale patterns of spatial variation of stream macroinvertebrate communities. *Ecol. Res.* 18, 365–379.

Bright, G.R. (1982). Secondary benthic production in a tropical island stream. *Limnol. Oceanogr.* 27, 472–480.

Brookshire, E.N.J., and Dwire, K.A. (2003). Controls on patterns of coarse organic particle retention in headwater streams. *J. N. Am. Benthol. Soc.* 22, 17–34.

Bunn, S.E. (1986). Origin and fate of organic matter in Australian upland streams. *In* "Limnology in Australia" (P. De Deckker and W.D. Williams, Eds), pp. 277–291, CSIRO/Dr W. Junk Publishers, Melbourne/Dordrecht.

Buzas, M.A., Collins, L.S., and Culver, S.J. (2002). Latitudinal differences in biodiversity caused by higher tropical rate of increase. *Proc. Nat. Acad. Sci.* 99, 7841–7843.

Caley, M.J., and Schluter, D. (1997). The relationship between local and regional diversity. *Ecology* 78, 70–80.

Carillo, M., Estrada, E., and Hazen, T.C. (1985). Survival and enumeration of the fecal indicators *Bifidobacterium adolescentis* and *Escherichia coli* in a tropical rainforest watershed. *Appl. Env. Microbiol.* 50, 468.

Cheshire, K., Boyero, L., and Pearson, R.G. (2005). Food webs in tropical Australian streams: shredders are not scarce. *Freshw. Biol.* 50, 748–769.

Chestnut, T.J., and McDowell, W.H. (2000). C and N dynamics in the riparian and hyporheic zones of a tropical stream, Luquillo Mountains, Puerto Rico. *J. N. Am. Benthol. Soc.* 19, 199–214.

Christidis, F. (2003). Systematics, Phylogeny and Ecology of Australian Leptophlebiidae (Ephemeroptera). Ph.D. Thesis, James Cook University, Townsville, Australia.

Coffman, W.P., and de la Rosa, C.L. (1998). Taxonomic composition and temporal organization of tropical and temperate species assemblages of lotic Chironomidae. *J. Kans. Entomol. Soc.* 71, 388–406.

Cotner, J.B., Montoya, J.V., Roelke, D.L., and Winemiller, K.O. (2006). Seasonally variable riverine production in the Venezuelan llanos. *J. N. Am. Benthol. Soc.* 25, 171–184.

Covich, A.P. (1988). Geographical and historical comparisons of Neotropical streams: biotic diversity and detrital processing in highly variable habitats. *J. N. Am. Benthol. Soc.* 7, 361–386.

Covich, A.P., Crowl, T.A., Johnson, S.L., Varza, D., and Certain, D.L. (1991). Post-hurricane Hugo increases in atyid shrimp abundance in a Puerto Rican montane stream. *Biotropica* 23, 448–454.

Covich, A.P., Crowl, T.A., and Scatena, F.N. (2003). Effects of extreme low flows on freshwater shrimps in a perennial tropical stream. *Freshw. Biol.* 48, 1199–1206.

Covich, A.P., Crowl, T.A., and Heartsill-Scally, T. (2006). Effects of drought and hurricane disturbances on headwater distributions of palaemonid river shrimp (*Macrobrachium* spp.) in the Luquillo Mountains, Puerto Rica. *J. N. Am. Benthol. Soc.* 25, 99–107.

Craig, D.A. (2003). Geomorphology, development of running water habitats, and evolution of black flies on Polynesian islands. *BioScience* 53, 1079–1093.

Cranston, P.S. (2000). August Thienemann's influence on modern chironomidology – an Australian perspective. *Verh. Internat. Verein. Limnol.* 27, 278–283.

Cranston, P.S., Cooper, P.D., Hardwick, R.A., Humphrey, C.L., and Dostine, P.L. (1997). Tropical acid streams – The chironomid (Diptera) response in northern Australia. *Freshw. Biol.* 37, 473–483.

Crow, G.E. (1993). Species diversity in aquatic angiosperms: latitudinal patterns. *Aquat. Bot.* 44, 229–258.

Crowl, T.A., and Covich, A.P. (1994). Responses of freshwater shrimp to chemical and tactile stimuli from large decapod predators: implications for habitat selection. *J. N. Am. Benthol. Soc.* 13, 291–298.

Crowl, T.A., McDowell, W.H., Covich, A.P., and Johnson, S.L. (2001). Freshwater shrimp effects on detrital processing and nutrients in a tropical headwater stream. *Ecology* 82, 775–783.

Cummins, K.W., and Klug, M.J. (1979). Feeding ecology of stream invertebrates. *Annu. Rev. Ecol. Syst.* 10, 147–172.

Dai, A.G., and Trenberth, K.E. (2002). Estimates of freshwater discharge from continents: latitudinal and seasonal variations. *J. Hydrometeorol.* 3, 660–687.

Davis, M. (2001). "Late Victorian Holocausts. El Niño Famines and the Making of the Third World." Verso, London, UK.

Dobson, M. (2004). Freshwater crabs in Africa. *Freshwat. Forum* 21, 3–26.

Dobson, M., Magana, A., Mathooko, J.M., and Ndegwa, F.K. (2002). Detritivores in Kenyan highland streams: more evidence for the paucity of shredders in the tropics? *Freshw. Biol.* 47, 909–919.

Dobson, M., Mathooko, J.M., Ndegwa, F.K., and M'Erimba, C. (2004). Leaf litter processing rates in a Kenyan highland stream, the Njoro River. *Hydrobiologia* 519, 207–210.

Dodds, W.K. (2002). "Freshwater Ecology: Concepts and Environmental Applications." Academic Press, San Diego, USA.

Donaldson, T.J., and Myers, R.F. (2002). Insular freshwater fish faunas of Micronesia: patterns of species richness and similarity. *Env. Biol. Fish.* 65, 139–149.

Douglas, M.M. (1999). Tropical Savannah Streams: Their Seasonal Dynamics and Response to Catchment Disturbance. Ph.D. Thesis, Monash University, Melbourne, Australia.

Douglas, M.M., Townsend, S.A., and Lake, P.S. (2003). Streams. *In* "Fire in Tropical Savannahs. The Kapalga Experiment" (A.N. Andersen, G.D. Cook and R.J. Williams, Eds), pp. 59–78, Springer–Verlag, New York, USA.

Downing, J.A., McClain, M., Twilley, R., Melack, J.M., Elser, J., Rabalais, N.N., Lewis, W.M. Jr., Turner, R.E., Corredor, J., Soto, D., Yanez-Arancibia, A., Kopaska, J.A., and Howarth, R.W. (1999). The impact of accelerating land-use change on the N-cycle of tropical aquatic ecosystems: current conditions and projected changes. *Biogeochemistry* **46**, 109–148.

Downum, K., Lee, D., Hallé, F., Quirke, M., and Towers, N. (2001). Plant secondary compounds in the canopy and understorey of a tropical rain forest in Gabon. *J. Trop. Ecol.* **17**, 477–481.

Dudgeon, D. (1988). The influence of riparian vegetation on macroinvertebrate community structure in four Hong Kong streams. *J. Zool. (Lond.)* **216**, 609–627.

Dudgeon, D. (1994). The influence of riparian vegetation on macroinvertebrate community structure and function in six New Guinea streams. *Hydrobiologia* **294**, 65–85.

Dudgeon, D. (1999a). "Tropical Asian streams: Zoobenthos, Ecology and Conservation." Hong Kong University Press, Hong Kong SAR, China.

Dudgeon, D. (1999b). Patterns of variation in secondary production in a tropical stream. *Arch. Hydrobiol.* **144**, 271–281.

Dudgeon, D. (2000). Large-scale hydrological changes in tropical Asia: prospects for riverine biodiversity. *BioScience* **50**, 793–806.

Dudgeon, D., and Wu, K.K.Y. (1999). Leaf litter in a tropical stream: food or substrate for macroinvertebrates? *Arch. Hydrobiol.* **146**, 65–82.

Dudgeon, D., Choowaew, S., and Ho, S.-C. (2000). River conservation in south-east Asia. *In* "Global Perspectives on River Conservation: Policy and Practice" (P.J. Boon, B.R. Davies and G.E. Petts, Eds), pp. 281–310, John Wiley & Sons, Chichester, UK.

Fairchild, G.W., Lowe, R.L., and Richardson, W.B. (1985). Algal periphyton growth on nutrient-diffusing substrata: an *in situ* bioassay. *Ecology* **66**, 465–472.

Faniran, A., and Areola, O. (1978). "Essentials of Soil Study with Special Reference to Tropical Areas". Heinemann, London, UK.

Flecker, A.S. (1992). Fish trophic guilds and the structure of a tropical stream – weak direct vs strong indirect effects. *Ecology* **73**, 927–940.

Flecker, A.S., and Feifarek, B. (1994). Disturbance and the temporal variability of invertebrate assemblages in 2 Andean streams. *Freshw. Biol.* **31**, 131–142.

Flecker, A.S., Feifarek, B.P., and Taylor, B.W. (1999). Ecosystem engineering by a tropical tadpole: density-dependent effects on habitat structure and larval growth rates. *Copeia* **1999**, 495–500.

Flowers, R.W. (1991). Diversity of stream-living insects in northwestern Panama. *J. N. Am. Benthol. Soc.* **10**, 322–334.

Freitag, H. (2004). Composition and longitudinal patterns of aquatic insect emergence in small rivers of Palawan Island, the Philippines. *Int. Rev. Hydrobiol.* **89**, 375–391.

Graça, M.A.S., Cressa, C., Gessner, M.O., Feio, M.J., Callies, K.A., and Barrios, C. (2001). Food quality, feeding preferences, survival and growth of shredders from temperate and tropical streams. *Freshw. Biol.* **46**, 947–957.

Greathouse, E.A., and Pringle, C.M. (2006). Does the river continuum concept apply on a tropical island? Longitudinal variation in a Puerto Rican stream. *Can. J. Fish. Aquat. Sci.* **63**, 134–152.

Grimm, N.B., and Fisher, S.G. (1989). Stability of periphyton and macroinvertebrates to disturbance by flash floods in a desert stream. *J. N. Am. Benthol. Soc.* **8**, 293–307.

Haines, A.T., Finlayson, B.L., and McMahon, T.A. (1988). A global classification of river regimes. *Appl. Geogr.* **8**, 255–272.

Hamada, N., McCreadie, J.W., and Adler, P.H. (2002). Species richness and spatial distribution of blackflies (Diptera: Simuliidae) in streams of Central Amazonia, Brazil. *Freshw. Biol.* **47**, 31–40.

Hawkins, B.A. (2001). Ecology's oldest pattern? *Trends Ecol. Evol.* **16**, 470.

Heino, J., Muotka, T., Paavola, R., Haemaelaeinen, H., and Koskenniemi, E. (2002). Correspondence between regional delineations and spatial patterns in macroinvertebrate assemblages of boreal headwater streams. *J. N. Am. Benthol. Soc.* **21**, 397–413.

Hill, W.R., and Knight, A.W. (1988). Nutrient and light limitation of algae in two northern California streams. *J. Phycol.* **24**, 125–132.

Ho, W.H., Hyde, K.D., and Hodgkiss, I.J. (2001). Fungal communities on submerged wood from streams in Brunei, Hong Kong, and Malaysia. *Mycol. Res.* **105**, 1492–1501.

Huryn, A.D., and Wallace, J.B. (2000). Life history and production of stream insects. *Ann. Rev. Entomol.* **45**, 83–110.

Hyde, K.D., and Goh, T.K. (1999). Fungi on submerged wood from the River Coln, England. *Mycol. Res.* **103**, 1561–1574.

Irons, J.G., Oswood, M.W., Stout, R.J., and Pringle, C.M. (1994). Latitudinal patterns in leaf-litter breakdown – is temperature really important? *Freshw. Biol.* **32**, 401–411.

Jackson, J.K., and Sweeney, B.W. (1995). Egg and larval development times for 35 species of tropical stream insects from Costa Rica. *J. N. Am. Benthol. Soc.* **14**, 115–130.

Jacobsen, D., and Terneus, E. (2001). Aquatic macrophytes in cool aseasonal and seasonal streams: a comparison between Ecuadorian highland and Danish lowland streams. *Aquat. Bot.* **71**, 281–295.

Jacobsen, D., Schultz, R., and Encalada, A. (1997). Structure and diversity of stream invertebrate assemblages: the influence of temperature with altitude and latitude. *Freshw. Biol.* **38**, 247–261.

Johnson, S.L., Covich, A.P., Crowl, T.A., Estrada-Pinto, A., Bithorn, J., and Wurtsbaugh, W.A. (1998). Do seasonality and disturbance influence reproduction in freshwater atyid shrimp in headwater streams (Puerto Rico)? *Verh. Internat. Verein. Limnol.* **26**, 2076–2081.

Jones, J.B., and Smock, L.A. (1991). Transport and retention of particulate organic matter in two low-gradient headwater streams. *J. N. Am. Benthol. Soc.* **10**, 115–126.

Junk, W.J., and Welcomme, R.L. (1990). Floodplains *In* "Wetlands and Shallow Continental Water Bodies" (B.C. Patten, S.E. Jorgensen and H. Dumont, Eds), pp. 491–524, SPB Academic, The Hague, The Netherlands.

Junk, W.J., and Wantzen, K.M. (2004). The flood pulse concept. New aspects, approaches, and applications – an update. *In* "Proceedings of the 2nd Large River Symposium (LARS), Phnom Penh, Cambodia" (R. Welcomme and T. Petr, Eds), pp. 117–149, Food and Agriculture Organization and Mekong River Commission, RAP Publication, 2004/16, Bangkok, Thailand.

Junk, W.J., Bayley, P.B., and Sparks, R.E. (1989). The flood pulse concept in river-floodplain systems. *Spec. Publ. Can. J. Fish. Aquat. Sci.* **106**, 110–127.

Jury, M.R., and Melice, J.L. (2000). Analysis of Durban rainfall and Nile river flow 1871–1999. *Theor. Appl. Climatol.* **67**, 161–169.

Kiffney, P.M., Volk, C.J., Beechie, T.J., Murray, G.L., Pess, G.R., and Edmonds, R.L. (2004). A high-severity disturbance event alters community and ecosystem properties in West Twin Creek, Olympic National Park, Washington, USA. *Am. Midl. Nat.* **152**, 286–303.

Kolar, C.S., and Lodge, D.M. (2001). Progress in invasion biology: predicting invaders. *Trends Ecol. Evol.* **16**, 199–204.

Lake, P.S., Barmuta, L.A., Boulton, A.J., Campbell, I.C., and St Clair, R.M. (1986). Australian streams and Northern Hemisphere stream ecology: comparisons and problems. *Proc. Ecol. Soc. Aust.* **14**, 61–82.

Lake, P.S., Schreiber, E.S.G., Milne, B.J., and Pearson, R.G. (1994). Species richness in streams: patterns over time, with stream size and with latitude). *Verh. Internat. Verein. Limnol.* **25**, 1822–1826.

Larned, S.T. (2000). Dynamics of coarse riparian detritus in a Hawaiian stream ecosystem: a comparison of drought and post-drought conditions. *J. N. Am. Benthol. Soc.* **19**, 215–234.

Larned, S.T., and Santos, S.R. (2000). Light- and nutrient-limited periphyton in low order streams of Oahu, Hawaii. *Hydrobiologia* **432**, 101–111.

Larned, S.T., Kinzie, R.A., Covich, A.P., and Chong, C.T. (2003). Detritus processing by endemic and non-native Hawaiian stream invertebrates: a microcosm study of species-specific effects. *Arch. Hydrobiol.* **156**, 241–254.

Layman, C.A., and Winemiller, K.O. (2004). Size-based responses of prey to piscivore exclusion in a species-rich Neotropical river. *Ecology* **85**, 1311–1320.

Levine, M. (2000). Species diversity and biological invasions: relating local process to community pattern. *Science* **288**, 852–854.

Lowe-McConnell, R.H. (1987). "Ecological Studies in Tropical Fish Communities." Cambridge University Press, New York, USA.

Lugo-Ortiz, C.R., and McCafferty, W.P. (1998). The *Centroptiloides* complex of Afrotropical small minnow mayflies (Ephemeroptera: Baetidae). *Ann. Entomol. Soc. Am.* **91**, 1–26.

Malmqvist, B. (1993). Interactions in stream leaf packs – effects of a stonefly predator on detritivores and organic-matter processing. *Oikos* **66**, 454–462.

Maltchik, L., and Pedro, F. (2001). Responses of aquatic macrophytes to disturbance by flash floods in a Brazilian semiarid intermittent stream. *Biotropica* **33**, 566–572.

Mantel, S.K., and Dudgeon, D. (2004). Growth and production of a tropical predatory shrimp, *Macrobrachium hainanense* (Palaemonidae), in two Hong Kong streams. *Freshw. Biol.* **49**, 1320–1336.

March, J.G., Benstead, J.P., Pringle, C.M., and Ruebel, M.W. (2001). Linking shrimp assemblages with rates of detrital processing along an elevational gradient in a tropical stream. *Can. J. Fish. Aquat. Sci.* **58**, 470–478.

Markewitz, D., Davidson, E.A., Figueiredo, R.D.O., Victoria, R.L., and Krusche, A.V. (2001). Control of cation concentrations in stream waters by surface soil processes in an Amazonian watershed. *Nature* **410**, 802–805.

Mathooko, J.M., Morara, G.M., and Leichtfried, M. (2001). Leaf litter transport and retention in a tropical Rift Valley stream: an experimental approach. *Hydrobiologia*, **443**, 9–18.

Mathuriau, C., and Chauvet, E. (2002). Breakdown of leaf litter in a Neotropical stream. *J. N. Am. Benthol. Soc.* **21**, 384–396.

McCreadie, J.W., Adler, P.H., and Hamada, N. (2005). Patterns of species richness for blackflies (Diptera: Simuliidae) in the Nearctic and Neotropical regions. *Ecol. Entomol.* **30**, 201–209.

McDowell, W.H. (1998). Internal nutrient fluxes in a Puerto Rican rain forest. *J. Trop. Ecol.* **14**, 521–536.

McDowell, W.H., McSwiney, C.P., and Bowden, W.B. (1996). Effects of hurricane disturbance on groundwater chemistry and riparian function in a tropical rain forest. *Biotropica* **28**, 577–584.

McIntyre, P., Jones, L.E., Flecker, A., and Vanni, M.J. (2007). Fish extinctions alter nutrient recycling in freshwaters. *PNAS* **104**, 4461–4466.

McKie, B.G., Pearson, R.G., and Cranston, P.S. (2005). Does biogeographic history matter? Diversity and distribution of lotic midges (Diptera: Chironomidae) in the Australian Wet Tropics. *Austral Ecol.* **30**, 1–13.

McMahon, T.A., Finlayson, B.L., Haines, A.T., and Srikanthan, R. (1992). "Global Runoff – Continental Comparisons of Annual Flows and Peak Discharges." Catena Verlag, Cremlingen–Destedt, Germany.

McNeeley, J.A., Miller, K.R., Reid, W.V., Mittermeier, R.A., and Werner, T.B. (1990). "Conserving the World's Biological Diversity." International Union for the Conservation of Nature, Gland, Switzerland.

Merriam, J.L., McDowell, W.H., Tank, J.L., Wollheim, W.M., Crenshaw, C.L., and Johnson, S.L. (2002). Characterizing nitrogen dynamics, retention and transport in a tropical rainforest stream using an in situ $^{15}$N addition. *Freshw. Biol.* **47**, 143–160.

Montoya, J.V., Roelke, D.L., Winemiller, K.O., Cotner, J.B., and Snider, J.A. (2006). Hydrological seasonality and benthic algal biomass in a neotropical floodplain river. *J. N. Am. Benthol. Soc.* **25**, 157–170.

Mosisch, T.D., Bunn, S.E., and Davies, P.M. (2001). The relative importance of shading and nutrients on algal production in subtropical streams. *Freshw. Biol.* **46**, 1269–1278.

Mulholland, P.J., Tank, J.L., Sanzone, D.M., Wollheim, W.M., Peterson, B.J., Webster, J.R., and Meyer, J.L. (2000). Nitrogen cycling in a forest stream determined by a $^{15}$N tracer addition. *Ecol. Monogr.* **70**, 471–493.

Mulholland, P.J., Fellows, C.S., Tank, J.L., Grimm, N.B., Webster, J.R., Hamilton, S.K., Marti, E., Ashkenas, L., Bowden, W.B., Dodds, W.K., McDowell, W.H., Paul, M.J., and Peterson, B.J. (2001). Inter-biome comparison of factors controlling stream metabolism. *Freshw. Biol.* **46**, 1503–1517.

Nolen, J.A., and Pearson, R.G. (1993). Processing of litter from an Australian tropical stream by *Anisocentropus kirramus* Neboiss (Trichoptera: Calamoceratidae). *Freshw. Biol.* **29**, 469–479.

Oberdorff, T., Guégan, J.-F., and Hugueny, B. (1995). Global scale patterns in fish species richness in rivers. *Ecography* **18**, 345–352.

Oberndorfer, R.Y., McArthur, J.V., Barnes, J.R., and Dixon, J. (1984). The effect of invertebrate predators on leaf litter processing in an alpine stream. *Ecology* **65**, 1325–1331.

Ortiz-Zayas, J.R., Lewis, W.M. Jr., Saunders, J.F. III, McCutchan, J.H., and Scatena, F.N. (2005). Metabolism of a tropical stream. *J. N. Am. Benthol. Soc.* **24**, 769–783.

Osterkamp, W.R. (2002). Geoindicators for river and river-valley monitoring in the humid tropics. *Environ. Geol.* **42**, 725–735.

Pan, Y.D., and Lowe, R.L. (1994). Independent and interactive effects of nutrients and grazers on benthic algal community structure. *Hydrobiologia* **291**, 201–209.

Patrick, R. (1964). A discussion of the results of the Catherwood expedition to the Peruvian headwaters of the Amazon. *Verh. Internat. Verein. Limnol.* **15**, 1084–1090.

Pearson, R.G. (2005). Biodiversity of the freshwater invertebrates of the Wet Tropics region of north-eastern Australia: patterns and possible determinants. *In* "Rainforests: Past Present and Future" (E. Bermingham, C.W. Dick and C. Moritz, Eds), pp. 470–485, University of Chicago Press, Chicago, USA.

Pearson, R.G., and Tobin, R.K. (1989). Litter consumption by invertebrates from an Australian tropical rainforest stream. *Arch. Hydrobiol.* **116**, 71–80.

Pearson, R.G., and Connolly, N.M. (2000). Nutrient enhancement, food quality and community dynamics in a tropical rainforest stream. *Freshw. Biol.* **43**, 31–42.

Pearson, R.G., Benson, L.J., and Smith, R.E.W. (1986). Diversity and abundance of the fauna in Yuccabine Creek, a tropical rainforest stream. *In* "Limnology in Australia" (P. De Deckker and W.D. Williams, Eds), pp. 329–342, CSIRO/Dr W. Junk, Melbourne/Dordrecht.

Pearson, R.G., Tobin, R.K., Benson, L.J., and Smith, R.E.W. (1989). Standing crop and processing of rainforest litter in a tropical Australian stream. *Arch. Hydrobiol.* **115**, 481–498.

Pereira, H.C. (1989). "Policy and Practice in the Management of Tropical Watersheds." Westview Press, San Francisco, USA.

Petersen, R.C., and Cummins, K.W. (1974). Leaf processing in a woodland stream. *Freshw. Biol.* **4**, 343–368.

Poff, N.L., and Ward, J.V. (1989). Implications of streamflow variability and predictability for lotic community structure: a regional analysis of streamflow patterns. *Can. J. Fish. Aquat. Sci.* **46**, 1805–1818.

Pringle, C.M., and Triska, F.J. (1991). Effect of geothermal groundwater on nutrient dynamics of a lowland Costa Rican stream. *Ecology* **72**, 951–965.

Pringle, C.M., and Hamazaki, T. (1998). The role of omnivory in a neotropical stream: separating diurnal and nocturnal effects. *Ecology* **79**, 269–280.

Pringle, C.M., Paaby-Hansen, P., Vaux, P.D., and Goldman, C.R. (1986). In situ nutrient assays of periphyton growth in a lowland Costa Rica stream. *Hydrobiologia* **134**, 207–213.

Pringle, C.M., Scatena, F.N., Paaby-Hansen, P., and Núñez-Ferrera, M. (2000a). River conservation in Latin America and the Caribbean. *In* "Global Perspectives on River Conservation: Policy and Practice" (P.J. Boon, B.R. Davies and G.E. Petts, Eds), pp. 41–77, John Wiley & Sons, Chichester, UK.

Pringle, C.M., Freeman, M.C., and Freeman, B.J. (2000b). Regional effects of hydrologic alterations on riverine macrobiota in the New World: tropical-temperate comparisons. *BioScience* **50**, 807–823.

Puckridge, J.T., Sheldon, F., Walker, K.F., and Boulton, A.J. (1998). Flow variability and the ecology of large rivers. *Mar. Freshwat. Res.* **49**, 55–72.

Ramírez, A., and Pringle, C.M. (1998). Structure and production of a benthic insect assemblage in a neotropical stream. *J. N. Am. Benthol. Soc.* **17**, 443–463.

Ramírez, A., and Hernandez-Cruz, L.R. (2004). Aquatic insect assemblages in shrimp-dominated tropical streams. *Biotropica* **36**, 259–266.

Ramírez, A., and Pringle, C.M. (2006). Fast growth and turnover of chironomid assemblages in response to stream phosphorus levels in a tropical lowland landscape. *Limnol. Oceanogr.* **51**, 189–196.

Ramírez, A., Pringle, C.M., and Molina, L. (2003). Effects of stream phosphorus levels on microbial respiration. *Freshw. Biol.* **48**, 88–97.

Ramírez, A., Pringle, C.M., and Douglas, M.M. (2006). Temporal and spatial patterns in stream physicochemistry and insect assemblages in tropical lowland streams. *J. N. Am. Benthol. Soc.* **25**, 108–125.

Ramos-Escobedo, M.G., and Vazquez, G. (2001). Major ions, nutrients and primary productivity in volcanic neotropical streams draining rainforest and pasture catchments at Los Tuxtlas, Veracruz, Mexico. *Hydrobiologia* **445**, 67–76.

Ranvestel, A.W., Lips, K.R., Pringle, C.M., Whiles, M.R., and Bixby, R.J. (2004). Neotropical tadpoles influence stream benthos: evidence for the ecological consequences of decline in amphibian populations. *Freshw. Biol.* **49**, 274–285.

Ribbink, A.J. (1994). Biodiversity and speciation of freshwater fishes with particular reference to African cichlids. *In* "Aquatic Ecology. Scale, Pattern and Process" (P.S. Giller, A.G. Hildrew and D.G. Raffaelli, Eds), pp. 261–288, Blackwell Science, Oxford, UK.

Robinet, T., Guyet, S., Marquet, G., Mounaix, B., Olivier, J.M., Tsukamoto, K., Valade, P., and Feunteun, E. (2003). Elver invasion, population structure and growth of marbled eels *Anguilla marmorata* in a tropical river on Reunion Island in the Indian Ocean. *Env. Biol. Fish.* **68**, 339–348.

Rohde, K. (1992). Latitudinal gradients in species diversity: the search for the primary cause. *Oikos* **65**, 514–527.

Rosemond, A.D., Pringle, C.M., and Ramírez, A. (1998). Macroconsumer effects on insect detritivores and detritus processing in a tropical stream. *Freshw. Biol.* **39**, 515–523.

Rosenzweig, M.L. (1997). Species diversity and latitudes: listening to area's signal. *Oikos* **80**, 172–176.

Rosser, Z., and Pearson, R.G. (1995). Responses of rock fauna to physical disturbance in two Australian tropical rainforest streams. *J. N. Am. Benthol. Soc.* **14**, 183–196.

Salas, M., and Dudgeon, D. (2002). Laboratory and field studies of mayfly growth in the tropics. *Arch. Hydrobiol.* **153**, 75–90.

Salas, M., and Dudgeon, D. (2003). Life histories, production dynamics and resource utilization of mayflies (Ephemeroptera ) in two tropical Asian forest streams. *Freshw. Biol.* **48**, 485–499.

Shivoga, W.A. (2001).The influence of hydrology on the structure of invertebrate communities in two streams flowing into Lake Nakuru, Kenya. *Hydrobiologia* **458**, 121–130.

Smith, R.E.W., and Pearson, R.G. (1987). The macro–invertebrate communities of temporary pools in an intermittent stream in tropical Queensland. *Hydrobiologia* **150**, 45–61.

Speaker, R.W., Moore, K., and Gregory, S. (1984). Analysis of the processes of retention of organic matter in stream ecosystems. *Verh. Internat. Verein. Limnol.* **22**, 1835–1841.

Stallard, R.F. (2001). Possible environmental factors underlying amphibian decline in eastern Puerto Rico: analysis of US government data archives. *Conserv. Biol.* **15**, 943–953.

Stallard, R.F. (2002). The biogeochemistry of the Amazon Basin: too little mud and perhaps too much water. *Hydrol. Proc.* **16**, 2042–2049.

Stanley, E.H., and Boulton, A.J. (2000). River size as a factor in river conservation. *In* "Global Perspectives on River Conservation: Policy and Practice" (P.J. Boon, B.R. Davies and G.E. Petts, Eds), pp. 399–409, John Wiley & Sons, Chichester, UK.

Stout, J., and Vandermeer, J. (1975). Comparison of species richness for stream-inhabiting insects in tropical and mid-latitude streams. *Am. Nat.* **109**, 263–280.

Tank, J.L., Meyer, J.L., Sanzone, D.M., Mulholland, P.J., Webster, J.R., Peterson, B.J., Wollheim, W.M., and Leonard, N.E. (2000). Analysis of nitrogen cycling in a forest stream during autumn using a N-15-tracer addition. *Limnol. Oceanogr.* **45**, 1013–1029.

Thorp, J.H., and Delong, M.D. (1994). The riverine productivity model: an heuristic view of carbon sources and organic processing in large river ecosystems. *Oikos* **70**, 305–308.

Thorp, J.H., and Delong, A.D. (2002). Dominance of autochthonous autotrophic carbon in food webs of heterotrophic rivers. *Oikos* **96**, 543–550.

Tonn, W.M., Magnuson, J.J., Rask, M., and Toivonen, J. (1990). Intercontinental comparison of small-lake fish assemblages: the balance between local and regional processes. *Am. Nat.* **136**, 345–375.

Townsend, C.R., and Hildrew, A.G. (1994). Species traits in relation to a habitat templet for river systems. *Freshw. Biol.* **31**, 265–275.

Tumwesigye, C., Yusuf, S.K., and Makanga, B. (2000). Structure and composition of benthic macroinvertebrates of a tropical forest stream, River Nyamweru, western Uganda. *Afr. J. Ecol.* **38**, 72–77.

Twinch, A.J. (1987). Phosphate exchange characteristics of wet and dried sediment samples from a hypereutrophic reservoir: implications for the measurement of sediment phosphorus status. *Water S. Afr.* **21**, 1225–1230.

Vannote, R.L., Minshall, G.W., Cummins, K.W., Sedell, J.R., and Cushing, C.E. (1980). The river continuum concept. *Can. J. Fish. Aquat. Sci.* **37**, 130–137.

Vinson, M.R., and Hawkins, C.P. (2003). Broad-scale geographical patterns in local stream insect genera richness. *Ecography* **26**, 751–767.

Walker, I. (1987). The biology of streams as part of Amazonian forest ecology. *Experientia* **43**, 279–287.

Walker, K., Neboiss, A., Dean, J., and Cartwright, D. (1995a). A preliminary investigation of the caddis-flies (Trichoptera) of the Queensland Wet Tropics. *Aust. Entomol.* **22**, 19–31.

Walker, K.F., Sheldon, F., and Puckridge, J.T. (1995b). A perspective on dryland river ecosystems. *Reg. Rivers: Res. Manag.* **11**, 85–104.

Wantzen, K.M., Wagner, R., Suetfeld, R., and Junk, W.J. (2002). How do plant-herbivore interactions of trees influence coarse detritus processing by shredders in aquatic ecosystems of different latitudes? *Verh. Internat. Verein. Limnol.* **28**, 815–821.

Wantzen, K.M., Ramírez, A., and Winemiller, K.O. (2006). New vistas in Neotropical stream ecology – Preface. *J. N. Am. Benthol. Soc.* **25**, 61–65.

Webster, J.R., Benfield, E.F., Golladay, S.W., Hill, B.H., Hornick, L.E., Kazmierczak, R.F., and Perry, W.B. (1987). Experimental studies of physical factors affecting seston transport in streams. *Limnol. Oceanogr.* **32**, 848–863.

Webster, J.R., Wallace, J.B., and Benfield, E.F. (1995). Organic processes in streams of the eastern United States. *In* "River and Stream Ecosystems" (C.E. Cushing, K.W. Cummins and G.W. Minshall, Eds), pp. 117–187. Elsevier Science, The Netherlands.

Webster, J.R., Mulholland, P.J., Tank, J.L., Valett, H.M., Dodds, W.K., Peterson, B.J., Bowden, W.B., Dahm, C.N., Findlay, S., Gregory, S.V., Grimm, N.B., Hamilton, S.K., Johnson, S.L., Marti, E., McDowell, W.H., Meyer, J.L., Morrall, D.D., Thomas, S.A., and Wollheim, W.M. (2003). Factors affecting ammonium uptake in streams – an inter-biome perspective. *Freshw. Biol.* **48**, 1329–1352.

Webster, P.J., Magada, V.O., Palmer, T.N., Shukla, J., Tomas, R.A., Yanai, M., and Yasunaris, T. (1998). Monsoons: processes, predictability and the prospects for prediction. *J. Geophys. Res.* **103**, 14451–14510.

Welcomme, R.L. (1985). River fisheries. *FAO Fish. Tech. Pap.* **262**, 1–330.

White, A.F., Blum, A.E., Schulz, M.S., Vivit, D.V., Stonestrom, D.A., Larsen, M., Murphy, S.F., and Eberl, D. (1998). Chemical weathering in a tropical watershed, Luquillo mountains, Puerto Rico: I. Long-term versus short-term weathering fluxes. *Geochim. Cosmochim. Acta.* **62**, 209–226.

Winterbourn, M.J., Rounick, J.S., and Cowie, B. (1981). Are New Zealand ecosystems really different? *N. Z. J. Mar. Freshwat. Res.* **15**, 321–328.

Woodward, G., and Hildrew, A.G. (2002). Foodweb structure in riverine landscapes. *Freshw. Biol.* **47**, 777–798.

Wootton, J.T., and Oemke, M.P. (1992). Latitudinal differences in fish community trophic structure, and the role of fish herbivory in a Costa Rican stream. *Env. Biol. Fish.* **35**, 311–319.

Wright, J.P., and Flecker, A.S. (2004). Deforesting the riverscape: the effects of wood on fish diversity in a Venezuelan piedmont stream. *Biol. Cons.* **120**, 439–447.

Wright, M.S., and Covich, A.P. (2005). The effect of macroinvertebrate exclusion on leaf breakdown rates in a tropical headwater stream. *Biotropica* **37**, 403–408.

Yam, R.S.W., and Dudgeon, D. (2006). Production dynamics and growth of atyid shrimps (Decapoda: *Caridina* spp.) in 4 Hong Kong streams: the effects of site, season, and species. *J. N. Am. Benthol. Soc.* **25**, 406–416.

Yule, C.M. (1996). Trophic relationships and food webs of the benthic invertebrate fauna of two aseasonal tropical streams on Bougainville Island, Papua New Guinea. *J. Trop. Ecol.* **12**, 517–534.

Yule, C.M., and Pearson, R.G. (1996). Aseasonality of benthic invertebrates in a tropical stream on Bougainville Island, Papua New Guinea. *Arch. Hydrobiol.* **137**, 95–117.

# 10

# *Tropical Stream Conservation*

Alonso Ramírez, Catherine M. Pringle, and Karl M. Wantzen

Tropical streams support diverse assemblages of plants and animals, including many species that still remain to be described by scientists. These rich and diverse ecosystems also provide valuable services to human populations in tropical countries. As a result stream conservation issues are often related to overfishing, excessive water removal, and pollution. The effects that human activities are having on tropical streams are poorly understood at best and the pace of stream deterioration in many regions exceeds the pace of scientific research to understand stream ecosystem responses. Conservation issues are complex and solutions require a clear understanding on how socioeconomic factors act as driving forces for stream degradation and how stream ecosystems respond to those forces. This chapter reviews the major factors impacting tropical streams: over-exploiting (mainly over-fishing), deforestation, water abstraction for irrigation and human consumption, pollution, and alterations in riverine connectivity. Major issues are also summarized for tropical regions (e.g. Latin America and the Caribbean, tropical Africa). Three case studies provide information on how conservation actions are helping protect tropical streams in Brazil, and how communication of the impacts of dams and socioeconomic changes in Puerto Rico affected threats to stream ecosystems as the island moved from an economy based on agriculture to one dependent upon industry. The complexity of the issues and the limited ecological information available highlights the need for further study of tropical streams in order to protect them effectively.

## I. INTRODUCTION

Streams in tropical regions often support diverse assemblages of plants and animals, including a high proportion of species that still remain to be described by scientists. Many of them have unique adaptations to specific habitats, microhabitats, or food sources, while others are more ubiquitous and capable of living in a wide range of conditions. The region we are referring as 'tropical' includes a variety of stream types, including small cold-water streams draining high-elevation snow-covered regions (e.g. the Andes: see also Chapter 8 of this volume), mountain streams with steep gradients and flashy hydrographs (e.g. island streams), and lowland streams draining expanses of low-elevation forest (e.g. Amazonian forest streams). Some of these stream systems are short, rapidly draining to the ocean, while others are components of some of the largest drainage basins on Earth. Similarly, diverse are the number of countries, cultures, and peoples inhabiting and modifying the margins and floodplains of streams and their catchments. This diversity of regions, habitat types, organisms, and peoples contributes to the complexity of environmental challenges facing tropical streams.

Major factors impacting tropical streams are over-exploitation (mainly over-fishing), deforestation, different types of pollution, water abstraction for irrigation and human consumption, and related alterations in longitudinal, lateral and vertical riverine connectivity (e.g. Dudgeon *et al.*, 2006). Although some of these factors cause local impacts within drainage basins, the capacity of streams to integrate over the landscape results in cumulative effects within the entire basin. Tropical regions support increasing numbers of people and many of the problems that will be discussed here are the direct result of human population growth, over-exploitation of resources, and lack of adequate planning by local governments. At the same time, the impacts of human activities on streams are poorly understood. The amount of information available for management is limited and baseline data, such as species inventories, are largely lacking in the tropics. Even more limited is the amount of time available to gather necessary information for conservation. The pace of stream deterioration in many regions exceeds the pace of scientific research to understand stream ecosystem structure and function and to gather baseline information on species diversity. Both scientific information and conservation efforts are not equal in all tropical regions, as has been evidenced by recent assessments of river conservation (Boon *et al.*, 2000; Dudgeon, 2000a, 2003; Pringle 2000; Pringle *et al.*, 2000a, b; Moulton and Wantzen, 2006).

This chapter focuses on conservation issues that are relevant to most tropical regions and stream types. Factors driving those conservation issues and the ecological response of streams are described, and an account of several major conservation issues that pose threats throughout the tropics is given. It is followed by a summary of available information on stream conservation in each major tropical region, although no claims are made for comprehensiveness. Stream conservation issues can be expected to change as countries implement different strategies toward improving the quality of life of their citizens and maximizing the use of natural resources. Examples from the Brazilian Cerrado and the island of Puerto Rico are used to explore how the biotic integrity of tropical streams changes as local policy and economies alter, and the chapter concludes by some speculation about what the future might hold for tropical stream ecosystems.

## II. WHAT DRIVES CONSERVATION ISSUES IN TROPICAL STREAMS?

Conservation issues affecting tropical streams are complex and result from particular combinations of socioeconomic factors and ecological responses that we are just beginning to understand. Socioeconomic realities of tropical regions result in resource overexploitation, water pollution, and overall degradation of stream ecosystems. At the same time, tropical stream ecosystems respond to degradation in ways that might not be expected based on our current understanding of stream ecology, which is mainly based on studies in temperate regions (see also Chapter 9 of this volume). Successful conservation strategies require a clear understanding of how socioeconomic factors act as driving forces, and how stream ecosystems respond to those forces.

### A. Socioeconomic Factors

Except for tropical Australia, the tropics are primarily composed of developing nations. There seems to be a general tendency by policy makers in developing countries to follow a 'develop now, clean up later' approach to try to gain economic stability or growth and solve short-term pressing issues (Dudgeon, 2000a). If resources are limited, conservation issues are commonly set aside and only those directly affecting human health are addressed. For example, expenditure of resources necessary for maintaining some level of water quality is driven by concerns about human health (Pringle, 2000). Water-treatment facilities are almost absent in many tropical countries and, where present, may be poorly maintained. Appropriate environmental legislation for the protection of aquatic systems is currently lacking in many tropical countries and, when present, it is poorly enforced (Dudgeon *et al.*, 2000; Wishart *et al.*, 2000).

Population growth is clearly a problem in developing countries. Most tropical regions have annual population growth rates of nearly 3% (Pringle, 2000). This rapid pace of growth is commonly followed by overexploitation of natural resources (Wishart *et al.*, 2000). Fisheries in tropical regions provide a clear example of such overexploitation. During 2002, nearly 9 million tons of fish were extracted from inland waters for human consumption and almost all natural populations of freshwater fish are showing signs of declines in abundance and body size (Allan *et al.*, 2005). Tropical developing countries accounted for nearly 70% of this catch, and the total haul is growing at a rate of 3% annually (Allan *et al.*, 2005). Overfishing of wild populations (i.e. excluding fish reared by aquaculture or artificially stocked lakes and rivers) is normally accompanied by other stress factors, such as habitat loss, flow alteration and pollution, and biodiversity losses of stream fishes are likely the result of combinations of different factors (e.g. Dudgeon, 2000c, 2005; Allan *et al.*, 2005; see also Chapter 5 of this volume). Intensive fishing may include practices that result in additional stream impacts. For example, stream poisoning is not uncommon as a fishing method in many tropical countries, and poisoning events in tropical island streams (which typically have small catchments) result in massive mortality of shrimps and fishes that can have consequential impacts on ecosystem function (Greathouse *et al.*, 2005). In addition, decreases in natural fish populations are likely to have impacts on higher trophic levels. For example, piscivorous mammals and birds can be out-competed by humans or may be hunted or otherwise exterminated to avoid competition. Available information on the responses of freshwater or semi-aquatic mammal populations to stream degradation are limited in tropical regions and, as many of them are globally-threatened or of conservation importance, there is an urgent need for research on this topic (see also Chapter 6 of this volume).

The problem of overexploitation of natural resources by developing countries is exacerbated by over-consumption in developed nations. Many tropical countries are under pressure to meet the steadily increasing demand for resources (e.g. agricultural products and minerals) from their trading partners in developed nations. Often, these goods are produced at a high environmental cost. For instance, as food crops commonly require 'zero' pest levels (van Emden and Peakall, 1996), and require the application of agrochemicals and large amounts of fertilizers and agrochemicals for pest control many of which are washed into streams and rivers. Extremely high rainfall in some tropical regions exacerbates this problem by moving pollutants farther away from their sources, and exacerbating erosion and soil loss from agricultural areas (Wantzen, 2006).

Marginalization of the rural poor in many tropical countries often coincides with complete degradation of all remaining riparian habitat remnants. Homeless people often colonize stream banks and flood-prone areas that, despite being state-owned, offer some agricultural potential, at least during dry season or low-flow periods; in doing, they expose themselves to water-borne diseases and flood hazards. Conservation initiatives must take account of such human concerns, especially in instances where the creation of state-owned forested areas is part of restoration or preservation plans for stream ecosystems.

## B. Ecological Factors

Tropical streams are diverse in nature, but certain common characteristics can be used to define the way in which they respond to human impacts. Understanding these characteristics is key for implementing management plans for tropical stream conservation. Pringle (2000) summarized how the particularities of the hydrological cycle (e.g. unpredictable high-rainfall events) and the warm prevailing water temperature that are characteristic of tropical regions can give rise to serious conservation problems. For example, the wet tropics are highly susceptible to erosion after land clearance for agriculture, and soil erosion rates of up to 200 t ha$^{-1}$ have been reported from cleared land in Costa Rica (Hartshorn *et al.*, 1982). In contrast, high evaporation rates in arid or seasonally-dry tropical environments can reduce the stability of vegetation cover, making land vulnerable to misuse (Pringle, 2000).

In large riverine systems, the importance of floodplains for the protection of biodiversity and ecosystem function and services has been generally recognized (e.g. Junk and Wantzen, 2004), but there is a need for more research on the spatial and temporal dynamics of floodplains and riparian vegetation (see Chapter 7 of this volume) and the resulting insights must be integrated into conservation, management, and rehabilitation plans (see also the discussion on deforestation below).

Tropical streams located in coastal areas and island streams tend to be dominated by migratory fishes and shrimps that must pass their immature stages in estuaries or coastal waters. Adult *Macrobrachium* (Palaemonidae) shrimps, for example, live in streams where they release their larvae into the current. The planktonic larvae drift downstream into estuarine environments where they develop into juveniles that migrate back upstream. Migratory life cycles make stream biota highly vulnerable to human impacts in different parts of the stream ecosystems and to human activities that disrupt the connectivity between streams and estuaries (March *et al.*, 2003). Similar examples exist among fishes in the Amazon basin (Barthem and Goulding, 1997) and diadromous neritid snails in coastal Puerto Rican streams (Blanco and Scatena, 2006).

## III. MAJOR CONSERVATION ISSUES

In the following section we restrict our discussion to four major issues affecting stream conservation in tropical regions: deforestation and erosion; agriculture; urban and industrial development, and alteration of hydrologic connectivity. These four issues are representative of issues common to tropical ecosystems across the globe. Since many more issues are locally important for specific regions, a summary of conservation information by major tropical region will be given after this section.

### A. Deforestation and Erosion

Deforestation is a major environmental problem in tropical regions, where rates of tropical forest loss currently exceed $1.25 \times 10^5$ km$^2$ yr$^{-1}$ (FAO, 1999). Rates of degradation or alteration of forest structure are almost certainly much higher. Every year, it is estimated that an additional 0.5 million kilometers of stream and river channel are affected by tropical deforestation (Benstead *et al.*, 2003b). In many regions, deforestation has reached alarming rates. For example, forest burning in the Amazon increased by 28% between 1996 and 1997, and 1994 deforestation estimates showed a 34% rise over 1991 (Schemo, 1996, 1997, 1998). Although deforestation estimates vary according to region, investigators and the precision of the satellites data used, they provide a basis for grave concern. By 2003, 648 500 km$^2$ of the Amazon had been deforested, with dramatic consequences for biodiversity and hydrology (Fearnside, 2005). Brazil is now losing more rainforest each year than any country in the world.

Deforestation has many direct impacts on stream and river ecosystems (see reviews by Dudgeon, 1992, 2000a, b; Pringle *et al.*, 2000b; Benstead *et al.*, 2003b; Benstead and Pringle,

2004). Removal of catchment vegetation, in particular riparian vegetation, alters water movement from land to stream, resulting in increases in erosion and sedimentation in the channel, alterations in discharge, increased light incidence on the water surface, increased water temperature, and changes in stream solute chemistry (Forti *et al.*, 2000; Neill *et al.*, 2001; Biggs *et al.*, 2005). The loss of allochthonous energy sources and structural components such as wood, which maintain habitat heterogeneity in the stream channel, also alter stream-ecosystem processes (see Chapters 3 and 7 in this volume). Deforestation can also have indirect effects on streams by facilitating the invasion of exotic species (Pringle and Benstead, 2001; Bunn and Arthington, 2002).

Deforestation of riparian buffers can have especially deleterious effects since many streams appear to be heterotrophic, and relying on allochthonous energy resources (for further discussion, see Chapters 3, 7, and 9). The loss of riparian vegetation can shift stream food webs from detritus-based to algal-based, and this change in basal food resources affects higher trophic levels. Benstead and Pringle (2004) found that most forest stream insects in Madagascar were unable to track shifts in their food resources caused by riparian vegetation removal, and consequently decline in abundance in deforested streams. Similarly, tropical fish assemblage composition in Australian streams is closely related to riparian zone integrity (Pusey and Arthington, 2003). In addition to providing allochthonous energy, riparian zones can also protect streams from changes in more distant parts of the catchment, as they intercept and remove or store sediments and nutrients from runoff. The association of wetlands, whether permanent or seasonal, with riparian buffers greatly enhances the role that they can play in maintaining stream integrity (see Chapter 7 in this volume).

Removal of protective plant cover over weathered and erosion-prone soils in tropical drainage basins increases stream sedimentation with consequences to the biota. Impacts include direct physiological stress of fine inorganic particles on fish and other organisms that use gills to breathe, disruption of food webs due to smothering of benthic communities, reduced visibility for predators, scouring of epilithic algal layers and organic debris, and loss of feeding and spawning sites (e.g. Waters, 1995; Wantzen, 1998, 2006; Fossati *et al.*, 2001; Mol and Ouboter, 2004). In seasonal Cerrado forest in Brazil, a single erosion gully has been reported to deliver up to 60 t sediments per day into a medium-sized stream (Wantzen, 2006). Impacts of gully erosion on stream biota are wide-ranging and can result in dramatic decreases in biodiversity and biomass of some taxa while favoring increases of a few taxa, such as sand-dwelling predators and riparian scavengers (Wantzen, 2006).

Sedimentation impacts in mountain regions can be particularly large in streams draining highly erosion-prone soils (see also Chapter 8 of this volume). These effects are evident in the rivers and streams in the central mountains of Madagascar, and can be observed many kilometers downstream of areas immediately affected by deforestation (Benstead *et al.*, 2000). In most cases, the fertile and fine-grained soil cover is lost and remaining sediments are easily displaced by groundwater flow from eroded gullies, surface runoff, or spates. Although natural recovery of eroded catchments is slow, rehabilitation and restoration strategies have been developed. These may involve some combination of damming erosion gullies, immobilizing sediments through bioengineering, or promoting plant growth by increasing soil fertility (e.g. adding fertilizer). The development of vegetated buffer strips can improve the situation, and may prevent further erosion of stream banks, and may even yield economic returns from non-wood products (Wantzen *et al.*, 2006).

At the landscape level, deforestation can alter regional patterns of the hydrological cycle. In the deforested parts of the Amazon, changes in precipitation are so large that tropical forests may be unable to reestablish (Shukla and Sellers, 1990), and changes in regional water budgets appear to be evident in the hydrology of the Amazon River itself (Gentry and Lopez-Parodi, 1980) although there is some dispute over the latter point (Richey *et al.*, 1989). Deforestation of lowlands has been found to decrease cloud cover over adjacent mountain ranges in Central America. The effect may be due to higher temperatures over deforested lowlands that lift the cloud cover above mountain tops with the consequence that water availability for small mountain streams is reduced (Lawton *et al.*, 2001).

Many countries in tropical regions have some type of environmental legislation in place to reduce negative effects of deforestation on the environment, including streams. For example, in Costa Rica, rivers and lakes must maintain a 15-m riparian buffer if they are located in rural areas, 10 m in urban areas, and 50 m in areas of steep slopes (Monika Springer, Biology School, University of Costa Rica, personal communication). Similar legislation in Brazil requires 50 m buffers in most states (see review on the development of Brazilian legislation in Wantzen *et al.*, 2006) but, as in the majority of tropical countries with environmental legislation in place, enforcement is weak. Meanwhile, the effects of deforestation on tropical stream biodiversity continue to be underestimated, and consequently it does not have sufficient weight in conservation policy and management plans (Benstead *et al.*, 2003a).

Conservation initiatives by several countries are worth mentioning. Madagascar has developed a framework for habitat conservation based on the notion that riparian vegetation contributes to the preservation of streams, riparian forests with multiple uses have been established to create economic benefits (e.g. wood production) and maintain ecological functions (Benstead *et al.*, 2003b). One such example is being developed in the Masoala National Park (Holloway, 2000). Similar efforts that are underway in Brazil and Puerto Rico include education programs and reestablishment of gallery forests (Ortíz-Zayas and Scatena, 2004; Wantzen *et al.*, 2006). Unfortunately, it is too early at present to properly evaluate their outcomes or effectiveness.

## B. Agriculture

Agriculture is the main economic activity of most tropical countries, with consequential negative effects on aquatic ecosystems, floodplains and riparian wetlands (Allan, 2004). Sheet erosion, due to heavy rainfall and thin soil layers that characterize many tropical regions, often limit the use of land for agriculture to only few years after forest removal. Once soil fertility is reduced, fertilizers are necessary to maintain crops at economically-profitable levels, especially those intended for export. The consumption of fertilizers in Latin America and the Caribbean increased by almost 50% between 1971 and 1973 and by another 50% between 1983 and 1985 (Postel, 1987). As a result of this intensive use of chemicals, impacts on streams arising from sedimentation are exacerbated by runoff loaded with fertilizers, pesticides, and herbicides (Pringle *et al.*, 2000b).

Determination of the presence and types of pesticides in streams requires sophisticated methods, analytical equipment, and well-trained personnel. This is probably the main reason for the limited data available on effects of agricultural pesticides and fertilizers on the biota of tropical streams and other inland waters (but see Sagardoy, 1993). Also, most pesticides were developed in temperate zones and we know little about the environmental consequences of application dosages or degradation times in tropical regions (Laabs *et al.*, 2002). A study of 12 river basins in the state of Paraná, Brazil, indicated that 91% of the samples obtained contained agrotoxin residues (Andreoli, 1993), while investigations of soil and water quality in the 15 000-ha irrigation district of Saldana, Colombia, reported significant traces (in ppb) of DDT in stream-bed sediments (Gomez-Sanchez, 1993). Other studies have shown the deleterious impacts of pesticides in agricultural runoff on benthic invertebrate assemblages in streams (Jergentz *et al.*, 2004). Pesticides carried by wind from distant regions have also been reported in streams (Laabs *et al.*, 2002) and in insect tissue (Standley and Sweeney, 1995). Empty pesticide cans and other such materials can be washed from dumpsites along stream margins, releasing high doses of pesticide over short periods (K.M. Wantzen, personal observations). The number of recorded human pesticide-related poisonings in Latin America provides a particularly illuminating indication of the potential environmental consequences of widespread chemical use: 18 000 pesticide poisonings occur per 6 000 000 persons each year in Central America, compared with an annual rate of only 1 per 600 000 persons in the United States (Pringle *et al.*, 2000b).

In tropical regions where agriculture requires large amounts of irrigation water, then there are usually has negative effects on streams serving as the source of water. Dams built for irrigation date back to prehistoric times in several parts of the tropics (e.g. India and China: Dudgeon, 1992, 2000c). The Parakrama Samudra Reservoir in Sri Lanka dates back to 386 AD (Gopal, 2000), and the first large irrigation dam in Latin America was built on the Saucillo River in Mexico in 1750 AD (Pringle *et al.*, 2000b). Conflicts over water use can arise in cases where rivers traverse national, state, or provincial boundaries. On a smaller scale, water abstraction from tropical streams is often uncontrolled, and some streams become intermittent in reaches downstream from water intakes. It will be obvious that even a temporary loss of permanent flow or surface water has severe negative consequences for native aquatic fauna, but more subtle effects include facilitation of subsequent invasions by exotic species (Bunn and Arthington, 2002). In addition to impacts arising from water extraction, irrigation affects stream ecosystems by way of salinization. Salinization is a major problem in coastal areas of Peru and Chile (Alva *et al.*, 1976; Peña-Torrealba, 1993) and elsewhere in Latin America and other tropical regions. Saltwater intrusion is affecting many aquifers as they are pumped for irrigation at rates that exceed natural rates of replacement. Salinization speeds the process of desertification, which claims 2250 km$^2$ of farmland in Mexico each year (Grainger, 1990).

## C. Urban and Industrial Development

Most urban areas in tropical countries are growing rapidly, with negative effects on water quality and quantity in their vicinity and streams receiving urban wastes. Walsh *et al.* (2005) proposed the 'urban stream syndrome' as an attempt to find generalities on the effects of urbanization on stream ecosystems, including those in the tropics. Urban stream catchments tend to include large areas of impervious surfaces that increase surface runoff, decrease the travel time of rainwater to the stream, and produce frequent high flood events (Walsh *et al.*, 2005). These hydrological changes result in increased erosion rates that lead to geomorphological changes in channel dimensions (Walsh *et al.*, 2005) and encourage the development of engineering responses such as channelization. Impervious surfaces also decrease water infiltration and result in lower water tables in urban streams, therefore, a decrease in stream baseflow is often observed (Walsh *et al.*, 2005). Stream organisms are affected by the associated decrease in suitable habitats and severe habitat reduction during dry periods, and tropical urban streams consequently support highly depleted and simplified invertebrate assemblages (Victor *et al.*, 1996; Cleto-Filho and Walker, 2001; Rebeca De Jesus, Department of Biology, University of Puerto Rico, and A. Ramírez, unpublished data).

In addition to changed flow regime, the water quality of tropical urban streams is impaired, since runoff carries contaminants (e.g. hydrocarbons, sediments, and nutrients) and dumping of untreated wastewater including common sewage and industrial wastes is common. Most industries in Asia discharge directly into rivers without proper wastewater treatment (Dudgeon, 2000a, b; Dudgeon *et al.*, 2000). Similarly, in Latin America, less than 2% of total urban sewage receives treatment before it is discharged into rivers (Pringle *et al.*, 2000b). Urban sewage not only results in expected decreases in oxygen and reduction of biodiversity (Daniel *et al.*, 2002), but such pollution can facilitate the spread of major human diseases (Pacini and Harper, 2000) and contaminate drinking water leading to a variety of health problems (Witt and Reiff, 1991).

An example of the severity of pollution in many urban areas is provided by the Tiete River, which flows 1120 km from the Atlantic Coast mountain range to the Parana River and subsequently empties into the Atlantic Ocean at Buenos Aires. The Tiete also passes through Sao Paulo (Brazil), a metropolitan area of more than 20 million people, and during the dry season an estimated 60% of the river's discharge consists of untreated residential wastes from the city. Industry adds another 4.5 t of chemical wastes and heavy metals each day (Switkes, 1995). Attempts to clean up the Tiete began in 1991 and were stimulated by public health concerns; they have since received international attention (Switkes, 1995). Measures include connecting 70% of urban residences to sewage treatment plants and control of industrial effluents, at an expected cost of over US $1 billion (Csillag, 2000).

More generally, urban stream management in tropical catchments requires the use of practices that have been established elsewhere, such as separation of wastewater discharge from rainfall runoff. In addition, there is a need to develop new strategies to deal with the large amount of runoff produced by tropical rainfall and can potentially carry vast quantities of contaminants to streams. There is a clear need for initiatives that incorporate environmental education to highlight the economic and social benefits of healthy streams and riparian zones (e.g. Pringle, 1997b).

## D. Alteration of Hydrologic Connectivity

Connectivity is a necessary component of ecosystem integrity in tropical rivers. Loss of longitudinal and lateral riverine connectivity is one of the main results of river regulation, damming, and water abstraction. Most lowland river biota rely on lateral connectivity with the floodplain, which is inundated periodically. Floodplain inundation provides resources for reproduction and growth for many species (Bunn and Arthington, 2002), and it is essential for riverine fisheries (Winemiller and Jepsen, 1998; Welcomme, 2000; Wantzen *et al.*, 2002; Junk and Wantzen, 2004; see also Chapter 5 of this volume). Migratory fauna (e.g. fishes, shrimps, snails) rely on longitudinal connectivity to move from one habitat to another as required by their life cycles (e.g. Dudgeon, 1992, 2000c; Winemiller and Jepsen, 2004). Disruptions to the movement of animals and matter along a river network and from the river to the floodplain can have profound negative implications for ecosystem integrity (Pringle, 1997a; Pringle *et al.*, 2000b; Bunn and Arthington, 2002; Agostinho *et al.*, 2004).

A synthesis paper by Pringle *et al.* (2000a) indicates that tropical stream ecosystems are very vulnerable to fragmentation by dams given their high degree of faunal endemism, the extent of faunal migratory behavior (e.g. potamodromy and amphidromy); the importance of seasonal inundation of floodplains for migration; and the adverse physical or chemical conditions often created in tropical reservoirs and tailwaters. The development of large dams in tropical latitudes is recent, relative to the era of large dams in north-temperate regions such as the United States. Over 70 large dams are planned for Brazil's Amazonian region alone (Fearnside, 1995). Effects on river ecology could be severe, and hydropower dams are considered as potentially the most dangerous human activity to Amazonian fisheries (Goulding *et al.*, 1996). Concern has been raised over the effects of huge dam arrays planned (and vast schemes now completed) on some of the great rivers of Asia (Dudgeon, 1992, 2000c, 2005).

Conservation research needs for dams planned for tropical rivers include pre-impoundment surveys and studies of aquatic biota, from headwaters to mouth, before dams are built, and evaluation of the applicability of hydropower technology developed for temperate regions to tropical regions (Pringle, 2000; Pringle *et al.*, 2000a). For example, assessment of the overall effects of dams on biodiversity has been hindered by a lack of pre-impoundment data: construction programs for the first five major dams in Amazonia did not include broad-scale investigations of fish migrations before the impoundments were closed (Goulding *et al.*, 1996).

Tropical dam builders have often assumed that fish pass facilities are not necessary and even when fishways are built they have often been based on the salmon (anadromous) fish pass model and are thus impassable for many native potamadromous species (Quiros, 1988; McCully, 1996; Dudgeon, 2000c). Other examples of the effects of dams are given in Section V-B.

## IV. REGIONAL CONSERVATION ISSUES

While several detailed summaries have been published on river conservation issues for specific tropical river catchments, most have focused on large river systems and issues related to stream conservation have been relatively neglected. Here, we briefly summarize key conservation issues for rivers and streams draining general geographic regions in the tropics.

### A. Latin America and the Caribbean Islands

Rivers and streams draining Latin America and the Caribbean islands contain valuable aquatic resources and provide important ecosystem services for humans. Several areas have been classified as biodiversity 'hot spots' (e.g. the Brazilian Atlantic Forest, Chilean Winter Rainfall–Valdivian Forest) and the region contains 'megadiverse' countries (e.g. Ecuador, Peru, Colombia: Mittermeier *et al.*, 1999). River ecosystems in Latin America in general, and the tropical streams of that region in particular, are facing threats from deforestation, agriculture, human population growth, and hydropower generation; the resulting conservation challenges are immense (see reviews by Pringle and Scatena, 1999; Pringle *et al.*, 2000b). A series of reports by the United Nations Economic Commission for Latin America and the Caribbean (ECLAC, 1990) also provides important information on water resources and river conservation problems in the region. Several authors have analyzed recent issues in river conservation in Latin American streams (Ometo *et al.*, 2000; Neill *et al.*, 2001; Branco and Pereira, 2002; Daniel *et al.*, 2002; Moulton and Magalhães, 2003; Gerhard *et al.*, 2004; Jergentz *et al.*, 2004; Biggs *et al.*, 2005; Blanco and Scatena, 2006; Wantzen, 2006; Wantzen *et al.*, 2006). Environmental problems in Latin America are mostly the result of rapid development in urban and agricultural areas at the expense of maintaining the conditions necessary for sustaining healthy ecosystems. Although a few conservation programs are underway, there is still much to be done to maintain the integrity of riverine systems in the region, and government initiatives are needed.

### B. Tropical Africa

Streams in tropical Africa are known for their high biodiversity. There are well over 3000 described species of freshwater fishes (Lévêque, 1997) and some expect that number to double (Ribbink, 1994). Invertebrate diversity has been less studied and its diversity remains to be completely assessed. Tropical Africa also has several biodiversity 'hot spots' (Mittermeier *et al.*, 1999). One of them, the Guinean Forest of West Africa, has a rich fish diversity, and some 35% of more than 500 species known from this area are endemic (http://www.biodiversityhotspots.org). River conservation issues in African countries were recently reviewed by Pacini and Harper (2000) and Davis and Wishart (2000), and Lévêque (1997) has described conservation concerns relevant to African freshwater fishes. Overall, conservation efforts for tropical African freshwaters have fallen behind efforts intended to protect forests and other terrestrial ecosystems. Major impacts have arisen from hydropower generation and dams for irrigation. Thus far, industrial and urban impacts remain localized, but pollution from agriculture on and the use of chemicals to control disease vectors are matters for concern. Overfishing has impacted many aquatic ecosystems, and the introduction of exotic species that compete with (and can extirpate) native species has contributed to significant biodiversity losses, especially in lakes (Lévêque, 1997).

## C. Madagascar

Madagascar is considered a 'hot spot of biodiversity' (McNeeley *et al.*, 1990), and the island's streams are characterized by a large number of endemic fishes, crustaceans, molluscs, and aquatic insects (Benstead and Pringle, 2004). Madagascan streams face large conservation challenges, as discussed in recent reviews by Benstead *et al.* (2000, 2003a). Deforestation, sedimentation, overfishing, and introduction of exotic species are among the major conservation concerns (Benstead *et al.*, 2000). Some of the main drivers behind stream degradation are rapid human population growth and a weak economy. Approximately 2.7% of Madagascar is included within several categories of protected areas (e.g. national parks and reserves), and the government has an aggressive plan to increase their extent over the next few years (http://www.biodiversityhotspots.org/).

## D. Tropical Asia-Pacific

Tropical Asia and adjacent Pacific islands are very diverse and species rich. Conservation International has identified at least seven biodiversity hot spots in the region (e.g. the Western Ghats and Sri Lanka, the Philippines: Mittermeier *et al.*, 1999). Stream diversity is rich: the Indonesian islands, for example, have a high number of endemic aquatic insects (e.g. Odonata), frogs, and fishes (Dudgeon *et al.*, 2000). Some streams in Asia were among the earliest ecosystems to experience human civilization and its impacts. Humans probably started to conduct agriculture in the region some 5000 years ago (and perhaps much earlier), and dams and irrigation systems were present in China over 4000 years ago. Several recent reviews of river conservation, with particular reference to biodiversity are available for Southeast Asia (Dudgeon, 2000a, b, c, 2005; Dudgeon *et al.*, 2000), India (Gopal, 2000), and Asia (Dudgeon *et al.*, 2000; Li *et al.*, 2000). Major conservation issues are related to deforestation, soil erosion, pollution, agricultural runoff, and river regulation (Gopal, 2000; Dudgeon *et al.*, 2000, Dudgeon, 2000a, c, 2005). Conservation efforts in the region are constrained by the lack of information available on stream ecosystems (e.g. Dudgeon, 2000b, 2003), but certain restoration actions and better ecosystem management are possible based on existing data and understanding of tropical Asian rivers given the necessary political will or social impetus (Dudgeon, 2003, 2005).

## E. Tropical Australia

Part of northern Australia is tropical, with strong dry-wet seasonality. Streams and rivers show marked flow periodicity and are rich in bivalve molluscs, insects, and other invertebrates (Outridge, 1987) and fishes (Herbert *et al.*, 1994). Stream conservation issues in Australia were summarized by Schofield *et al.* (2000). Major problems for tropical Australian streams are related to deforestation and sedimentation, irrigation and salinization, and contamination with pesticides in agricultural runoff (Schofield *et al.*, 2000).

## V. CASE STUDIES

### A. Erosion-Prone Soils in the Brazilian Cerrado

The Brazilian Cerrado is a large biome supporting a highly-diverse flora and fauna that is adapted to marked changes between dry and wet seasons, intrusion by cold air masses, and recurring fire events (Gottsberger and Silberbauer-Gottsberger, 2006). Cerrado vegetation once covered about 2 million km² or 20% of the Brazilian territory. After aggressive government-led development programs in the 1970s and 1980s, more than half of the Cerrado was destroyed and converted into agriculture (Mittermeier *et al.*, 1999), mainly for soy bean, corn, sugar cane, and cotton. Streams and their riparian zones are now protected by law, but enforcement is lacking. Human impacts are evident and include selective logging, poaching, invasion of cattle and goats, construction of aquaculture ponds, dams and irrigation, and pesticide spills. An overriding problem is stream siltation due to gully erosion from dirt roads and sediments from gold and diamond mining (Wantzen, 2006; see also Section III-A). Erosion gullies drain riparian wetlands, change the vegetation structure and release large amounts of carbon from drying soils (K.M. Wantzen, unpublished observations). All rivers and streams flowing toward the Pantanal, which is the largest wetland in the world, carry excess sediment loads that impact riparian vegetation, destroy spawning habitats for fishes, and block secondary channels that connect main channels to floodplain lakes.

Attempts are being made to mitigate siltation impacts on streams. State governments in Brazil are now requiring farmers to prove that they are applying all possible techniques to reduce erosion. Some suggested technical solutions are expensive and unlikely to be used widely in impoverished areas, but farmers are developing innovative solutions. Construction of small dams along gullies is one option, creating impoundments that can serve as fish ponds. At the same time, the reservoir increases soil fertility around the gully and allows the reestablishment of native vegetation reducing erosion. The economic return from the fishpond often compensates for the investment in dam construction (Wantzen *et al.*, 2006).

Erosion from areas of intense agriculture can be a serious problem for stream conservation (see Section III-B), but a viable response seems to be reforestation of riparian zones to buffer the impact of agriculture on streams. In addition to stream protection, riparian reforestation can allow reconnection of isolated 'islands' of Cerrado vegetation into integrated corridors. The chances of success of this approach will be enhanced by combining reforestation with the selective use of non-wood products by human populations, providing added economic incentives for stakeholders (Wantzen *et al.*, 2006).

## B. The Effects of Dams on Puerto Rican Streams

Streams draining Puerto Rico are fragmented by many dams, including 25 exceeding 15 m in height (Fig. 1), and are hydrologically altered by low-head (2–3 m) dams and associated water withdrawals. All but one of the nine stream drainages within the Caribbean National Forest (CNF) in northeastern Puerto Rico have low-head dams and water intakes on their main channels. A recent water budget indicates that, on an average day, up to 70% of the water draining the CNF is withdrawn for municipal water supplies (Crook, 2005). Water withdrawals can lead to direct mortality of the aquatic biota, while dams affect their dispersal to varying degrees (Holmquist *et al.*, 1998; March *et al.*, 2003), with their combined impacts potentially affecting the ecological integrity of the forest and stream ecosystem processes (Pringle 1997a; Greathouse *et al.*, 2006). Large dams with no spillway discharge (i.e. water flow over the face of the dam) are complete barriers to the migration of native shrimps and fishes in Puerto Rico; low-head dams are also problematic since water extraction for municipal water leads to direct mortality of drifting shrimp larvae, whereas the concrete barrier and low water flows behind the dam impedes upstream migration of juvenile shrimps (Benstead *et al.*, 1999; Greathouse *et al.*, 2006).

Loss of connectivity and stream fragmentation can even result from what appear to be minor channel modifications. For example, adults of *Neritina* (Neritidae) snails that live in Puerto Rican streams, and other tropical coastal streams, have planktonic larvae that must develop

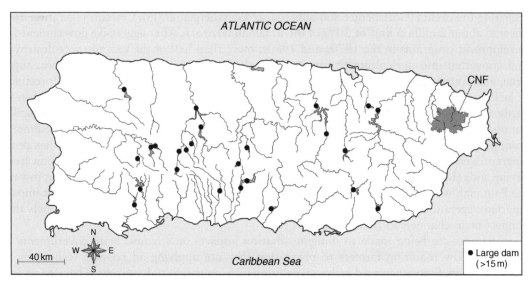

*FIGURE 1* The location of large (>15 m) dams in Puerto Rico: CNF = Caribbean National Forest.

in marine environments. Upstream migration of juveniles is essential to maintain freshwater populations of neritid snails (Blanco and Scatena, 2006). Juvenile snails prefer to use areas of shallow, fast-flowing water during upstream migrations, and channel modifications for flood control that transform whole reaches into a single, deep run can prevent upstream movements (Blanco, 2005).

Effective communication by ecologists and environmental groups of the results of research showing the impacts of dam migratory biota, stream connectivity, and ecosystem integrity have produced some responses. The Puerto Rican Aqueduct and Sewage Authority has altered the design of two new water withdrawal systems to minimize mortality of migrating stream animals and maintain baseflow. Water withdrawal from an intake on one river has been prohibited during peak period of downstream drift of shrimp larvae (1900–2300 h) and a fish ladder has been installed (March *et al.*, 2003). Additional conservation measures are necessary to protect the biotic integrity of streams draining Puerto Rico (examples are given by March *et al.*, 2003), especially since climate-change scenarios predict reductions in the island's rainfall (Wang *et al.*, 2003) that are liable to lead to increased extraction of stream water and further threat to streams on Puerto Rico.

## C. Forest Cover Increases in Puerto Rico

The history of deforestation in Puerto Rico is similar to that observed in other tropical regions (Section III-A); by the late 1940s, only around 7% of the island remained under forest with agriculture comprising the main economic activity (Grau *et al.*, 2003). However, a series of socioeconomic changes initiated soon after led to an increase in industrialization. The result was considerable movement of people from the country to the cities, and abandonment of marginal agricultural lands. Such lands became secondary forest and forest cover in Puerto Rico forest cover has been steadily increasing overt the last six decades (Grau *et al.*, 2003).

The increase in forest cover is proving positive for stream ecosystems and human populations. The effect is particularly evident during the hurricane season, when frequent heavy tropical storm systems impact the island. Forested catchments hold more water during and after rainstorms, and streams draining them tend to have less extreme hydrographs, reducing the incidence and intensity of flash floods and their impacts on humans and stream ecosystems. The beneficial effect of newly-established forests was clearly evident during a storm that impacted Puerto Rico and Hispaniola during May 2004. The storm moved over both islands with similar strength producing similar amounts of precipitation, but had devastating effects only in Hispaniola where there is much less forest and where flooding resulted in the loss of human lives and property. In Puerto Rico, the intense rains resulted in increased stream flow and also caused some flooding, but the impacts were more limited and localized (Aide and Grau, 2004). Although we still have limited information on the impacts of the newly-established forest on stream ecosystems in Puerto Rico, a reduction in flash flooding is likely to increase retention of nutrients and organic matter in streams (Biggs *et al.*, 2002).

The abandonment of agriculture in Puerto Rico has also had a positive effect on water quality, as chemically-loaded runoff into streams has decreased. The Fajardo catchment, located in northeast Puerto Rico, is a good example with records of water quality dating back to the 1970s. Land use changed from sugar cane plantations in the 1950 to urban uses and forest in 2000. Over approximately the same period (1950–2000), human populations grew by 182%, increasing demands on stream water (Jorge Ortiz-Zayas, Institute for Tropical Ecosystem Studies, University of Puerto Rico, unpublished information). Despite the growing population, the changes in land use were reflected by improved stream water quality. Total nitrogen and phosphorus concentrations (Fig. 2) and fecal coliform bacteria (Fig. 3) steadily decreased from 1973 to 1998 (Jorge Ortíz-Zayas, unpublished data). Planned changes in waste and drinking

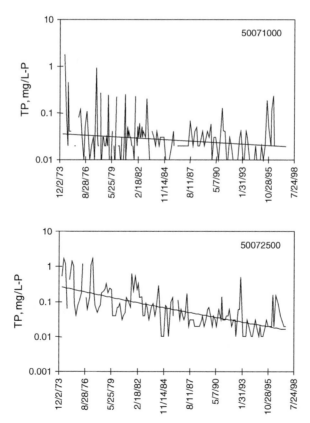

FIGURE 2 Total phosphorus (TP) concentrations at two sites along the Fajardo River, Puerto Rico, from 1973 to 1998. Steeper declines have occurred at the lower site (lower graph) where land use was formerly dominated by sugar cane cultivation.

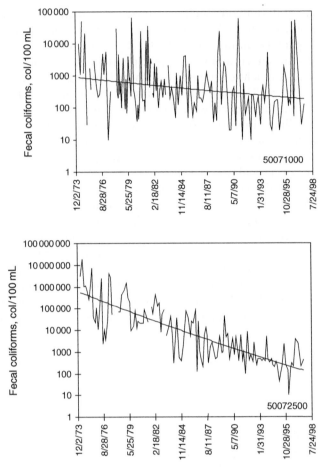

FIGURE 3 Concentrations of fecal coliform bacteria at two sites in the Fajardo River, Puerto Rico, from 1973 to 1998. Steeper declines in the lower course (lower graph) appear to be related to abandonment of cattle ranching in the catchment.

water treatment in the area promises further benefits to the Fajardo catchment, and it is hoped to increase the amounts of water flowing along stream channels during periods of low discharge.

Environmental improvements in Puerto Rico are possible due to the presence of relevant legislation and the willingness of the government to spend money and allocate resources. The association of Puerto Rico with the United States was a key component in the industrialization of the island, and plays an important role in the enforcement of environmental legislation. This is a rather unusual situation and thus does not represent a model applicable to other tropical regions. Nonetheless, changes in Puerto Rico have provided valuable information on how tropical streams respond to changes in land use, water treatment facilities, and to the distribution of humans in the landscape, and indicate how tropical streams might respond to improvements in water quality and quantity.

## VI. SUMMARY – WHAT MIGHT THE FUTURE HOLD?

Environmental problems facing tropical streams are not easy to address, mainly because most countries lack the necessary resources and/or political will to protect these vulnerable ecosystems. Conservation is normally seen as a luxury and a process that should be started once countries reach a certain level of development. As has occurred during the history of developed countries, stream conservation is seen as an activity that will take place at some point in the future, but at present, it is not something that most tropical countries can afford (Dudgeon, 2000b, 2003; Pringle *et al.*, 2000b). Given the complex drinking water and health problems that will become widespread if pollution and stream degradation continue at their current pace, investment in the protection of stream ecosystems might be a cheaper alternative that 'fixes' to these problems in the long term. A few tropical countries have begun to start protecting their streams (see case studies in Section V), and while specific economic and strategic formulae are beyond the scope of this chapter, their experiences provide useful information about possible future scenarios for tropical streams.

There is certainly a need to conduct additional research to understand how humans are impacting tropical stream ecosystems, but conservation efforts could benefit from better communication of effective solutions to problems encountered different parts of the tropics; better exchange of existing ecological information is critically needed also. For example, the development of an inter-tropical database on studies of anthropogenic impacts on streams, together with information on mitigation or restoration methods applied, would be a useful first step. In addition, there is a need to enforce existing legislation to protect riparian zones and streams. 'Decriminalization' of farmers and the rural poor, who are often considered as willing and active destroyers of natural habitats, is necessary given that such people are usually driven by economic necessity. Improvements in environmental education might prove successful in changing attitudes toward stream ecosystems, as would a better appreciation of the economic values of intact riparian zones or forest buffers. Valuation in terms of carbon units may offer an additional economic incentive, because carbon additions to the atmosphere are especially high from the type of soils that are most sought after for agriculture (Bellamy *et al.*, 2005).

There are several key knowledge gaps evident in our understanding of how tropical stream ecosystems respond to human impacts that need to be addressed. Firstly, as economic growth is a primary goal of all tropical nations, there is a need to search for commonalities on the impacts that humans have on streams as nations strive to become wealthier. Such commonalities might prove useful in developing effective conservation strategies and in learning from others experiences if they were communicated effectively. Secondly, there is a need to look into ecological concepts or theories relating to streams, as they are mainly based on studies undertaken in temperate regions. While tropical streams similarly might respond to environmental change, some aspects of their ecology may differ from their temperate counterparts. This matter is considered in more detail in Chapter 9 of this volume. Some examples of likely differences that are likely to have important implications for conservation and management include the fact that the range of allochthonous food resources in tropical streams seems to be wider than in temperate latitudes (see Chapter 3 of this volume). In addition, the drivers of stream seasonality change from day length and temperature in temperate latitudes to hydrological periodicity in the tropics (Junk and Wantzen, 2004). Furthermore, rainfall and runoff appear to be more intense in some tropical climates, causing severe conservation problems where they are associated with deforestation, soil erosion, and urbanization. A third need is to make more use of the tropical biota as indicators of stream health. Biotic indicators have been used to successfully monitor stream health or integrity in many temperate regions, and form part of some environmental education programs. Most tropical countries lack geographically-relevant biotic indices and bioindicators are not employed widely. If indices from temperate streams are adopted in the tropics, critical evaluation of their performance will be essential, and appropriate adjustments made to suit the composition of the local biota. The use of reference sites to evaluate individual streams might be an appropriate model for tropical countries intending to monitor stream health. This technique has been applied in Australia for some time (Bailey *et al.*, 2003) and has proven useful when a local reference condition (i.e. the 'natural' condition) is known. This approach to biomonitoring does not require information about the pollution tolerance or life history of organisms, and thus holds promise for tropical streams where the ecology and habitat of most of the biota are very incompletely known.

## ACKNOWLEDGMENTS

Alonso Ramírez and C. M. Pringle were partially funded by the Luquillo LTER program (NSF-DEB 0218039). Karl M. Wantzen was partially funded by the Deutsche Forschungsgemeinschaft (DFG WA 1612) and received travel funds from the DLR (BRA 02/26). The authors are grateful to two anonymous reviewers for comments on the manuscript.

## REFERENCES

Agostinho, A. A., Thomaz, S. M., and Gomes, L. C. (2004). Threats for biodiversity in the floodplain of the Upper Parana River: effects of hydrological regulation by dams. *Ecohydrology and Hydrobiology* 4, 267–280.

Aide, M. T., and Grau, H. R. (2004). Globalization, migration, and Latin American ecosystems. *Science* 305, 1915–1916.

Allan, J. D. (2004). Landscapes and riverscapes: the influence of land use on stream ecosystems. *Annual Review of Ecology and Systematics* 35, 257–284.

Allan, J. D., Abell, R., Hogan, Z., Revenga, C., Taylor, B. W., Welcomme, R. L., and Winemiller, K. O. (2005). Overfishing of inland waters. *BioScience* 55, 1041–1051.

Alva, C. A., van Alphen, J. G., de la Torre, A., and Manrique, L. (1976). "Problemas de Drenaje y Salinidad den la Costa Peruana." International Institute for Land Reclamation and Improvement, Wageningen, The Netherlands.

Andreoli, C. (1993). The influence of agriculture on water quality. *In* "Prevention of Water Pollution by Agriculture and Related Activities" (J. A. Sagardoy, Ed.), pp. 53–65. FAO, Rome, Italy.

Bailey, R. C., Norris, R. H., and Reynoldson, T. B. (2003). "Bioassessment of Freshwater Ecosystems: Using the Reference Condition Approach." Springer-Verlag, New York, USA.

Barthem, R. B., and Goulding, M. (1997). "The Catfish Connections: Ecology, Migration, and Conservation of Amazon Predators." Columbia University Press, New York, USA.

Bellamy, P. H., Loveland, P. J., Bradley, R. I., Lark, R. M., and Kirk, G. J. D. (2005). Carbon losses from all soils across England and Wales 1978–2003. *Nature* 437, 245–248.

Benstead, J. P., and Pringle, C. M. (2004). Deforestation alters the resource base and biomass of endemic stream insects in eastern Madagascar. *Freshwater Biology* 49, 490–501.

Benstead, J. P., March, J. G., Pringle, C. M., and Scatena, F. N. (1999). Effects of a low-head dam and water abstraction on migratory tropical stream biota. *Ecological Applications* 9, 656–668.

Benstead, J. P., Stiassny, M. L. J., Loiselle, P. V., Riseng, K. J., and Raminosoa, N. (2000). River conservation in Madagascar. *In* "Global Perspectives on River Conservation: Science, Policy and Practice" (P. J. Boon, B. R. Davis, and G. E. Petts, Eds), pp. 205–231. John Wiley and Sons Ltd., New York, USA.

Benstead, J. P., De Rham, P. H., Gattolliat, J. L., Gibon, F. M., Loiselle, P. V., Sartori, M., Sparks, J. S., and Stiassny, M. L. J. (2003a). Conserving Madagascar's freshwater biodiversity. *BioScience* 53, 1101–1111.

Benstead, J. P., Douglas, M. M., and Pringle, C. M. (2003b). Relationships of stream invertebrate communities to deforestation in eastern Madagascar. *Ecological Applications* 13, 1473–1490.

Biggs, T. W., Dunne, T., Domingues, T. F., and Martinelli, L. A. (2002). Relative importance of natural watershed properties and human disturbance on stream solute concentration in the southwestern Brazilian Amazon basin. *Water Resources Research* 38, 1150, DOI: 1110.1029/2001 WR000271.

Biggs, T. W., Dunne, T., and Martinelli, L. A. (2005). Natural controls and human impacts on stream nutrient concentrations in a deforested region of the Brazilian Amazon basin. *Biogeochemistry* 68, 227–257.

Blanco, J. F. (2005). "Physical Habitat, Disturbances, and the Population Ecology of the Migratory Snail *Neritina virginea* (Gastropoda: Neritidae) in Streams of Puerto Rico." Unpublished Ph.D. Thesis, University of Puerto Rico, Puerto Rico.

Blanco, J. F., and Scatena, F. N. (2006). Hierarchical contribution of river–ocean connectivity, water chemistry, hydraulics and substrate to the distribution of diadromous snails in Puerto Rico streams. *Journal of the North American Benthological Society* 25, 82–98.

Boon, P. J., Davis, B. R., and Petts, G. E. (2000). "Global Perspectives on River Conservation: Science, Policy and Practice." John Wiley and Sons Ltd., New York, USA.

Branco, L. H. Z., and Pereira, J. L. (2002). Evaluation of seasonal dynamics and bioindication potential of macroalgal communities in a polluted tropical stream. *Archive für Hydrobiologie* 155, 147–161.

Bunn, S. E., and Arthington, A. H. (2002). Basic principles and ecological consequences of altered flow regimes for aquatic biodiversity. *Environmental Management* 30, 492–507.

Cleto-Filho, S. E. N., and Walker, I. (2001). Effects of urban occupation on the aquatic macroinvertebrate from a small stream of Manaus, Amazonas State, Brazil. *Acta Amazonica* 31, 69–89.

Crook, K. E. (2005). "Quantifying the Effects of Water Withdrawal on Streams Draining the Caribbean National Forest, Puerto Rico." Unpublished Master's Thesis, The University of Georgia, GA, USA.

Csillag, C. (2000). Environmental health in Brazil. *Environmental Health Perspectives* 108, 504–511.

Daniel, M. H. B., Montebelo, A. A., Bernardes, M. C., Ometto, J. P. H. B., De Camargo, P. B., Krusche, A. V., Ballester, M. V., Victoria, R. L., and Martinelli, L. A. (2002). Effects of urban sewage on dissolved oxygen, dissolved inorganic and organic carbon, and electrical conductivity of small streams along a gradient of urbanization in the Piracicaba River basin. *Water, Air, and Soil Pollution* 136, 189–206.

Davis, B. R., and Wishart, M. J. (2000). River conservation in the countries of the Southern African Development Community (SADC). *In* "Global Perspectives on River Conservation: Science, Policy and Practice" (P. J. Boon, B. R. Davis, and G. E. Petts, Eds), pp. 179–204. John Wiley and Sons Ltd., New York, USA.

Dudgeon, D. (1992). Endangered ecosystems: a review of the conservation status of tropical Asian rivers. *Hydrobiologia* 248, 167–191.

Dudgeon, D. (2000a). The ecology of tropical Asian rivers and streams in relation to biodiversity conservation. *Annual Review of Ecology and Systematics* 31, 239–263.

Dudgeon, D. (2000b). Riverine biodiversity in Asia: a challenge for conservation biology. *Hydrobiologia* 418, 1–13.

Dudgeon, D. (2000c). Large-scale hydrological alterations in tropical Asia: prospects for riverine biodiversity. *BioScience* 50, 793–806.

Dudgeon, D. (2003). The contribution of scientific information to the conservation and management of freshwater biodiversity in tropical Asia. *Hydrobiologia* 500, 295–314.

Dudgeon, D. (2005). River rehabilitation for conservation of fish biodiversity in monsoonal Asia. *Ecology and Society* 10, 15. http://www.ecologyandsociety.org/vol10/iss2/art15/.

Dudgeon, D., Choowaew, S., and Ho, S. C. (2000). River conservation in south-east Asia. *In* "Global Perspectives on River Conservation: Science, Policy and Practice" (P. J. Boon, B. R. Davis, and G. E. Petts, Eds), pp. 281–310. John Wiley and Sons Ltd., New York, USA.

Dudgeon, D., Arthington, A. H., Gessner, M. O, Kawabata, Z., Knowler, D., Lévêque, C., Naiman, R. J., Prieur-Richard, A.-H., Soto, D., Stiassny, M. L. J., and Sullivan, C. A. (2006). Freshwater biodiversity: importance, threats, status and conservation challenges. *Biological Reviews* 81, 163–182.

ECLAC (1990). "Latin America and the Caribbean: Inventory of Water Resources and their Use. Volume I: Mexico, Central America, and the Caribbean." United Nations Economic Commission for Latin America and the Caribbean (ECLAC), Santiago, Chile.

FAO (Food and Agriculture Organization of the United Nations) (1999). "State of the World's Forests 1999." FAO, Rome, Italy.

Fearnside, P. M. (1995). Hydroelectric dams in the Brazilian Amazon as sources of 'greenhouse' gases. *Environmental Conservation* 22, 7–19.

Fearnside, P. M. (2005). Deforestation in Brazilian Amazonia: history, rates, and consequences. *Conservation Biology* 19, 680–688.

Forti, M. C., Boulet, R., Melfi, A. J., and Neal, C. (2000). Hydrogeochemistry of a small catchment in Northeastern Amazonia: a comparison between natural with deforested parts of the catchment (Serra do Navio, Amapa State, Brazil). *Water, Air, and Soil Pollution* 118, 263–279.

Fossati, O., Wasson, J. G., Cecile, H., Salinas, G., and Marin, R. (2001). Impact of sediment releases on water chemistry and macroinvertebrate communities in clear water Andean streams (Bolivia). *Archive für Hydrobiologie* 151, 33–50.

Gentry, A. H., and Lopez-Parodi, J. (1980). Deforestation and increased flooding of the upper Amazon. *Science* 210, 1354–1356.

Gerhard, P., Moraes, R., and Molander, S. (2004). Stream fish communities and their associations to habitat variables in a rain forest reserve in southeastern Brazil. *Environmental Biology of Fishes* 71, 321–340.

Gomez-Sanchez, C. E. (1993). The influence of the agriculture on water quality in Colombia. *In* "Prevention of Water Pollution by Agriculture and Related Activities" (J. A. Sagardoy, Ed.), pp. 93–101. FAO, Rome, Italy.

Gopal, B. (2000). River conservation in the Indian sub-continent. *In* "Global Perspectives on River Conservation: Science, Policy and Practice" (P. J. Boon, B. R. Davis, and G. E. Petts, Eds), pp. 233–261. John Wiley and Sons Ltd., New York, USA.

Gottsberger, G., and Silberbauer-Gottsberger, I. (2006). "Life in the Cerrado." Reta Verlag, Ulm, Germany.

Goulding, M., Smith, N. J. H., and Majar, D. J. (1996). "Floods of Fortune: Ecology and Economy along the Amazon." Columbia University Press, New York, USA.

Grainger, A. (1990). "The Threatening Desert." Earthscan Publications Ltd., London, UK.

Grau, H. R., Aide, T. M., Zimmerman, J. K., Thomlinson, J. R., Helmer, E., and Zou, X. (2003). The ecological consequences of socioeconomic and land-use changes in postagriculture Puerto Rico. *BioScience* 53, 1159–1168.

Greathouse, E. A., March, J. G., and Pringle, C. M. (2005). Recovery of a tropical stream after a harvest-related chlorine poisoning event. *Freshwater Biology* 50, 603–615.

Greathouse, E. A., Pringle, C. M., McDowell, W. H., and Holmquist, J. G. (2006). Indirect upstream effects of dams: consequences of migratory consumer extirpation in Puerto Rico. *Ecological Applications* 16, 339–352.

Hartshorn, G. S., Hartshorn, L., Atmeila, A., Gomez, L. D., Mata, A., Matta, L., Morales, R., Ocampo, R., Pool, O., Quesada, C., Solera, C., Solorzano, R., Stiles, G., Tosi, J., Umana, A., Villalibos, C., and Wells, R. (1982). "Costa Rica Country Environmental Profile: A Field of Study." Tropical Science Center, San Jose, Costa Rica.

Herbert, B., Peeters, J., Graham, P., and Hogan, A. (1994). "Natural Resources Assessment Program – Fish Fauna Survey." Project NR10, CYPLUS Report, Queensland Department of Primary Industry, Brisbane, Australia.

Holloway, L. (2000). Catalysing rainforest restoration in Madagascar. *In* "Diversite et Endemisme a Madagascar" (W. R. Lourenco and S. M. Goodman, Eds), pp. 115–124. Societe de Biogeographie, Paris, France.

Holmquist, J. G., Schmidt-Gengenbach, J. M., and Yoshioka, B. B. (1998). High dams and marine-freshwater linkages: effects on native and introduced fauna in the Caribbean. *Conservation Biology* 12, 621–630.

Jergentz, S., Mugni, H., Bonetto, C., and Schulz, R. (2004). Runoff-related endosulfan contamination and aquatic macroinvertebrate response in rural basins near Buenos Aires, Argentina. *Archives of Environmental Contamination and Toxicology* 46, 345–352.

Junk, W. J., and Wantzen, K. M. (2004). The Flood Pulse Concept. New aspects, approaches, and applications – an update. *In* "Proceedings of the 2nd Large River Symposium (LARS), Phnom Penh, Cambodia" (R. Welcomme and T. Petr, Eds), pp. 117–149. Food and Agriculture Organization and Mekong River Commission, RAP Publication, 2004/16, Bangkok, Thailand.

Laabs, V. W., Amelung, A. A., Pinto, M., Wantzen, K. M., da Silva, C. J., and Zech, W. (2002). Pesticides in surface water, sediments and rainfall of the north-eastern Pantanal basin, Brazil. *Journal of Environmental Quality* 31, 1636–1648.

Lawton, R. O., Nair, U. S., Pielke, R. A., and Welch, R. M. (2001). Climatic impact of tropical lowland deforestation on nearby montane cloud forest. *Science* 294, 584–587.

Lévêque, C. (1997). "Biodiversity Dynamics and Conservation: The Freshwater Fish of Tropical Africa." Cambridge University Press, Cambridge, UK.

Li, L., Liu, C., and Mou, H. (2000). River conservation in central and eastern Asia. *In* "Global Perspectives on River Conservation: Science, Policy and Practice" (P. J. Boon, B. R. Davis, and G. E. Petts, Eds), pp. 263–279. John Wiley and Sons Ltd., New York, USA.

March, J. G., Benstead, J. P., Pringle, C. M., and Scatena, F. N. (2003). Damming tropical island streams: problems, solutions, and alternatives. *Bioscience* 53, 1069–1078.

McCully, P. (1996). "Silenced Rivers: The Ecology and Politics of Large Dams." Zed Books Ltd., London, UK.

McNeeley, J. A., Miller, K. R., Reid, W. V., Mittermeiner, R. A., and Wener, T. B. (1990). "Conseving the World's Biological Diversity." International Union for the Conservation of Nature (IUCN), Gland, Switzerland.

Mittermeier, R. A., Myers, N., and Mittermeier, C. G. (1999). "Hot spots – Earths Biologically Richest and Most Endangered Terrestrial Ecoregions." CEMEX, Conservation International, New York, USA.

Mol, J. H., and Ouboter, P. E. (2004). Downstream effects of erosion from small-scale gold mining on the instream habitat and fish community of a small Neotropical rainforest stream. *Conservation Biology* 18, 201–214.

Moulton, T. P., and Magalhães, S. A. P. (2003). Responses of leaf processing to impacts in streams in Atlantic Rain Forest, Rio de Janeiro, Brazil – a test of the biodiversity-ecosystem functioning relationship? *Brazilian Journal of Biology* 63, 87–95.

Moulton, T. P., and Wantzen, K. M. (2006). Conservation of tropical streams – special questions or conventional paradigms? *Aquatic Conservation* 16, 659–663.

Neill, C., Deegan, L. A., Thomas, S. M., and Cerri, C. C. (2001). Deforestation for pasture alters nitrogen and phosphorus in small Amazonian streams. *Ecological Applications* 11, 1817–1828.

Ometo, J. P. H. B., Martinelli, L. A., Ballester, M. V., Gessner, A., Krusche, A. V., Victória, R. L., and Williams, M. (2000). Effects of land use on water chemistry and macroinvertebrates in two streams of the Piracicaba River Basin, South-East Brazil. *Freshwater Biology* 44, 327–337.

Ortíz-Zayas, J. R., and Scatena, F. N. (2004). Integrated water resources management in the Luquillo Mountains: an evolving process. *International Journal of Water Resources Development* 20, 387–398.

Outridge, P. M. (1987). Possible causes of high species diversity in tropical Australian freshwater macrobenthic communities. *Hydrobiologia* 150, 95–107.

Pacini, N., and Harper, D. M. (2000). River conservation in central and tropical Africa. *In* "Global Perspectives on River Conservation: Science, Policy and Practice" (P. J. Boon, B. R. Davis, and G. E. Petts, Eds), pp. 155–178. John Wiley and Sons Ltd., New York, USA.

Peña-Torrealba, H. (1993). Natural water quality and agricultural pollution in Chile. *In* "Prevention of Water Pollution by Agriculture and Related Activities" (J. A. Sagardoy, Ed.), pp. 67–76. FAO, Rome, Italy.

Postel, S. (1987). "Defusing the Toxics Threat: Controlling Pesticides and Industrial Waste." Worldwatch Institute, Washington, DC, USA.

Pringle, C. M. (1997a). Exploring how disturbance is transmitted upstream: going against the flow. *Journal of the North American Benthological Society* **16**, 425–438.

Pringle, C. M. (1997b). Expanding scientific research programs to address conservation challenges in freshwater ecosystems. *In* "Enhancing the ecological basis of conservation: Heterogeneity, ecosystem function and biodiversity" (S. T. A. Pickett, R. S. Ostfeld, M. Shachak, and G. E. Likens, Eds), pp. 305–319. Chapman & Hall, New York, USA.

Pringle, C. M. (2000). Riverine conservation in tropical versus temperate regions: Ecological and socioeconomic considerations. *In* "Global Perspectives on River Conservation: Science, Policy and Practice" (P. J. Boon, B. R. Davis, and G. E. Petts, Eds), pp. 367–378. John Wiley and Sons Ltd., New York, USA.

Pringle, C. M., and Benstead, J. P. (2001). The effects of logging on tropical river ecosystems. *In* "The Cutting Edge: Conserving Wildlife in Logged Tropical Forests" (R. A. Fimbel, A. Grajal, and J. G. Robinson, Eds), pp. 305–325. Columbia University Press, New York, USA.

Pringle, C. M., and Scatena, F. N. (1999). Aquatic ecosystem deterioration in Latin America and the Caribbean. *In* "Managed Ecosystems: The Mesoamerican Experience" (U. Hatch and M. E. Swisher, Eds), pp. 104–113. Oxford University Press, New York, USA.

Pringle, C. M., Freeman, M., and Freeman, B. (2000a). Regional effects of hydrologic alterations on riverine macrobiota in the new world: tropical-temperate comparisons. *BioScience* **50**, 807–823.

Pringle, C. M., Scatena, F. N., Paaby-Hansen, P., and Nuñez-Ferrera, M. (2000b). River conservation in Latin America and the Caribbean. *In* "Global Perspectives on River Conservation: Science, Policy and Practice" (P. J. Boon, B. R. Davis, and G. E. Petts, Eds), pp. 41–77. John Wiley and Sons Ltd., New York, USA.

Pusey, J. P., and Arthington, A. H. (2003). Importance of the riparian zone to the conservation and management of freshwater fish: a review. *Marine and Freshwater Research* **54**, 1–16.

Quiros, R. (1988). Structures assisting the migrations of non-salmonid fish: Latin America. *Food and Agriculture Organization Technical Paper* **5**, 1–41.

Ribbink, A. J. (1994). Biodiversity and speciation of freshwater fishes with particular reference to African cichlids. *In* "Aquatic Ecology: Scale, Pattern and Process" (P. S. Giller, A. G. Hildrew, and D. G. Raffaelli, Eds), pp. 261–288. Blackwell Scientific Publishers, Oxford, UK.

Richey, J. E., Norbe, C., and Deser, C. (1989). Amazon River discharge and climate variability; 1903–1985. *Science* **246**, 01–103.

Sagardoy, J. A. (1993). "Prevention of Water Pollution by Agriculture and Related Activities." FAO, Rome, Italy.

Schemo, D. J. (1996). Burning of Amazon picks up pace, with vast areas lost. *The New York Times*, 12 September.

Schemo, D. J. (1997). More fires by farmers raise threat to Amazon. *The New York Times*, 2 November.

Schemo, D. J. (1998). Brazil Says recent burning of Amazon is worst ever. *The New York Times*, 27 January.

Schofield, N. J., Collier, K. J., Quinn, J., Sheldon, F., and Thoms, M. C. (2000). River conservation in Australia and New Zealand. *In* "Global Perspectives on River Conservation: Science, Policy and Practice" (P. J. Boon, B. R. Davis, and G. E. Petts, Eds), pp. 311–333. John Wiley and Sons Ltd., New York, USA.

Shukla, J. C. N., and Sellers, P. (1990). Amazon deforestation and climate change. *Science* **247**, 1322–1325.

Standley, L. J., and Sweeney, B. W. (1995). Organochlorine pesticides in stream mayflies and terrestrial vegetation of undisturbed tropical catchments exposed to long-range atmospheric transport. *Journal of the North American Benthological Society* **14**, 38–49.

Switkes, G. (1995). Tiete River cleanup: IDB flushes hundreds of millions down the drain in 'environmental boondoggle. *World Rivers Review* **10**, 1.

van Emden, H., and Peakall, D. B. (1996). "Beyond Silent Spring: Integrated Pest Management and Chemical Safety". Chapman & Hall, New York, USA.

Victor, R., Onomivbori, O., Schiemer, F., and Boland, K. T. (1996). The effects of urban perturbations on the benthic macroinvertebrates of a southern Nigerian stream. *In* "Perspectives in Tropical Limnology" (F. Schiemer and K. T. Boland, Eds), pp. 223–238. SPB Academic Publishing, Amsterdam, The Netherlands.

Walsh, C. J., Roy, A. H., Feminella, J. W., Cottingham, P. D., Groffman, P. M., and Morgan, R. P. (2005). The urban stream syndrome: current knowledge and the search for a cure. *Journal of the North American Benthological Society* **24**, 706–723.

Wang, H., Hall, C. A. S., Scatena, F. N., Fetcher, N., and Wu, W. (2003). Modeling the spatial and temporal variability in climate and primary productivity across the Luquillo Mountains, Puerto Rico. *Forest Ecology and Management* **179**, 69–94.

Wantzen, K. M. (1998). Effects of siltation on benthic communities in clear water streams in Mato Grosso, Brazil. *Verhandlungen der Internationalen Vereinigung fur Theoretische und Angewandte Limnologie* **26**, 1155–1159.

Wantzen, K. M. (2006). Physical pollution: effects of gully erosion in a tropical clear-water stream. *Aquatic Conservation* **16**, 733–749.

Wantzen, K. M., Machado, F. A., Voss, M., Boriss, H., and Junk, W. J. (2002). Floodpulse-induced isotopic changes in fish of the Pantanal wetland, Brazil. *Aquatic Sciences* **64**, 239–251.

Wantzen, K. M., Sá, M. F. P., Siqueira, A., and Nunes da Cunha, C. (2006). Conservation scheme for forest-stream-ecosystems of the Brazilian Cerrado and similar biomes in the seasonal tropics. *Aquatic Conservation* **16**, 713–732.

Waters, T. F. (1995). "Sediment in Streams: Sources, Biological Effects, and Control." American Fisheries Society, Bethesda, MD, USA.

Welcomme, R. L. (2000). Biodiversity in floodplains and their associated rivers. *In* "Biodiversity in Wetlands: Assessment, Function and Conservation. Volume 1" (B. Gopal, W. J. Junk, and J. Davis, Eds), pp. 61–87. Backhuys Publishers, Leiden, The Netherlands.

Winemiller, K. O., and Jepsen, D. B. (1998). Effects of seasonality and fish movement on tropical river food webs. *Journal of Fish Biology* **53**, 267–296.

Winemiller, K. O., and Jepsen, D. B. (2004). Migratory Neotropical fish subsidize food webs of oligotrophic blackwater rivers. *In* "Food Webs at the Landscape Level" (G. A. Polis, M. E. Power, and G. R. Huxel, Eds), pp. 115–132. University of Chicago Press, Chicago, USA.

Wishart, M. J., Davis, B. R., Boon, P. J., and Pringle, C. M. (2000). Global disparities in river conservation: 'First World' values and 'Third World' realities. *In* "Global Perspectives on River Conservation: Science, Policy and Practice" (P. J. Boon, B. R. Davis, and G. E. Petts, Eds), pp. 353–369. John Wiley and Sons Ltd., New York, USA.

Witt, V. M., and Reiff, F. M. (1991). Environmental health conditions and cholera vulnerability in Latin America and the Caribbean. *Journal of Public Health Policy* **12**, 450–463.

# Index

Printed and bound by CPI Group (UK) Ltd, Croydon, CR0 4YY

03/10/2024

01040316-0010